Modern RF and Microwave Measurement Techniques

This comprehensive, hands-on review of the most up-to-date techniques in RF and microwave measurement combines microwave circuit theory and metrology, in-depth analysis of advanced modern instrumentation, methods and systems, and practical advice for professional RF and microwave engineers and researchers.

Topics covered include microwave instrumentation, such as network analyzers, real-time spectrum analyzers, and microwave synthesizers; linear measurements, such as VNA calibrations, noise figure measurements, time domain reflectometry and multiport measurements; and nonlinear measurements, such as load- and source-pull techniques, broadband signal measurements, and nonlinear NVNAs.

Each technique is discussed in detail, and accompanied by state-of-the-art solutions to the unique technical challenges associated with its deployment. With each chapter delivered by internationally recognized experts in the field, this is an invaluable resource for researchers and professionals involved with microwave measurements.

Valeria Teppati is a Researcher in the Millimeter Wave Electronics Group of the Department of Information Technology and Electrical Engineering at ETH Zürich, developing innovative solutions to aspects of linear and nonlinear measurement techniques.

Andrea Ferrero is a Professor in the RF, Microwave and Computational Electronics group of the Department of Electronics and Telecommunications at Politecnico di Torino. He is a Distinguished Microwave Lecturer of the IEEE Microwave Theory and Techniques Society, and a Fellow of the IEEE.

Mohamed Sayed is the Principal Consultant for Microwave and Millimeter Wave Solutions, and has nearly thirty years' experience of developing microwave and millimeter wave systems for Hewlett-Packard Co. and Agilent Technologies Inc.

The Cambridge RF and Microwave Engineering Series

Series Editor
Steve C. Cripps, Distinguished Research Professor, Cardiff University

Peter Aaen, Jaime Plá, and John Wood, *Modeling and Characterization of RF and Microwave Power FETs*
Dominique Schreurs, Máirtín O'Droma, Anthony A. Goacher, and Michael Gadringer (Eds), *RF Amplifier Behavioral Modeling*
Fan Yang and Yahya Rahmat-Samii, *Electromagnetic Band Gap Structures in Antenna Engineering*
Enrico Rubiola, *Phase Noise and Frequency Stability in Oscillators*
Earl McCune, *Practical Digital Wireless Signals*
Stepan Lucyszyn (Ed.), *Advanced RF MEMS*
Patrick Roblin, *Nonlinear FR Circuits and the Large-Signal Network Analyzer*
Matthias Rudolph, Christian Fager, and David E. Root (Eds), *Nonlinear Transistor Model Parameter Extraction Techniques*
John L. B. Walker (Ed.), *Handbook of RF and Microwave Solid-State Power Amplifiers*
Anh-Vu H. Pham, Morgan J. Chen, and Kunia Aihara, *LCP for Microwave Packages and Modules*
Sorin Voinigescu, *High-Frequency Integrated Circuits*
Richard Collier, *Transmission Lines*
Valeria Teppati, Andrea Ferrero, and Mohamed Sayed (Eds), *Modern RF and Microwave Measurement Techniques*

Forthcoming
David E. Root, Jason Horn, Jan Verspecht, and Mihai Marcu, *X-Parameters*
Richard Carter, *Theory and Design of Microwave Tubes*
Nuno Borges Carvalho and Dominique Schreurs, *Microwave and Wireless Measurement Techniques*

Modern RF and Microwave Measurement Techniques

Edited by

VALERIA TEPPATI
ETH Zürich

ANDREA FERRERO
Politecnico di Torino

MOHAMED SAYED
Microwave and Millimeter Wave Solutions

CAMBRIDGE
UNIVERSITY PRESS

University Printing House, Cambridge CB2 8BS, United Kingdom

One Liberty Plaza, 20th Floor, New York, NY 10006, USA

477 Williamstown Road, Port Melbourne, VIC 3207, Australia

314-321, 3rd Floor, Plot 3, Splendor Forum, Jasola District Centre, New Delhi - 110025, India

79 Anson Road, #06-04/06, Singapore 079906

Cambridge University Press is part of the University of Cambridge.

It furthers the University's mission by disseminating knowledge in the pursuit of education, learning and research at the highest international levels of excellence.

www.cambridge.org
Information on this title: www.cambridge.org/9781107036413

First published 2013
3rd printing 2014

A catalogue record for this publication is available from the British Library

Library of Congress Cataloging in Publication data
Modern RF and microwave measurement techniques / [edited by] Valeria Teppati, Andrea Ferrero, Mohamed Sayed.
pages cm. – (The cambridge RF and microwave engineering series)
Includes bibliographical references and index.
ISBN 978-1-107-03641-3 (hardback)
1. Radio measurements. 2. Microwave measurements. 3. Radio circuits.
I. Teppati, Valeria, 1974– editor of compilation.
TK6552.5.M63 2013
621.382028′7–dc23 2013000790

ISBN 978-1-107-03641-3 Hardback

This book is dedicated to the memory of our colleague Dr. Roger D. Pollard, innovator, educator, contributor and friend.

Contents

Preface

In the last few years, the field of microwave testing has been evolving rapidly with the development and introduction of digital techniques and microprocessor based instruments, and reaching higher and higher frequencies. Nevertheless, the basic underlying concepts, such as frequency synthesis, network analysis and calibration, and spectrum analysis, still constrain even the more modern equipment.

In recent years, microwave instrumentation has had to meet new testing requirements, from 3G and now LTE wireless networks, for millimeter wave and THz applications. Thus instrumentation and measurement techniques have evolved from traditional instruments, such as vector network analyzers (VNAs), to increasingly more complex multifunction platforms, managing time and frequency domains in a unified, extensive approach.

We can identify two main directions of evolution:

- linear measurements, essentially S-parameter techniques;
- nonlinear measurements, for high power and nonlinear device characterization.

S-parameter measurements have been moving towards the multiport and millimeter wave fields. The first to characterize multi-channel transmission structures such as digital buses, and the latter for space or short-range radio communication or security scanner applications. New calibrations and instrument architectures have been introduced to improve accuracy, versatility and speed.

Nonlinear applications have also evolved. Traditional high power transistor characterization by load-pull techniques now also typically includes time domain waveform measurements under nonlinear conditions. These techniques can nowadays also handle the broadband signals used in most communication links, or pulsed signals. Moreover, even nonlinear measurements had to evolve to multiport, with differential and common mode impedance tuning, due to the spreading of amplifiers and devices exploiting differential configuration.

The idea of a comprehensive book on microwave measurement was born when we noticed that the knowledge of these modern instrumentation and measurement techniques was scattered inside different books or papers, sometimes dealing more with design or modeling than with the measurement itself or the metrological aspects, and there was no recent book covering these topics extensively.

We thus tried to make an effort to produce a book that could:

- give an overview of modern techniques for measurements at microwave frequencies;
- be as complete and comprehensive as possible, giving general concepts in a unitary way;
- treat *modern* techniques, i.e. the state of the art and all the most recent developments.

As editors of the book, we have been honored to work with several international experts in the field, who contributed their invaluable experience to the various chapters of this book. This multi-author approach should guarantee the reader a deep understanding of such a complex and sophisticated matter as microwave measurements.

The book is structured in four main sections:

1. general concepts
2. microwave instrumentation
3. linear measurement techniques
4. nonlinear measurement techniques.

An already expert reader may directly jump to a specific topic, to read about innovative instruments or techniques, such as synthesizers, modular RF instruments, multiport VNAs or broadband load-pull techniques, or follow the book's organization that will guide him/her through the development of the instruments and their applications.

Fifteen chapters form the body of the four book sections. Two of them describe fundamentals, from the theory behind the S-parameters to the interconnections; five chapters are then devoted to microwave instrumentation: synthesizers, network and spectrum analyzers, power meters, up to modern microwave modular instrumentation. The third section on linear measurements covers traditional two-port S-parameter calibration, multiport S-parameter techniques, noise measurements and time domain reflectometry techniques. Finally the last section on nonlinear measurements describes nonlinear VNAs, load-pull, broadband load-pull, and concludes with pulsed measurements.

All the content is correlated with details on metrological aspects whenever possible, and with some examples of typical use, though we have tried to be as independent as possible of a specific device under test and to concentrate on the measurement technique rather than the particular application.

Contributors

Jin Bains
National Instruments Corp., USA

Alexander Chenakin
Phase Matrix, Inc., USA

Juan-Mari Collantes
University of the Basque Country (UPV/EHU), Spain

Kaviyesh Doshi
Teledyne LeCroy, USA

Andrea Ferrero
Politecnico di Torino, Italy

Ronald Ginley
NIST, USA

Leonard Hayden
Teledyne LeCroy, USA

Gian Luigi Madonna
ABB Corporate Research, Baden, Switzerland

Mauro Marchetti
Anteverta Microwave B.V., the Netherlands

Jon Martens
Anritsu Company, USA

Nerea Otegi
University of the Basque Country (UPV/EHU), Spain

Anthony Parker
Macquarie University, Sydney, Australia

Roger Pollard
Agilent Technologies, USA and University of Leeds, United Kingdom

Peter J. Pupalaikis
Teledyne LeCroy, USA

Yves Rolain
Vrije Universiteit Brussel, Belgium

Mohamed Sayed
Microwave and Millimeter Wave Solutions, USA

Maarten Schoukens
Vrije Universiteit Brussel, Belgium

Marcus Da Silva
Tektronix Inc., USA

Marco Spirito
Delft University of Technology, the Netherlands

Valeria Teppati
ETH Zürich, Switzerland

Gerd Vandersteen
Vrije Universiteit Brussel, Belgium

Abbreviations

ACLR	Adjacent Channel Leakage Ratio
ACPR	Adjacent Channel Power Ratio
AD	Analog-to-Digital
ADC	Analog-to-Digital Converter
ADS	Advanced Design System
ALC	Automatic Level Control
AM	Amplitude Modulation
ASB	All-Side-Band
AWG	Arbitrary Waveform Generator
BER	Bit Error Rate
BWO	Backward Wave Oscillators
CDMA	Code Division Multiple Access
CIS	Coherent Interleaved Sampling
CMOS	Complementary Metal-Oxide Semiconductor
CPU	Central Processing Unit
CPW	CoPlanar Waveguide
CRT	Cathode Ray Tube
CW	Continuous Wave
CZT	Chirp Zeta Transform
DAC	Digital-to-Analog Converter
DDC	Digital Down Converter
DDS	Direct Digital Synthesizer
DFT	Discrete Fourier Transform
DMM	Digital Multimeter
DPD	Digital Pre-Distortion – used to linearize power amplifiers
DPX	Digital Phosphor Processing – Tektronix implementation of a variable persistence spectrum display.
DSA	Discrete-time Spectrum Analyzer
DSB	Double Side Band
DSP	Digital Signal Processing
DTC	Digital to Time Converter
DTFT	Discrete-Time Fourier Transform
DUT	Device Under Test
DVM	Digital Volt Meter

DWT	Discrete Wavelet Transform
EDA	Electronic Design & Automation
EDGE	Enhanced Data rates for GSM Evolution
eLRRM™	enhanced Line-Reflect-Reflect Match (Cascade Microtech)
ENR	Excess Noise Ratio
EVM	Error Vector Magnitude
FEM	Finite Element Method
FET	Field Effect Transistor
FEXT	Far End CrossTalk
FFT	Fast Fourier transform
FM	Frequency Modulation
FMT	Frequency Mask Trigger
FOM	Figure of Merit
FPGA	Field-Programmable Gate Arrays
GMSK	Gaussian Minimum Shift Keyed
GPIB	General Purpose Interface Bus
GPU	Graphics Processing Unit
GSG	Ground-Signal-Ground
GSM	Global System for Mobile Communications
GUI	Graphical User Interface
HBT	Heterojunction Bipolar Transistor
HEMT	High Electron Mobility Transistor
HW	HardWare
I	In-phase component of vector modulation
IC	Integrated Circuit
IDE	Integrated Drive Electronics
IDFT	Inverse Discrete Fourier transform
IDWT	Discrete Wavelet Transform
IEEE	Institute of Electrical and Electronics Engineers
IF	Intermediate Frequency
IFFT	Inverse Fast Fourier Transform
IL	Insertion Loss
IM_3	Third-order Inter Modulation distortion
IM_5	Fifth-order Inter Modulation distortion
IMD	Inter Modulation Distortion
IP	Intellectual Property
IQ	Cartesian vector modulation format – In-phase and Quadrature.
IVI	Interchangeable Virtual Instruments
JTFA	Joint Time-Frequency Analysis
KCL	Kirchhoff's Current Law
KVL	Kirchhoff's Voltage Law
LAN	Local Area Network
LCD	Liquid Crystal Display
LDMOS	Laterally Diffused Metal Oxide Semiconductor

LO	Local Oscillator
LPF	Low Pass Filter
LRL	Line Reflect Line
LRM	Line Reflect Match
LSNA	Large Signal Network Analyzer
LTE	Long Term Evolution
LTI	Linear Time Invariant
LUT	Look Up Table
LXI	LAN eXtensions for Instrumentation
MESFET	MEtal-Semiconductor Field Effect Transistor
MIMO	Multiple Input Multiple Output
MMIC	Monolithic Microwave Integrated Circuit
MTA	Microwave Transition Analyzer
NCO	Numerically Controlled Oscillator
NEXT	Near End CrossTalk
NF	Noise Figure
NFA	Noise Figure Analyzer
NIST	National Institute of Standards and Technology
NVNA	Nonlinear Vector Network Analyzer
OFDM	Orthogonal Frequency-Division Multiplexing
OIP_2	Output Second Order Intercept Point
OIP_3	Output Third Order Intercept Point
ORFS	Output RF Spectrum
OS	Operating System
P2P	Peer-to-Peer
PA	Power Amplifier
PAE	Power Added Efficiency
PC	Personal Computer
PCB	Printed Circuit Board
PCI	Peripheral Component Interconnect
PCMCIA	Personal Computer Memory Card International Association
PDF	Probability Density Function
PFER	Phase and Frequency Error
pHEMT	pseudomorphic High Electron Mobility Transistor
PICMG	PCI Industrial Computer Manufacturers Group
PISPO	Periodic In, Same Period Out
PLL	Phase Locked Loop
PM	Phase Modulation
PMC	PCI Mezzanine Card
PVT	Power Versus Time
PXI	PCI eXtensions for Instrumentation
PXISA	PXI Systems Alliance
Q	Quadrature component of vector modulation
QMF	Quadrature Mirror Filter

QSOLT	Quick Short Open Load Thru
RAID	Redundant Array of Inexpensive Disks
RAM	Random Access Memory
RBW	Resolution BandWidth- The minimum bandwidth that can be resolved in a spectrum analyzer display.
RF	Radio Frequency
RL	Return Loss
RMS	Root Mean Square
ROM	Read Only Memory
RSA	Tektronix nomenclature for its real-time signal analyzer family
RSS	Root Sum of Squares
RTSA	Real-Time Signal Analyzer
SA	Spectrum Analyzer
SATA	Serial Advanced Technology Attachment
SI	Signal Integrity
SMVR	R&S nomenclature for its real-time signal analyzer family
SNR	Signal to Noise Ratio
SOL	Short Open Load
SOLT	Short Open Load Thru
SSB	Single-Side Band
SW	SoftWare
SWR	Standing Wave Ratio
TDEMI	Gauss Instruments nomenclature for its real-time spectrum analyzers targeted at emissions measurement
TDMA	Time Division Multiple Access
TDR	Time Domain Reflectometry
TDT	Time Domain Transmission
TE	Transverse Electric
TEM	Transverse ElectroMagnetic
TL	Transmission Line
TM	Transverse Magnetic
TOI	Third Order Intercept
TRL	Thru Reflect Line
TSD	Thru Short Delay
TXP	Transmit Power
UHF	Ultra High Frequency
UML	Universal Modeling Language
USB	Universal Serial Bus
VCO	Voltage Controlled Oscillator
VISA	Virtual Instrument Software Architecture
VNA	Vector Network Analyzer
VPN	Virtual Private Network
VSA	Vector Signal Analyzer
VSG	Vector Signal Generator

VSWR	Voltage Standing Wave Ratio
VXI	VMEbus eXtensions for Instrumentation
W-CDMA	Wideband Code Division Multiple Access
WLAN	Wireless Local Area Network
YIG	Yttrium-Iron-Garnet

Part I

General concepts

1 Transmission lines and scattering parameters

Roger Pollard and Mohamed Sayed

1.1 Introduction

This chapter introduces the reader to the topics presented in the rest of the book, and serves as a quick guide to the basic concepts of wave propagation and scattering parameters.

Understanding these concepts becomes very important when dealing with RF and microwave frequencies, as is shown in Section 1.2, where a simplified formulation for the transmission line theory is given.

Section 1.3 provides the definition of the scattering matrix or S-matrix, the key element to describe networks at RF, microwaves, and higher frequencies.

Section 1.4 deals with the most important component in microwave measurements, the directional coupler, while Section 1.5 revises a common way to represent quantities in the RF domain, the Smith Chart.

Finally, in Appendix A signal flow graphs, a typical way to represent simple linear algebra operations, are presented, while Appendix B summarizes the various types of transmission lines cited in this book.

1.2 Fundamentals of transmission lines, models and equations

1.2.1 Introduction

Electromagnetic waves travel at about the speed of light ($c = 299\ 792\ 458$ m/s) in air. Using the relationship

$$\nu = f\lambda, \tag{1.1}$$

where ν is velocity ($= c$ in air), f is frequency and λ is wavelength, the wavelength of a 100 GHz wave is about 3 mm. If a simple connection on a circuit is of the order of magnitude of a wavelength, it is then necessary to consider its behavior as distributed and regard it as a transmission line. In fact, propagation phenomena already appear for lengths of $1/10^{\text{th}}$ of a wavelength.

Let's clarify this concept by a simple example. When a source of electrical power is connected to a load, as shown in Figure 1.1, the voltage appears at the load instantaneously over a short distance.

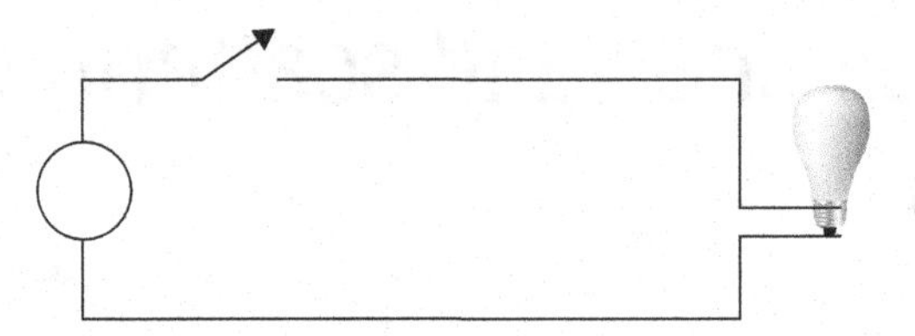

Fig. 1.1 Connection of a light bulb close to the source of electrical power.

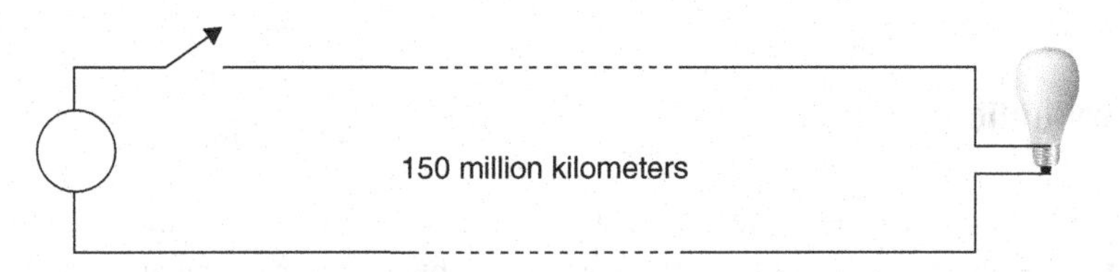

Fig. 1.2 Connection of a light bulb at 150 million km from the source of electrical power.

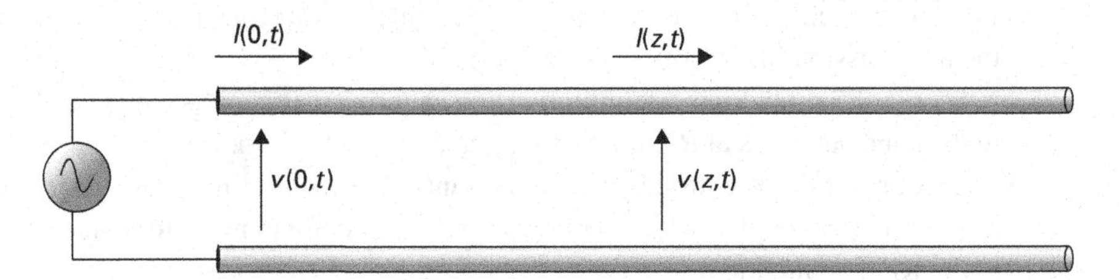

Fig. 1.3 Two-wire line.

However, if the connection wiring is very long, as shown in Figure 1.2, it takes time for the signal to propagate to the load. In this example, using the approximate distance from the sun, the bulb would light some 8 minutes after the switch is closed.

This means that the connection cannot be modeled with a short circuit anymore, since the voltage and current (or electric and magnetic fields) are now functions of both time and position.

Let us consider a two-wire line, as shown in Figure 1.3.

Here both the voltage and current are functions of position and time. Now, if we model the line as an infinite number of very short sections, each element can be considered as a series inductance and shunt capacitance with associated losses, as shown in Figure 1.4. This model can actually be applied to any kind of transmission line (waveguide, coaxial, microstrip, etc.; see Appendix B for a brief description of the most common types of transmission lines referred to in this book).

1.2.2 Propagation and characteristic impedance

In a two-conductor line, the model may be explained physically. The wire properties and skin effect generate the inductance, the two conductors the capacitance, and leakage

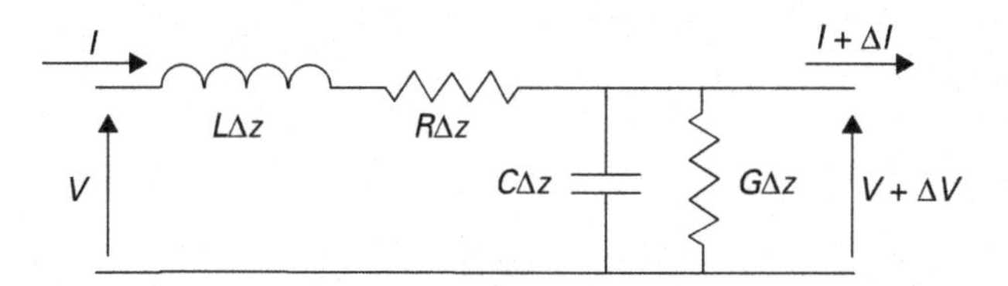

Fig. 1.4 Lumped-element model of a section of the two-wire line of Fig. 1.3.

and losses produce the parasitic resistances. These model elements are also functions of frequency.

Solving the model circuit for the voltage and current, yields

$$\Delta V(z,t) = (R\Delta z + j\omega L\Delta z)I(z,t) \tag{1.2}$$

and

$$\Delta I(z,t) = (G\Delta z + j\omega C\Delta z)V(z,t). \tag{1.3}$$

Taking Δz as infinitely short, the partial derivatives of voltage and current with respect to the z coordinate appear as:

$$\frac{\partial V(z,t)}{\partial z} = -(R + j\omega L)\, I(z,t) \tag{1.4}$$

$$\frac{\partial I(z,t)}{\partial z} = -(G + j\omega C)\, V(z,t). \tag{1.5}$$

Then, by differentiating (1.4) again with respect to z and substituting (1.5) in the obtained equation (and vice versa) one gets:

$$\frac{\partial^2 V(z,t)}{dz^2} = \gamma^2 V(z,t) \quad \text{and} \quad \frac{\partial^2 I(z,t)}{dz^2} = \gamma^2 (z,t) I(z,t), \tag{1.6}$$

where $\gamma = \sqrt{(R + j\omega L)(G + j\omega C)} = \alpha + j\beta$ is the ***propagation constant***.

The equations have exponential solutions of the form

$$V = V_1 e^{-j\gamma z} + V_2 e^{+j\gamma z}, \tag{1.7}$$

where the first part of the solution ($V^+ = V_1 e^{-j\gamma z}$) is referred to as an ***incident wave***, and the second part ($V^- = V_2 e^{+j\gamma z}$) as a ***reflected wave***.

In the same way, one can write the solution for the current as

$$I = I_1 e^{-j\gamma z} + I_2 e^{+j\gamma z}. \tag{1.8}$$

By substituting (1.7) and (1.8) inside (1.4) and (1.5) one can find the relationship between I_1-V_1 and I_2-V_2, which are:

$$V_1 = Z_0 I_1 \tag{1.9}$$

and

$$V_2 = -Z_0 I_2 \tag{1.10}$$

with

$$Z_0 = \frac{\sqrt{R + j\omega L}}{\sqrt{G + j\omega C}}, \tag{1.11}$$

where Z_0 is referred to as the *characteristic impedance* of the transmission line. Note that the ***wave number*** β can be expressed as a function of v_p, the so-called ***phase velocity***, or of the wavelength (λ):

$$\beta = \frac{\omega}{v_p} = \frac{2\pi}{\lambda}. \tag{1.12}$$

The time dependence of the voltage and current can be made explicit in this way

$$V(z,t) = V(z)e^{j\omega t} \quad I(z,t) = I(z)e^{j\omega t} \tag{1.13}$$

and the circuit equations rewritten as

$$\frac{\partial V}{\partial z} = -\left(RI + L\frac{\partial I}{\partial t}\right) \quad \text{and} \quad \frac{\partial I}{\partial z} = -\left(GV + C\frac{\partial V}{\partial t}\right). \tag{1.14}$$

Again, differentiating gives

$$\frac{\partial^2 V}{\partial z^2} = R\left(GV + C\frac{\partial V}{\partial t}\right) + L\left(G\frac{\partial V}{\partial t} + C\frac{\partial^2 V}{\partial t^2}\right) \tag{1.15}$$

or

$$\frac{\partial^2 V}{dz^2} = -(RC + LG)\frac{\partial V}{\partial t} - LC\frac{\partial^2 V}{dt^2} - RGV = 0. \tag{1.16}$$

Note that the current I satisfies an identical equation.

In the case of lossless transmissions lines with $R = G = 0$, the propagation constant and the characteristic impedance simplify to the trivial

$$\gamma = j\beta = j\omega\sqrt{LC} \quad \text{and} \quad Z_0 = \sqrt{\frac{L}{C}}. \tag{1.17}$$

For most practical purposes, however, especially in a hollow pipe waveguide, the low-loss case ($R = \omega L,\ G = \omega C$) provides accurate values:

$$\gamma \approx \alpha + j\beta = j\omega\sqrt{LC} + \frac{1}{2}\sqrt{LC}\left(\frac{R}{L} + \frac{G}{C}\right) \tag{1.18}$$

with

$$\alpha = \frac{1}{2}\sqrt{LC}\left(\frac{R}{L} + \frac{G}{C}\right) = \frac{1}{2}(RY_0 + GZ_0) \tag{1.19}$$

1.2.3 Terminations, reflection coefficient, SWR, return loss

We have seen how the total voltage on a transmission line is the vector sum of the incident and reflected voltages and the phase relationship between the waves depends on the position along the line. The nature of a discontinuity determines the phase relationship of the incident and reflected waves at that point on the line and that phase relationship is repeated at points that are multiples of a half-wavelength (180°).

The classical example is when the line is terminated with a load impedance Z_L that is not the characteristic impedance. Some of the incident energy may be absorbed by the load and the rest is reflected. The maximum and minimum values of the standing wave voltage and the positions of these maxima and minima are related to Z_L. The maximum occurs where the incident and reflected voltages are in phase, the minimum where they are 180° out of phase.

$$E_{\max} = |V_{incident}| + |V_{reflected}| \quad \text{and} \quad E_{\min} = |V_{incident}| - |V_{reflected}| \tag{1.20}$$

with $V_{incident}$ a constant and $V_{reflected}$ a function of Z_L, $\frac{E_{\max}}{E_{\min}}$ is the **Voltage Standing Wave Ratio**, abbreviated VSWR or SWR and is a way of describing the discontinuity at the plane of the load. The SWR is 1 when the load termination is equal to the characteristic impedance of the line, since $V_{reflected} = 0$, and infinite when a lossless reflective termination (short circuit, open circuit, capacitance, etc.) is connected, since $V_{reflected} = V_{incident}$ in that case. SWR is commonly used as a specification for components, most commonly loads and attenuators.

For a finite Z_L, the magnitude and phase of the reflected signal depends on the ratio of Z_L/Z_0. Since the total voltage (and current) across Z_L is the vector sum of the incident and reflected voltages (and currents) we have

$$Z_L = \frac{V_L}{I_L} = \frac{V_{incident} + V_{reflected}}{I_{incident} + I_{reflected}}. \tag{1.21}$$

The voltage and current in each of the waves on the transmission line are related by the characteristic impedance, as already shown in (1.9) and (1.10)

$$\frac{V_{incident}}{I_{incident}} = Z_0 \quad \text{and} \quad \frac{V_{reflected}}{I_{reflected}} = -Z_0 \tag{1.22}$$

so

$$Z_L = \frac{V_{incident} + V_{reflected}}{\frac{V_{incident}}{Z_0} - \frac{V_{refelected}}{Z_0}} = Z_0 \frac{1 + \frac{V_{reflected}}{V_{incident}}}{1 - \frac{V_{reflected}}{V_{incident}}} = Z_0 \frac{1+\Gamma}{1-\Gamma} \tag{1.23}$$

where Γ is the **reflection coefficient**, a complex value with magnitude and phase. The magnitude of Γ is usually denoted by the symbol ρ and its phase by θ. The values of ρ vary from zero to one. It is common practice to refer to the magnitude of the reflection coefficient as the **return loss** ($20\log_{10}\rho$).

Note that ρ, the magnitude of Γ, remains constant as the observation point is moved along a lossless transmission line. In this case, the phase θ changes and thus the complex value of Γ rotates around a circle on a polar plot. Since, at the plane of the load

$$\Gamma = \frac{Z_L - Z_0}{Z_L + Z_0} \tag{1.24}$$

the value of the impedance seen looking into the transmission line at any point is readily calculated by rotating Γ by the electrical length (a function of the signal frequency, $360° = \lambda/2$) between the plane of the load and the point of observation. Thus, for example, at a quarter-wavelength distance (180° electrical length) from the plane of a short circuit, the impedance appears as an open circuit. The same impedance repeats at multiples of a half-wavelength.

1.2.4 Power transfer to load

The maximum power transfer from sources with source impedance of R_s to load impedance of R_L occurs at the value of R_s equal to R_L. For complex impedances, the maximum power transfer occurs when $Z_L = R_L + jX_L$, $Z_s = R_s - jX_s$ and $R_s = R_L$, and $X_L = X_s$, otherwise there will be a mismatch and standing wave ratio.

1.3 Scattering parameters

A key assumption when making measurements is that networks can be completely characterized by quantities measured at the network terminals (ports) regardless of the contents of the networks. Once the parameters of a (linear) n-port network have been determined, its behavior in any external environment can be predicted.

At low frequencies, typical choices of network parameters to be measured and handled are Z-parameters or Y-parameters, i.e. the impedance or admittance matrix, respectively. In microwave design, S-parameters are the natural choice because they are easier to measure and work with at high frequencies than other kinds of parameters. They are conceptually simple, analytically convenient, and capable of providing a great insight into a measurement or design problem.

Similarly to when light interacts with a lens, and a part of the light incident is reflected while the rest is transmitted, scattering parameters are measures of reflection and transmission of voltage waves through an electrical network.

Let us now focus on the generic n-port network, shown in Figure 1.5

To characterize the performance of such a network, as we said, any of several parameter sets can be used, each of which has certain advantages. Each parameter set is related to a set of $2n$ variables associated with the n-port model. Of these variables, n represents the excitation of the network (independent variables), and the remaining n represents the response of the network to the excitation (dependent variables). The network of Figure 1.5, assuming it has a linear behavior, can be represented by its

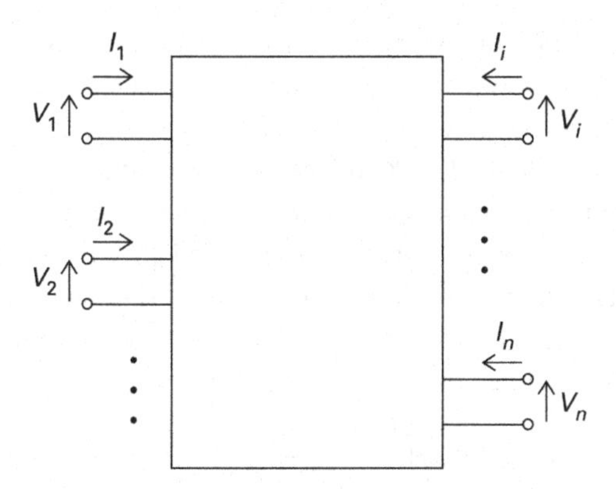

Fig. 1.5 Generic n-port network.

Z-matrix (impedance matrix):

$$\begin{bmatrix} V_1 \\ V_2 \\ \vdots \\ V_n \end{bmatrix} = \begin{bmatrix} Z_{11} & Z_{12} & \cdots & Z_{1n} \\ Z_{21} & Z_{22} & \cdots & Z_{2n} \\ \vdots & \vdots & \ddots & \vdots \\ Z_{n1} & Z_{n2} & \cdots & Z_{nn} \end{bmatrix} \begin{bmatrix} I_1 \\ I_2 \\ \vdots \\ I_n \end{bmatrix}, \tag{1.25}$$

where V_1-V_n are the node voltages and I_1-I_n are the node currents. Alternatively, one can use the dual representation:

$$\begin{bmatrix} I_1 \\ I_2 \\ \vdots \\ I_n \end{bmatrix} = \begin{bmatrix} Y_{11} & Y_{12} & \cdots & Y_{1n} \\ Y_{21} & Y_{22} & \cdots & Y_{2n} \\ \vdots & \vdots & \ddots & \vdots \\ Y_{n1} & Y_{n2} & \cdots & Y_{nn} \end{bmatrix} \begin{bmatrix} V_1 \\ V_2 \\ \vdots \\ V_n \end{bmatrix}. \tag{1.26}$$

Here, port voltages are the independent variables and port currents are the dependent variables; the relating parameters are the short-circuit admittance parameters, or Y-parameters. In the absence of additional information, n^2 measurements are required to determine the n^2 Y-parameter. Each measurement is made with one port of the network excited by a voltage source while all the other ports are short-circuited. For example, Y_{21}, the forward trans-admittance, is the ratio of the current at port 2 to the voltage at port 1, when all other ports are short-circuited:

$$Y_{21} = \left.\frac{I_2}{V_1}\right|_{V_2=\ldots=V_n=0}. \tag{1.27}$$

If other independent and dependent variables had been chosen, the network would have been described, as before, by n linear equations similar to (1.24), except that the variables and the parameters describing their relationships would be different. However, all parameter sets contain the same information about a network, and it is always possible to calculate any set in terms of any other set [1,2].

"Scattering parameters," which are commonly referred to as S-parameters, are a parameter set that relates to the traveling waves that are scattered or reflected when an n-port network is inserted into a transmission line.

Scattering parameters were first defined by Kurokawa [3], where the assumption was to have real and positive reference impedances Z_i. For complex reference impedances, Marks and Williams [4] addressed the general case in 1992 and gave a comprehensive solution to it. They describe the interrelationships of a new set of variables, the ***pseudo-waves*** a_i, b_i, which are the normalized complex voltage waves incident on and reflected from the ith port of the network, defined as:

$$a_i = \alpha\sqrt{\Re\{Z_i\}}\frac{V_i + Z_i I_i}{2|Z_i|}.$$

$$b_i = \alpha\sqrt{\Re\{Z_i\}}\frac{V_i - Z_i I_i}{2|Z_i|}. \tag{1.28}$$

where voltage V_i and I_i are the terminal voltages and currents, Z_i are arbitrary (complex) reference impedances and α is a free parameter whose only constraint is to have unitary modulus, from now on assumed to be 1.

The linear equations describing the n-port network are therefore:

$$\begin{bmatrix} b_1 \\ b_2 \\ \vdots \\ b_n \end{bmatrix} = \begin{bmatrix} S_{11} & S_{12} & \cdots & S_{1n} \\ S_{21} & S_{22} & \cdots & S_{2n} \\ \vdots & \vdots & \ddots & \vdots \\ S_{n1} & S_{n2} & \cdots & S_{nn} \end{bmatrix} \begin{bmatrix} a_1 \\ a_2 \\ \vdots \\ a_n \end{bmatrix} \tag{1.29}$$

where by definition

$$S_{ij} = \left.\frac{b_j}{a_i}\right|_{a_2=\ldots=a_n=0}. \tag{1.30}$$

Note that in principle each port can use a different reference Z_i, and they need not be related to any physical characteristic impedance.

The ease with which scattering parameters can be measured makes them especially well suited for describing transistors and other active devices. Measuring most other parameters calls for the input and output of the device to be successively opened and short-circuited. This can be hard to do, especially at RF frequencies where lead inductance and capacitance make short and open circuits difficult to obtain. At higher frequencies these measurements typically require tuning stubs, separately adjusted at each measurement frequency, to reflect short or open circuit conditions to the device terminals. Not only is this inconvenient and tedious, but a tuning stub shunting the input or output may cause a transistor to oscillate, making the measurement invalid.

S-parameters, on the other hand, are usually measured with the device embedded between a matched load and source, and there is very little chance for oscillations to occur. Another important advantage of S-parameters stems from the fact that traveling waves, unlike terminal voltages and currents, do not vary in magnitude at points along a lossless transmission line. This means that scattering parameters can be measured on a device located at some distance from the measurement transducers, provided that the measuring device and the transducers are connected by low-loss transmission lines.

The relationship between some of the most commonly used parameters can be found in [1], which is valid for real reference impedances. When dealing with complex reference impedances, then the corrections of [2] should be taken into account.

1.4 Microwave directional coupler

1.4.1 General concepts

Probably the most important passive component in the microwave measurement field, the ***directional coupler*** [5], [6], is a device that can separate the incident and reflected waves, which were described in Section 1.2.2 of this chapter.

A sketch of a generic directional coupler is shown in Figure 1.6(a). Independent of the typology and the coupling strategy, any directional coupler is made of two transmission lines, respectively the ***main line*** (line 1–2 in Figure 1.6(a)) and the ***coupled line*** (line 3–4 in Figure 1.6(a)).

The directional coupler properties can be understood by inspecting its S-matrix:

$$\begin{bmatrix} S_{11} & S_{12} & S_{13} & S_{14} \\ S_{21} & S_{22} & S_{23} & S_{24} \\ S_{31} & S_{32} & S_{33} & S_{34} \\ S_{41} & S_{42} & S_{43} & S_{44} \end{bmatrix} = \begin{bmatrix} \rho_1 & l_1 & k & \iota \\ l_1 & \rho_1 & \iota & k \\ k & \iota & \rho_2 & l_2 \\ \iota & k & l_2 & \rho_2 \end{bmatrix}. \tag{1.31}$$

Here we assume that the device is passive (reciprocal) and perfectly symmetrical, thus the return losses of the main line and the coupled line are $S_{11} = S_{22} = \rho_1$ and $S_{33} = S_{44} = \rho_2$. If the device is well designed (well matched to the reference impedance) then $\rho_1 \approx \rho_2 \approx 0$. The terms l_1 and l_2 represent the losses of the two transmission lines.

The term k in the S-matrix is the ***coupling factor*** (expressed in linear units). When a signal source is connected to port 1 of the coupler and port 2 is connected to non-matched load Γ_L, if port 3 and 4 are well matched (i.e. $a_3 \approx a_4 \approx 0$) we have

$$b_3 = ka_1 + \iota a_2 \tag{1.32}$$

$$b_4 = \iota a_1 + ka_2 \tag{1.33}$$

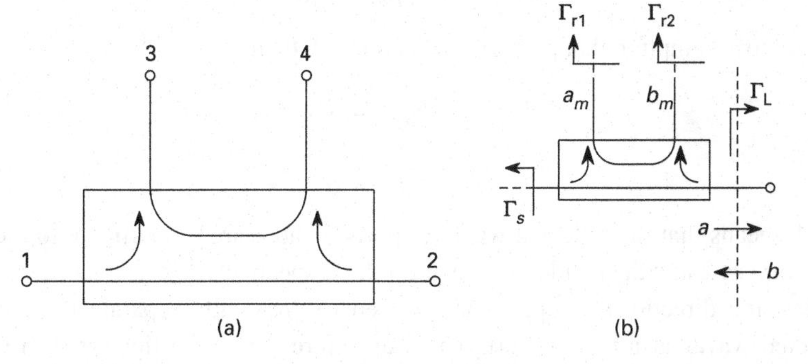

Fig. 1.6 Sketch of a generic directional coupler (a) and a directional coupler used as a reflectometer (b).

where a_i and b_i are, respectively, the incident and reflected waves at each i-port of the coupler. As long as ι, the ***isolation factor***, is kept small, b_3 is proportional to a_1 through the coefficient k and b_4 is proportional to a_2, through the same factor.

Coupling and isolation factors are typically expressed in dB. The ***directivity*** expresses the ratio ι/k, in other words how much the coupler is capable of separating the incident and reflected waves.

The different typologies of directional couplers available depend on the type of the transmission line (see Appendix B for a quick overview) used for the main and the coupled lines [7].

For example, in microstrip technology, the coupling between the two lines can be realized by progressively reducing the distance between the main and coupled line or in a "branch line" configuration (with two parallel microstrips physically coupled together with two or more branch lines between them, placed at proper distance).

In a waveguide, coupling is typically realized with single or multiple holes along one side or other of the guide. Coaxial couplers can be realized by manufacturing holes in the external shields of the coaxial lines. Mixed-technology couplers are also possible, such as waveguide-coaxial, coaxial-microstrip, etc.

In all cases the design involves finding the proper physical dimensions in order to achieve the desired performances, in terms of:

- coupling factor,
- directivity,
- insertion losses,
- frequency bandwidth.

Depending on the application, it is also common to find 3-port directional couplers, where one of the coupled ports (3 or 4) is typically physically terminated with a matched load.

1.4.2 The reflectometer

The directional coupler can be used as a ***reflectometer***, as shown in Figure 1.6(b). From (1.31), a_1 and b_2 are related through

$$b_2 = l_1 a_1 + \rho_1 a_2. \tag{1.34}$$

Under the assumption that $\rho_1 \approx \rho_2 \approx 0$ and $\iota = 0$, then

$$b_3 = \frac{k}{l_1} b_2 \tag{1.35}$$

$$b_4 = k a_2 \tag{1.36}$$

which means that the reflected waves at ports 3 and 4 are proportional to the reflected wave at port 2 and the incident wave at port 2, respectively.

Thus the directional coupler can be used to physically separate the incident and reflected waves at a certain port. There are more details on this topic in Chapter 8, which also considers the case of a non-ideal directional coupler.

1.5 Smith Chart

The Smith Chart, shown in Figure 1.7, is a graph of the reflection coefficient Γ in the polar plane. Phillip H. Smith invented this chart in the 1930s [8]. Using the Smith Chart it's very easy to convert impedances to reflection coefficients and vice versa.

The Smith Chart represents the bilinear conformal transformation $\Gamma = (z-1)/(z+1)$ where z is the normalized impedance (Z/Z_0). In other words, the Smith Chart is the transformation of the right part of the Z complex plane (only positive real parts of Z are considered) into a circle, where the infinite values for the real and imaginary parts of Z converge to the point (1, 0) on the transformed plane.

The normalization to Z_0 of the Smith Chart implies, for example, that an impedance of $(30+\mathrm{j}10)\ \Omega$ will be plotted as $(0.6+\mathrm{j}\ 0.2)$ on the Smith Chart, normalized to 50 Ω.

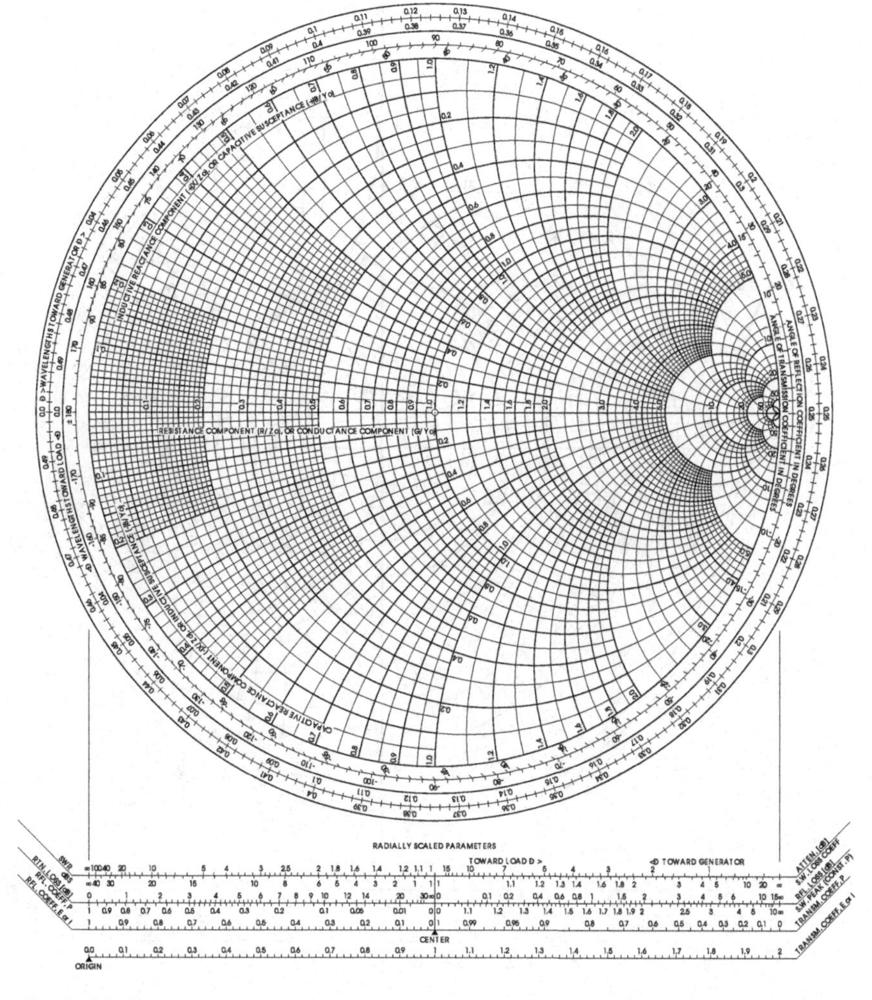

Fig. 1.7 The Smith Chart.

The lines of the Smith chart define the loci of the constant real part of the impedance or constant imaginary part, as shown in Figure 1.8. Constant resistance maps to circles and constant reactance maps to arcs. In the Z plane, these would simply be vertical and horizontal lines, respectively. Note that since the transformation is conformal, the 90° angles formed between these lines are also maintained in the transformed plane.

A perfect load (equal to Z_0) occurs when Γ equals zero, which is the center of the Smith Chart. An open load will have a Γ of unity and 0° (point (1,0) on the Smith Chart) and a short load will have a Γ of unity and 180° (point $(-1{,}0)$ on the Smith Chart).

Figure 1.8 shows a Smith Chart with the constant VSWR contours. A constant VSWR corresponds to a constant $|\Gamma|$.

The constant impedance magnitude and phase can also be plotted on the Smith Chart as shown in Figures 1.9(a) and (b).

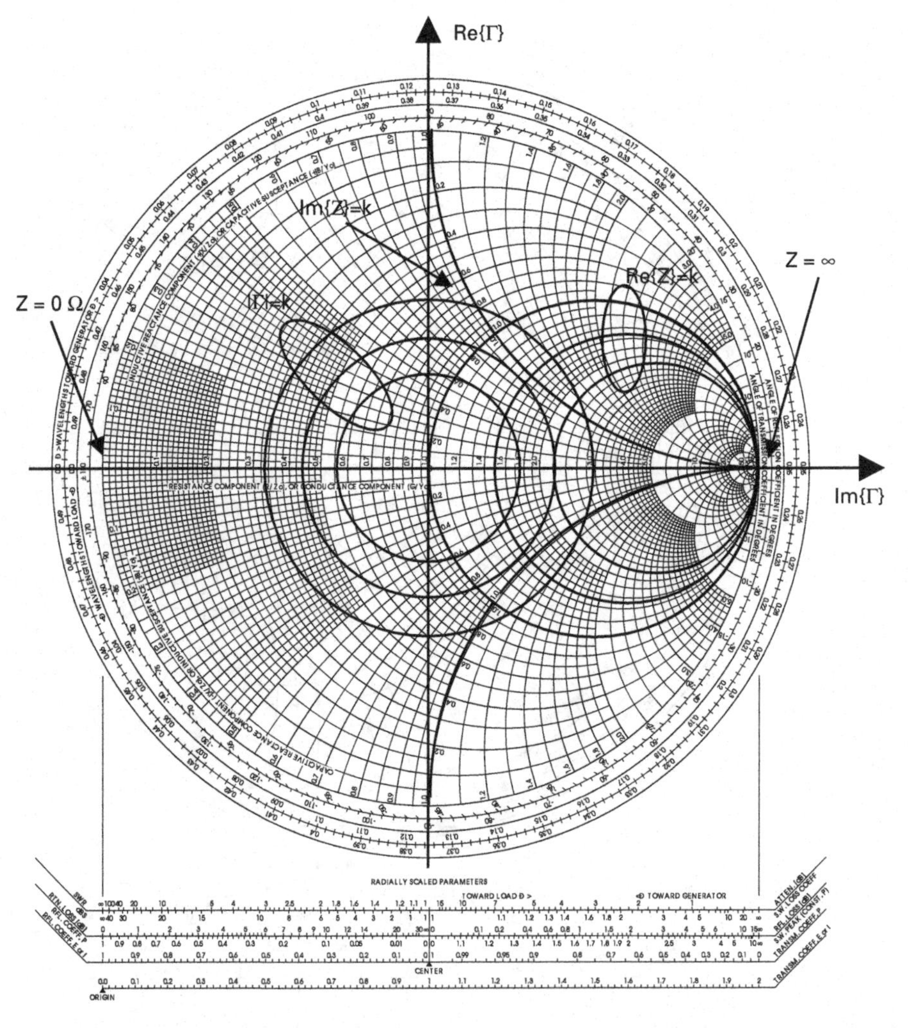

Fig. 1.8 Constant resistance and reactance lines on the Smith Chart.

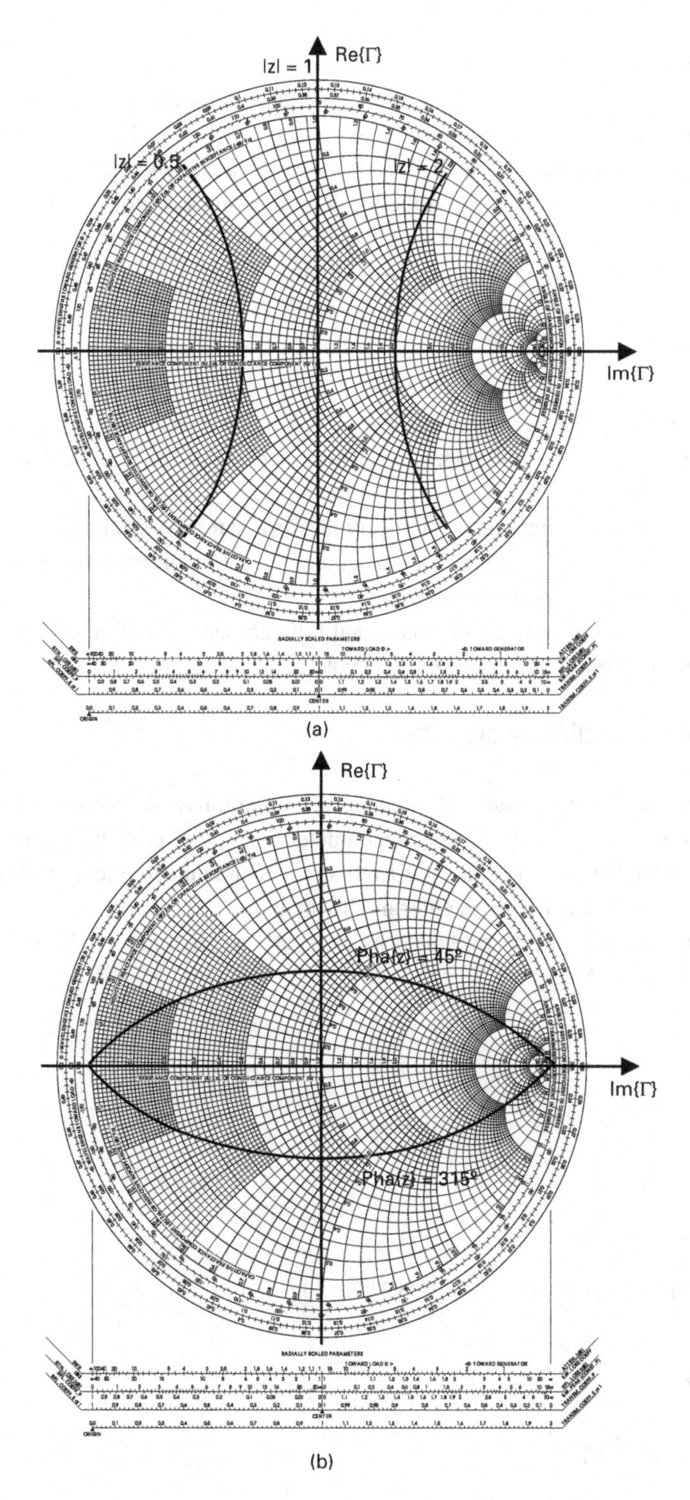

Fig. 1.9 Constant impedance magnitude (a) and constant impedance phase (b) represented on the Smith Chart.

The Smith Chart is typically used to map the Γ (and the impedance) across the length of the transmission line. In the absence of losses, $\Gamma(l) = \Gamma_0 e^{j2\beta l}$, where $2\beta l = 4\pi l/\lambda$. So while moving along a transmission line, the Γ moves on a circle with constant $|\Gamma| = \Gamma_0$, changing only its phase. Clockwise direction represents moving towards the generator and counter clockwise represents moving towards the load. Moving from short circuit to open circuit represents a quarter of a wavelength.

1.6 Conclusions

In this chapter the concepts of wave propagation along a transmission line, which are important when the excitation signal frequency increases to the RF and microwave regions, were revised.

The representation of a generic linear n-port network in terms of scattering parameters was presented and the most important passive component for microwave measurements, the directional coupler, was described.

There follow two Appendices, one on signal flow graphs and the other summarizing the types of transmission lines cited in this book.

Appendix A – Signal flow graphs

Microwave networks can be analyzed using signal flow graphs and scattering parameters. Each variable becomes a node, and each parameter becomes a branch. A branch enters a dependent variable node and emanates from an independent variable node. Each node is equal to the sum of the branches entering the node.

A two-port network can be presented as two parts a_1, b_1, and a_2, b_2 as shown in Figures 1.A.1. and 1.A.2.

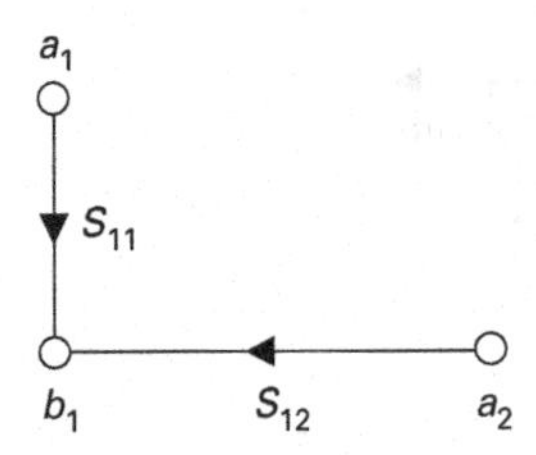

Fig. 1.A.1 Signal flow graph describing the scattering equation: $b_1 = S_{11}a_1 + S_{12}a_2$.

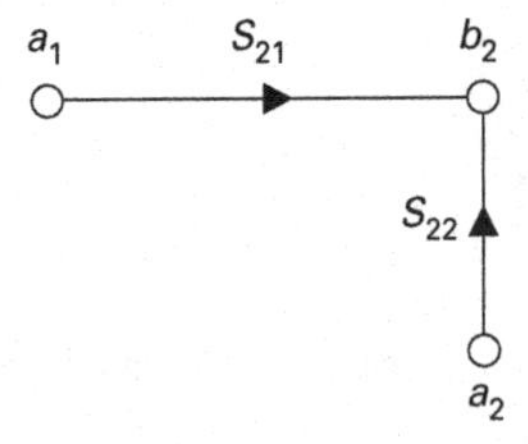

Fig. 1.A.2 Signal flow graph describing the scattering equation: $b_2 = S_{21}a_1 + S_{22}a_2$

Thus, the complete two-port flow graph is shown in Figure 1.A.3.

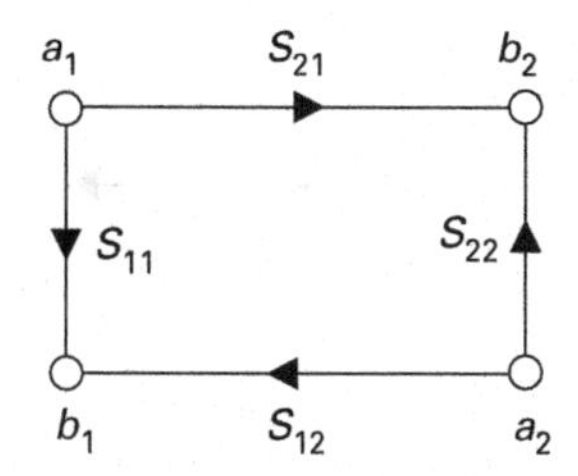

Fig. 1.A.3 Signal flow graph describing a set of two-port scattering equations.

The generator and load add more nodes and branches as shown in Figure 1.A.4.

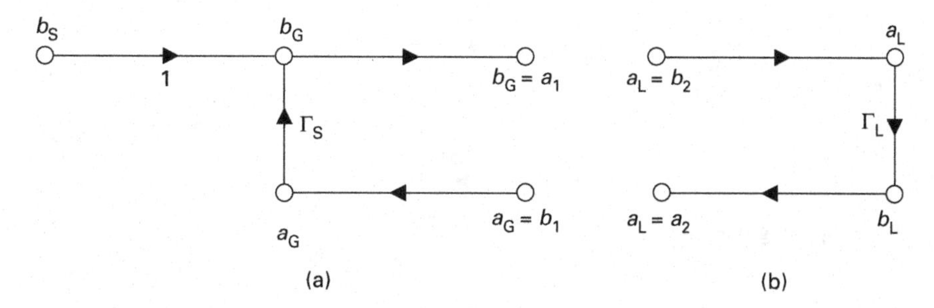

Fig. 1.A.4 Generator (a) and load (b) representation with signal flow graphs.

Thus the overall flow graph can be combined as shown in Figure 1.A.5.

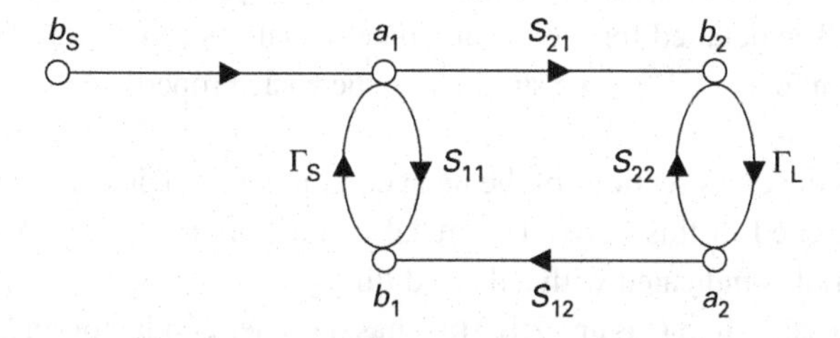

Fig. 1.A.5 Full representation of a microwave source and load connected to an S-matrix.

Finally, some basic rules for the nodes are described in Figures 1.A.6–1.A.9.

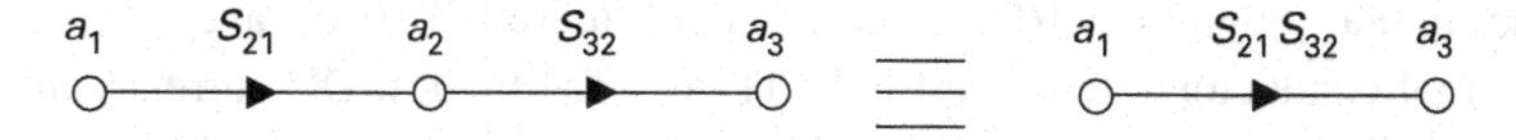

Fig. 1.A.6 Series rule.

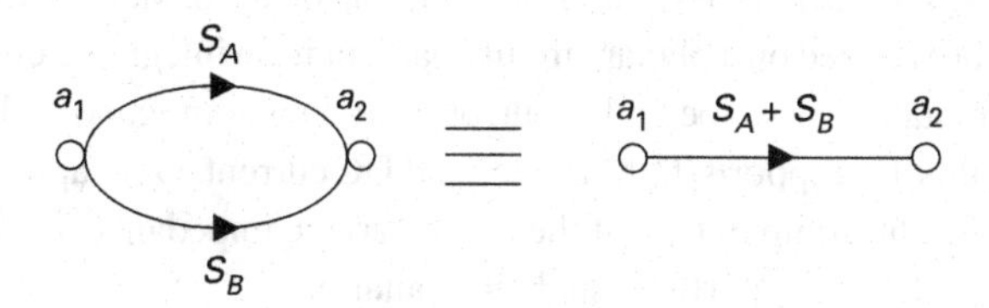

Fig. 1.A.7 Parallel rule.

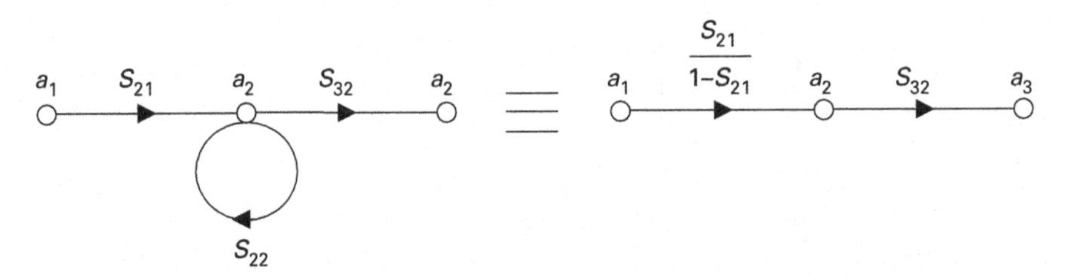

Fig. 1.A.8 Self-loop rule.

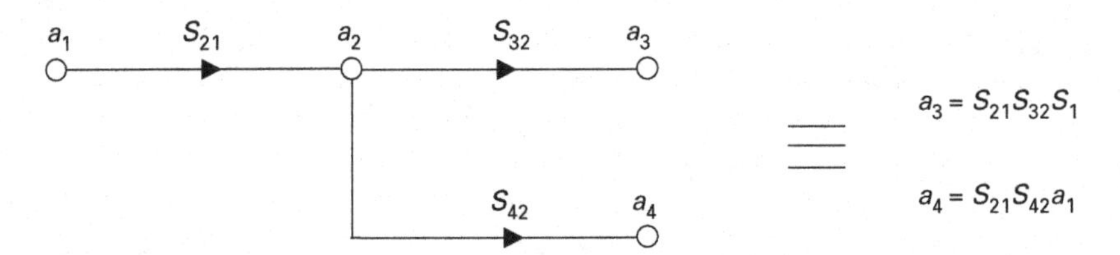

Fig. 1.A.9 Splitting rule.

Appendix B – Transmission lines types

Complete information and description of transmission line typologies is far from the purpose of this book. The detailed formalism and modal analysis can be found in many other books, as for example [6]. Here we summarize the basic properties of some of the most used.

Figure 1.B.1 shows the cross sections of the most common typologies of transmission lines, some of which cited in this book. The metal conductors are depicted in black, while dielectric material is indicated with a dashed filling.

The coaxial transmission line (Figure 1.B.1(a)) has an inner conductor and an outer ground shield. It supports TEM modes if the dielectric is homogenous and a DC current can flow through such a transmission line. Closed formulas for the computation of the characteristic impedance from the physical dimensions are available.

The circular (Figure 1.B.1(b)) or rectangular (Figure 1.B.1(d)) waveguides do not have an internal conductor and support only TE and TM modes; DC current can flow. The typical medium within the metal shield is air; this keeps the losses of a waveguide very low, typically much lower than those of coaxial cable of the same length.

The stripline (Figure 1.B.1(c)) is the natural evolution of a coaxial cable when a transmission line must be realized on a planar circuit board, or in an integrated circuit. The inner conductor has a rectangular shape and is surrounded by a homogenous dielectric. Like the coaxial cable, this TL supports TEM modes and DC current. Only approximated formulas are available for the computation of the characteristic impedance; nevertheless modern simulators (e.g. FEM) can perform such computations.

The microstrip (Figure 1.B.1(e)) is also typical of integrated circuits or PCBs; it's very easy to fabricate since the strip does not need to be embedded in the circuit but can be

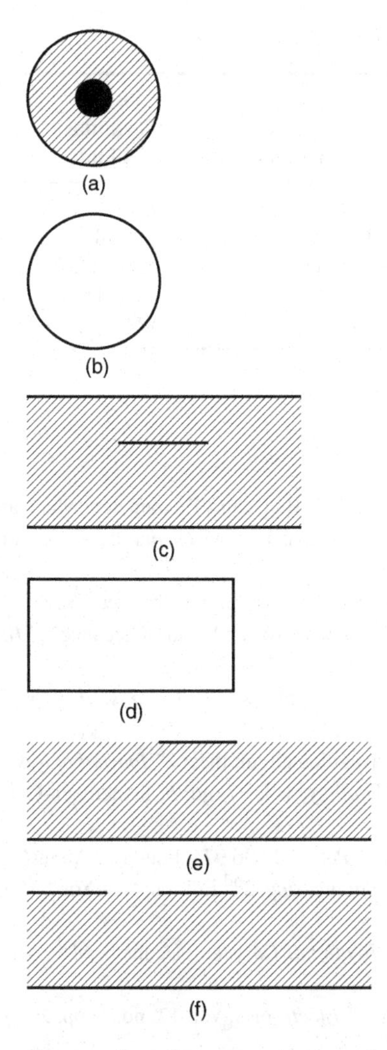

Fig. 1.B.1 Cross section of common use transmission lines: coaxial (a), circular waveguide (b), stripline (c), rectangular waveguide (d), microstrip (e), coplanar waveguide (f). Dashed lines represent a dielectric material.

fabricated by the classical exposure-etch methods. As the medium is non-homogenous, the supported modes are quasi-TEM.

Finally, the coplanar waveguide (Figure 1.B.1(f)), also typical of PCBs and ICs, has two ground shields placed at the side of the main central line. The ground potential is achieved by drilling some via-holes, to reach the bottom ground conductor. CPWs reduce the crosstalk between different lines on the same circuit. The drawback is that the top ground shields must be kept at the same zero potential along all the line length: via-holes must be then placed with proper spacing all along the line length.

Table 1.B.1 summarizes the basic properties of these TLs.

Table 1.B.1 Basic properties of the most common transmission line types

Name	Figure	Type	DC supported	Closed Formulas Available
Coaxial	1.B.1 (a)	TEM	Yes	Yes
Waveguide	1.B.1 (b–d)	non-TEM	No	Yes [6]
Microstrip	1.B.1 (e)	quasi-TEM	Yes	No [9], [10]
Stripline	1.B.1 (c)	TEM	Yes	No [11]
Coplanar waveguide	1.B.1 (f)	quasi-TEM	Yes	Yes [6], [10], [12]

References

[1] D. A. Frickey, "Conversions between S, Z, Y, h, ABCD, and T parameters which are valid for complex source and load impedances," *IEEE Trans. Microw. Theory and Tech.*, vol. MTT-42, no. 2, February 1994.

[2] R. B. Marks and D. F. Williams, "Comments on 'Conversions between S, Z, Y, h, ABCD, and T parameters which are valid for complex source and load impedances'," *IEEE Trans. Microw. Theory and Tech.*, vol. 43, no. 4, April 1995.

[3] K. Kurokawa, "Power waves and the scattering matrix," *IEEE Trans. Microw. Theory and Tech.*, vol. MTT-13, no. 2, March 1965.

[4] R. Marks and D. Williams, "A general waveguide circuit theory," *Journal of Research of the National Institute of Standards and Technology*, vol. 97, no. 5, September–October 1992, pp. 533–562.

[5] H. A. Wheeler, "Directional Coupler," U.S. Patent 2 606 974, issued 12 August 1952.

[6] R. E. Collin, *Foundations for Microwave Engineering*. 2nd Edition, New York: McGraw-Hill, 1992.

[7] L. Young, *Parallel Coupled Lines and Directional Couplers*. Dedham, MA: Artech House, 1972.

[8] P. H. Smith, "Transmission line calculator," *Electronics*, vol. 12, no. 1, pp. 20–31, January 1939.

[9] H. A. Wheeler, "Transmission-line properties of a strip on a dielectric sheet on a plane," *IEEE Trans. Microw. Theory Tech.*, vol. MTT-25, pp. 631–647, Aug. 1977.

[10] K. C. Gupta, R. Garg, I. J. Bahl, and P. Bhartia, *Microstrip Lines and Slotlines*, 2nd Edition. Dedham, MA: Artech House, 1996.

[11] H. Howe, *Stripline Circuit Design*, Dedham, MA: Artech House, 1974.

[12] T. Q. Deng, M. S. Leong, and P. S. Kooi, "Accurate and simple closed-form formulas for coplanar waveguide synthesis," *Electronics Letters*, vol. 31, is. 23, pp. 2017–2019, November 1995.

2 Microwave interconnections, probing, and fixturing

Leonard Hayden

2.1 Introduction

In this chapter concepts related to connecting test equipment to a device-under-test are explored. Application-specific definitions of device boundaries and measures for signal path and power-ground performance are introduced. Practical measurement system accuracy implications of fixture losses are examined, with the surprising result that sometimes more fixture loss can be beneficial to measurement precision. An introduction to the basic elements of microwave probing and probing applications concludes the discussion.

2.2 Device boundaries and measurement reference planes

It is necessary to clearly define the boundary of the target of a measurement (known generally as the Device Under Test, or DUT), to distinguish it from the test system – fixture, probes, or other interconnections. The DUT can take many forms. It could be a functional block in a housing with connectors or waveguides for the inputs and outputs. Or a circuit DUT could be an embeddable semiconductor functional design element with a standard interface point such as a microstripline or other transmission lines. At the other extreme, the DUT could be a circuit component such as a transistor, inductor, capacitor, or resistor with no interface elements other than the constituent electrical contacts. Somewhere in between is the fully distributed circuit element. In all cases, the measurement *reference planes* define the boundaries of the DUT; see Figure 2.1.

For the purposes of discussion let us broadly and perhaps arbitrarily assign DUTs into three categories: Devices, Transmission Lines, and Circuits.

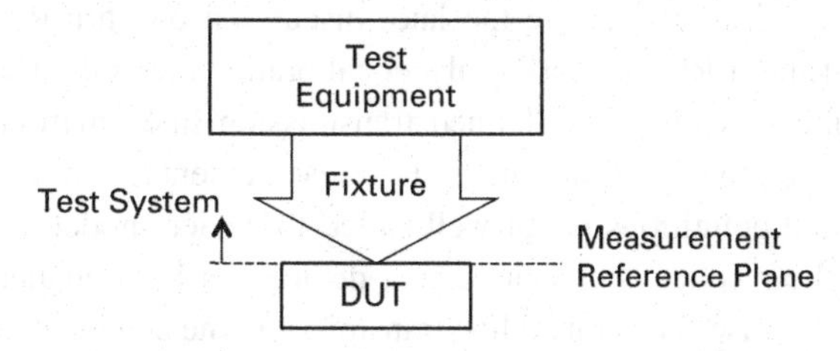

Fig. 2.1 The measurement reference plane is the dividing line between the test system, including test equipment and fixturing, and the Device Under Test.

2.2.1 Devices

This category of DUT includes the components making up, for example, an integrated circuit design. When electrically small, these devices may have essentially lumped element behavior, changing to lumped element with parasitics, and to fully distributed behavior as the electrical size grows with frequency. Often, it is desirable to consider devices with distributed behavior in the transmission line category of DUT.

Device models describe the behavior of the circuit component either with a functional black-box with a network parameter description, through an equivalent circuit made up of a topology of ideal components arranged to mimic the device behavior, or some hybrid of the two. A so-called "compact" circuit model of a CMOS transistor can easily exceed 100 parameters defining the functioning behavior and combines lumped element topology modeling with special mathematical expressions.

The goal of a device model is to predict the performance of a circuit from a theoretical array (in a circuit simulator) of the devices and topology of the design. The measure of success of a device model is the degree of its success *for the required application.* In modeling applications, the generality of a model is compromised for efficiency and the model is always created with a context or range of applicability in mind. Device modeling measurements, likewise, are scaled and evaluated based upon the application needs.

2.2.2 Transmission lines

Transmission lines, distributed circuits, and other interconnection elements, such as adapters, pose a particular challenge for a measurement system. The desired electrical behavior can approach the ideal with close to no insertion loss and no reflections. Measurement attempts are often limited by the residual uncertainty of the system itself. For example the mismatch measured at the input of an ideal transmission line is the mismatch of the termination; or, in calibrated systems, the residual error in the characterization of the termination mismatch.

The location of the measurement reference planes is critical for transmission lines. In modeling a transmission line, one of the most fundamental properties is the propagation constant – the attenuation and delay/phase per-unit-length. For accurate calculation of the per-unit-length normalization, the physical distance between the measurement reference planes must be known accurately.

For this class of measurement, the reference planes must be in the middle of uniform sections of the transmission line: see the later discussion of Thru-Reflect-Line (TRL) and Line-Reflect-Line (LRL) network analyzer calibration methods. These methods support a DUT embedded in a well-defined transmission line environment that effectively becomes part of the DUT as far as the measurement is concerned. But using such line-surrounded definitions is not well suited to device modeling. Moving the reference planes to the physical boundaries of a device can be a challenging de-embedding or modeling exercise. The embedding transmission line behavior can change as it nears a discontinuous transition to the device due to the complex electromagnetic interaction.

2.2.3 Circuits

The simplest conceptual measurement case is the basic amplifier (or passive two-port device) with a coaxial connector at the input and output; see Figure 2.2. The reason these circuits seem simple is that they directly connect to instrumentation and, in the simplest cases, the only measurement concern is transmission gain or attenuation in a 50 ohm environment, either as a frequency response or pulse response in the time domain. This measurement can be made either with a pulse source and an oscilloscope or with a swept frequency signal generator and a power meter. A direct connection of the input to the output provides the input reference excitation that can be removed/normalized from the measured response to isolate the DUT behavior.

Adding one level of complexity raises concern about the input and output match, perhaps characterized as a standing wave ratio (SWR); see Figure 2.3. When a circuit is always tested and used in an environment supplying perfect 50 ohm terminations, then DUT reflection behavior is simply a contributor to the transmission response. A scalar value of the mismatches allows an estimation of the uncertainty bounds on the transmission behavior associated with the DUT and system or test fixture mismatches interacting with each other. Adding a reflectometer or SWR bridge to the scalar measurement transmission test system facilitates scalar mismatch measurement.

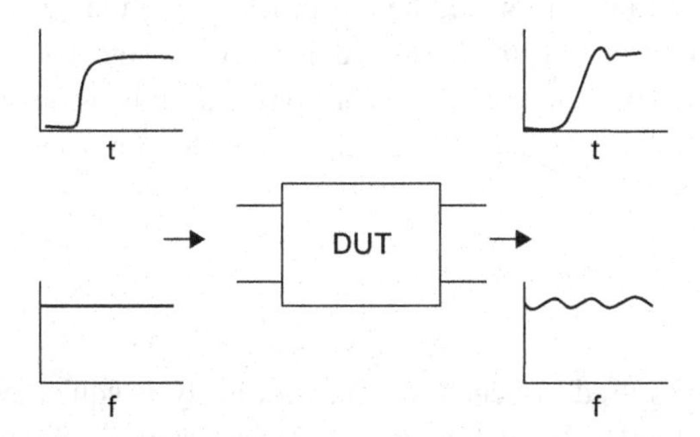

Fig. 2.2 Simple circuit transmission behavior.

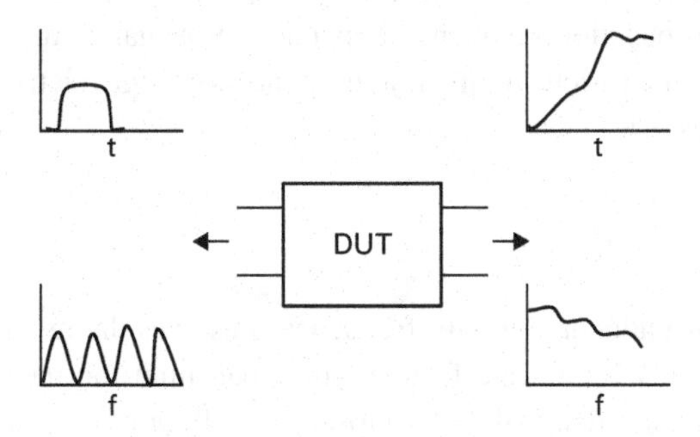

Fig. 2.3 Circuit behavior with reflections.

Adding yet another level of complexity, we can consider the signal distortion in wide bandwidth signals caused by frequency-dependent delay (dispersion/deviation from linear phase). Often the absolute delay is not a concern, but dispersion can significantly change a pulse shape through increased rise-time and the ringing of a step response.

For this case, adding a vector voltmeter to our previously scalar measurement system would work, but the more general answer is to use a Vector Network Analyzer (VNA) with two or more port signal switching, directional couplers (or bridges), wide dynamic range receivers, and "calibration" software. These features allow a VNA to clearly define the measurement reference planes and computationally remove the fixture behavior from the measurement.

For well-behaved linear systems, the VNA measurement is capable of fully characterizing the DUT behavior at the VNA calibration reference planes, independent of the non-idealities of the measurement system. While it is a very powerful instrument, the VNA has limitations and there are measurement situations that add further complexity: multi-mode excitations, nonlinear DUT gain sensitivity to port terminations, frequency conversion and intermodulation products, etc.

2.3 Signal-path fixture performance measures

At the most fundamental level, a fixture would be most ideal if it could be electrically represented as a node, i.e. a connection with zero loss, delay, or other signal or impedance impacts. This is a useful concept in the regime where physical dimensions are small compared to the electrical wavelengths and lumped element approximations are usable for circuit work.

2.3.1 Delay

The roughly meter-long cables used in bench-top network analysis equipment fail to behave as nodes for all but the lowest of frequencies. Delay or phase-shifts become a behavior that must be considered. Linearly increasing phase-shift with frequency, or constant-delay, has the least impact, as signals can propagate without distortion – a complex waveform will retain its shape from input to output of the interconnection. The alternative dispersive propagation, frequency-dependent propagation delay, causes changes in the wave-shape as the various frequency components change relative position in time due to unequal delays.

2.3.2 Loss

Loss in the form of a uniform attenuation with frequency preserves the relative signal wave-shape (it is modified only by a scale factor), but it can limit the measurement system dynamic range as attenuated signals grow closer to the floor of uncertainty, due to various types of noise or correlated or uncorrelated interfering signals.

Frequency-dependent loss is present in interconnections when the skin-depth, i.e. the conductivity and frequency-dependent dimension of the electromagnetic penetration of current distribution in a metal, becomes smaller than the cross-sectional dimensions of the interconnection. For the lower range of frequencies where the skin-depth exceeds the conductor dimensions, the series resistance and commensurate attenuation are essentially constant as the entire cross-sectional area carries current. At increasing frequencies, the depth of penetration effectively limits the cross-sectional conductor area carrying current, resulting in series resistance that is proportional to the square root of the frequency. This behavior is known as skin-effect loss.

The skin-effect loss behavior occurs for frequencies above which the conductors are no longer fully penetrated by the current distribution. The transition to skin-effect loss behavior may be observed in the MHz frequency range for larger dimension structures like coaxial cables used in instrumentation. However, for small dimension interconnections such as thin-film traces on ceramic or semiconductor substrates, the conductors are very thin and the transition to partial penetration may occur as high as several GHz.

Dielectric loss is also a component of interconnection loss, particularly for lower quality materials such as common FR-4 circuit boards. Measurement systems tend to avoid using material with significant dielectric loss contribution, frequency-dependence, or resonant behavior over the frequencies of use.

2.3.3 Mismatch

Even a lossless, constant-delay transmission line can contribute to signal attenuation and distortion. In any distributed system, impedancc mismatch will cause signal reflections, and pairs of mismatches work to cause a multiply reflected signal path to combine with the direct signal path resulting in a frequency dependence of loss (as the path-length varies between constructive and destructive interference) and a phase-shift (when the path-length combines to cause signal leading or lagging). Example reflection magnitude values are shown in Table 2.1.

The single mismatch section example suggests that a severe mismatch is required to cause a significant non-ideality. Indeed, for a significantly mismatched 40 ohm section in a 50 ohm environment, the attenuation ripple amplitude is only about 0.2 dB, the deviation from the linear phase is less than 1 degree, and the peaks in return loss exceed 30 dB.

However, a real interconnection system may have several regions of mismatch each contributing such a response. With the phase coherence of constructive interference, the reflections add as voltages, so the worst-case combination of two 30 dB return loss non-idealities is 6 dB higher or a 24 dB return loss. As the number of transitions increases, the performance can rapidly degrade with each contributor of non-ideal match (all combinations of impedance discontinuity interactions).

Estimation of lumped parasitic values – a practical tool

Transmission line theory can be helpful in estimating the behavior of electrically short sections of interconnections such as bond-wires, trace-overlap regions, or socket elements. While transmission line equations represent l and c on a per-unit-length basis,

Table 2.1 Impedance vs. reflection coefficient, SWR, and return loss relative to a 50 ohm environment.

Impedance (ohms)	Γ	SWR	Return Loss (dB)
200	0.60	4.00	−4.4
150	0.50	3.00	−6.0
120	0.41	2.40	−7.7
100	0.33	2.00	−9.5
80	0.23	1.60	−12.7
75	0.20	1.50	−14.0
70	0.17	1.40	−15.6
65	0.13	1.30	−17.7
60	0.09	1.20	−20.8
55	0.05	1.10	−26.4
54	0.04	1.08	−28.3
53	0.03	1.06	−30.7
52	0.02	1.04	−34.2
51	0.01	1.02	−40.1
50	0.00	1.00	–
49	−0.01	1.02	−39.9
48	−0.02	1.04	−33.8
47	−0.03	1.06	−30.2
46	−0.04	1.09	−30.2
45	−0.05	1.11	−25.6
40	−0.11	1.25	−19.1
35	−0.18	1.43	−15.1
30	−0.25	1.67	−12.0
25	−0.33	2.00	−9.5
20	−0.43	2.50	−7.4
10	−0.67	5.00	−3.5
5	−0.82	10.00	−1.7

with a constraint of electrically short regions we can use these same equations for total inductance L and total capacitance C. The characteristic impedance and total delay of the electrically short section are then given by:

$$Z_0 = \sqrt{\frac{L}{C}} \tag{2.1}$$

$$T = \sqrt{L \cdot C} \tag{2.2}$$

and the total inductance and capacitance in terms of impedance and total delay are then:

$$L = T \cdot Z_0 \tag{2.3}$$

$$C = \frac{T}{Z_0} \tag{2.4}$$

For any particular trace region, an estimate of the physical length and approximate dielectric coefficient is enough to determine a surprisingly accurate estimate of the total delay T. Transmission lines created using coax, stripline, or microstrip require extremely small or large physical dimensions to realize particularly high or low impedances. Going much above 100 ohms or much below 10 ohms becomes difficult and the impedance extreme is readily apparent from the extreme dimensions. It becomes possible to look at structures and have a good idea of the impedance value to better than an order of magnitude – or even as close as a factor of 2 or 3. For example, a very wide trace over a ground plane is likely to be close to an estimate of 10 ohms, while a thin wire in air far from ground might be approximated by 100 ohms. Using estimated delay and impedance values, the total inductance and capacitance are computed using (2.3) and (2.4) with reasonable accuracy.

For example, consider a narrow bond wire 200 μm above a conductive plane and 1.2 mm long in air. Air dielectric has a propagation velocity of 3×10^8 m/s, or 300 μm/ps for convenience at this scale. The total delay of the 1200 μm length is then 4 ps to high precision. Characteristic impedance equations or a cross-sectional simulator could be used for best accuracy, but a safe guess for the impedance of the bond wire is of the order of 100 ohms. This estimate is expected to be well within a factor of two of the actual value. The impedance is safely above 50 ohms, since a typical 25 μm diameter bond wire in air would have to be much closer than 25 μm to realize 50 ohms. And, as a bond wire becomes very far from a ground plane, the impedance curve flattens out making 200 ohms a likely maximum to achieve. The total inductance estimate is then 400 pH and the less significant capacitance is 40 fF. These values should be within about a factor of two of the actual values, providing an easy to obtain and often very useful estimate.

2.3.4 Crosstalk

Ideally, multiple interconnections do not electrically interact and closed transmission line structures such as coaxial cables exhibit very low crosstalk. However, open interconnection structures such as parallel microstriplines on circuit boards, integrated circuit packaging, or other interconnections will exhibit crosstalk. Crosstalk effects are cumulative and grow with the complexity of the circuit (e.g. number of inputs and outputs), but even a single input, single output circuit such as a transistor test fixture can be sensitive to input/output port coupling when used for device modeling or critical performance metrics such as F_{max} derived from Mason's gain.

Crosstalk is particularly difficult to remove from a measurement because crosstalk mechanisms in test fixtures can have electrical behavior dependent on a DUT. The thought experiment of a parasitic capacitive coupling between input and output in a two-port measurement system easily demonstrates this. The current through the capacitor depends on the dynamic voltage. If the DUT measured is low-impedance, low voltages are present and capacitive crosstalk is small. A high-impedance DUT maximizes the capacitive crosstalk.

Particularly subtle and often difficult to identify is the measurement "suckout" which is often related to crosstalk. A suckout is a frequency response magnitude aberration with a characteristically narrow and small dip in transmission – often of the order of 0.1–1 dB in depth. One cause for a suckout is when an otherwise clean transmission line is lightly coupled to an adjacent transmission line without terminations. The adjacent open-ended line acts as a high-Q, half-wavelength resonator that is only lightly loaded by the coupling to the primary signal path. Energy is sucked away from the signal path at the peak of the resonance.

2.3.5 Multiple-modes

Measurement systems often expect interconnections to only allow a single propagating mode at a network port. Network analysis and S-parameter theory depend upon this and when a physical interconnection has significant energy in another mode, then this mode must be mathematically separated and considered as effectively an additional network port.

The problem with multiple-modes at a reference plane is with the transmission line definition of the port. Non-degenerate modes propagate at different velocities, creating an interaction pattern with distance that is not compatible with a propagation constant description. Over electrically short transitions this cannot happen and the effect does not have an impact.

The consistent summation of all mode behavior at a location may be used to instead define a port voltage and current allowing network modeling where this non-distributed behavior is appropriate. This is the situation for an abrupt transition – there is no single propagation behavior; multiple-modes or even higher-order electromagnetic coupling exists, but the region with this behavior is electrically short allowing voltage-current based circuit modeling to effectively describe the transition non-idealities.

The conductor-backed finite ground coplanar waveguide is an example of such a multi-mode path; see Figure 2.4. The three conductors along with the backside ground plane allow three modes of propagation: the desired *coplanar waveguide mode* with outbound

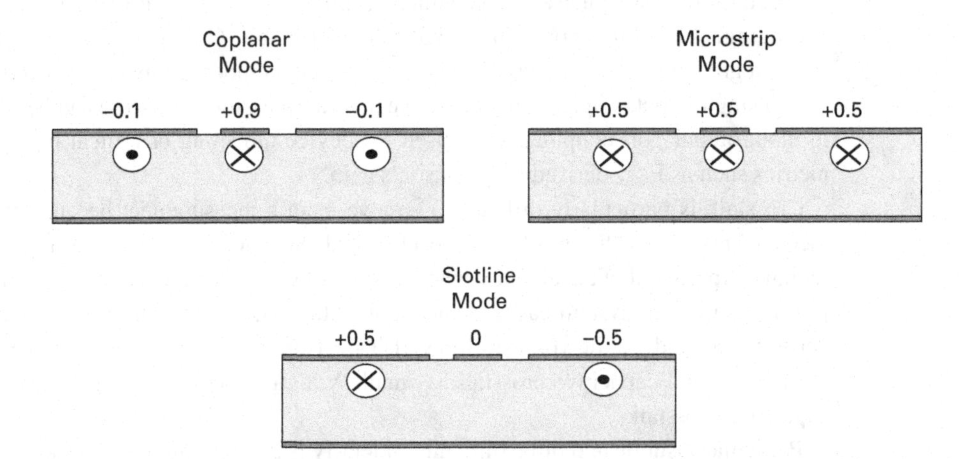

Fig. 2.4 Propagating modes in conductor-backed coplanar waveguide with finite ground conductor width.

current in the signal trace and equally split return currents in the ground traces; *microstrip mode* where all traces carry outbound symmetric currents and the return current is in the ground plane; and *slot-line mode* where the signal trace is current-free and the ground traces carry equal currents in opposing directions.

2.3.6 Electromagnetic discontinuity

In an electrical transition between two different transmission lines, such as between a coaxial cable and a microstripline, the discussion above shows that maintaining a constant impedance is necessary to avoid mismatch losses. But even if every cross section along the transition has the same equivalent impedance, does this guarantee optimal performance?

When approaching this problem from the electrical circuit or even distributed circuit approach the answer would seem to be yes, but these are only approximations to the physical world modeled by Maxwell's electromagnetic equations. Examining the electric field patterns for coaxial cable and microstrip shows two radically different shapes. An abrupt transition can be made between the two but it will incur a mismatch associated with the mismatched fields. The simple transmission line approximation fails to predict this.

Eisenhart [1] used the concept of an electromagnetic transition as a way to optimize behavior and minimize mismatch effects. A continuously varying cross section is used that progressively shifts the electrical field patterns from the radial pattern of the coax case to the vertical with fringing microstrip field. When done over a distance that is electrically long at frequencies where the electrical discontinuity would otherwise be of significance, an optimal transition is created; see Figure 2.5.

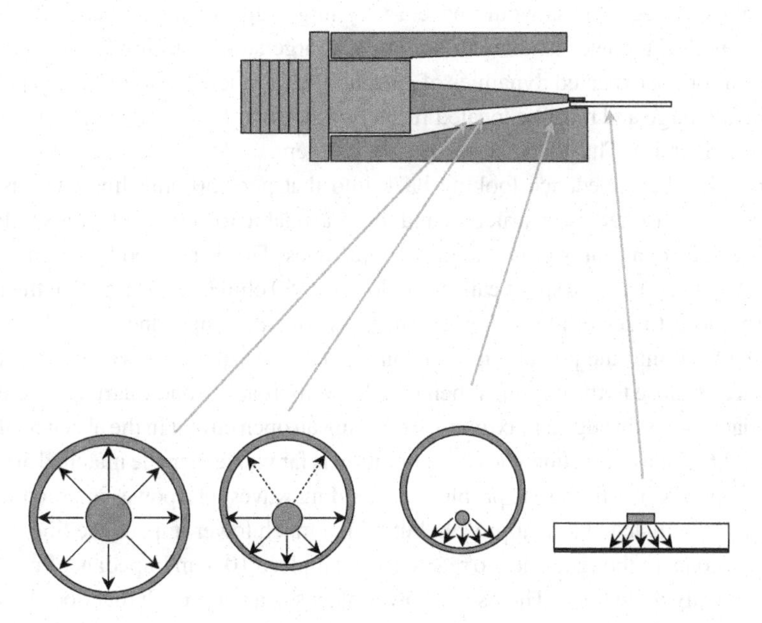

Fig. 2.5 The Eisenhart launcher creates a continuous field transition between coaxial cable and microstripline, minimizing mismatches.

2.4 Power-ground fixture performance measures

Power and ground path performance often has greater impact on a measurement of circuit performance than non-ideal signal paths. Circuits expect a "stiff" or low-impedance supply with a constant voltage independent of the dynamic current draw. Ground is expected to be an equipotential reference everywhere it is used.

De-embedding or compensation of the impact of a non-ideal power or ground is not something that has been demonstrated. Unlike the signal path where correction may be possible, it is essential that the power and ground paths are optimized as much as practical in our measurement system. In some cases the best test is obtained when the power and ground connections are identical to what would be used in application. Often this means using the original application circuit with most connections retained, but modified to allow microwave measurement launches to the signal input and output.

2.4.1 Non-ideal power

If circuits could function nominally when supplied DC power from a 50 ohm source, then this would be a very short topic. But many active circuits will suffer degraded gain, dynamic range, linearity, and increased tendency to oscillate with a high-resistance power supply. The latter concern requires not just a low and controlled-impedance power supply over the frequencies where there is signal energy, but the impedance must be well controlled anywhere there is gain and the possibility of oscillation.

Bypass capacitors are the key components for reducing power bounce, as they have the tendency to stabilize voltage by acting as a reservoir of charge to supply surges in current. A perfect bypass would have a very large capacitance immediately connected across the DUT power leads with sufficient charge storage to hold a constant voltage across it for the expected dynamics of current. But in practice, large value capacitors are physically large and must be located further away from, for example, the tiny pads of an integrated circuit. This distance can pose a problem.

Consider the impedance looking back into that poor 50 ohm line used as a power source. For reasonable construction and to a reasonable tolerance, 50 ohms will be seen, and it will be constant over low to high frequencies. This is not good enough, so the first thought is to add a large shunt capacitor along our 50 ohm line so large that the reactance is very close to zero even at low frequencies. But the impedance seen by the circuit looking back into the power supply is only small at frequencies where the capacitor is electrically close to the circuit. When this delay approaches one-quarter wavelength, the impedance goes through a maximum, becoming an open circuit in the absence of loss; see Figure 2.6. For these frequencies this condition is far worse than the matched 50 ohm line.

One approach to fixing the problem of standing waves on a power interconnect would be to use the matched line approach, but with a much lower impedance line. Feasibility can be a concern though, since to reach, for example, a 10 ohm impedance can require an impractically wide trace. The use of power planes is an approach that does benefit from this concept, but isn't available when considering sockets, integrated circuit packages, or wafer probing measurement cases.

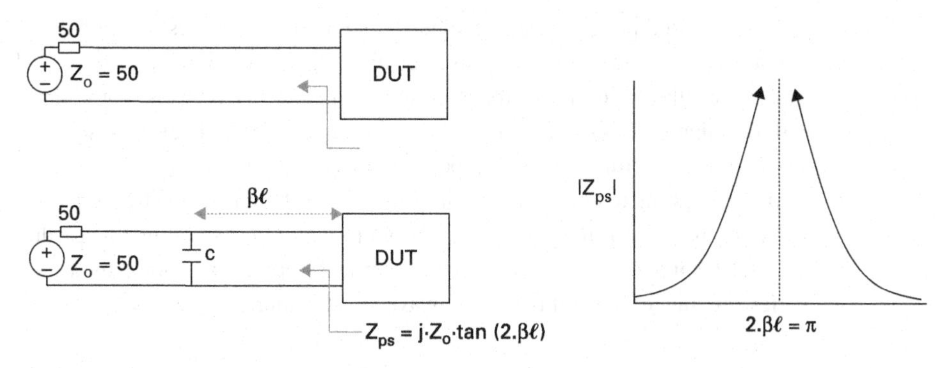

Fig. 2.6 When electrically far from a bypass capacitor undesired extremes of impedance occur.

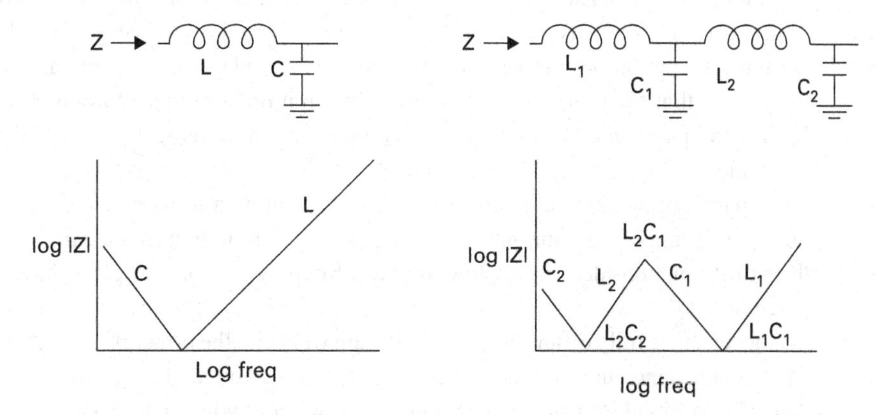

Fig. 2.7 A distributed power system makes use of progressively larger capacitors at progressively greater distances from the DUT.

A distributed power system is often the answer to situations with significant regions of non-zero impedance. Instead of using a single, very large capacitor (value and size) a set of progressively larger valued capacitors is used – with a very small value very close to the DUT, a moderately valued capacitor at a moderate distance from the DUT, and a large valued capacitor at a large distance from the DUT until finally the low-frequency behavior is controlled by the feedback circuits in the power supply that maintain an arbitrarily low impedance.

Careful study of Figure 2.7 leads to insight into the behavior of the power system. In each section the transmission line behavior is modeled well as an inductor, since they are electrically short for the frequency components that can get past the capacitors closer to the DUT.

At the highest frequencies, C_1 behaves as a short circuit and the equivalent behavior is simply that of L_1. As the frequency is lowered, a series resonance of L_1 and C_1 occurs creating an impedance minimum. Below this minimum the impedance grows with the behavior dominated by C_1. Looking past C_1 we see L_2 terminated by C_2. C_2 is large for this frequency range and acts as a short.

An undesired parallel resonance between L_2 and C_1 creates an impedance maximum limited only by losses. Adding resistance in series with C_1 will load the parallel resonance and reduce the magnitude of the maximum impedance. Below the parallel resonance, L_2 will dominate the response to a series resonance with C_2 causing a second impedance minimum and dominating the response at even lower frequencies.

It is interesting to note that with this kind of distributed power structure, loss is required in the transmission lines or the capacitors to avoid peaks of impedance due to high-Q parallel resonance. Bypass capacitor manufacturers often feature their extremely low series resistance, when this is not always an advantage.

2.4.2 Non-ideal ground

Grounding in circuits has been one of the black arts of analog RF and microwave circuit design [2]. Conceptually a ground is a common potential reference, independent of the current distribution present in it. In practice a ground plane must be of sufficiently low impedance that the return currents imposed upon it do not impact circuit performance. If this condition cannot be met, then the equipotential requirement of a ground reference is not met.

In many applications a ground plane is sufficient as a ground reference. However, high power amplifiers running low-voltage swing and high currents can easily exhibit the effects of non-ideal grounding where different spots on the plane have different potentials.

Conversely, a fully differential circuit design works on the principle of a virtual ground where net ground currents sum to zero. For differential circuits the behavior is independent of the ground impedance. Zero current means zero voltage drop and the ground has a single potential level.

Evaluating the impedance of a ground path is particularly troublesome. For a socket, integrated circuit package, or probing environment a reasonable approach can be taken by using a transmission measurement where a signal transmission path is shorted by the imperfect ground connection; see Figure 2.8.

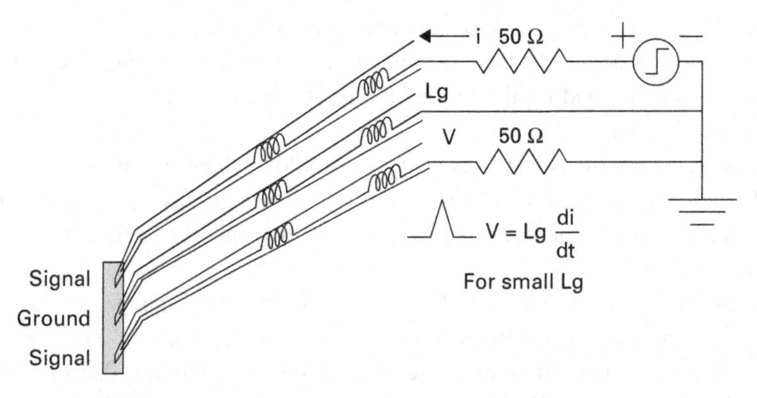

Fig. 2.8 Direct measurement of ground impedance is a difficult problem; the shunt imperfect short method shown allows a reasonable inference of the equivalent ground inductance.

2.5 Fixture loss performance and measurement accuracy

As a general rule it is not bad to assume that higher performance fixtures make for more accurate measurement results. This is true when the measurement and DUT reference planes don't coincide due to limitations in our ability to calibrate our system. In these cases de-embedding or modeling techniques are used to identify and remove the intervening interconnection behavior. Usually, the more ideal this element, the easier it is to identify.

In the ideal case the losses and mismatches of the fixture are small enough that they may be ignored, but in all cases interconnection delay (even if distortion free) is present and may need to be identified. Practical interconnections have losses. Loss is usually split between conductor and dielectric loss. Other loss mechanisms are possible, but effects like radiating loss are considered something to avoid in a measurement system, since the energy may be going to unpredictable locations.

Conductor losses are often dominant in test fixtures, particularly when good quality dielectrics are used (e.g. air, alumina, semi-insulating GaAs, SiO_2, etc) and when conductor cross-sectional dimensions are small and resistance is high. Both propagation constant and characteristic impedance may vary with extreme loss values, but in low-loss cases the mismatch effect may be small enough to ignore. Fundamental transmission line theory [3] tells us how the line impedance is determined from per-unit-length r, l, g, and c (resistance, inductance, admittance, and capacitance, respectively).

$$Z_o = \sqrt{\frac{r + j\omega l}{g + j\omega c}}, \tag{2.5}$$

where ω is the radian frequency given by $2\pi f$.

When g is small compared to $\omega{\cdot}c$, the case for low dielectric loss to surprisingly low frequencies, then at higher frequencies where r is small compared to $\omega{\cdot}l$ we find that Z_o simplifies to the real and often constant high-frequency approximation:

$$Z_o = \sqrt{\frac{l}{c}}. \tag{2.6}$$

In this region, the loss of the interconnection no longer contributes to mismatch, as the inductive reactance and capacitive susceptance dominate the behavior.

In some measurement systems loss can be a stabilizing factor. In vector network analysis, the measurement system dynamic range is often sufficient to tolerate some signal loss without significant impact, and calibration effectively corrects for the fixture loss in the signal path. When an imperfect measurement system has electrically separated elements with high reflections, the interaction resonance errors may be noticeable in the frequency response. Similarly, a suckout caused by coupling to an adjacent unterminated interconnection can create a sharp resonance response.

Corrections need repeatability in the measurement system to be applicable. A small temperature change will cause physical and electrical length changes in the metal interconnect conductors, moving the resonances and invalidating the calibration; see

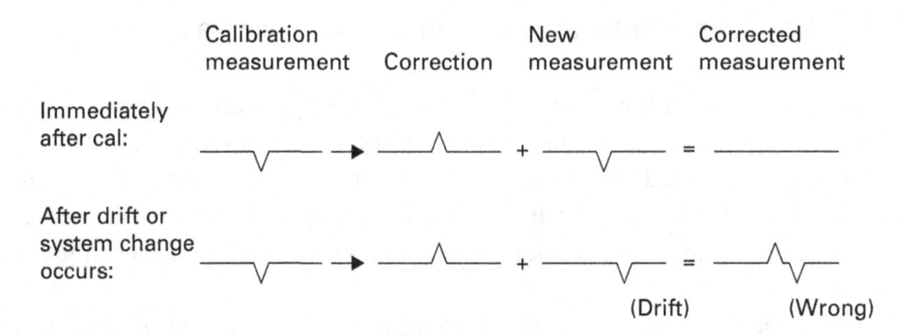

Fig. 2.9 A small change in fixture electrical length can cause any resonances to move in frequency, invalidating the calibration.

Figure 2.9. The presence of loss between the interacting reflections or in the coupled adjacent line will load the resonance causing the effect to be shallower and broader. The calibration will be imperfect but still helpful over a broader temperature range.

Mechanisms other than temperature change will contribute to frequency shifts in the small bumps inherently present in a non-ideal measurement system. Sensitivity to changes from cable bending or twisting is also mitigated by the presence of loss.

Adding loss to a VNA port has a significant drawback, however. The port directivity, the ability to distinguish between the incident and reflected waves, is reduced by twice the attenuation added to the port. For any specific measurement system the solution to the trade-off for optimal loss will differ. A well-matched, suck out-free system with high-quality phase stable cables with minimal bending and twisting during use will work best with minimal fixture loss; while, as reported in [4], a broadband probing measurement system using a poorly matched combiner (between a coaxial low-band and a waveguide high-band), benefits from some degree of loss to stabilize the response from the probing discontinuity and combiner interaction. Loss improves the raw source and load match of the system, and the impact on directivity is not important since the waveguide couplers have very high natural directivity and some degradation can be tolerated.

2.6 Microwave probing

Before the concept of a monolithic microwave integrated circuit became viable hybrid circuits were common practice. These used passive components patterned on a ceramic substrate and individual transistors connected via bond wires. Larger value resistors and capacitors were attached through bonding or direct attach (epoxy or solder). Test fixtures were often of a similar concept using coax-to-microstrip launches and bond wires connecting the DUT to the microstrips; see Figure 2.10. For best performance, the electromagnetically tapered launch developed by Eisenhart was used.

Microwave probes [5] were significant to the development of microwave integrated circuits. Probes enabled much more accurate transistor models due to smaller and more consistent launch structures and the small tip geometry facilitates precise calibration. The

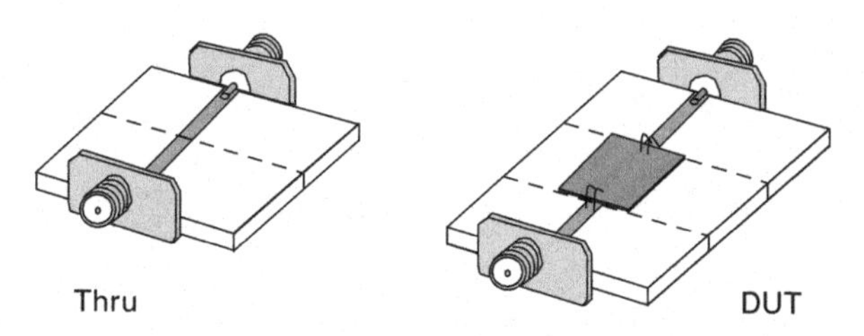

Fig. 2.10 Use of microstrip launchers to characterize a microwave component. Hybrid circuits bonded transistors and capacitors to ceramic substrates and connected them using thin-film interconnects and resistors.

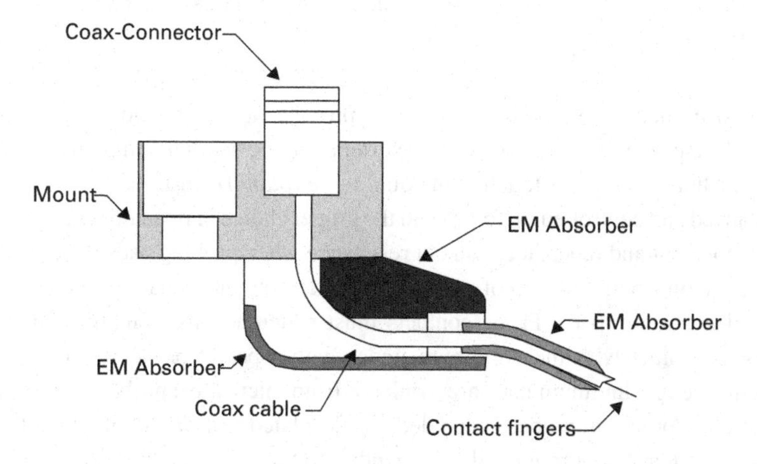

Fig. 2.11 Probe cross section showing typical microwave probe features.

ability to probe devices, test patterns, and circuits while on an undiced, unpackaged wafer reduces the design, fabricate, test cycle time speeding up iterations for improvement.

2.6.1 Probing system elements

A typical microwave probe consists of a coaxial connector, a coaxial cable, and a tip made up of multiple contact fingers; see Figure 2.11. The probe design is optimized to make the size transition from the 1–3 mm diameter coax connector to the 100–500 μm tip width while maintaining a constant 50 ohm impedance.

For best electrical repeatability, the probe tip is designed with contacts that maintain a fixed pitch (center-to-center spacing); see Figure 2.12. Signal path shielding is optimized when ground contacts are immediately adjacent on both sides. Electromagnetic fields are most confined in this configuration and crosstalk to nearby traces or probes is also minimized.

Unlike simple needles, a microwave probe tip is an array of points making contact with multiple pads simultaneously. These pads can be as small as 50 μm square so a precision positioning system is required. A typical microwave probe station uses a high power microscope for viewing and the probe is driven in x-y-z by a three-axis

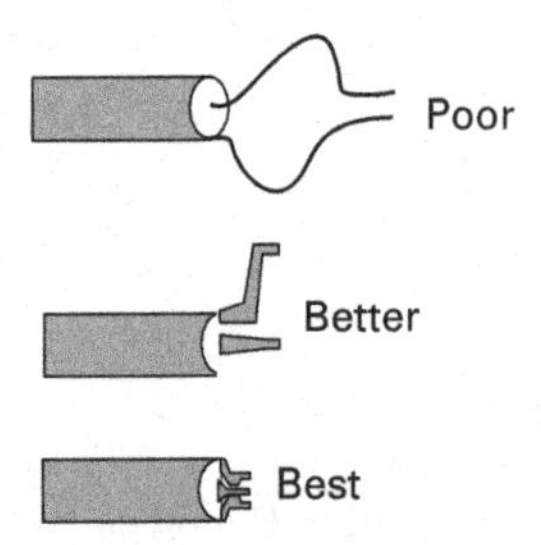

Fig. 2.12 High-performance microwave probes used precisely fabricated fixed spacing contacts in a ground-signal-ground (GSG) configuration. The double ground contact provides better electromagnetic shielding of the signal path and the shorter fingers minimize the impedance and field discontinuity.

micro-positioner. Often the positioner arm provides a roll axis adjustment so that the plane of the probe contact fingers can be oriented to be parallel to the plane of the probe pads – so that all contacts touch with equal force (*planarizing*).

Balanced and controlled tip forces and the proper choice of tip material are necessary to ensure constant and repeatable contact resistance when probing pads with an aluminum surface. A thin barrier layer of aluminum oxide, known as native oxide, forms over exposed aluminum pads. Probe contacts must penetrate this insulating layer to make electrical contact with the aluminum underneath. Special probe versions optimizing performance on aluminum pads are available from microwave probe vendors. The extra cost of these premium probes is avoided if gold-plated pads are available in the process (common in GaAs, but rare in silicon foundries).

2.6.2 VNA calibration of a probing system

Calibration standards for probes can be relatively simple to fabricate. A nearly ideal low-inductance short-circuit is made by a metal shorting bar for the contacts. A probe elevated in air or landed on open pads behaves like it is capacitively terminated (with negative and positive value, respectively). Constant value lumped models (inductor or capacitor) are often used; these are much simpler than the polynomial of frequency dependence combined with delay offset used to model coaxial short and open calibration standards.

A match termination for the GSG configuration uses equal 100 ohm thin-film resistors from the signal to each ground in parallel to provide a well-defined 50 ohm load. These resistors are symmetrically trimmed to tight tolerance with a laser or other mechanism. The electrical model of these standards is simply 50 ohms in series with a lumped, frequency-independent inductor. Coaxial standards normally assume perfectly matched loads, but a probe load requires an extra parameter describing the inductance of the load.

Unlike the case for complementary gender coaxial connectors, no direct thru connection of probes is possible. Instead, a transmission line of known electrical behavior is fabricated on the calibration substrate. By keeping the length of this structure very short electrically (e.g. 1 ps) uncertainty in the impedance or accuracy of the delay parameter has minimal impact on the measured S-parameters of a DUT.

Actual standard electrical behavior varies with specific positional placement of the probes on the standards. A change in probe to pad overlap of even a few microns is discernible in the calibration result. Alignment marks are fabricated and help the probing operator to dock probes into a precise separation. With the aid of the reference and high-power optics, better than 5 μm placement repeatability can usually be achieved. Using a probe station separation stroke the probes are lifted together and landed on simultaneous shorts, loads, or a thru standard. Visible symmetry helps to ensure that both probes overlap equally.

The calibration method plays a part in the sensitivity of the calibration to probe placement. Traditional Short-Open-Load-Thru (SOLT) calibration requires fully known electrical behavior of the standards and is most sensitive to probe placement error caused by variation from the definitions.

The enhanced Line-Reflect-Reflect-Match (eLRRM) with automatic determination of load inductance calibration method [6] uses the same physical standards as SOLT, but does not require the shorts and opens to be anything other than symmetric on the two ports. The inductance of the load/match standard (only one port's match measurement is required) is determined with redundancy in the calibration data. The load need only be modeled well by an R-L equivalent circuit (where R is the known low-frequency value, usually 50 ohms). The eLRRM calibration is far less sensitive to imprecise probe placement than the SOLT calibration method.

In probing two-port standards, such as thru or line structures, the structure orientation and geometry must match the probes. Having a straight thru standard is of no help if the probes are oriented orthogonally or on the same side of a DUT.

This requirement of geometric compatibility between two-port standards and ports is not a requirement in coaxial setups, where the cables may be reoriented as needed to complete the standard measurement. Conversely, the coaxial setup does have to worry about connector gender. These differences in important issues, along with the desire to automate calibration steps, has created a need for specialized calibration software for microwave probing [7]. This software goes beyond the mainly coaxial and rectangular waveguide calibration support provided in most commercial vector network analyzers.

2.6.3 Probing applications – in situ test

While microwave probes were developed for the specific needs of testing devices and circuits while still in wafer form, their advantages can also be applied in areas more familiar to the days of the hybrid circuit. Small ceramic elements containing a GSG probe interface with a transition to microstrip are commercially available [8]. These probe launches can be bonded to the input and output of a compatible circuit that is otherwise in its native environment (power, ground, and other connections).

2.6.4 Probing applications – transistor characterization

A measurement of a transistor (or other small device) has, as its goal, the behavior of the intrinsic device (the device without pads). Measurement after probe tip calibration provides the extrinsic device behavior (the device with pads). Separate measurement and modeling of the device pads allows their mathematical removal. The most commonly used version of this process uses two-steps and is known as Y-Z de-embedding.

In the first step of Y-Z de-embedding, the lumped, frequency-dependent Pi-model Y-parameters of the parasitic shunt parasitics are determined from a measurement of the open pads (device removed). In the second step, the frequency-dependent T-model Z-parameters of the series parasitics are determined from a measurement of the shorted pads (device replace by a short of both signals to ground). Simple subtractions of Y and Z parameters facilitate the de-embedding [9].

2.7 Conclusion

This chapter has provided some key concepts and tools related to device fixturing and performance and boundary determination. There are many more topics that could be explored in much greater depth – particularly as the unique needs of specific applications are considered.

References

[1] R. L. Eisenhart, "A Better Microstrip Connector," *Microwave Symposium Digest, 1978 IEEE-MTT-S International*, pp. 318–320, 27–29 June 1978.

[2] E. Holzman, *Essentials of RF and Microwave Grounding*. Norwood, MA: Artech House, 2006.

[3] Philip C. Magnusson, *et al.*, *Transmission Lines and Wave Propagation*, 4th ed., Boca Raton, FL: CRC Press, 2001.

[4] *Agilent Technologies application note 5989–1941* [Online]. Available: www.agilent.com.

[5] Eric Strid, "A History of Microwave Wafer Probing," *ARFTG Conference Digest-Fall, 50th*, vol. 32, pp. 27–34, Dec. 1997.

[6] L. Hayden, "An enhanced Line-Reflect-Reflect-Match calibration," *ARFTG Conference, 2006 67th*, pp. 143–149, 16 June 2006.

[7] *WinCal^{TM} Calibration Software* [Online]. Available: www.cascademicrotech.com.

[8] *ProbePoint^{TM} Adapter Substrates* [Online]. Available: www.jmicrotechnology.com.

[9] M. C. A. M. Koolen, *et al.*, "An improved de-embedding technique for on-wafer high-frequency characterization," *Bipolar Circuits and Technology Meeting, 1991, Proceedings of the 1991*, pp. 188–191, 9–10 Sep. 1991.

Part II

Microwave instrumentation

3 Microwave synthesizers

Alexander Chenakin

3.1 Introduction

A frequency synthesizer is the most versatile piece of microwave equipment. Synthesizers come in a variety of forms ranging from tiny chips to complex instruments. Single-chip synthesizers are available in a die form or as surface-mount ICs. They include the key elements (such as RF and reference dividers, phase detector, and lock indicator) required to build a simple single-loop synthesizer. Such ICs are mounted on a PCB with additional circuitry. The PCB-based modules range from small, surface-mount, "oscillator-like" designs to more complex connectorized assemblies. Such PCB assemblies can be packaged into a metal housing and are presented as stand-alone, complete synthesizer modules. Connectorized synthesizer modules can be used to build larger instruments such as signal generators for test-and-measurement applications.

Not surprisingly, frequency synthesizers are among the most challenging of high-frequency designs. Many approaches have been developed to generate clean output signals [1–17]. This chapter presents a brief overview of today's microwave synthesizer technologies. It starts with general synthesizer characteristics followed by a review of the main architectures. Direct analog, direct digital, and indirect techniques are compared in terms of performance, circuit complexity, and cost. Synthesizer parameters can be further improved in hybrid designs by combining these main technologies and taking advantage of the best aspects of each. Finally, sophisticated test-and-measurement signal generator solutions are reviewed. The signal generators come with high-end technical characteristics and extended functionality including output power calibration and control, frequency and power sweep, various modulation modes, built-in modulation sources, and many other functions.

3.2 Synthesizer characteristics

A frequency synthesizer can be treated as a black box that translates one (or more) input frequency (usually called *reference*) to a number of output frequencies as shown in Fig. 3.1. This black box contains various components such as VCOs, frequency multipliers, dividers, mixers, and phase detectors, which being properly connected, perform the desired translation function. Although all synthesizers exhibit significant differences as a result of specific applications, they share basic core characteristics depicting their frequency and timing, spectral purity, and output power parameters. These core characteristics are reviewed below. Other specifications (not listed here) may include AC or DC power consumption, control interface, mechanical and environmental characteristics as well as some special functions such as modulation and output power control.

3.2.1 Frequency and timing

Operating frequency range denotes the range of frequencies that can be generated by the synthesizer. It is specified in the units of Hz (MHz, GHz) by indicating the minimum and maximum frequencies generated by the synthesizer.

Frequency resolution or *step size* is the maximum frequency difference between two successive output frequencies, indicated in Fig. 3.1 as $\Delta f = f_{n+1} - f_n$. The operating frequency range and frequency resolution are fundamental synthesizer specifications set by a particular application. Some applications (e.g. test-and-measurement) require very wide coverage and fine frequency resolution while others need a relatively narrow bandwidth with a rough step size or just a single fixed frequency.

Frequency accuracy indicates the maximum deviation between the synthesizer's set output frequency and its actual output. Frequency accuracy is normally determined by the reference signal, which can be internal or external to the synthesizer. Frequency synthesizers usually employ a crystal oscillator as an internal reference. The crystal oscillator's *temperature stability* and *aging* are important characteristics that define the synthesizer's frequency accuracy. Temperature stability denotes the maximum frequency drift over the operating temperature range and is usually expressed in ppm (the term ppm is an acronym for parts-per-million – a dimensionless coefficient equal to 10^{-6}). Aging is a change in frequency over time that occurs because of changes in the resonator material or a buildup of foreign material on the crystal. It is also specified in ppm over a certain period of time. Aging leads to a permanent frequency error; thus, it is good practice to

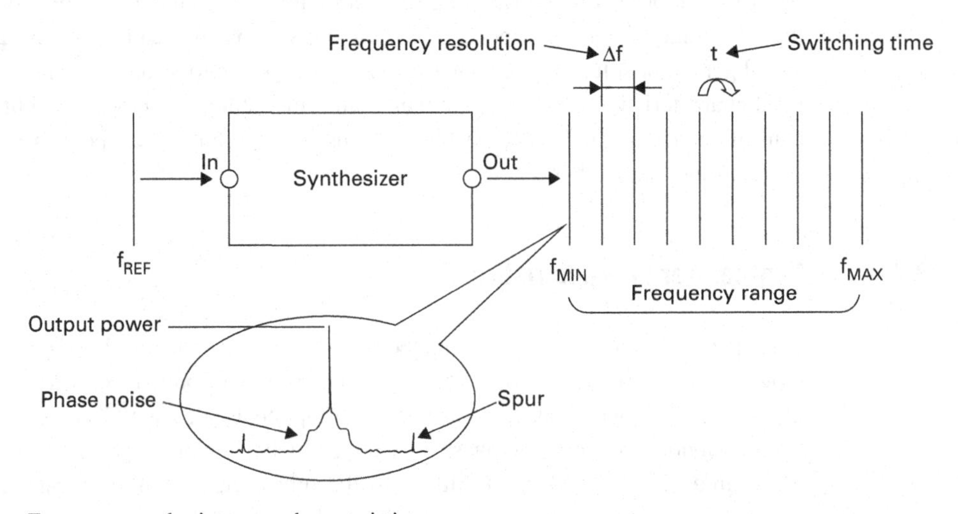

Fig. 3.1 Frequency synthesizer core characteristics.

use mechanical or electronic frequency adjustment means to compensate for internal reference aging.

Switching or *tuning speed* determines how fast the synthesizer transitions from one desired frequency to another and is defined as time spent by the synthesizer between these two states. Thus, the *switching time* is a more proper term (indicated as t in Figure 3.1). The switching time is calculated from the time when the synthesizer receives a command to the time it approaches the desired frequency with a specified accuracy.

3.2.2 Spectral purity

Harmonics appear in the synthesizer spectrum as integer multiples of the output frequency because of signal distortion in nonlinear components. For example, if the fundamental frequency is represented by f, the frequencies of the harmonics would be represented by $2f$, $3f$, etc. Harmonics are expressed in dBc (decibels relative to the carrier) and represent the power ratio of a harmonic to a carrier signal, as depicted in Figure 3.2. For narrow-band synthesizer harmonics can be easily controlled by adding a simple low-pass filter. However, in test-and-measurements applications, a tunable filter or a switched filter bank is required since the synthesizer bandwidth normally exceeds an octave.

Sub-harmonics appear at frequencies that are "sub-harmonically" related to the main signal such as $f/2$, $f/3$, etc. Propagating through nonlinear components, these signals exhibit their own harmonics. Thus, in a more general case, the sub-harmonics are considered as products appearing at N/K of the output frequency, where N and K are integers. Sub-harmonics can be created in some nonlinear devices such as a frequency doubler. The doubler generates a number of harmonics of the incoming signal. It usually employs a balanced scheme that intends to suppress odd products, as shown in Figure 3.3. Since the second harmonic now becomes the main signal, all the odd products (which are not completely suppressed) do not meet the harmonic relationship with respect to the desired output and are, therefore, treated as sub-harmonics.

Spurious signals or *spurs* are undesired artifacts created by the synthesizer at some discrete frequencies that are not harmonically related to the output signal. As a typical example, Figure 3.4 shows reference spurs often created in PLL synthesizers. Other spurs can come from many other sources such as mixer intermodulation products,

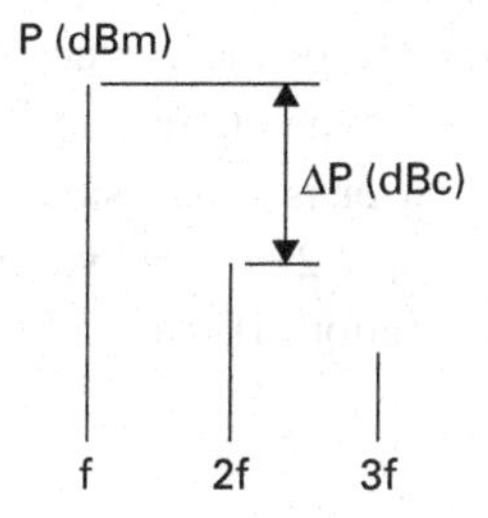

Fig. 3.2 Harmonics appear at integer multiples of the output frequency.

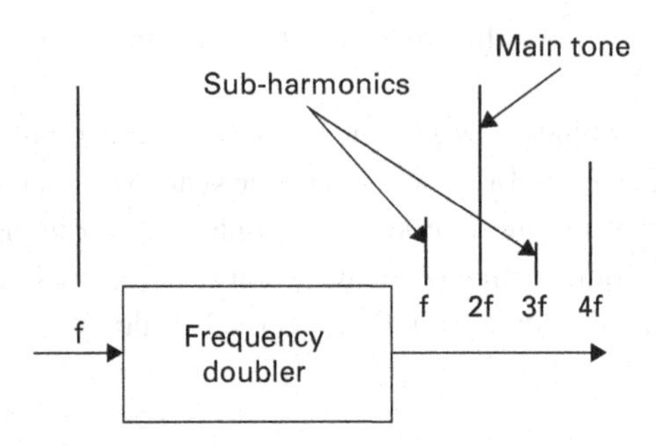

Fig. 3.3 Sub-harmonic products.

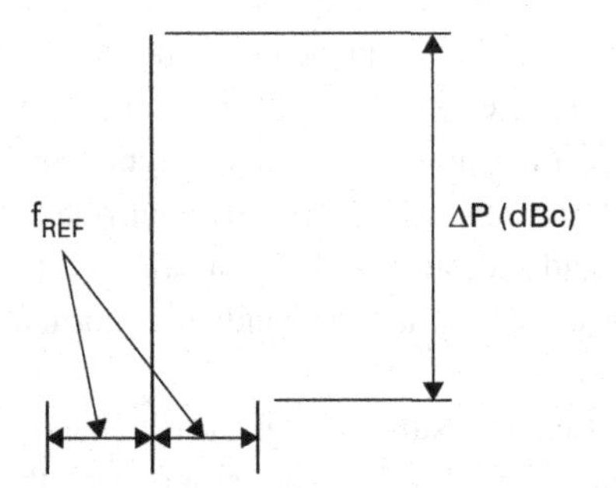

Fig. 3.4 Reference spurs of a PLL synthesizer.

local oscillator leakages, and external signals coming through the bias or control interface. In contrast to harmonics, the spurs are much more troublesome products that can limit the ability of receiving systems to resolve and process a desired signal. Spurs can sit very close to the main tone and in many cases cannot be filtered. Thus, the spurious level has to be minimized, typically to −60 dBc relative to the main signal, although many applications require bringing this level even lower. This presents a certain design challenge, especially if a small step size is required. A different concern is mechanically induced spurs, usually referred to as "microphonics." These spurs appear due to the sensitivity of certain synthesizer components to external mechanical perturbations.

Phase noise is one of the major parameters that ultimately limits the performance of microwave systems. In general, phase noise is a measure of the synthesizer's short-term frequency instability, which manifests itself as random frequency fluctuations around the desired tone. The output of an ideal synthesizer is a pure sine-wave signal $V_{OUT} = A_0 \sin \omega_0 t$ with amplitude A_0 and frequency $\omega_0 = 2\pi f_0$. However, in reality the output signal demonstrates amplitude and phase variations (Figure 3.5), which can be represented as follows

$$V_{OUT} = (A_0 + a(t)) \sin(\omega_0 t + \varphi(t)), \tag{3.1}$$

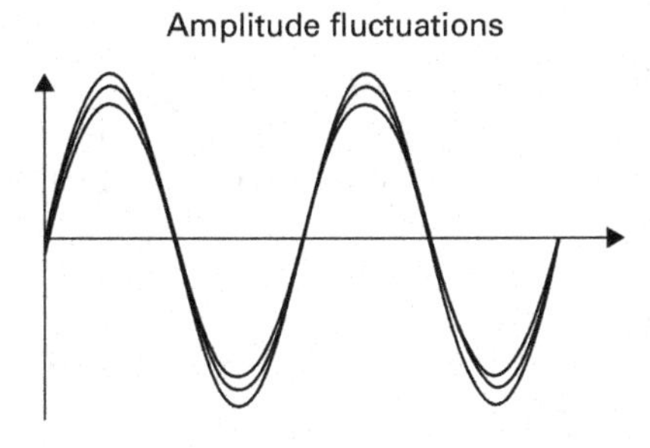

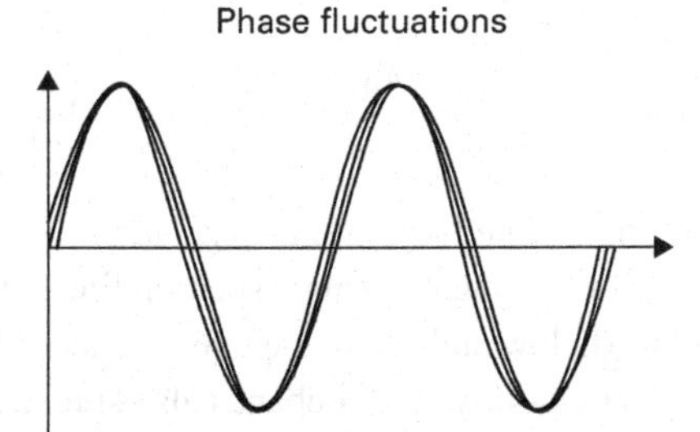

Fig. 3.5 Amplitude and phase fluctuations.

where $a(t)$ and $\varphi(t)$ are the amplitude and phase fluctuations, respectively. Amplitude noise is rarely as critical as phase noise. The amplitude variations can be easily reduced by balanced mixers, amplifiers in compression, diode limiters, or an automatic level control circuit. Hence, the phase effects generally dominate, reducing (3.1) to $V_{OUT} = A_0 \sin(\omega_0 t + \varphi(t))$.

These phase fluctuations result in uncertainty in the signal zero-crossing, which in the time domain is referred as *jitter*. Assuming that the phase fluctuations $\varphi(t)$ are caused by an unwanted fixed-frequency signal $\omega_m = 2\pi f_m$ that modulates the synthesizer output frequency and is expressed as $\varphi(t) = A_m \sin \omega_m t$, then the output signal can be described by

$$V_{OUT} = A_0 \sin(\omega_0 t + A_m \sin \omega_m t)$$

and using the well-known trigonometric identity $\sin(\alpha + \beta) = \sin\alpha \cos\beta + \cos\alpha \sin\beta$ is further transformed to

$$V_{OUT} = A_0 [\sin \omega_0 t \cos(A_m \sin \omega_m t) + \cos \omega_0 t \sin(A_m \sin \omega_m t)]. \quad (3.2)$$

Assuming that the amplitude of the modulating signal A_m is small, we can simplify the corresponding terms of (3.2) to

$$\cos(A_m \sin \omega_m t) \approx 1$$
$$\sin(A_m \sin \omega_m t) \approx A_m \sin \omega_m t$$

reducing (3.2) to

$$V_{OUT} \approx A_0 (\sin\omega_0 t + A_m \cos\omega_0 t \sin\omega_m t). \tag{3.3}$$

Using another elementary trigonometric formula, $\sin\alpha\cos\beta = \frac{1}{2}[\sin(\alpha+\beta) + \sin(\alpha-\beta)]$, (3.3) is further modified to

$$V_{OUT} \approx A_0 \{\sin\omega_0 t + \frac{A_m}{2}[\sin(\omega_m t + \omega_0 t) + \sin(\omega_m t - \omega_0 t)]\}$$

and finally

$$V_{OUT} \approx A_0 \sin\omega_0 t + \frac{A_0 A_m}{2}\sin(\omega_0 + \omega_m)t - \frac{A_0 A_m}{2}\sin(\omega_0 - \omega_m)t. \tag{3.4}$$

Note that (3.4) has three sinusoidal terms related to ω_0, $\omega_0 - \omega_m$, and $\omega_0 + \omega_m$. Thus, in the frequency domain, the output signal is no longer a single spectral line but adds two spurious sidebands equally spaced by f_m (below and above the main signal). Obviously, if f_m is not a fixed frequency but changes randomly, the sidebands also spread randomly over frequencies both above and below the nominal signal frequency. Phase noise can be quantified by measuring the output power at many frequencies away from the nominal frequency and comparing it to the power at the nominal frequency, as illustrated in Figure 3.6. This leads to a quantitative definition of the phase noise as the ratio of the noise power found in a 1-Hz bandwidth at a certain frequency offset Δf to the total power at the carrier frequency f_0, which can be written as

$$\mathbf{L} = 10\log\left(\frac{P^{1Hz}_{f_0+\Delta f}}{P_{f_0}}\right) \tag{3.5}$$

This ratio is normally taken in the logarithmic scale; hence, the phase noise is expressed in units of dBc/Hz (dBc per hertz) at various offsets from the carrier frequency and is usually specified by a table or as a graphic representation.

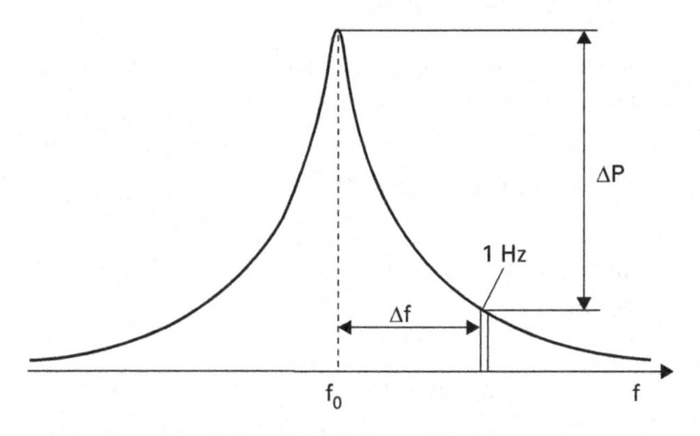

Fig. 3.6 Phase noise is quantified by measuring the output power at many frequencies away from the nominal frequency and comparing it to the power at the nominal frequency.

3.2.3 Output power

Output power is a measure of the synthesizer output signal strength specified in units of watt or, more frequently, in dBm. The term dBm refers to the ratio in decibels of the measured power referenced to one milliwatt. The relationship between these two units is expressed by

$$P_{dBm} = 10\log P_{mWatt}. \tag{3.6}$$

A simple synthesizer usually delivers a fixed power level that cannot be changed. More complex designs provide an ability to control the output power in a specified range. In the latter case, the *output power control range* (i.e. the minimum and maximum values between which power can be set) and the *power step size* (i.e. the minimum change between two consecutive power settings) are specified as well. Note that output power can differ from its set value. This discrepancy is described by the *output power accuracy* that defines the absolute maximum variance between programmed and actual (i.e. measured) power values.

3.3 Synthesizer architectures

Synthesizer characteristics depend heavily on a particular architecture. Synthesizer architectures can be classified into a few main groups, as shown in Figure 3.7. The direct architectures are intended to create the output signal directly from the available input frequency signals either by manipulating and combining them in the frequency domain (direct analog synthesis) or by constructing the output waveform in the time domain (direct digital synthesis). The indirect methods assume that the output signal is regenerated inside the synthesizer in such a manner that the output frequency relates (e.g. is phase-locked) to the input reference signal. A practical synthesizer, however, is usually a hybrid design that combines various techniques to achieve specific design goals.

3.3.1 Direct analog synthesizers

The direct analog synthesizer is today's most advanced technique, offering exceptional tuning speed and phase-noise characteristics. The output signal is obtained by mixing input frequencies followed by switched filters, as shown in Figure 3.8. These input frequencies can be created by mixing, dividing, and multiplying the output of low-noise fixed-frequency oscillators [1–5]. The output frequency change is accomplished by

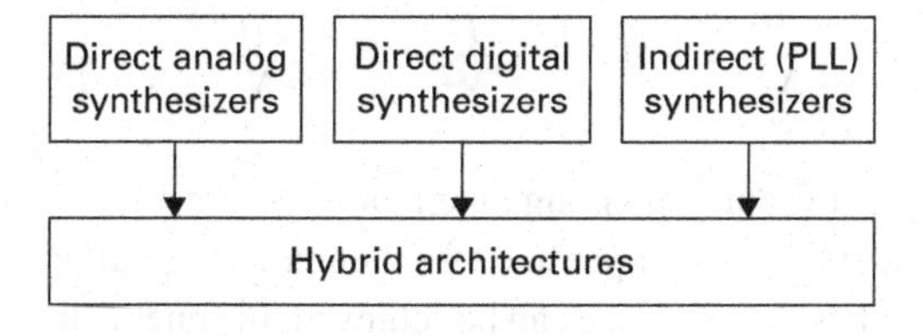

Fig. 3.7 Synthesizer architectures.

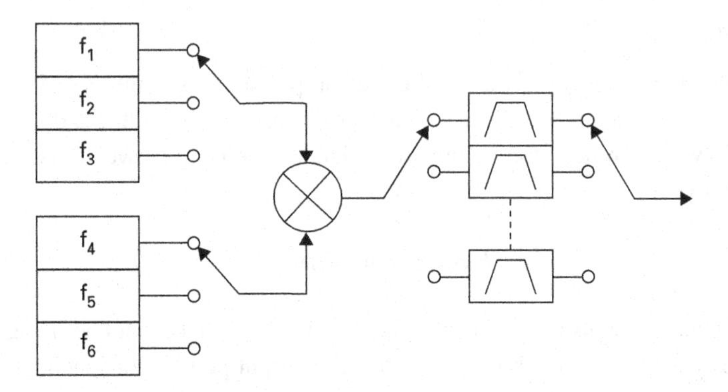

Fig. 3.8 Basic block diagram of a direct analog synthesizer.

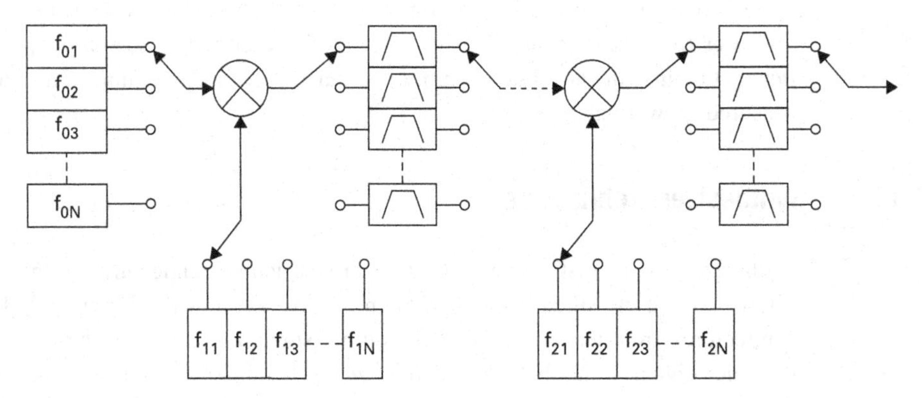

Fig. 3.9 The number of output frequencies is increased by cascading individual mixer stages.

switching appropriate input frequencies; thus, tuning speed is only limited by propagation delays inserted by the switches and their control circuits. Phase noise mainly depends on the noise of the available fixed-frequency sources and can potentially be very low. The main disadvantage of this simple scheme is the limited frequency coverage and step size. In our example, only eighteen output frequencies can be generated (even by utilizing both mixer sidebands). The number of output frequencies can be increased by cascading individual mixer stages, as shown in Figure 3.9. However, this rapidly increases the design complexity and overall component count.

The frequency resolution can also be improved by repeatedly mixing and dividing the input frequencies, as conceptually shown in Figure 3.10. The synthesizer contains a chain of frequency mixer-divider cells that generate a signal at

$$f_{OUT} = \sum_{i=0} \frac{f_i}{N^i} = f_0 + \frac{f_1}{N} + \frac{f_2}{N^2} + \cdots + \frac{f_i}{N^i}, \tag{3.7}$$

where f_i is a frequency driving the corresponding mixer and N is the division coefficient of the utilized frequency dividers. Using proper fixed frequencies and a sufficient number of individual cells, an arbitrarily small step size can be achieved. In general, the frequency

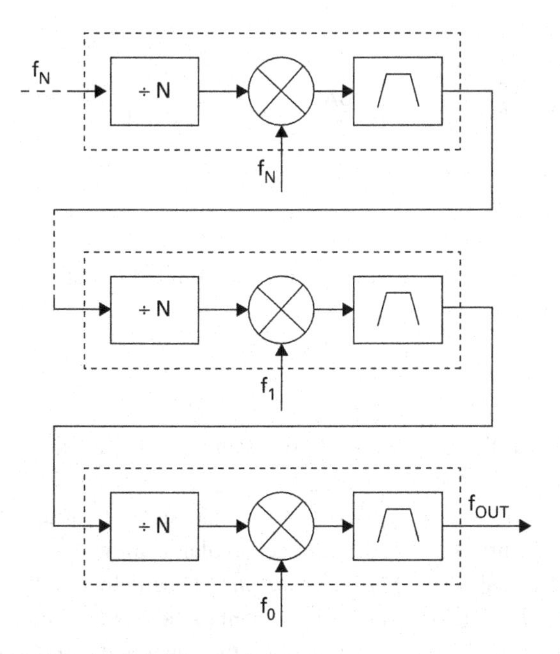

Fig. 3.10 Mixing and dividing technique.

division coefficients can also be arbitrary; however, $N = 10$ is the most commonly used scenario that leads to

$$f_{OUT} = f_0 + \frac{f_1}{10} + \frac{f_2}{100} + \cdots + \frac{f_i}{10^i}. \tag{3.8}$$

The frequencies f_i are usually generated from a common reference F_0 by utilizing its harmonics, i.e. $f_i = A_i F_0$, where A_i is an integer between 1 and 9. This allows rewriting the synthesizer tuning formula to

$$f_{OUT} = F_0 \left(A_0 + \frac{A_1}{10} + \frac{A_2}{100} + \cdots + \frac{A_i}{10^i} \right), \tag{3.9}$$

where the decimal coefficients A_i simply show which harmonic is chosen. Therefore, the output frequency is conveniently represented in a decimal form by setting corresponding digits. Similarly, the synthesizer can be constructed using different frequency division coefficients to represent its output frequency in a binary or any other desired form, or a combination thereof.

The main disadvantage of the direct analog synthesizers is the large number of mixing products that have to be filtered. These include the undesired mixer sideband, LO leakage, and intermodulation products. Depending on a particular frequency plan, filtering close-in spurs can be a challenging task. Another serious issue is cross-coupling between individual filter channels and whole stages. Although a large variety of mixing and filtering schemes are possible, they tend to be hardware intensive if a small frequency step and wide coverage are required. Therefore, while direct analog synthesis offers excellent

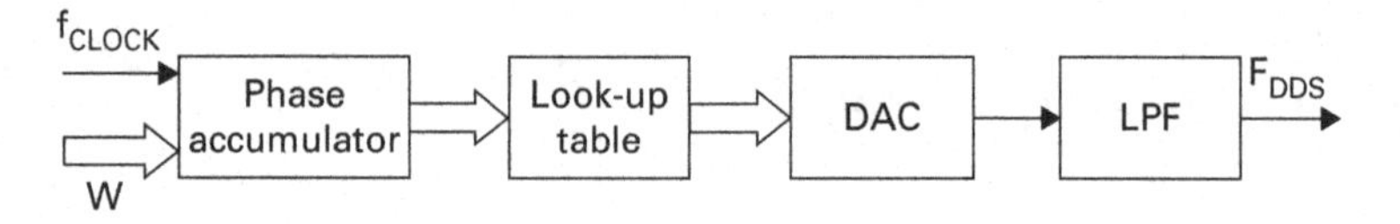

Fig. 3.11 Direct digital synthesizer concept.

tuning speed and phase-noise characteristics, its usage is limited to applications where a fairly high cost can be tolerated.

3.3.2 Direct digital synthesizers

Direct digital synthesizers utilize digital signal processing to construct an output signal waveform in the time domain piece-by-piece from an input (called *clock*) signal [1–6]. The direct digital synthesizer includes a phase accumulator, digital look-up table, DAC, and LPF as depicted in Figure 3.11. The phase accumulator allows the entering and storing of a digital word W called the *phase increment.* At each clock pulse, the phase accumulator adds (i.e. accumulates) the phase increment to the previously stored digital value that represents an instantaneous digital phase of the generated signal. This digital phase is continually updated until it reaches the capacity of the accumulator.

For an N-bit accumulator and the smallest increment of one least significant bit, it will take 2^N clock cycles to fill up the accumulator. Then the accumulator resets and the process starts over again. Hence, the lowest generated frequency is given by

$$f_{MIN} = \frac{f_{CLK}}{2^N} \tag{3.10}$$

that also equals the smallest frequency step. With a larger phase increment W, the phase accumulator obviously fills up faster and the DDS output frequency increases to

$$f_{DDS} = \frac{W}{2^N} f_{CLK}. \tag{3.11}$$

Therefore, frequency tuning is accomplished by changing the phase increment word. This word defines the DDS output frequency and can be loaded into the accumulator through either a serial or parallel interface. The tuning process has essentially no settling time delays other than what is inserted by the digital interface. The frequency can be changed in very small steps determined by the length of the phase accumulator. For example, assuming that f_{CLK} is 100 MHz and N equals 32, we can calculate f_{MIN} to approximately 0.023 Hz. The length of the phase accumulator can be further increased; thus, millihertz or even microhertz steps are easily achievable.

The next step is to convert the digital phase value into a digital representation of the signal waveform. This is accomplished with a look-up table. It uses a ROM to store a digital code that sets a proper address on the DAC's bus, and consequently, its output voltage. In general, any desired waveform can be created; however, the sine wave is most commonly used. The waveform construction process completes with a low-pass

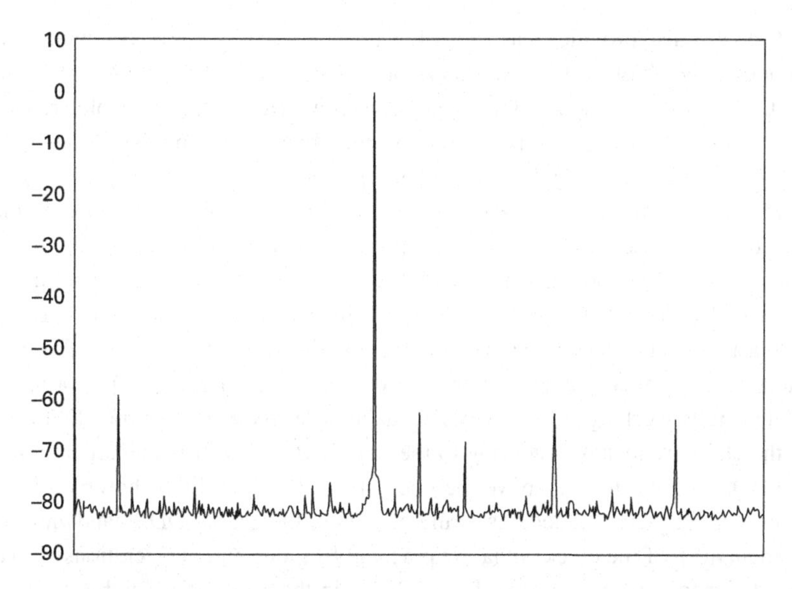

Fig. 3.12 DDS output contains a number of spurious signals.

filter required to remove some unwanted spurious components because of the imperfect approximation of the desired waveform.

Practical realization of this concept brings further modifications. For example, the length of the phase accumulator, required to achieve the necessary resolution, can exceed practical limits for ROM and DAC devices. Due to the sine function's symmetry, only one-fourth of the cycle needs to be stored, thus, greatly reducing the required memory capacity. Furthermore, the DAC usually utilizes a smaller number of bits available from the phase accumulator. This reduction in DAC resolution is called *phase truncation* and leads to increased spurious levels.

The DDS output contains a number of spurious signals (Figure 3.12) as a result of truncation, amplitude quantization, and DAC nonlinearities. However, the most significant ones are aliased images of the output signal that appear on either side of the clock frequency and its multiples because of the sampling nature of digital signal synthesis. From this point of view, the DDS works as a frequency mixer producing spurs at

$$f_{SPUR} = \pm n\, f_{CLK} \pm m\, f_{DDS}, \tag{3.12}$$

where n and m are integers. Similar to mixer intermodulation products, these spurs require careful frequency planning, since they can be very close to the output signal and, therefore, cannot be filtered. While spur location in the frequency domain can be easily determined, its amplitude is much less predictable. As a general rule, lower-order spurs are the strongest; although, fairly high-order spurs can still be harmful and must be taken into account. Typical DDS spurious levels are −50 to −60 dBc for output signal ranges between a few tens to a few hundreds of megahertz.

The DDS also provides reasonably low phase-noise levels, even showing an improvement over the phase noise of the clock source itself. From the phase-noise point of view, the DDS works as a fractional frequency divider with a very fine, variable, frequency division coefficient. The phase-noise improvement is described by the $20\log(f_{CLK}/f_{DDS})$ function and is limited by the residual noise floor.

The DDS is currently available as a tiny, yet highly integrated, surface-mount IC that includes the phase accumulator, look-up table, and DAC in a single chip. It needs only a few external components (LPF and bias circuitry) to build a powerful and versatile module. The most valuable DDS feature is its exceptionally fine frequency resolution and fast switching speed that is comparable to direct analog schemes. The main disadvantages are limited usable bandwidth and relatively poor spurious performance. While a DDS starts working from nearly DC, its highest frequency is limited within one half of the clock frequency according to the sampling theory. It is theoretically possible to use DDS aliased images above the one half of the clock limit; however, the spurious content is further degraded. As a rule of thumb, the usable DDS bandwidth is limited to about 40% of the clock signal by practical LPF design considerations. Typical clock speeds for today's commercial DDS ICs are in the range of a few hundred megahertz to a few gigahertz. The DDS technique is rarely used alone at microwave frequencies because of the previously mentioned bandwidth and spurious limitations. Rather DDS is used as a fine-frequency-resolution block in conjunction with direct analog and indirect architectures.

3.3.3 Indirect synthesizers

Indirect frequency synthesizers utilize phase-lock loop techniques offering a smaller step size and lower complexity in comparison with direct analog schemes [9–17]. A typical single-loop PLL synthesizer (Figure 3.13) includes a tunable voltage-controlled oscillator that generates a signal in a desired frequency range. This signal is fed back to a phase detector through a frequency divider with a variable frequency division ratio N. The other input of the phase detector is a reference signal equal to a desirable step size. The phase detector compares the signals at both inputs and generates an error voltage, which following filtering and optional amplification, slews the VCO until it acquires the lock frequency given by $f_{OUT} = N f_{PD}$, where f_{PD} is the comparison frequency at the phase detector inputs.

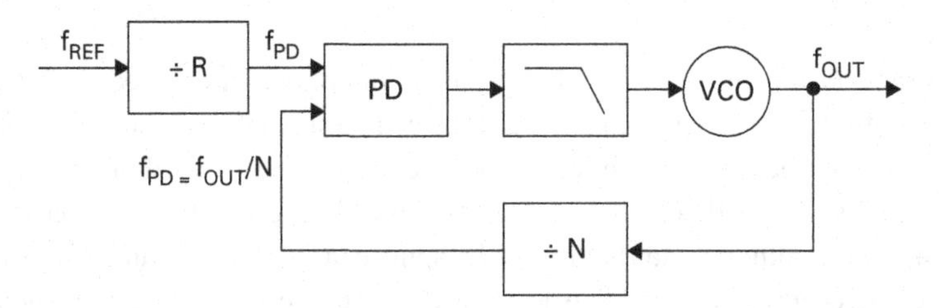

Fig. 3.13 Single-loop PLL synthesizer.

The frequency tuning is achieved in discrete frequency steps equal to f_{PD} by changing the division coefficient N. The available reference frequency can be divided down by another divider to reduce the step size. If the division coefficient of the reference divider is R, then the output frequency is set by

$$f_{OUT} = \frac{N}{R} f_{REF}. \tag{3.13}$$

Since the output signal of the PLL synthesizer is generated at microwave frequencies, all spurs associated with the direct architectures are generally absent. The only source of the spurs in the PLL block diagram shown in Fig. 3.13 is the reference signal itself. The reference signal and its harmonics modulate the VCO tuning port and create sidebands both above and below the main signal. The loop filter bandwidth has to be significantly lower than f_{PD} (usually ten times or more) to keep the reference spurs at a reasonable level. However, the loop bandwidth is inversely proportional to the settling time. Thus, achieving fine frequency resolution, low spurs, and fast switching is an arduous task as it means balancing mutually exclusive terms.

Another important consideration and design tradeoff is phase noise. The noise outside the PLL filter bandwidth is mainly determined by the VCO's free-running noise. The phase noise within the loop filter bandwidth is given by

$$\mathbf{L}_{PLL} = \mathbf{L}_{\Sigma PD} + 20 \log N, \tag{3.14}$$

where $L_{\Sigma PD}$ is the cumulative phase noise of the reference signal, reference and feedback dividers, phase detector, LPF, and loop amplifier recalculated to the phase detector input. The phase noise generated by PLL components is degraded by the large division ratios required to provide a high-frequency output with a fine resolution. Moreover, programmable dividers are usually not available at high frequencies; thus, an additional, fixed-division-coefficient divider (called a *prescaler*) is required. In this case, the total division ratio increases by the prescaler division coefficient resulting in further phase noise degradation. At high frequency offsets, the VCO's free-running noise can be (and normally is) better than the multiplied PLL noise. The optimal phase-noise profile is achieved by choosing the loop bandwidth at the cross point of the multiplied PLL noise and VCO free-running noise curves, as depicted in Figure 3.14. Clearly, by utilizing a low noise VCO and narrower loop bandwidth, it is possible to mask some excessive PLL noise at high-frequency offsets. However, this results in a slower switching speed. Alternatively, a good PLL design can suppress VCO noise at higher-frequency offsets and also provide faster tuning.

Overall, the major advantages of the PLL schemes are reduced levels of spurious signals resulting from the low-pass filter action of the loop and a much less complex compared to the direct analog architectures. In fact, all key PLL components can be integrated into a single chip that leads to low-cost, miniature designs. The main disadvantages are slower tuning, limited step size, and considerably higher phase noise compared to direct analog and direct digital architectures.

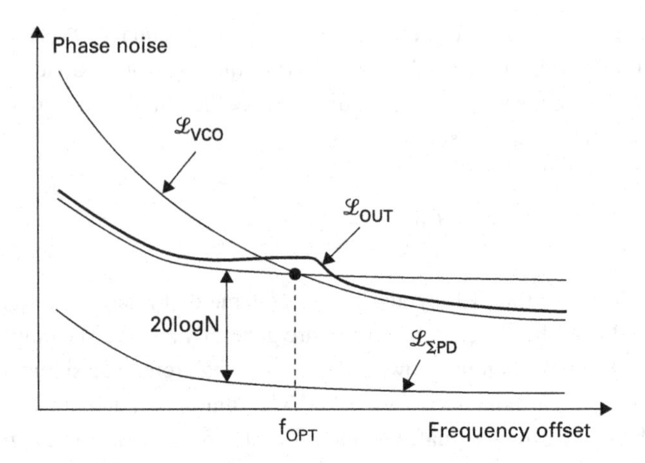

Fig. 3.14 Phase noise of a single-loop PLL synthesizer.

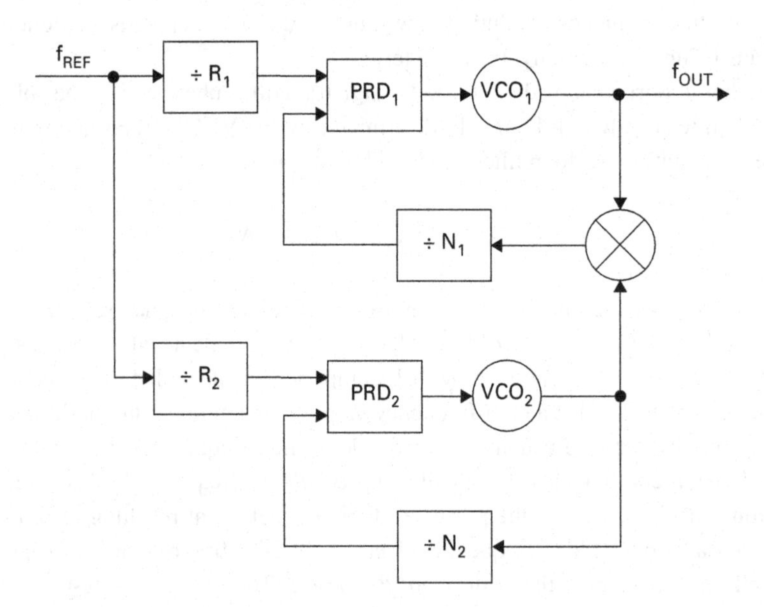

Fig. 3.15 Multiloop synthesizer concept.

3.3.4 Hybrid architectures

A practical synthesizer is usually a hybrid design that combines various techniques to achieve specific design goals. A good example is a multiloop synthesizer that is essentially a combination of direct analog (frequency mixing) and indirect (PLL) techniques. The idea is to convert the VCO output to a much lower frequency with the aid of a mixer and an offset frequency source, as illustrated in Figure 3.15. The second PLL generates an auxiliary signal, which is used as an offset signal for the first loop. Splitting the design in two loops allows for the reduction division coefficients in both loops, thus, improving the overall phase-noise performance.

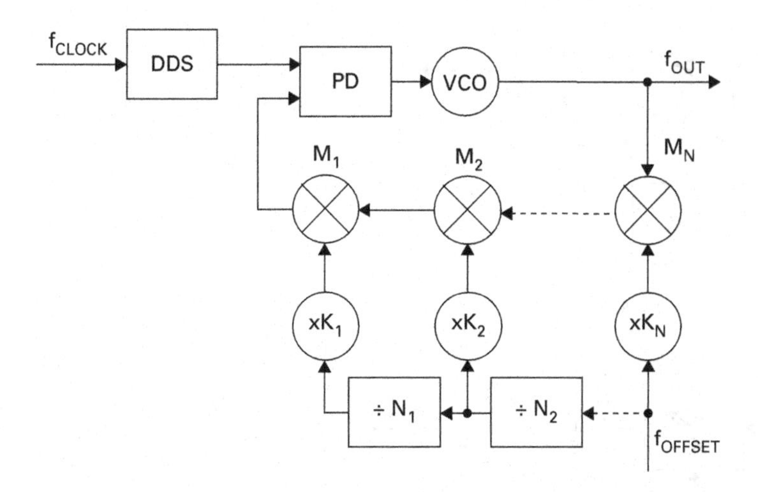

Fig. 3.16 Hybrid design combines three frequency synthesis techniques.

Another interesting design employs a chain of mixers converting the VCO signal to a lower frequency, as depicted in Figure 3.16. This scheme allows for the minimization, or even complete removal, of the frequency divider from the PLL feedback path that results in very low phase noise and low spurious performance [18]. Note that the local oscillator offset signals are created from a common source utilizing frequency mixing, division, and multiplication. In other words, these offset signals are created by direct analog synthesis means. The fine frequency step is achieved by adding a DDS. Thus, this design combines all three main frequency synthesis techniques (i.e. direct analog, direct digital, and indirect) to achieve high performance and extended functionality [19].

3.4 Signal generators

Although synthesizers can be found in virtually any microwave system, test-and-measurement is probably the most challenging application that calls for advanced synthesizer solutions. Broadband operation, very fine frequency resolution, low spurs, and low phase noise are the key specifications for signal generator instruments. Another important parameter that impacts overall test-and-measurement system throughput is the switching speed. The time spent by the synthesizer transitioning between frequencies becomes increasingly valuable since it cannot be used for measurement and data processing. Besides these key parameters, signal generators feature extended functionality that includes output power calibration and control, frequency and power sweep, list mode, and various modulation functions [20].

3.4.1 Power calibration and control

For a simple synthesizer design, the output power can vary across the operating frequency range because of individual component gain variations. More sophisticated designs bring

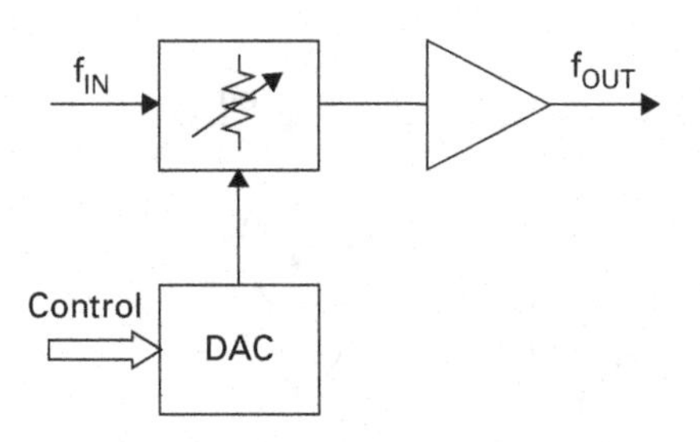

Fig. 3.17 Open-loop power control.

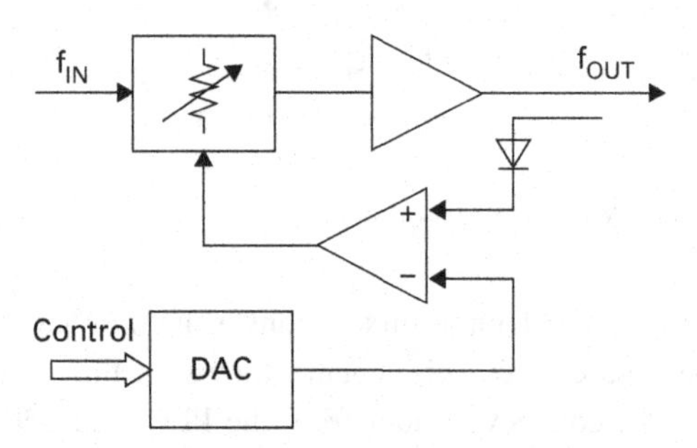

Fig. 3.18 Better performance is achieved with a closed-loop control.

the ability to equalize the output power response and also change the power level as needed. A synthesizer's output power can be controlled in many different ways, for example, using an open-loop technique illustrated in Figure 3.17. The amplitude control circuit includes an RF attenuator and a DAC. The DAC generates a proper voltage for any given frequency to ensure a flat output response across the entire operating frequency range. The DAC values are generated during a calibration routine and are stored in a look-up table. The output power can be changed within certain limits (set by the available attenuator dynamic range), adding one more dimension to the calibration table. Furthermore, the synthesizer output circuit may include many devices that exhibit temperature variations. Thus, the synthesizer may also include a temperature sensor to provide further correction if required. By employing a sophisticated interpolation routine, this technique provides reasonably flat and repeatable output power characteristics across operating frequency and temperature ranges. Note that the output power is set almost instantaneously. Therefore, the open-loop method is well suited for fast switching applications. The main disadvantage of this method is limited accuracy caused by component temperature variations. The output power delivered to an external load also depends on how well the synthesizer and load impedances are matched.

Better performance can be achieved with a closed-loop ALC method. The output power is sampled with a directional coupler and routed to an RF detector, as depicted in Figure 3.18. The detector generates a voltage proportional to the output power. This voltage is compared to a reference voltage generated by a DAC. An error signal controls the attenuator, thus, closing the loop. In other words, the RF detector continuously measures and adjusts the output power to a value set by the DAC. This configuration ensures a precise output power level regardless of the load mismatch. Furthermore, temperature variations of the synthesizer components are also taken into account. The only significant source of temperature instability is the RF detector itself (and – to a smaller degree – the directional coupler). Temperature variations of the detector are further reduced by controlling (i.e. stabilizing) its temperature. The power control range can be further extended by adding an electromechanical step attenuator.

3.4.2 Frequency and power sweep

It is often desired to linearly change the output frequency within certain limits. This function is called *frequency sweep* and is widely used in test-and-measurement applications for characterizing various devices across their operating frequency range. The frequency sweep is defined by setting start frequency, stop frequency, and sweep time. There are two basic modes – continuous and stepped sweeps. The continuous (analog) sweep is realized by changing the VCO tuning voltage directly (i.e. breaking the phase-lock-loop) with a sawtooth generator or a DAC. This mode requires linear and repeatable tuning characteristics; however, a frequency error is always present since the oscillator remains unlocked during the sweep. The stepped (also called discrete or digital) frequency sweep is realized by changing the synthesizer output frequency in discrete frequency increments (steps) as illustrated in Figure 3.19. It assumes that the synthesizer is locked at every discrete point across the sweep range. Hence, this mode provides significantly better frequency accuracy compared to the analog sweep.

Similarly, synthesizer output power can be swept between desired power levels. This function (called *power sweep*) is used in the characterization of output power and linearity characteristics of various devices such as transistors, amplifiers, mixers, and many others. The synthesizer output power can be swept continuously or in steps.

The frequency (or power) sweep mode normally assumes that the synthesizer steps linearly in equal increments. However, frequency and power can be set to any desired value. A list mode provides better flexibility. The idea is to create a table (list) of frequencies and

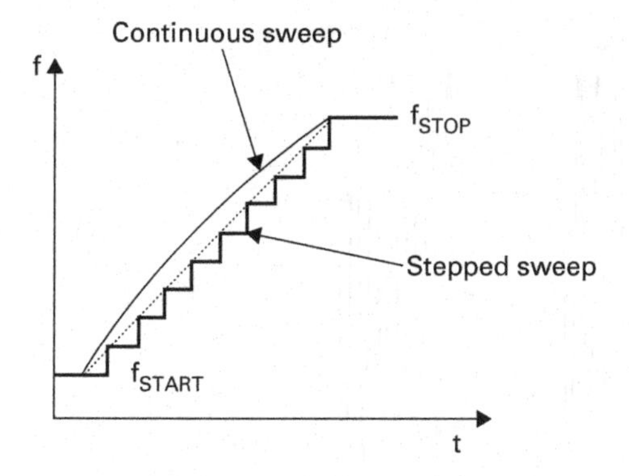

Fig. 3.19 Frequency sweep methods.

store it in the synthesizer's memory. The list is executed by sending a proper command or by applying a trigger signal (which is a voltage pulse) to a corresponding control line. Once the synthesizer's control circuit detects a trigger pulse, it commands the synthesizer to move from one frequency to another according to the programmed list. Alternatively, the synthesizer can go to the next frequency, stop there and wait for the next trigger pulse; then, the process repeats. One of the advantages of the list mode is a significant throughput improvement compared to normal programming, since it is possible to precalculate and memorize all necessary parameters required to control individual components of the synthesizer.

3.4.3 Modulation

Signal generators utilize various modulation forms ranging from simple pulse, amplitude, frequency, and phase modulation to complex digital modulation formats. The most commonly used modulation modes are reviewed below. Further details on modulation theory and implementation techniques can be found in [1], [9], [20], [21].

Pulse modulation

Pulse modulation is achieved by switching the output signal on and off in accordance with the applied modulating pulses. The result is a sequence of RF pulses that replicate (or tend to replicate) the input modulating signal as shown in Figure 3.20. The minimum RF pulse width, rise time, fall time, and overshoot are important characteristics that define how well the modulating signal is replicated. Typical rise time and fall time numbers required are in the order of ten nanoseconds. The pulse modulation on/off ratio is another critical parameter. A typical specification is 80 dB or higher. The modulating signal frequency (also called *rate*) can be between DC and several megahertz.

Amplitude modulation

Amplitude modulation historically has been one of the most popular methods for carrying information via RF frequencies. It is realized by varying the output signal amplitude in accordance with an applied modulating signal, as indicated in Figure 3.21. The simplest

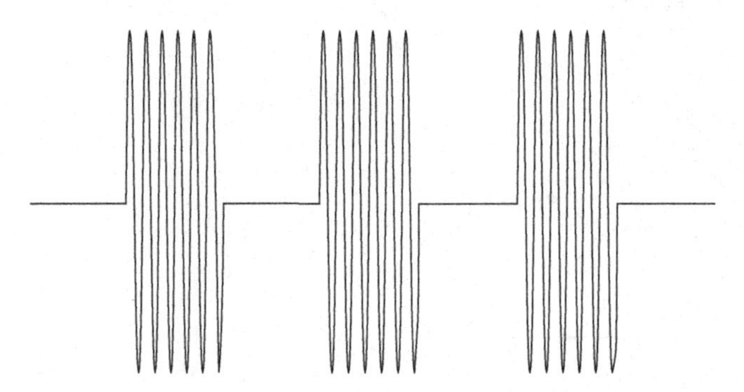

Fig. 3.20 Pulse modulated signal.

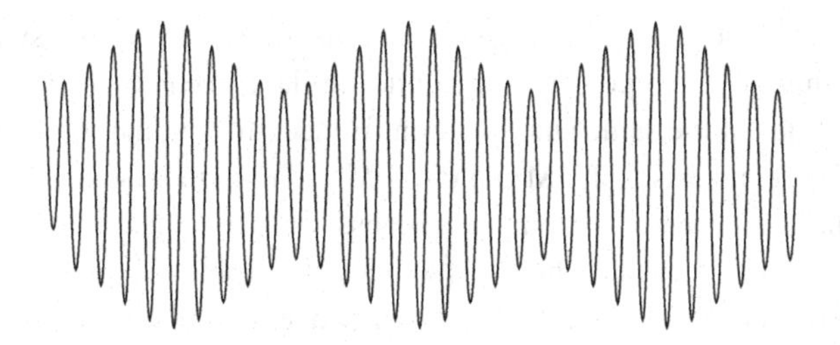

Fig. 3.21 Amplitude-modulated signal.

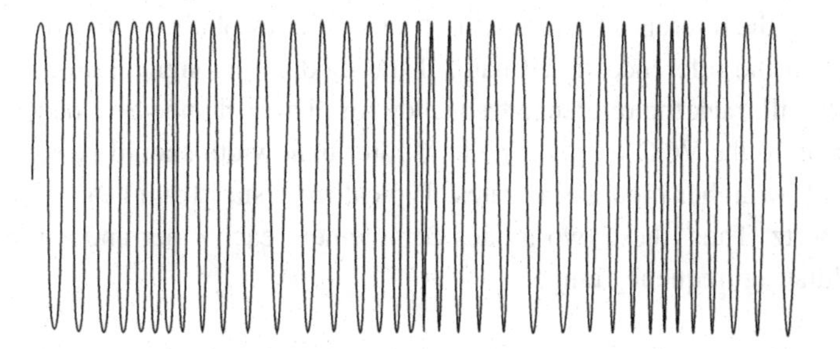

Fig. 3.22 Frequency-modulated signal.

way to implement AM is to control the insertion loss of an attenuator inserted into the synthesizer output circuit. The maximum power variation (which can also be expressed in terms of *modulation index* or *depth*) is achieved by setting the output power level in the middle of its control range. Another important requirement is linearity, since the modulator must translate the modulating signal with minimal distortion. This may further limit a realizable modulation depth. Various linearization techniques can be applied to minimize AM signal distortion.

Alternatively, amplitude modulation can be implemented by summing the modulating signal into the ALC loop. In general, the ALC-based amplitude modulation offers better linearity and repeatability characteristics. However, the modulation depth may be limited by the available ALC dynamic range which, in turn, depends on the detector that is used. The maximum modulating signal rate is also lower compared to the open-loop alternative because of the settling time of the closed-loop ALC system.

Frequency and phase modulation

Frequency modulation is another popular form of analog modulation that offers better signal immunity than AM. The process of producing a frequency-modulated signal involves the variation of the synthesizer output frequency in accordance with the modulating signal, as shown in Figure 3.22. The frequency bandwidth where the synthesized signal fluctuates is proportional to the peak amplitude of the modulating signal and is called frequency *deviation*. FM can also be described by the *modulation index*, which is the ratio of the maximum frequency deviation to the frequency of the modulating signal.

Note that we can vary not only the frequency but also the phase of the synthesized signal, thus, producing PM. Both processes are quite similar since in both cases we vary the argument (the angle) of the same sine function. Hence, the angular modulation is a more general case that represents both FM and PM. FM and PM modulated signals can be produced in many different ways. For example, it is possible to modulate the synthesizer's VCO tuning voltage around the value where it is settled. The problem, however, is that the PLL will tend to correct any voltage change. For proper operation, the modulating signal rate has to be well above the PLL filter bandwidth. Typical achievable modulating rates range from a few kilohertz to tens of megahertz. An alternative solution is to modulate not the VCO but the reference oscillator. Furthermore, a higher deviation can be achieved by changing not the reference frequency itself but rather its phase by inserting a variable phase-shifter in the reference signal path. If the modulating signal rate is sufficiently low, the PLL will track the reference frequency (or phase) change and, hence, translate the modulation to the VCO output. The loop filter bandwidth should be set as wide as possible to allow higher modulating rates. Typical rates start from nearly DC to a few tens of kilohertz. Thus, these two modes complement each other and can extend the overall modulating frequency range.

Digital modulation

Modern communication systems migrate from simple analog modulation to more sophisticated digital modulation techniques. Note that more effective modulation forms are possible by simultaneously varying both amplitude and phase. The simplest way to visualize such a complex signal is to draw it as a vector on a polar diagram. The amplitude and phase are represented as the length and the angle of the vector, as shown in Figure 3.23. In digital communication systems, such a signal is expressed in I (in-phase) and Q (quadrature) terms, which are projections of the signal vector on a corresponding orthogonal axis. Therefore, the amplitude and phase modulation assumes the change of the signal vector, which can be conveniently accomplished by varying two

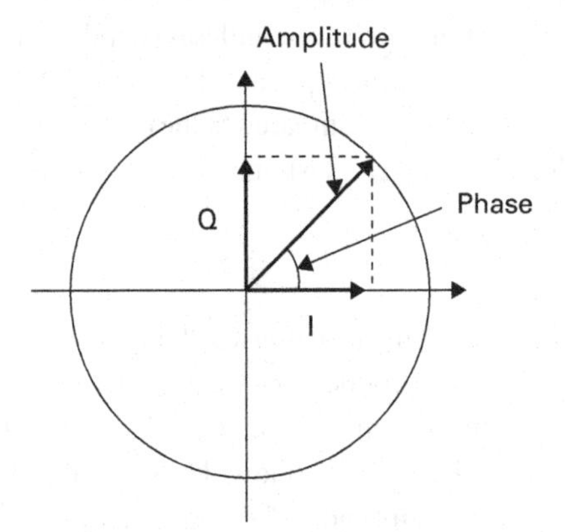

Fig. 3.23 More effective modulation forms are possible by simultaneously varying both amplitude and phase.

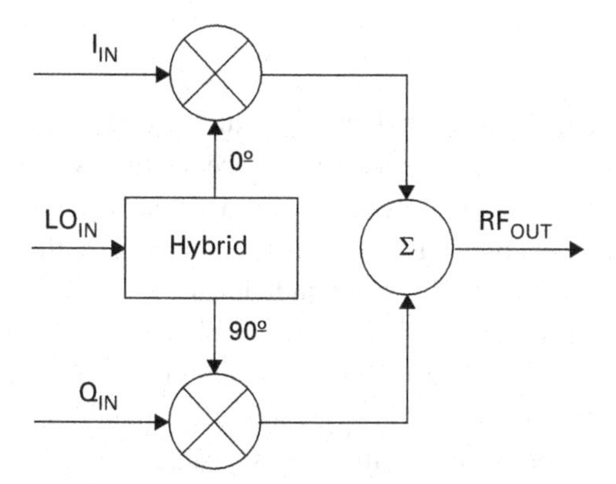

Fig. 3.24 Basic diagram of an IQ-modulator.

independent IQ-components. Hence, such a complex modulation is also called vector or IQ-modulation.

Vector modulation can be realized with an IQ-modulator. It consists of two identical mixers driven with a 90-degree phase-shift at their LO ports, as shown in Figure 3.24. The base-band data signals are applied directly to the IF ports, upconverted and summed together with no phase-shift between them. The resulting output is an IQ-modulated signal at the same carrier frequency as the LO. The quality of the synthesized signal can be tested by applying two base-band signals of the same frequency and amplitude with a 90-degree phase-shift with respect to each other. For a perfect modulator, only one sideband should be present. However, in reality, the output signal contains another sideband because of imperfect amplitude and phase balance. Furthermore, an LO leakage also takes place. The undesired sideband can be suppressed (cancelled) by adjusting the amplitude and phase of the applied IQ signals. The LO leakage can be controlled by adjusting DC offset voltages for the diodes used in the balanced mixers. Therefore, it is generally possible to calibrate the modulator characteristics to a degree where it can be practically used. The difficulty is that this calibration has to be implemented at many frequencies across the entire operating range. Moreover, the calibration has to survive over time and temperature changes. Thus, achieving a good image and LO leakage suppression for a broadband, high-frequency, direct IQ modulator is a challenging task.

An alternative solution is to create a desired IQ-modulated signal at a lower, fixed frequency and upconvert it to microwave frequencies. Obviously, it is much easier to achieve better cancellation of undesired products at a single (and lower-frequency) point. However, the difficulty is now to remove undesired up-conversion products in a wide frequency range. This can be accomplished with a YIG-tuned filter. The disadvantage of the YIG filter is slow tuning speed and relatively narrow passband that can be insufficient in certain applications. A switched filter bank offers better tuning characteristics. However, it usually requires a large number of channels and, hence, is hardware extensive. Nevertheless, IQ-modulation is a very desirable function in modern signal generator instruments.

3.5 Conclusions

Frequency synthesizers are among the most challenging of high-frequency devices. The industry feels persistent pressure to deliver higher-performance designs. Broadband operation, fine frequency resolution, low spurs, low phase noise, and fast switching speed are desirable characteristics. Another challenge is size and cost reduction. In the past, complex microwave synthesizers were often built using individual connectorized modules connected with coaxial cables. The designer could easily isolate and refine individual blocks to make them perfect. These days, such complex assemblies have to be made on a common PCB using tiny surface-mount parts. A great effort is required to minimize interactions between individual components sitting on the same board. Furthermore, many parts are reused to accomplish different functions, which are distributed through the whole assembly. The net result is a significant increase in "design density," meaning both component count and functionality per square inch. All these factors drastically complicate the design process. Nevertheless, this seems to be a "must" approach these days.

References

[1] V. Manassewitsch, *Frequency Synthesizers: Theory and Design*, 3rd ed. NJ: Wiley, 2005.

[2] V. F. Kroupa, *Frequency Synthesis: Theory, Design and Applications*. New York: Wiley, 1973.

[3] V. Reinhardt, *et al.*, "A Short Survey of Frequency Synthesizer Techniques," *Proc. 40th Annual Symposium on Frequency Control*, May 1986, pp. 355–365.

[4] R. R. Stone Jr., "Frequency Synthesizers," *Proc. 21st Annual Symposium on Frequency Control*, April 1967, pp. 294–307.

[5] Z. Galani, and R. A. Campbell, "An overview of frequency synthesizers for radars," *IEEE Trans. Microw. Theory and Tech.*, vol. 39, no. 5, May 1991, pp. 782–790.

[6] V. F. Kroupa (ed.), *Direct Digital Frequency Synthesizers*. New York: IEEE Press, 1999.

[7] A. Chenakin, "Frequency synthesis: Current solutions and new trends," *Microwave Journal*, May 2007, pp. 256–266.

[8] A. Chenakin, *Frequency Synthesizers. Concept to Product*. Norwood, MA: Artech House, 2010.

[9] J. A. Crawford, *Advanced Phase-Lock Techniques*. MA: Artech House, 2008.

[10] R. E. Best, *Phase-Locked Loops: Theory, Design and Applications*. New York: McGraw-Hill, 1984.

[11] W. F. Egan, *Phase-Lock Basics*, 2nd ed. NJ: Wiley, 2007.

[12] W. F. Egan, *Frequency Synthesis by Phase Lock*, 2nd ed. New York: Wiley, 1999.

[13] F. M. Gardner, *Phaselock Techniques*, 3rd ed. NJ: Wiley, 2005.

[14] J. Klapper and J. T. Frankle, *Phased-Locked and Frequency-Feedback Systems*. New York: Academic Press, 1972.

[15] U. L. Rohde, *Digital PLL Frequency Synthesizers: Theory and Design*. NJ: Prentice-Hall, 1983.

[16] V. F. Kroupa, *Phase Lock Loops and Frequency Synthesis*. NJ: Wiley, 2003.

[17] S. J. Goldman, *Phase-Locked Loop Engineering Handbook for Integrated Circuits*. MA: Artech House, 2007.

[18] A. Chenakin, "Low phase noise PLL synthesizer," *US Patent No. 7 701 299*, April 2010.

[19] A. Chenakin, “A compact synthesizer module offers instrument-grade performance and functionality,” *Microwave Journal,* February 2011, pp. 34–38.
[20] C. F. Coombs, Jr., (ed.), *Electronic Instrument Handbook*, 3rd ed. New York: McGraw-Hill, 1999.
[21] F. E. Terman, *Electronic and Radio Engineering*. New York: McGraw-Hill, 1955.

4 Real-time spectrum analysis and time-correlated measurements applied to nonlinear system characterization

Marcus Da Silva

Nonlinear performance characterization and measurements have become an important consideration for today's modern technologies as digital clock rates increase and wireless communication systems attempt to keep up with the dynamic delivery of voice, video, and data over a finite wireless spectrum. Transient anomalies caused by the interaction of various digital and analog signals within a system, nonlinear behavior and device memory effects, can all degrade system performance, causing EMI/EMC regulatory violations, lost calls, packet errors, and system inefficiency.

This chapter explores the techniques used in real-time spectrum and signal analysis and presents applications where real-time technologies can be applied to offer a modern approach to the detection, discovery, and analysis of transient behavior, including those caused by nonlinear memory effects and interactions between the digital and analog parts of modern embedded systems.

4.1 Introduction

The class of instruments called spectrum analyzers has evolved with the uses of the electromagnetic spectrum and with the available technology. Early instruments, then called *Wave Analyzers*, were manually tuned receivers that measured the signal level at the frequency to which they were tuned. The addition of sweep tuning and a CRT enabled a two-dimensional display of amplitude versus frequency and engendered the Swept Tuned *Spectrum Analyzer* (SA). The advent of digital modulation techniques and the availability of precision Analog-to-Digital Converters (ADCs), coupled with enough computing power for Digital Signal Processing (DSP), brought forth the *Vector Signal Analyzer* (VSA). The explosion of digital communications and the need to maximize the ever-increasing amount of information that must be transferred across a limited spectrum created a need for techniques that separate signals in the time domain as well as the frequency domain and the ability to observe and measure signals that happen far too fast for traditional analyzers. This need for speed and the need to correlate time and frequency in the analysis of RF signals led to the creation of the Real-Time Signal Analyzer (RTSA).

This chapter describes the architecture of real-time signal and spectrum analyzers; explores some of the theoretical implications of the techniques used; and provides some examples of RTSA applications. It also covers methods of sequentially applying Discrete

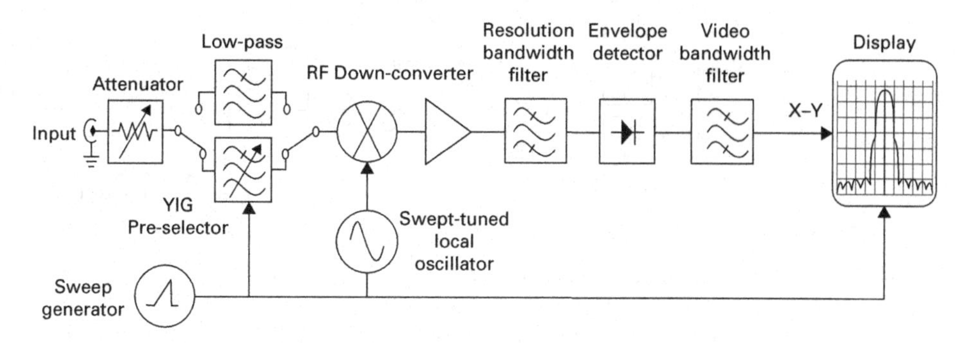

Fig. 4.1 Swept-tuned spectrum analyzer (SA) architecture.

Fourier Transforms to a continuous stream of time domain samples in order to perform real-time spectrum analysis. Methods of analyzing and displaying the fast-changing spectral events captured by real-time spectrum analysis are explored as well. Applications of the technique to identify and measure transient events caused by interactions between digital modulation and nonlinear circuit effects are also presented.

4.1.1 Types of spectrum analyzers

Swept-tuned spectrum analyzer

The swept-tuned, super-heterodyne spectrum analyzer shown in Figure 4.1 is the traditional architecture that first enabled engineers to make frequency domain measurements several decades ago. The swept SA has since evolved along with the applications that it serves. Current generation swept SAs include digital elements such as ADCs, DSPs, and microprocessors. The basic swept approach, however, remains largely the same and is best suited for observing signals that change slowly relative to the sweep speed.

The swept SA makes power vs. frequency measurements by down-converting the signal of interest and sweeping it through the pass-band of a resolution bandwidth (RBW) filter. The RBW filter is followed by a detector that calculates the amplitude at each frequency point in the selected span. This approach of effectively sweeping a filter across the span of interest is based on the assumption that the signal being measured does not change during the time it takes the analyzer to complete one sweep. Measurements are valid for relatively stable, unchanging input signals. The assumption that signals are stable for the duration of a sweep is certainly valid in many cases, but becomes an impediment when analyzing the more dynamic signals prevalent in modern digital communications, imaging, and radar applications.

Vector signal analyzer

Analyzing signals carrying digital modulation requires *vector* measurements that provide both magnitude and phase information. The capability of performing vector measurements gave the Vector Signal Analyzer, shown in Figure 4.2, its name.

A VSA digitizes all the RF signals within the pass-band of the instrument and puts the digitized waveform into memory. The waveform in memory contains both the magnitude

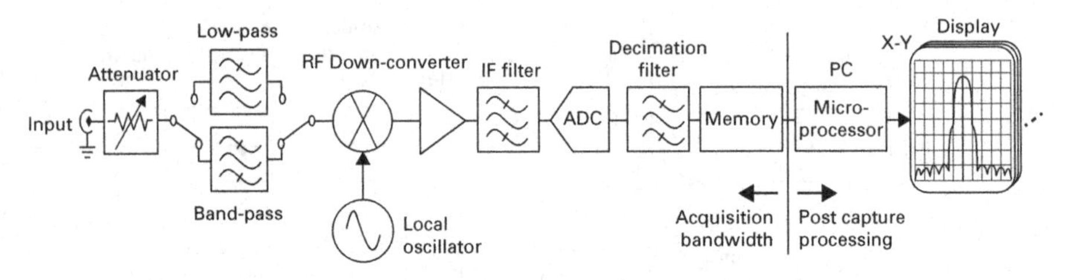

Fig. 4.2 Vector signal analyzer (VSA) architecture.

and phase information which can be used by DSP for demodulation, measurements, or display processing. Transformation from the time domain to the frequency domain is done using Discrete Fourier Transform (DFT) algorithms. Digital signal processing is also used to provide a variety of other functions including the measurement of modulation parameters, channel power, power versus time, frequency versus time, phase versus time, and others.

While the VSA uses DSP to greatly expand its signal analysis capability, it is limited in its ability to analyze transient events. Signals that are acquired must be stored in memory before being processed. The serial nature of this batch processing means that the instrument is effectively blind to events that occur between acquisitions. Single events or events with low repetition rates cannot be reliably captured into memory unless a trigger is available to isolate the event in time. The dynamic nature of modern RF signals is not always accurately portrayed due to the relatively slow cycle time for acquisition and analysis.

Real-time signal analyzer

The RTSA shown in Figure 4.3 is designed with dynamic and transient signals in mind. Like the VSA, a wide pass-band of interest is down-converted to an IF and digitized. The time domain samples are continuously digitally converted to a baseband data stream composed of a sequence of I (in-phase) and Q (quadrature) samples. This digital down-conversion is done in real-time, allowing no gaps in the time record. The same IQ samples that are fed to the real-time engine[12] can also be simultaneously stored in memory for subsequent analysis using batch mode digital signal processing.

The real-time processing Engine[3] is a combination of hardware and software optimized to perform computations at a rate that keeps up with the incoming data stream. DFTs are sequentially performed on segments of the IQ record, generating a mathematical representation of frequency occupancy over time. RTSAs generate spectrum data at rates that are much too fast for the human senses. Visual data compression techniques must be used to generate meaningful displays. Persistence Spectrum[4] and *DPX®Spectrum*[5] (Digital Phosphor Spectrum) are techniques that provide an intuitive "live" view of complex and dynamic spectrum activity. The real-time processing engine can also be used to generate a *trigger* signal based on specific occurrences within the input signal. These occurrences can be frequency domain patterns, time domain events, or modulation events. The trigger signal can be used to store specific segments of the IQ time record for further analysis using batch-mode DSP.

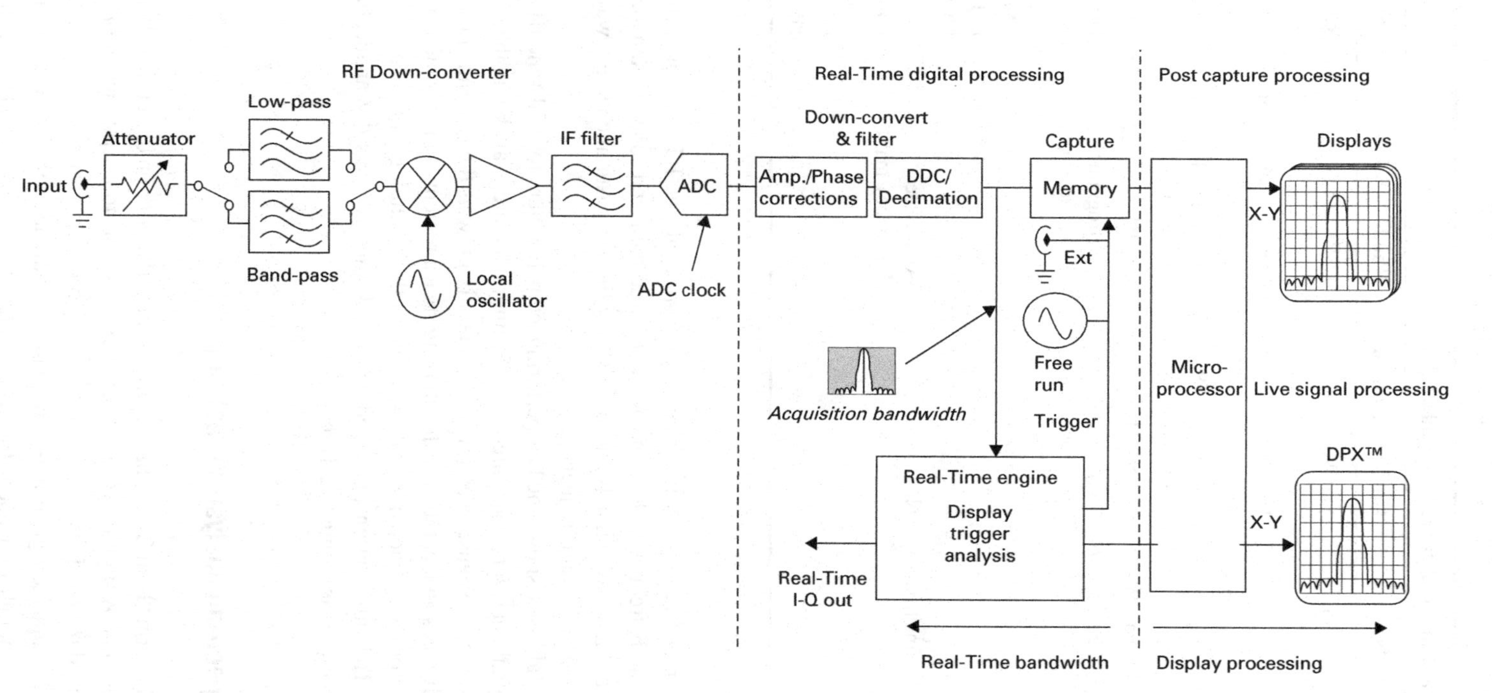

Fig. 4.3 Real-time signal analyzer architecture.

Table 4.1 Swept-tuned spectrum analyzer (SA) architecture

Real-time spectrum and signal analyzers			
Manufacturer	**Gauss instruments**	**Rhode & Schwarz**	**Tektronix**
Product family	TDEMI series	FSVR series	RSA 5000 series RSA 6000 series
Purpose	EMI Measurements	Spectrum and signal analysis	Spectrum and signal analysis
Frequency range	0.03, 1, 3, 6, 18, 26.5, 40 GHz	7, 13.6, 30, 40 GHz Extendable to 110 GHz	3, 6.2, 14, 20 GHz
Real-time BW	162.5 MHz	40 MHz	25, 40, 85, 110 MHz
Real-time displays	Weighted spectrogram	Spectrogram Persistence spectrum	DPX Zero-span Spectrogram AM FM Phase
Real-time triggers	N/A	Amplitude Frequency mask	Runt Power (Amplitude) Frequency edge (FM) Frequency mask DPX Density trigger

In addition to the traditional spectrum analysis, RTSAs can perform multiple time domain, frequency domain, modulation-domain, and code-domain measurements on RF and microwave signals and can display these measurements in a way that is correlated in both time and frequency.

Table 4.1 shows the key performance parameters for real-time RF/uW spectrum and signal analyzers from three manufacturers that are available at the time of writing. The list is not exhaustive and will undoubtedly grow as the real-time processing technology advances and as the need for making measurements in real-time becomes critical. The performance quoted will also change as the technology advances.

The information presented in this chapter is generally applicable to all real-time spectrum and signal analyzers.

4.2 Spectrum analysis in real-time

The term "Real-Time" has its origins in early work on digital simulations of mechanical systems. A digital system *simulation* was said to operate in *real-time* if its operating speed matched that of the *real system* which it was simulating.

To analyze signals in real-time means that the analysis operations must be performed fast enough to account for all the relevant signal components in the frequency band of interest. This definition, when applied to Fourier analysis, implies that a real-time spectrum analyzer that is based on sequential DFTs, as shown in Figure 4.3, must sample the analog IF signal fast enough to satisfy the Nyquist criteria. This means that the sampling frequency must ***exceed*** twice the bandwidth of interest. It must also perform all computations continuously at a fast enough rate that the output analysis keeps up with the changes in the input signal. At the time of publication, RTSAs can process bandwidths exceeding 160 MHz and generate in the order of 300 K spectrums per second.

4.2.1 Real-time criteria

The statement that a real-time system must keep up with changes in the incoming signal requires a more precise definition. There are two useful criteria based on scenarios in which RTSAs are used to perform spectrum analysis.

1. **Transient detection and measurement:** An RTSA is often used to observe a section of spectrum and to detect and measure short-duration events that happen with unknown timing. A useful measure of performance is the minimum *single-event* duration that can be detected and measured with 100% probability, at the specified accuracy.
2. **Spectrum monitoring with no loss of information:** An RTSA is often used to analyze unknown signals in spectrum management and surveillance applications. It is desirable that all information contained in the signal of interest be included in the analysis, with no gaps or lost content.

An elaboration of the two definitions requires us to explore the basic RTSA operation of performing sequential DFTs on a continuous input stream of time domain samples.

4.2.2 Theoretical background

Sequential discrete Fourier transforms

The heart of an RTSA is a DFT engine capable of analyzing frequency behavior over time. Figure 4.4 shows a simplified schematic of a real-time processing engine sequentially computing DFTs. The frequency domain behavior over time can be visualized as a spectrogram, shown in Figure 4.5, where frequency is plotted horizontally, time is plotted vertically, and the amplitude is represented as a color, intensity, or shade. A swept spectrum analyzer, in contrast, is tuned to a single frequency at any given point in time. The diagonal line in Figure 4.5 traces the time-frequency trajectory taken by a sweep. The slope of the line becomes steeper as the sweep slows, so that the function of a spectrum analyzer in *zero-span* can be represented as a vertical line indicating that the instrument is tuned to a single frequency as time advances. Figure 4.5 also shows how a sweep can miss transient events such as the single frequency hop depicted in the spectrogram.

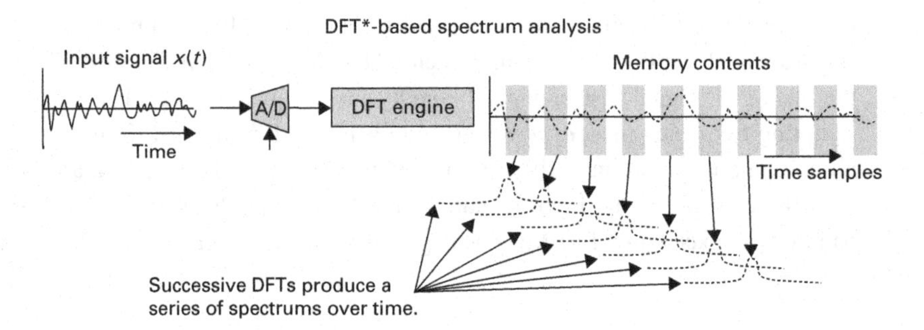

Fig. 4.4 A DFT-based spectrum analyzer computing a series of transforms over time.

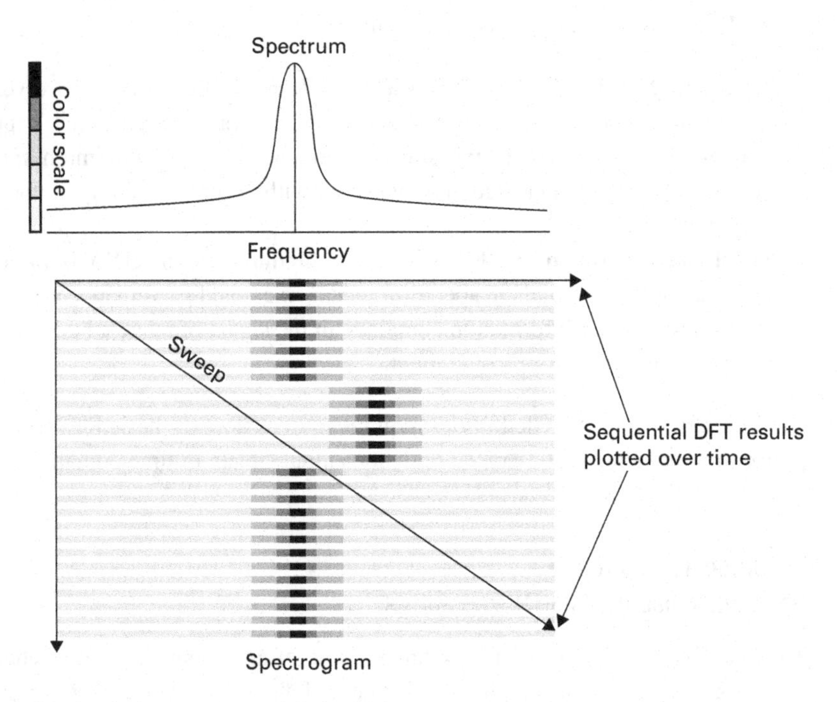

Fig. 4.5 Spectrum, Spectrogram and their relation to sweep.

4.3 Spectrum analysis using discrete Fourier transforms

4.3.1 The Fourier transform for discrete-time signals

Consider an input signal $x(t)$ as shown in Figure 4.4 that is digitized with a sampling period T_S resulting in the sampled representation of the input signal given by (4.1).

$$x[n] = x(nT_s), \quad \text{for} \quad n = 0, 1, 2, 3, 4, 5, \ldots \tag{4.1}$$

The Fourier integral, computed for N samples starting at sample n_m becomes the sum

$$X(\omega) = \sum_{n=n_{\mathrm{m}}}^{n_m+N-1} x(nT_s)e^{-j\omega nT_s}T_s. \tag{4.2}$$

Normalizing the frequency variable and scaling the magnitude by T_S, leads us to the Discrete Time Fourier Transform (DTFT) for a finite time interval[6].

$$X(\Omega) = \sum_{n=n_{\mathrm{m}}}^{n_m+N-1} x[n]e^{-j\Omega n}. \tag{4.3}$$

The DTFT has N time domain samples at its input and generates a continuous function of frequency, $X(\Omega)$. $X(\Omega)$ is periodic in Ω with a period of 2π. The time interval represented by the N samples covered in the summation, starting at sample n_m and ending at $n_m + N - 1$, is called a *frame*.

Many DTFT algorithms provide an output that is sampled in the frequency domain. Assuming regularly spaced frequency sampling, the output of each DTFT can be denoted as

$$X[k] = \sum_{n=n_{\mathrm{m}}}^{n_m+N-1} x[n]\mathrm{e}^{-jk\frac{2\pi}{K}n}, \tag{4.4}$$

where $\frac{2\pi}{K}$ is the spacing between frequency domain samples.

The formula in (4.4) can be recognized as the Discrete Fourier Transform or DFT[7] performed at an arbitrary starting point. It should be noted that, although many algorithms have an equal number of input and output points (Cooley-Tukey FFT for example), the number of samples in the frequency domain output does not need to be equal to the number of time domain samples in the input. (4.4) simply samples the continuous frequency function $X(\Omega)$.

4.3.2 Regularly spaced sequential DFTs

One can extend the above formula for a sequence of transforms by using the index m to denote the transform number in a sequence. Consider now the continuous flow of *regularly spaced* DFTs shown in Figure 4.4. We have a new frequency domain output being generated every L samples. The time-frequency output can be expressed as a sampled function of time and frequency:

$$X[k,m] = \sum_{n=n_{\mathrm{m}}}^{n_m+N-1} x[n]\mathrm{e}^{-\mathrm{jk}\frac{2\pi}{\mathrm{K}}n}, \tag{4.5}$$

where K represents the number of frequency points in the output of each transform, k represents the frequency sample index, L represents the number of time samples between the start of each successive DFT frames, and N represents the number of time domain samples in each DFT frame. The indices for frequency, time, and frame number are k, n,

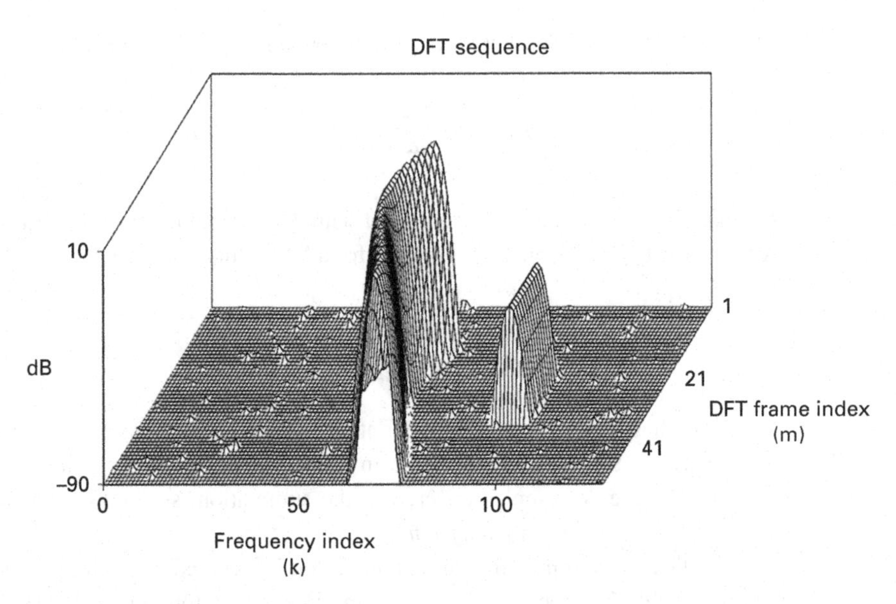

Fig. 4.6 A 3-dimensional view of a sequence of DFTs taken over time.

and m, respectively. Each value of m represents a new DFT frame with N time domain samples and produces a frequency spectrum with K frequency domain samples as shown in Figure 4.6.

There are three relevant cases to consider:

1. $L > N$: The spacing between frames is greater than the frame duration. There is a gap between frames. The portion of the input that lies in the gap is ignored. Data is lost.
2. $L = N$: The first sample of a frame is the sample immediately following the last sample of the previous frame. There is no gap. The frames are back-to back. Every sample of the input is included.
3. $L < N$: The spacing between frames is smaller than the frame length. The frames overlap. Not only are all input samples included but a given frame shares some of the samples with frames that precede it and with frames that follow it.

4.4 Windowing and resolution bandwidth (RBW)

The mathematics of DFTs assumes that the data to be processed is a single period of a periodically repeating signal. The upper graph in Figure 4.7 depicts a series of time domain samples. When DFT processing is applied to the 64 samples starting at sample 32 in Figure 4.7, the periodic extension is made to the signal as shown in the lower graph. The resulting discontinuities can generate spectral artifacts that are not present in the original signal. This effect produces an inaccurate representation of the signal and is called *spectral leakage*[8]. Spectral leakage not only creates signals in the output that

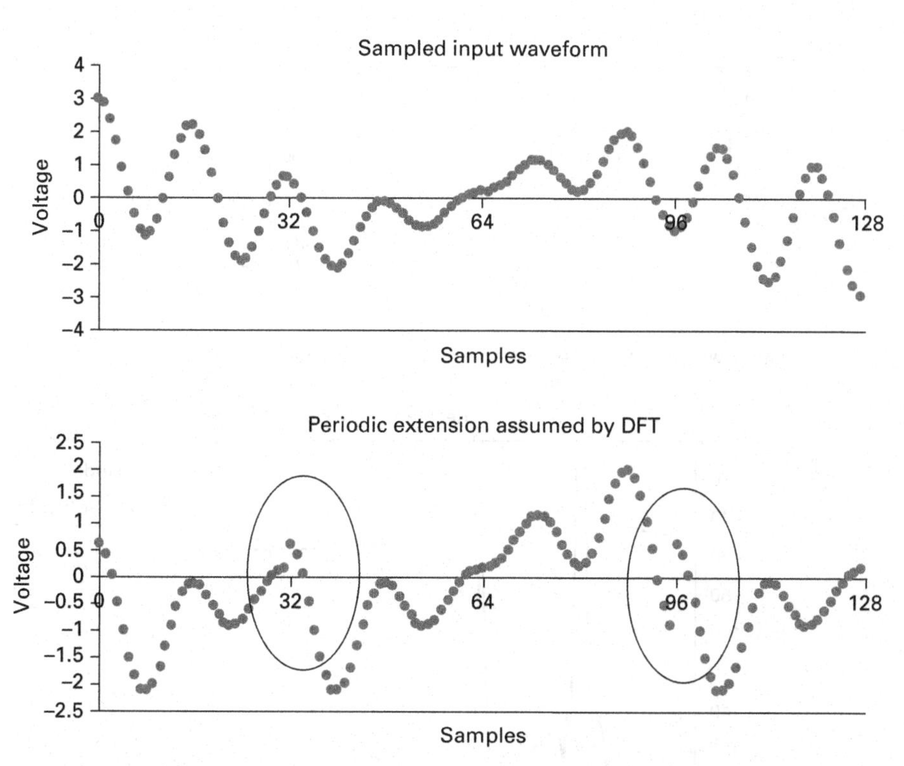

Fig. 4.7 Time domain samples and the discontinuities caused by periodic extension of samples in a single frame.

were not present in the input, but also reduces *dynamic range*, the ability to observe small signals in the presence of nearby large ones.

Windowing is a technique that is commonly used to reduce the effects of spectral leakage and to improve the resulting dynamic range. Before performing the DFT, the DFT *frame* is multiplied by a *window function* with the same length sample by sample. The window functions usually have a bell shape, reducing or eliminating the discontinuities at the ends of the DFT frame. Figure 4.8 shows an example of a Kaiser window and its Fourier transform.

Figure 4.9 shows the effects of spectral leakage on dynamic range and how windowing can be used to reduce its effects. The input signal in Figure 4.9 contains two pure sinusoids, one at full amplitude with a frequency of $1/13^{th}$ of the sampling rate; the second signal is at $1/7^{th}$ of the sample rate and has an amplitude 1000 times lower (−60 dB). The trace in black shows the magnitude in dB of a 1024-point DFT with a rectangular window (un-windowed). Spectral leakage reduces the dynamic range so that the weaker of the two signals is not visible. The application of a window similar to the one shown in Figure 4.9 increases the dynamic range so that the weaker signal is easily visible, as shown in the lower trace.

The choice of window function depends on its frequency response characteristics such as side-lobe level, equivalent noise bandwidth, and amplitude error. The window shape

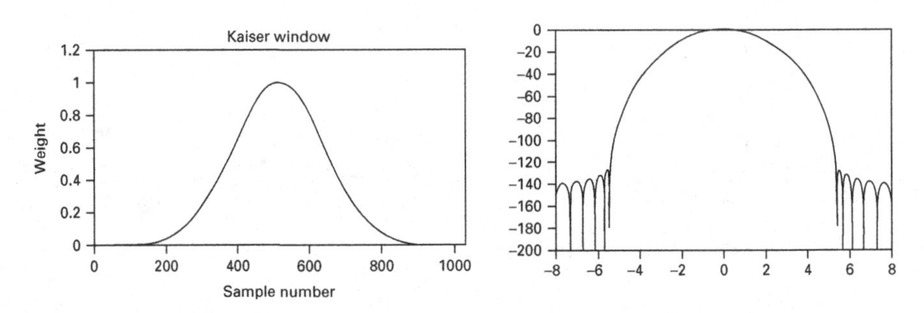

Fig. 4.8 Kaiser window (beta=16.7) and its Fourier transform.

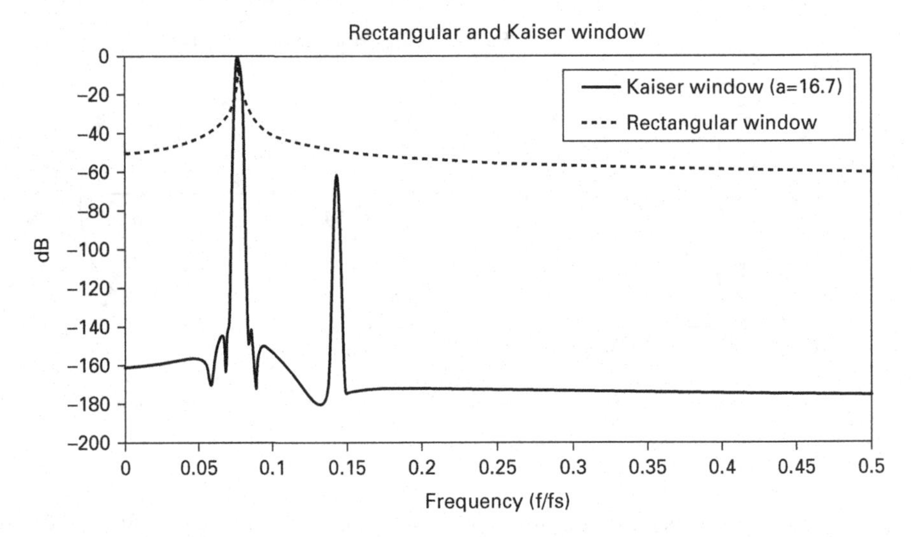

Fig. 4.9 Dynamic range improvement with windowing.

also determines the effective *resolution bandwidth* (RBW)[9]. Figure 4.9 shows a widening of the line-width in the spectrum. Many DFT-based spectrum analyzers vary window parameters as a means of allowing a selectable resolution bandwidth or RBW. RTSAs offer either user selectable RBW or a choice of several popular window functions.

4.4.1 Windowing considerations

Consider a window function as shown in Figure 4.8. The mth frame of the input $x(n)$ signal is multiplied by the window to generate the windowed signal, $x_W[n,m]$. The DFT of the mth frame is then taken as

$$X_W[k,m] = \sum_{n=mL}^{mL+N-1} x[n]\,W[n-mL]e^{-jk\Omega_s n} = \sum_{n=mL}^{mL+N-1} x_W[n,m]\,e^{-jk\Omega_s n}. \quad (4.6)$$

The windowed function, $x_W[n,m]$, has a value of zero at $n = mL$ and at $n = mL + N - 1$, the beginning and end of each summation. Discontinuities, like those shown in

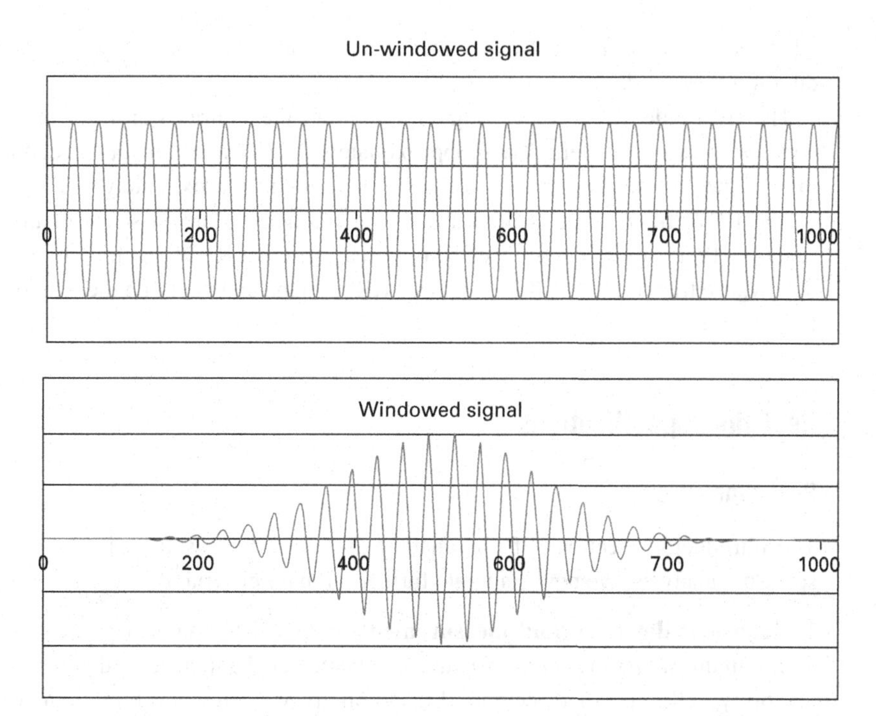

Fig. 4.10 Un-windowed and windowed signal $x[n]$ and $x_W[n,m]$.

Figure 4.7, are thus attenuated. Figure 4.10 shows a frame of a sinusoidal input signal and the same frame after windowing.

4.4.2 Resolution bandwidth (RBW)

The term “Resolution bandwidth” (RBW) was coined in the days when spectrum analyzers actually swept a signal through a physical filter as shown in Figure 4.1. Resolution bandwidth is defined as the smallest bandwidth that can be resolved in a spectrum analyzer. A swept analyzer, when presented with a pure sinusoidal tone at its input, traces the RBW filter shape on its display.

RTSAs use windowed DFTs to generate spectrum displays. The windowing operation involves multiplying the window shape with the incoming signal. The multiplication of the window function $W(t)$ and the signal to be analyzed $x(t)$ implies that the DFT of the time domain product is the convolution of the frequency domain functions.

In commercially available spectrum analyzers, the RBW filter defines the spectrum shape traced by a perfect sinusoid presented at the input. Therefore, the RBW of a DFT-based spectrum analyzer is defined as the convolution of a spectral impulse with the Fourier transform of the window function. The effective RBW filter shape is the *Fourier transform of the window function.*

The 3 dB bandwidth of the RBW filter is given by

$$RBW = \frac{K_W}{D_{Frame}}, \tag{4.7}$$

where K_W is a coefficient that is related to a particular window and D_{Frame} is the frame duration in seconds.

The right side of Figure 4.8 Figure 4.1 shows the Fourier transform of the Kaiser window shown on its left. K_W is approximately 2.23. The *shape factor*, defined as the ratio of the 3dB bandwidth to the 60 dB bandwidth, is approximately 4:1.

The RBW filter shape is the same as the frequency domain shape of the window function. Performing spectrum analysis with a particular RBW requires choosing a window whose transform yields the required RBW shape and applying the window to each DFT frame.

4.5 Real-time specifications

4.5.1 Real-time criteria

The introductory sections of this chapter described two useful criteria for real-time spectrum analysis. We now elaborate further on these criteria.

1. **Transient detection and measurement:** A useful measure of performance is the minimum *single-event duration* that can be detected and measured with 100% probability, at the specified accuracy. For simplicity, we define this minimum event as the narrowest *rectangular RF burst* that can be detected with 100% probability and have its underlying RF signal measured without degradation of accuracy.
2. **Spectrum monitoring with no loss of information:** It is desirable that all information contained in the signal of interest be included in the analysis, with no gaps or lost content. A useful test for no information loss in a sequence of DFTs is that input time domain data can be recovered from the frequency domain output.

An exploration of the two criteria requires us to look at the sequence of DFTs and the relationship between the DFT length, the spacing between successive DFTs, and the effects of windowing.

4.5.2 Minimum event duration for 100% probability of intercept at the specified accuracy

Consider a rectangular burst of RF that is down-converted and sampled. The stream of samples can be expressed as

$$x[n] = a[n]e^{j\omega_0 n T_s}\left[u(n-n_0) - u(n-n_0-n_p)\right], \tag{4.8}$$

where $a(n)$ is the sampled complex envelope of the RF signal within the burst, ω_0 is the RF carrier frequency after down-conversion, n_0 is the starting sample of the pulse, and n_P is the number of samples contained within the pulse. A representation of such an RF burst is shown in Figure 4.11.

The DFT for any particular *frame* will be the same as that for a continuous signal as long as *that frame* is completely contained within the pulse. The spectrum of the RF

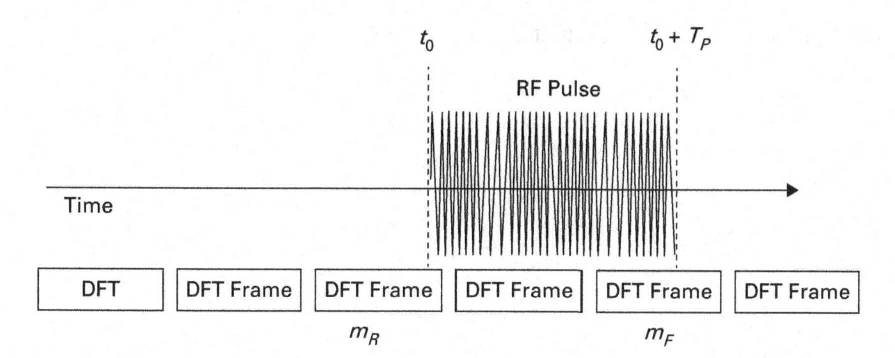

Fig. 4.11 A DFT frame must be contained within the pulse for an accurate representation of the RF signal contained in the pulse.

signal within the pulse will be accurately represented as long as the pulse contains at least one entire DFT frame. Frames that contain either a rising edge or a falling edge will reflect the presence of those edges in their spectrums and will not accurately depict the underlying RF signal.

Consider now performing successive DFTs on the single RF burst shown in Figure 4.11. The starting point of the burst is *unknown* and *not synchronous* with the sequential DFT operation.

$$X_W[k,m] = \sum_{n=mL}^{mL+N-1} W[n-mL]a(n)e^{j\omega_0 nT_s}[u(n-n_0)-u(n-n_0-n_p)]e^{-jk\Omega_s n}. \tag{4.9}$$

If the **rising** edge is contained in frame m_R and the falling edge in frame m_F then the above expression can be expanded to include all frames that contain some of the RF burst.

The frames that occur before the beginning and after the end of the burst contain no signal. The frames that contain the rising and falling edges have a truncated summation. Their DFT will show the spectral effects of the rise and fall.

$$X_W[k,m_R] = \sum_{n=n_0}^{m_R L+N-1} W[n-m_R L]a(n)e^{j\omega_0 nT_s}e^{-jk\Omega_s n}$$

$$X_W[k,m_F] = \sum_{n=m_F L}^{n_0+n_p} W[n-m_F L]a(n)e^{j\omega_0 nT_s}e^{-jk\Omega_s n}. \tag{4.10}$$

The DFT for frames that fall completely within the burst will be indistinguishable from those of a continuous signal. These frames do not contain the rising and falling edges and have a complete summation. Their spectrum shows a faithful representation of the signal inside the burst. Any modulation, distortion, or other spectral effects present in the signal will be accurately represented.

$$
\begin{aligned}
X_W[k, m_R+1] &= \sum_{n=(m_R+1)L}^{(m_R+1)L+N-1} W[n-(m_R+1)L]a(n)e^{j\omega_0 n T_s}e^{-jk\Omega_S n} \\
X_W[k, m_R+2] &= \sum_{n=(m_R+2)L}^{(m_R+2)L+N-1} W[n-(m_R+2)L]a(n)e^{j\omega_0 n T_s}e^{-jk\Omega_S n} \\
&\vdots \\
X_W[k, m_F-1] &= \sum_{n=(m_F-1)L}^{(m_F-1)L+N-1} W[n-(m_F-1)L]a(n)e^{j\omega_0 n T_s}e^{-jk\Omega_S n}.
\end{aligned}
\tag{4.11}
$$

Pulses and DFT frames are, in the general case, asynchronous. Figure 4.12 illustrates the case where the DFT frames overlap and the case where there is a gap between frames. Figure 4.12 also shows that a burst must be at least as wide as the time it takes for two consecutive frames to be acquired in order to have a 100% probability of containing a complete frame, considering the arbitrary timing for the RF burst.

Let $T_{P\,\min}$ be the minimum pulse duration required for a pulse to be captured with a 100% probability with full accuracy. Let T_{Frame} and T_{Gap} be the frame duration and gap duration, respectively. Then

$$T_{P\,\min} = 2T_{Frame} + T_{Gap}. \tag{4.12}$$

The requirements for discovering, capturing, and analyzing transients are:

- Enough **capture bandwidth** to support the signal of interest.
- A high enough ADC clock rate to exceed the **Nyquist criteria** for the capture bandwidth.
- A **frame duration** long enough to support the narrowest resolution bandwidth (RBW) of interest as defined by the Fourier transform of the window function.
- A fast enough **DFT transform rate** to support the minimum event duration as given by (4.12). It should be noted that overlapping frames reduce the minimum time by making T_{Gap} negative.

4.5.3 Comparison with swept analyzers

The definitions for $T_{P\,\min}$ can apply to any spectrum analyzer, including the traditional swept spectrum analyzers. If we replace T_{Frame} by the sweep time, T_{SW}, and T_{Gap} by the retrace time, $T_{Retrace}$, then (4.12) becomes

$$T_{P\,\min} = 2T_{SW} + T_{\mathrm{Retrace}}. \tag{4.13}$$

The minimum time for 100% probability of capture and measurement with full accuracy for a swept analyzer is twice the sweep time plus the retrace time. $T_{P\,\min}$ for a

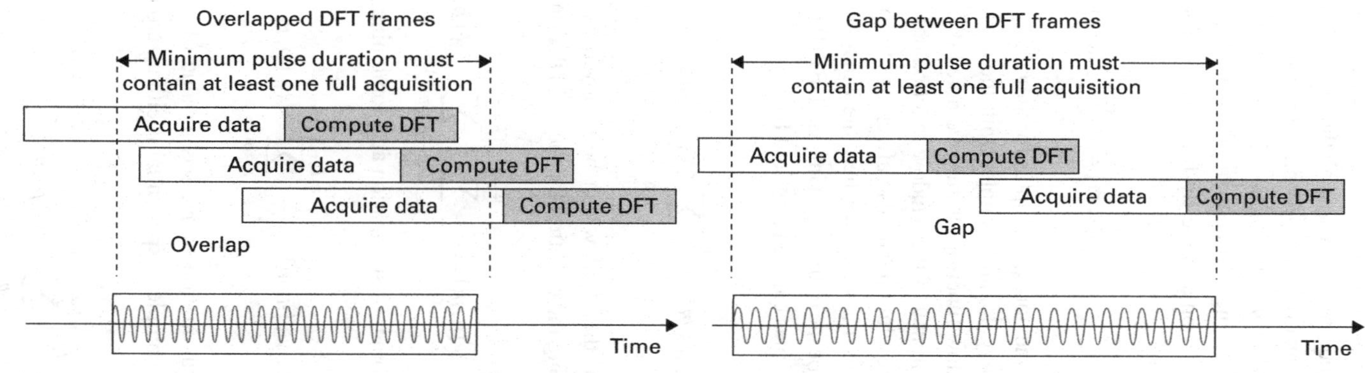

Fig. 4.12 Overlapped DFT frames and frames with a time gap.

swept analyzer observing a 100 MHz span with a 1 MHz RBW is in the order of many milliseconds. An RTSA can achieve $T_{P\,\mathrm{min}} < 6\ \mu\mathrm{s}$ for the same settings.

4.5.4 Processing all information within a signal with no loss of information

It is sometimes required to process all the information contained within a signal, making sure that no information is lost. As mentioned in the introduction to this section, a useful test for *no loss of information* is that the input data can be recovered from the output.

Consider performing consecutive DFTs on a sampled time domain signal $x[n]$ that has been windowed on a frame-by-frame basis. The DFT of the mth frame is related to its time samples by

$$X[k,m] = \sum_{n=mL}^{mL+N-1} x[n]\mathrm{e}^{-\mathrm{j}k\frac{2\pi}{K}n}. \tag{4.14}$$

The original time domain samples for the mth frame can, in general, be reconstructed from the frequency domain samples with the application of an inverse DFT and division by the window function. It must be noted that the value of a time domain sample becomes indeterminate if the window function has a weight of zero at a particular sample. Numerical resolution can also affect the ability to accurately recover time domain data, especially for the samples where the window has small values.

Let $y_W[n,m]$ be the output of an inverse DFT (IDFT) for the mth frame. The windowed time domain samples can be recovered by

$$y_W[n,m] = \frac{1}{N}\sum_{k=-\frac{K}{2}}^{\frac{K}{2}-1} X_W[k,m]e^{jk\frac{2\pi}{K}n}, \quad mL \le n \le N-1. \tag{4.15}$$

The effects of the window can be removed by dividing $y_W[n,m]$ by the window function. Applying the equation for the mth DFT and exchanging the index n for p, we have

$$y_W[n,m] = \frac{1}{N}\sum_{k=-\frac{K}{2}}^{\frac{K}{2}-1}\ \sum_{p=mL}^{mL+N-1} x_W[p,m]e^{-jk\Omega_S p}e^{jk\frac{2\pi}{K}n}. \tag{4.16}$$

Inverting the order of summation and combining the exponentials, we get

$$y_W[n,m] = \frac{1}{N}\sum_{p=mL}^{mL+N-1} x_W[p,m]\sum_{k=-\frac{K}{2}}^{\frac{K}{2}-1} e^{jk\frac{2\pi}{K}(n-p)}. \tag{4.17}$$

The second summation above has a value of zero except for the cases where the argument of the complex exponential is either zero or a multiple of 2π, where its value is K.

$$y_W[n,m] = \frac{K}{N}\sum_{p=mL}^{mL+N-1} x_W[p,m]\partial[p-n+iK], \quad \text{where } i \text{ is an integer.} \tag{4.18}$$

We must now remember that $x_W[p,m]$ is non-zero only for the N samples in the window, $mL \leq p < mL + N$. We must also remember that the DFT is periodic with period K.

Consider the following cases:

1. $K \geqq N$: **Data recovery is possible.** The number of frequency domain points is greater than or equal to the number of time domain points. The only value that falls inside the summation limits is $i = 0$. The result is an exact reproduction of the original windowed function for the mth frame.

$$y_W[n,m] = \frac{K}{N} \sum_{p=mL}^{mL+N-1} x_W[p,m]\partial[p-n]$$

$$y_W[n,m] = \frac{K}{N} x_W[n,m]. \tag{4.19}$$

2. $K < N$: **Aliasing occurs.** The number of frequency domain points is smaller than the number of time domain points. There are more than K points in the summation. Non-zero values of i fall within the summation limits. The recovered output contains contributions from multiple periods. The frequency domain signal is said to be under-sampled. The time domain samples cannot be uniquely recovered from the DFT output.

$$y_W[n,m] = \frac{K}{N} \left\{ x_W[n,m] + \sum_{i=1}^{\text{int}\left(\frac{N}{K}\right)} x_W[n-iK,m] \right\}. \tag{4.20}$$

4.5.5 Windowing and overlap

The derivations in (4.18) – (4.20) show that the original time domain data for a particular frame can be recovered from its DFT provided that the number of frequency samples is at least as great as the number of time samples and that the window has non-zero values across the DFT frame. The recovery of all data can still be achieved with DFT frame overlap even if the window function goes to zero at the beginning and end of each frame.

Consider now a sequence of windowed DFTs taken at regular intervals. Taking inverse DFTs can recover the windowed time domain samples for each frame. There are three cases to consider.

1. $L > N$: The spacing between each DFT is greater than the DFT length. There is a gap between consecutive DFT frames. Any variations in the signal that happen inside the gap are lost.
2. $L = N$: Consecutive DFTs begin with the sample immediately following the last sample of the previous one. There are no gaps between successive DFTs. Every time domain sample is considered. The application of inverse DFTs can, in theory, recover each and every sample of the input. There are, however, practical considerations:

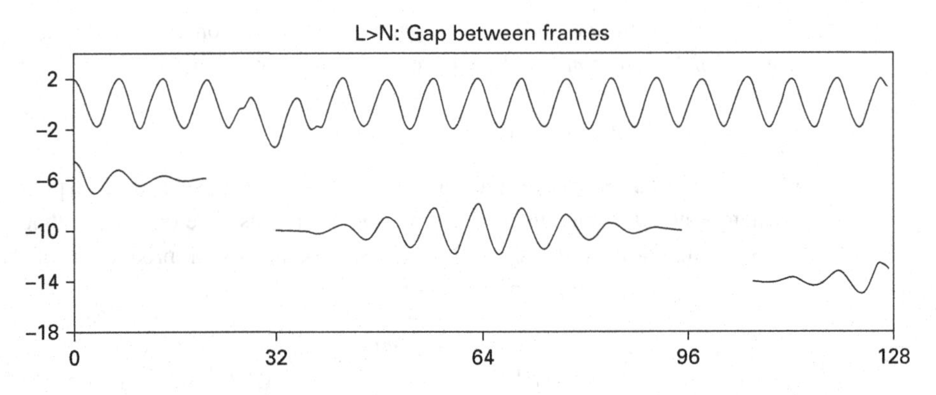

Fig. 4.13 L > N – Gap between DFT frames.

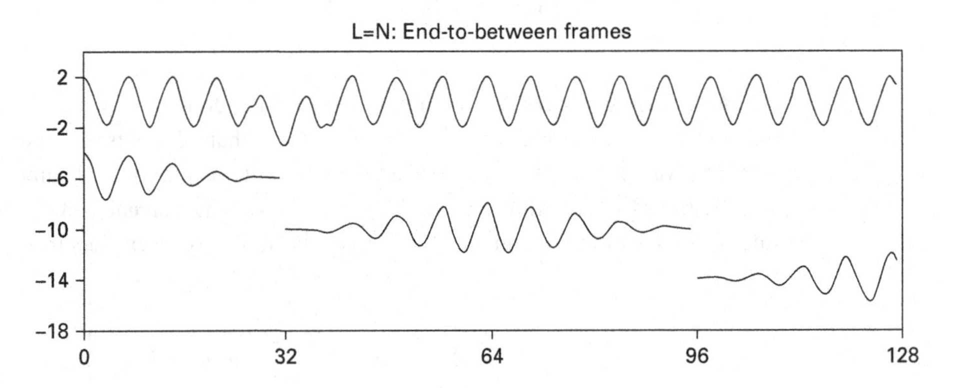

Fig. 4.14 L = N- Frame arranged end-to-end, with no gap.

- The window cannot be zero at any point in the frame.
- Numerical precision may cause an unrecoverable degradation of data at points where the window value is small.

Figure 4.14 shows the same input signal, now with an end-to-end frame arrangement. Limited numerical precision makes the recovery of samples near the ends of the frames impractical. Figure 4.14 illustrates this point since the anomaly around sample 32 is attenuated to a value very near zero by the window.

3. L<N. Consecutive DFTs overlap. All DFT frames share some samples with both their successors and predecessors. Successive inverse DFTs can uniquely reconstruct the input signal from the output as long as the windowing function is known and the zeroes of overlapping windows don't occur at the same place. Bell-shaped windows, such as the ones typically used in spectrum analysis, allow reconstruction when the amount of overlap is enough to overcome any numerical precision issues. Information is not lost.

Figure 4.15 shows the same input signal as in the previous two examples, this time processed with 50% frame overlap. Note that the effects of the anomaly near sample

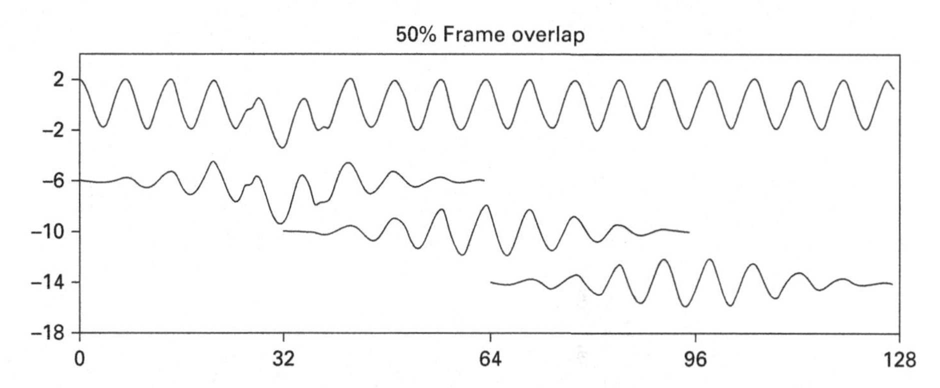

Fig. 4.15 L=N – DFT frames with 50% overlap.

32 are included in the first and second frames. An inverse DFT (IDFT) provides two estimates of the input signal for samples lying within the gap. The input signal can be recovered with high confidence by choosing the best of the two estimates or by optimally processing the two results.

It must be noted here that the proceeding discussion assumes that both magnitude and phase information are available from the DFT outputs. Although this is inherently true for DFT computations, most spectrum analyzers typically only display the magnitude. The test that the time domain data be recoverable is still useful for us to determine if *all information* present in a signal is included in the analysis and reflected in its results.

4.5.6 Sequential DFTs as a parallel bank of filters

Figure 4.16 illustrates an intuitive interpretation of the process used in real-time spectrum analysis. Taking sequential equally spaced DFTs over time, as shown in the upper part of Figure 4.16, is conceptually equivalent to the filtering and sampling system shown in the lower part of the figure.

Consider passing the input signal though a bank of K band-pass filters whose frequency spacing is the same as the spacing between DFT bins and whose bandwidth and frequency response is the same as the RBW shape described in Figure 4.8. The resulting K analog waveforms are then sampled and digitized in magnitude and phase (or I and Q) at the rate and timing with which the DFTs are computed. The resulting output is indistinguishable from the result of the computations in the upper part of Figure 4.12 Overlapped DFT frames and frames with a time gap.[10]

Consider one of the band-pass filter paths. Slow signal variations, those with frequency components contained within one RBW, result in changes in the level and phase of that path over time. A faithful representation of all signals present within the RBW requires that the magnitude and phase be sampled at a fast enough rate to meet the Nyquist criteria, at least *one complex sample* (a complex sample contains two samples) for every Hz of bandwidth. The required time between frames is related to the RBW by

$$T_{Frame} \leq \frac{1}{RBW}. \tag{4.21}$$

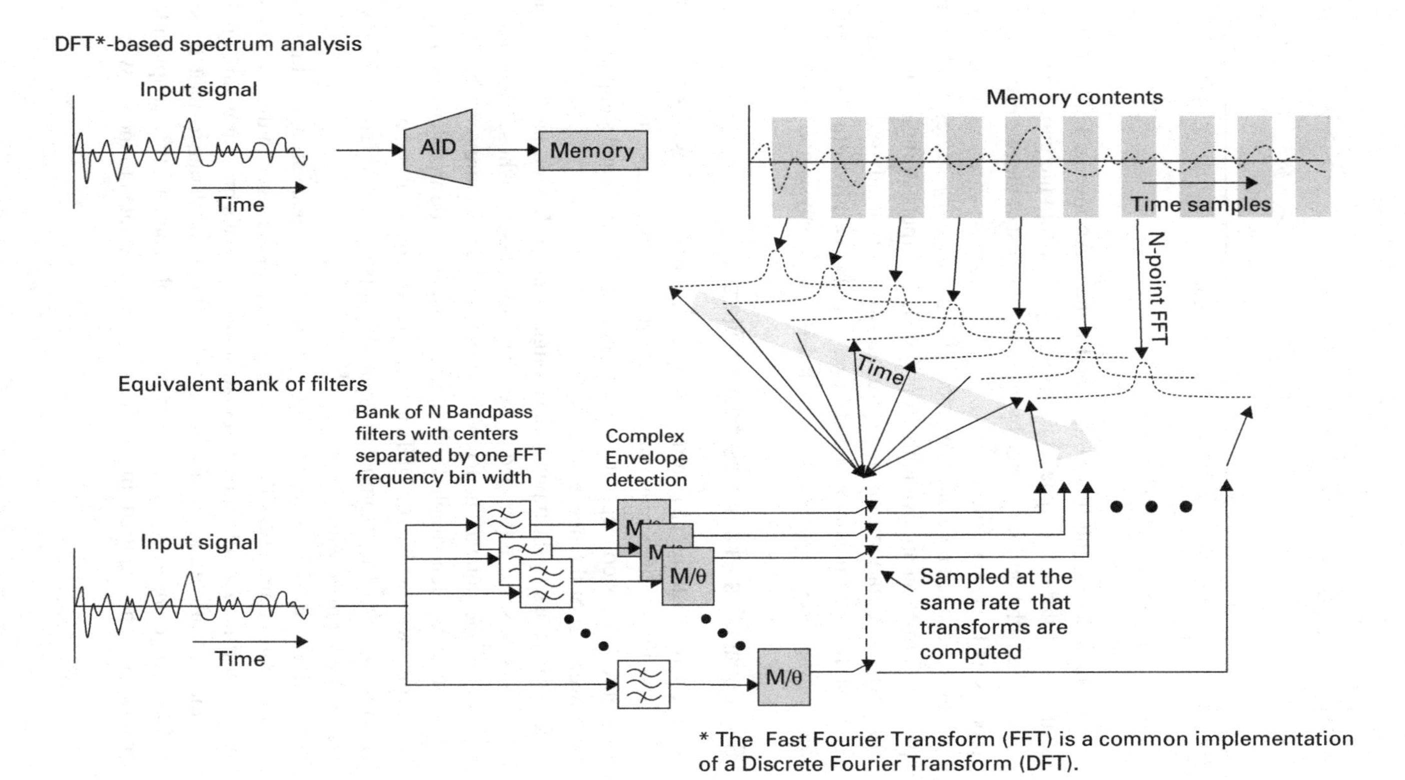

Fig. 4.16 DFTs can be interpreted as a parallel bank of band-pass filters.

Signal variations that are too fast to be contained within one RBW affect the magnitudes and phases of adjacent paths and are resolved in the spectrum graph.

4.5.7 Relating frame rate, frame overlap, and RBW

Equation (4.7) relates the frame duration to RBW. Combining this with (4.21) gives us

$$\frac{D_{Frame}}{T_{Frame}} \geq K_W. \tag{4.22}$$

The duration of the frame must exceed the time between consecutive frames. Consecutive frames must overlap in time. Each contains some samples in common with both the previous and the next frames. For the Kaiser window shown in Figure 4.8, $K_W = 2.23$. This window requires an overlap of at least 55% to ensure that all data contained in the time domain input signal is included in the three-dimensional output of spectrum versus time.

4.5.8 Criteria for processing all signals in the input waveform with no loss of information

In order to take all the information contained in a time domain waveform and transform it into a frequency domain representation with no loss of information in *real-time* requires several important signal processing requirements:

- Enough **capture bandwidth** to support analysis of the signal of interest.
- A high enough ADC clock rate to exceed the **Nyquist criteria** for the capture bandwidth.
- A long enough **frame duration** to support the narrowest resolution bandwidth (RBW) of interest as defined by the Fourier transform of the window function
- **Overlapping DFT frames**. The amount of overlap depends on the window or BRW used (4.22).

4.6 Applications of real-time spectrum analysis

4.6.1 Displaying real-time spectrum analysis data

RTSAs generate spectrum data at rates that are far too fast for the human eye to see. A means must be employed to compress the available data so that the user can observe dynamic, time-varying signals in a live manner. Several methods are used in the industry to compress the data into a form that is observable. These include three-dimensional displays (level, frequency, and time) like the Weighted Spectrogram[11] offered by Gauss industries, as well as ways to display the statistics of frequency occupancy over time such as Persistence Spectrum[12] and DPX®[13], offered by R&S and Tektronix, respectively. This text focuses on the Tektronix DPX method.

4.6.2 Digital persistence displays

The names "Digital Phosphor" and "Persistence Spectrum" are derived from the phosphor coating on the inside of cathode ray tubes (CRTs) used in older oscilloscopes and spectrum analyzers, where the electron beam is directly controlled by the waveform to be displayed. When the phosphor is excited by an electron beam, it fluoresces, lighting up the path drawn by the stream of electrons. The phosphors had the properties of **persistence** and **proportionality**.

Persistence is the property that the phosphor continues to glow even after the electron beam has passed by. Persistence allows the human eye to see events that would otherwise occur too fast to be seen. Proportionality means that the brightness of a phosphor is proportional to the number of electrons that hit a target point in the phosphor. The brightness of the spot increases as the electron beam hits it more frequently.

Digital Phosphor (DPX)[14] technology was developed by Tektronix to bring the analog benefits of a variable persistence CRT to modern digital instruments. DPX includes digital enhancements such as intensity grading, selectable color schemes, and statistical traces that communicate more information in less time. In the DPX display, both color and brightness provide z-axis emphasis.

4.6.3 The DPX spectrum display engine

A real-time signal analyzer computes hundreds of thousands of spectrums every second. This high transform rate is the key to detecting infrequent events, but it is far too fast for the liquid-crystal display to keep up with, and well beyond what human eyes can perceive. The DPX engine writes the incoming spectrums into a bitmap database at full speed then transfers the resulting image to the screen at a viewable rate. The bitmap database can be viewed as a dense grid created by dividing a spectrum graph into rows representing trace amplitude and columns for points on the frequency axis. Each cell in this grid contains the count of how many times it was occupied by an incoming spectrum. The result is a three-dimensional database where each point in the grid is represented by its x-axis (frequency), y-axis (amplitude) and z-axis (number of occurrences).

Persistence

Persistence is obtained by accumulating the contents of many DFTs and storing the results in a bitmap where the x and y axes correspond to frequency and amplitude, respectively and the z-axis, usually represented as a color, is an indication of how often a particular point is occupied.

The 11×10 matrix shown in Figure 4.17 illustrates the concept. The bitmap is computed by adding the contents of consecutive grids, each grid corresponding to a DFT frame. The picture on the left of Figure 4.17 shows what the database cells might contain after a single spectrum is mapped into it. Blank cells contain the value zero, meaning that no points from a spectrum have fallen into them yet. The grid on the right shows values that our simplified database might contain after an additional eight spectral transforms have been performed and their results stored in the cells. One of the nine spectrums

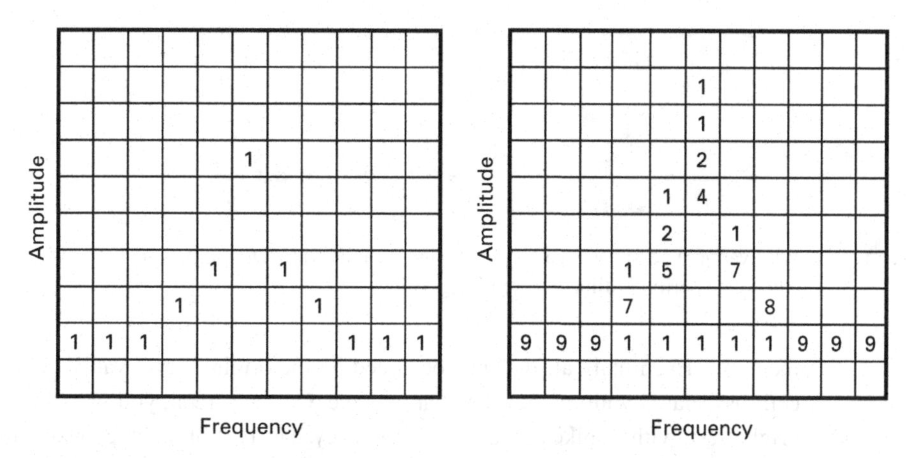

Fig. 4.17 Example 3-D bitmap database after 1 (left) and 9 (right) updates. Note that each column contains the same total number of "hits."

happened to be computed at a time during which the signal was absent, as indicated by the string of "1" values at the noise floor.

The DPX engine receives spectrum information at the full DFT rate and then accumulates the bitmap for a large number of DFTs, passing the bitmap to the display at a much slower rate. Each accumulated bitmap that is passed to the display is called a *display frame*. The display frame is computed by accumulating the contents of the bitmap. The Tektronix RSA6100A series of RTSAs, for example, performs nearly 292 000 DFT operations per second and updates the display frame at 20 times per second.

Variable persistence occurs when only a fraction of each count is carried over to the next display. Adjusting the fraction changes the length of time it takes for a signal event to decay from the database, and thus fade from the display. Raising the fraction to unity provides infinite persistence, where each point in the bitmap contains a histogram of the number of times it was hit since the process was started. A value of zero means that there is no persistence and it is freshly updated each display frame.

Color mapping and persistence

The display processing system then maps the *number of occurrences* to a color scale. Warmer colors (red, orange, yellow) indicate more occurrences. Cooler colors (blues and purples) indicate fewer occurrences. Black indicates no occurrences. Other intensity-grading schemes can also be used.

Imagine a 20 dB step change in RF level happening at the end of a display frame. Assume that the initial level was present for all 14,600 of the spectrum updates in one *display frame* and that the variable persistence factor causes 25% attenuation after each frame ($\alpha = 0.75$). After the initial frame, the cells affected by the initial level would start out with a value of 14,600 and be displayed at full force (Red). One display frame later, the values of those cells would drop by 25% (10,950), while those occupied by the new level would be at 25% (3650). After the next frame, the cells occupied by the initial level decrease by another 25%, while the value for the cells occupied by the new level

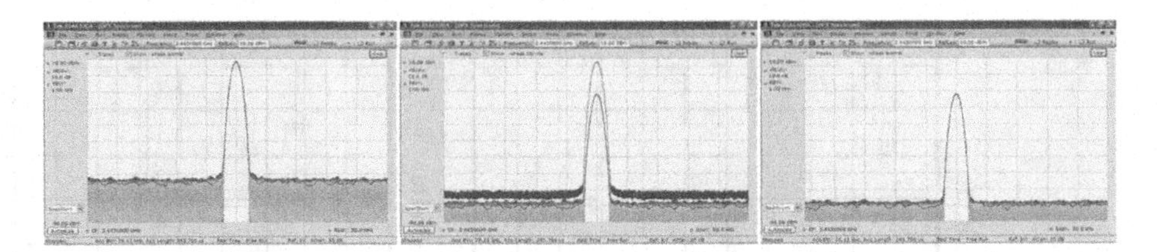

Fig. 4.18 With variable persistence, a brief CW signal captured by DPX remains in the display for an adjustable period of time before fading away.

increases. Eventually all the cells occupied by the original level vanish and only those cells associated with the new level are visible. On the screen, you would initially see a bright trace with a spike at the signal frequency. The part of the trace where the original signal occurred fades away. During this time, the pixels start to brighten at the new level below the fading signal. In the end, there is only the new trace in the display, as shown in Figure 4.18. Infinite persistence can catch even a single occurrence of a spectrum event. Variable persistence can provide an insight into dynamic signal behavior as it happens.

DPX line traces

The DPX engine can also produce line traces. The +Peak and −Peak traces show signal maxima and minima instantly and clearly. The average trace finds the mean level for the signal at each frequency point.

Using Hold on the DPX + Peak trace is almost exactly the same as the Max Hold trace on a typical spectrum analyzer, with the important difference that the DPX trace's update rate is *orders of magnitude* faster. The +Peak and −Peak traces show signal maxima and minima over arbitrarily long time periods. The +Peak trace displays a single occurrence of an RF burst lasting a few microseconds in an observation period of many hours or even days. Similarly, the −Peak trace shows a momentary gap in a signal lasting as little as a few microseconds.

4.7 Triggering in the frequency domain

Triggering is critical to capturing time domain information. Triggering allows a user to concentrate attention and analysis on a window of time where an event of interest happens. RTSAs offer unique trigger functionality, providing the ability to trigger on a *frequency mask, modulation and signal statistics* as well as power in a bandwidth. Other trigger functions including external triggering and gated triggers are included as well. This section explores the frequency mask trigger (FMT), beginning with an overview of triggering in general.

4.7.1 Digital triggering

Triggering was originally implemented as a way to stabilize oscilloscope displays. In traditional analog oscilloscopes, the signal to be observed is fed to one input while the

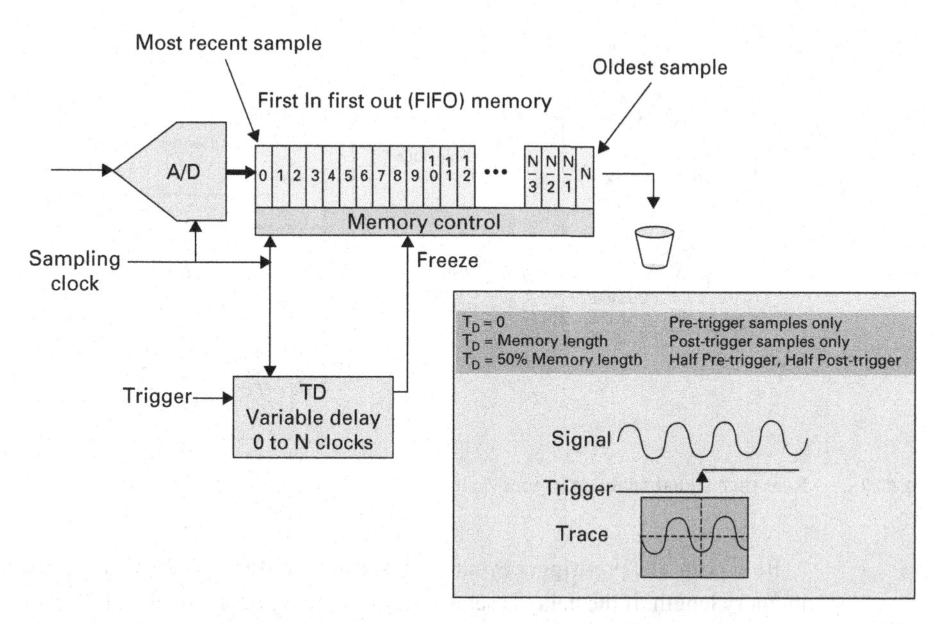

Fig. 4.19 Triggering in digital acquisition systems.

trigger is fed to another. The trigger event causes the start of a horizontal sweep while the amplitude of the signal is shown as a vertical displacement superimposed on a calibrated grid. In its simplest form, analog triggering allows events that happen *after* the trigger to be observed.

4.7.2 Triggering in systems with digital acquisition

The ability to represent and process signals digitally, when coupled with large memory capacity, allows the capture of events that happen before the trigger as well as after it. Digital acquisition systems of the type used in Tektronix RTSAs use an analog-to-digital converter (ADC) to fill a deep memory with time samples of the received signal. Conceptually, new samples are continuously fed to the memory, while the oldest samples fall off. The example shown in Figure 4.19 shows a memory configured to store N samples. The arrival of a trigger stops the acquisition, freezing the contents of the memory. The addition of a variable delay in the path of the trigger signal allows events that happen before a trigger, as well as those that come after it, to be captured.

Consider a case in which there is no delay. The trigger event causes the memory to freeze immediately after a sample concurrent with the trigger is stored. The memory then contains the sample at the time of the trigger as well as $N-1$ samples that occurred before the trigger. Only **pre-trigger** events are stored.

Consider now the case in which the delay is set to match exactly the length of the memory. N samples are then allowed to come into the memory after the trigger occurrence before the memory is frozen. The memory then contains N samples of signal activity after the trigger. Only **post-trigger** events are stored.

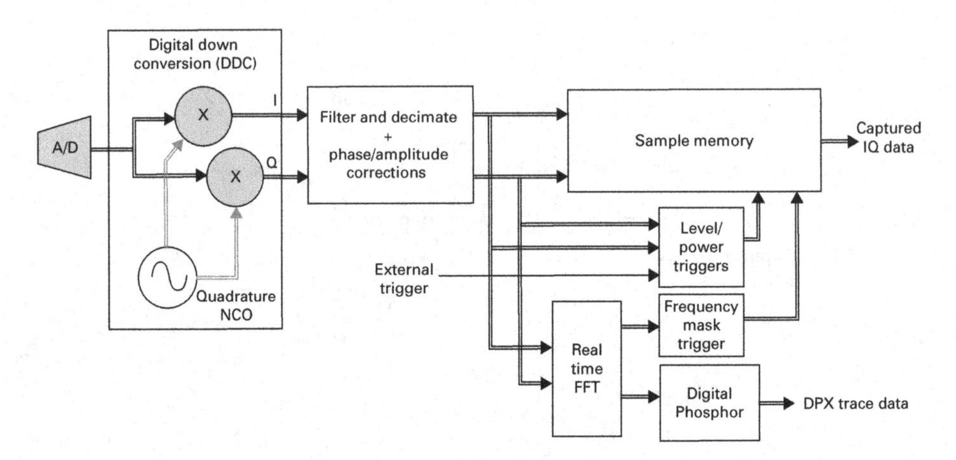

Fig. 4.20 Real-time signal analyzer trigger system.

Both post- and pre-trigger events can be captured if the delay is set to a fraction of the memory length. If the delay is set to half the memory depth, then half the stored samples are those that preceded the trigger and half are the stored samples that followed it.

Once data is stored in memory, it is available for further analysis using a DSP. The stored signal data can subsequently be analyzed in the frequency, time, and modulation domains. This is a powerful tool for applications such as signal monitoring and device troubleshooting.

4.7.3 RTSA trigger sources

RTSAs provide several methods of internal and external triggering, as shown in Figure 4.17. Triggers can come from an external source, from a computation of signal power, from a real-time demodulation of the input signal or from the frequency domain content of the input signal in a frequency mask trigger (FMT).

4.7.4 Frequency mask trigger (FMT)

Frequency mask triggering compares the spectrum shape to a user-defined mask. This powerful technique allows changes in a spectrum shape to trigger an acquisition. Frequency mask triggers can reliably detect signals far below full-scale even in the presence of other signals at much higher levels. This ability to trigger on weak signals in the presence of strong ones is critical for detecting intermittent signals, the presence of intermodulation products, transient spectrum containment violations, and much more. Trigger events are determined in the frequency domain using a dedicated hardware DFT processor, as shown in the block diagram in Figure 4.20. A full DFT is required to compare a signal to a mask, requiring a complete frame.

Consider a sequence of log-magnitude traces generated by consecutive DFTs. Each of these spectrums is compared with a stored mask. A trigger is generated whenever one of the incoming spectrums violates the mask conditions.

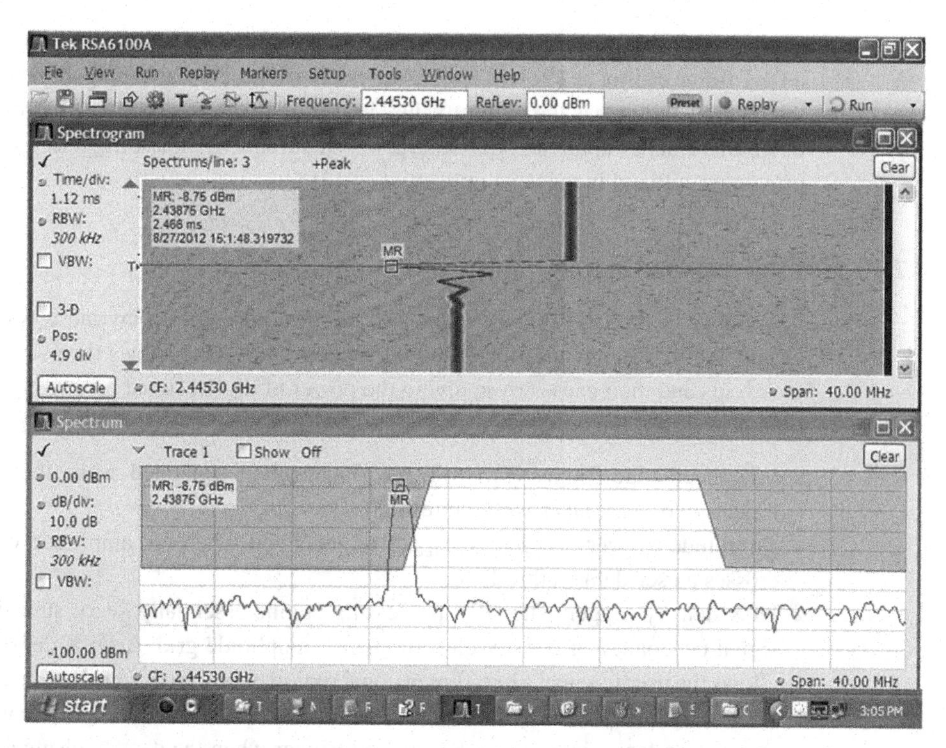

Fig. 4.21 Frequency mask trigger operation.

Figure 4.21 shows frequency mask triggering applied to a frequency hop. The frequency mask, shown as a shaded area, is drawn to exclude the normal hop behavior and to catch a frequency overshoot. The spectrogram in the upper part of Figure 4.21 shows that there is a momentary transient as the frequency hops from its original setting just to the right of center to a new one about 6 MHz lower. The overshoot violated the frequency mask and generated the trigger for the acquisition. Trigger parameters were set to display events that preceded the trigger as well as events that happened after the trigger, as shown in the spectrogram in the lower part of Figure 4.21.

Markers can be used to display the spectrum corresponding to each horizontal line in the spectrogram, pinpointing both the time and frequency of a spectrum feature. The full time record represented by the spectrogram is also stored and can be subjected to analysis using the many other analytic functions of the RTSA.

4.7.5 Frequency mask trigger time resolution and time alignment

Triggering in the time domain as illustrated in Figure 4.19 can uniquely locate an event to a resolution of a single time sample (some oscilloscope trigger systems use interpolation to achieve sub-sample time resolution). This is the case for external, level, and power triggers in RTSAs. Frequency mask triggering, however, compares the output of a sequence of DFTs with a frequency domain mask. Each spectrum is computed for an entire DFT frame containing a large number of samples. The location of an event within a DFT frame cannot be known. The *center* of the frame, corresponding to the sample for which the window has the highest value, is chosen by convention. The narrowest burst of RF that can be captured at full accuracy with a frequency mask triggering follows the principles outlined in the development of (4.13).

4.7.6 Other real-time triggers

The ability to digitally process signals in real-time provides a myriad ways to detect events within a signal and isolate those events in time. This allows the user to analyze the events and their causes by applying the power of digital signal processing to a short time period near the occurrence of interest. A list of these triggers includes:

- Frequency-edge trigger (FM trigger): A trigger is generated when the frequency trajectory of a signal crosses a threshold.
- Amplitude (power) trigger: A trigger is generated when the amplitude of a signal crosses a user-defined threshold.
- DPX density trigger: The DPX processing engine essentially keeps statistics of the signal occupancy or density on a frequency-amplitude grid. A DPX density trigger allows the user to select a two-dimensional region or box on this grid and trigger when the average density over the box exceeds a user-selectable level. Alternatively, the system can learn the average density and trigger when the density changes from the "normal" value.

4.8 Application examples: using real-time technologies to solve nonlinear challenges

4.8.1 Discovering transient signals

Behaviors that create transient nonlinear effects have traditionally been difficult to identify and to troubleshoot. These events often cause non-stationary behavior that is causally linked to other parts of a complex system. Some examples of nonlinear behavior caused by coupling between independent parts of complex systems include:

- Systems that combine digital and analog circuitry can exhibit power supply voltage dips caused by variations in current draw as computations are performed that create momentary clipping in the analog signal path.
- Systems involving DSP can have software errors that cause incorrect filter values to be momentarily applied.
- Phase hits can be caused by the physical effects of component heating.
- Systems that use RF bursts can exhibit nonlinear effects that change in time due to thermal transients in amplifying devices.

Discovering the root cause of transient events often involves mixed domain analysis and multiple instruments. The tools that are used to correlate events require precise triggering

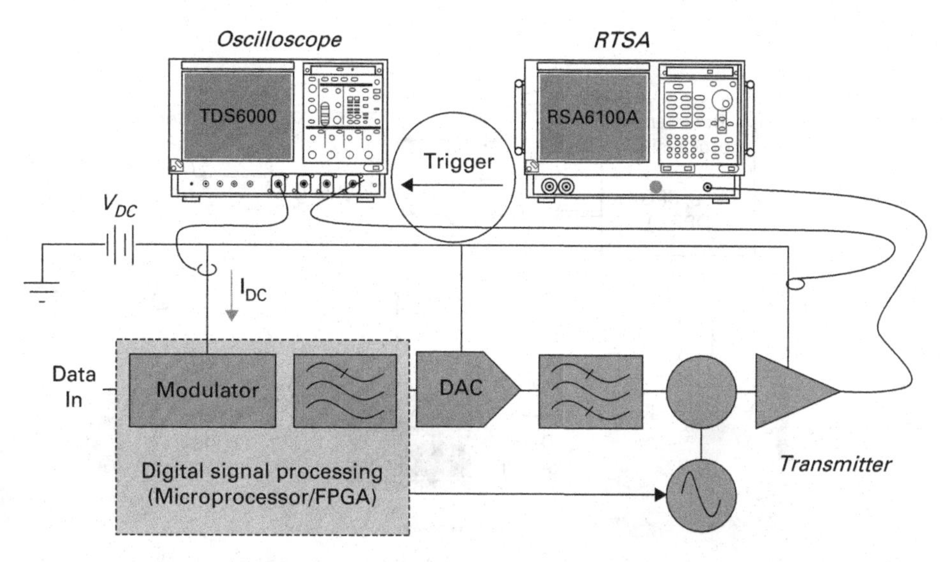

Fig. 4.22 Frequency mask triggering enables precise event capture.

and time alignment across multiple measurement domains. The real-time process of frequency mask triggering provides a unique method of transient event isolation.

4.8.2 Adjacent channel power (ACP) violation caused by power supply fluctuations

Consider a case where a power supply current spike is causing a spectrum anomaly in a transmitter system. Figure 4.22 demonstrates an approach to resolving the root cause of transient events.

A frequency mask trigger (FMT) is used to trigger on spectrum anomalies such as excessive **adjacent channel power (ACP)**. The trigger is then sent to the oscilloscope to reveal the simultaneous behavior of the power supply drain current. The real-time triggering functionality on the RTSA can not only enable a trigger output to cross-trigger other instruments: it can also capture the IQ representation of the RF signal containing the transient event into internal memory. Once captured into memory, the RF signal can be analyzed for power fluctuations, modulation errors, phase stability, or any other RF parameter of interest.

4.8.3 Software errors affecting RF performance

Software errors are often the root cause of a transient spectrum anomaly. Tracking down a software problem often requires the simultaneous observation of signals in many parts of a complex system, often crossing the RF, analog baseband, digital, and software domains.

Figure 4.23 illustrates what might be needed to track an RF signal integrity problem across the various building blocks of a modern RF transmitter. The RTSA is used to trigger the oscilloscope and logic analyzer when a transient event appears in the spectrum.

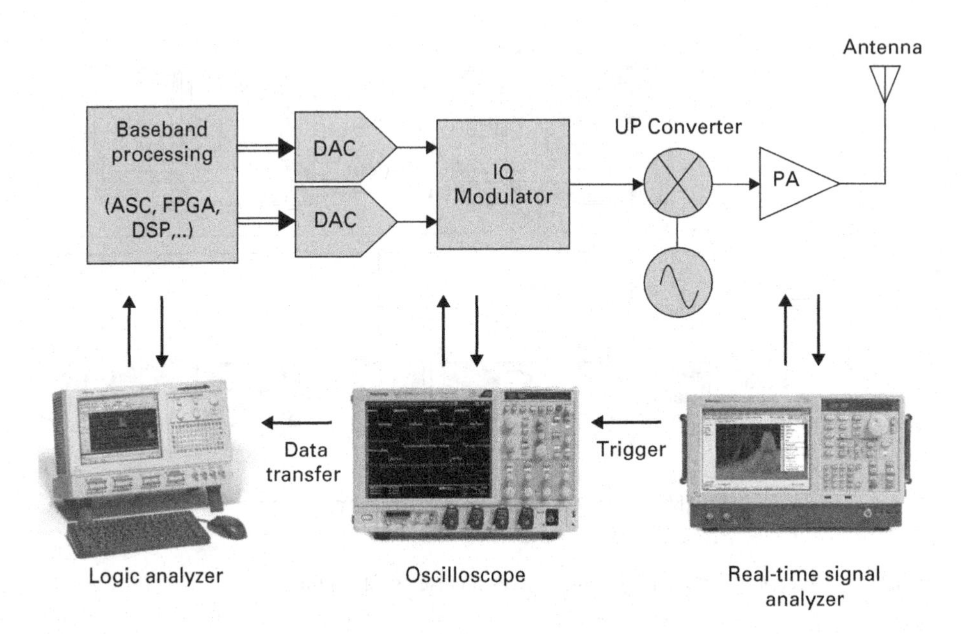

Fig. 4.23 Multi-domain event analysis enable by real-time triggering.

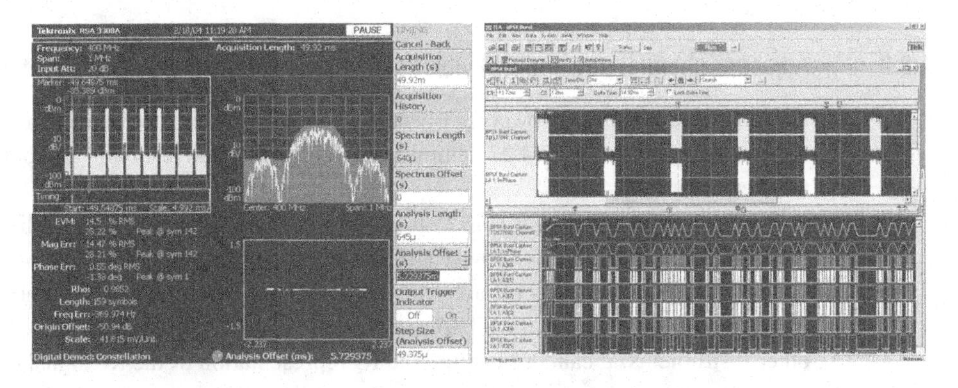

Fig. 4.24 Frequency mask trigger used to troubleshoot analog and digital errors.

The external trigger output from the RTSA is then fed to the oscilloscope and the logic analyzer. The time-correlated baseband analog signal and the time-correlated digital signal can be simultaneously captured and displayed on an integrated view. The screen shot on the left of Figure 4.24 shows a spectrum mask violation in a BPSK signal that was used to trigger the scope and logic analyzer shown in Figure 4.23. The IQ baseband signal from the oscilloscope and the digital bus that time correlates to the trigger event are shown in the screen shot on the right of Figure 4.24. With common source code debug tools, a trace can be put on the real-time hardware and real-time instructions being executed, so the line of code being executed at the time of the spectrum event can be isolated.

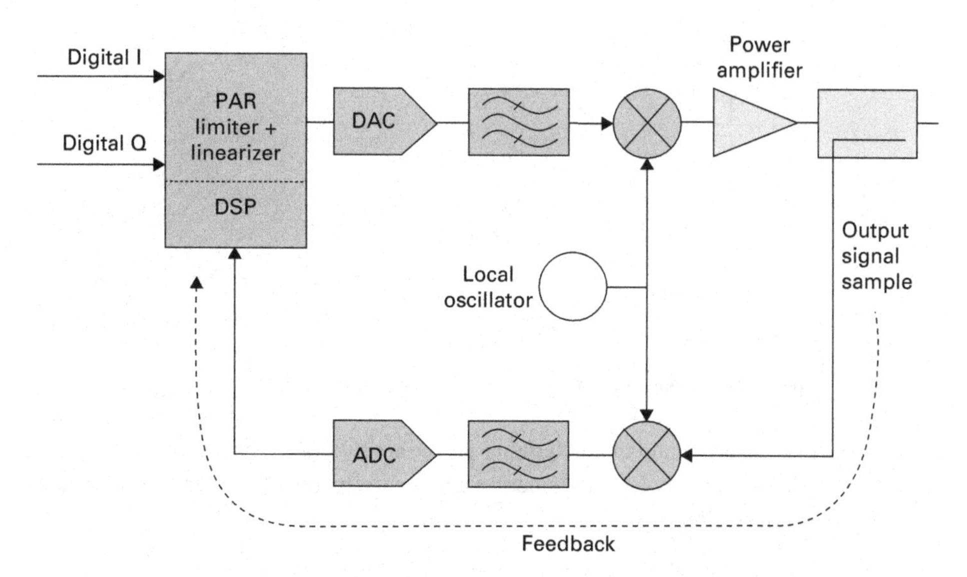

Fig. 4.25 Digital pre-distortion with feedback.

4.8.4 Memory effects in digitally pre-distorted (DPD) amplifiers

RF power amplifiers used in digital communications systems need to be linear to minimize out-of-band spurious emissions and distortion. They also need to be efficient to minimize heat dissipation and extend battery life. Various adaptive digital pre-distortion (DPD) techniques have emerged as promising linearization methods and are being applied to applications ranging from cell phones to radars. DPD employs digital baseband pre-distortion ahead of the amplifier to compensate for the nonlinear distortion contributed by PAs as they operate in high-efficiency yet nonlinear regions. Compensation levels are monitored and adapted based on a sensing feedback loop, as shown in Figure 4.25.

Memory effects have been identified as a major source of degradation of DPD performance in PA systems primarily due to the finite delay in the sensing and adapting of the feedback system. Memory effects can be defined as the frequency-dependent distortion in RF power amplifiers caused by circuit elements with memory. Memory effects in PAs[15] are generally categorized into two groups: electrical effects and thermal effects. Electrical effects are usually caused by the source and/or load impedance variations over the modulation bandwidth. Thermal memory effects are caused by electro-thermal coupling. Since PAs always exhibit some sort of thermal inertia and most of the electrical parameters of the transistors are functions of temperature, thermal effects are unavoidable.

The left side of Figure 4.26 shows an example of a traditional spectrum display of a power amplifier output after the DPD correction has been applied. The right side shows a DPX display of the same signal. The DPX display shows the effects of infrequent spectral regrowth. These infrequent events go undetected on the traditional spectrum analyzer. Adjacent channel power (ACP) measurements and other conventional ways

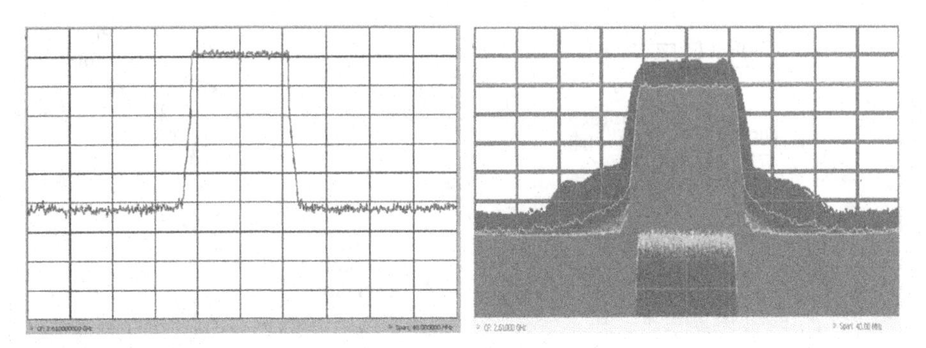

Fig. 4.26 Spectrum of a power amplifier after DPD correction has been applied.

of measuring and characterizing power amplifiers are based on averaging signal power over a relatively long time period and are incapable of showing the transients caused by memory effects.

4.9 Conclusions

Spectrum and signal analyzers have evolved along with the many novel ways we've learned to use the electromagnetic spectrum and with the changes in technology. Modern digital communications formats have created a need to simultaneously observe events in the time as well as the frequency domain and to measure dynamic signals that change far too fast for traditional analyzers.

Real-time spectrum analysis provides a methodology to observe events that are far too brief to be detected with traditional equipment. Advanced trigger functions allow the pinpoint capture of these rare RF events and a way to time-correlate them to their causes.

Nonlinear effects in RF devices such as power amplifiers often display anomalous spectrum behaviors that coincide with occurrences in the digital portions of a system or as a result of modulation anomalies (occasional power peaks for example) that are carried on the RF. Errors in programming, noise from switching power supplies, and RF energy radiated from nearby digital circuitry can also cause spectrum abnormalities that are rare and of short duration. Real-time spectrum analysis, with its ability to perform gapless analysis and to detect single events lasting as little as a few microseconds, provides a tool to isolate these rare events in an RF signal and track down their causes.

End notes

1. R&S White Paper "Implementation of Real-Time Spectrum Analysis," page 6.
2. Tektronix, "Fundamentals of real-time Spectrum Analysis."
3. Implementations differ among various equipment manufacturers. The discussion in this book focuses on the Tektronix implementation.
4. R&S FSVR Data Sheet, page 16.
5. Tektronix, "Fundamentals of real-time Spectrum Analysis."
6. Stremler [5], p. 135.

7. Oppenheim [6], p. 520.
8. We have effectively used a "rectangular window [6]."
9. Oppenheim [7], p. 717.
10. Rabiner and Gold [7], p. 386.
11. Gauss Instruments TDEMI 40G Data Sheet.
12. R&S White Paper "Implementation of Real-Time Spectrum Analysis", p. 19.
13. Tektronix, "Fundamentals of real-time Spectrum Analysis."
14. US Patent # 7,216,046.
15. Y. He, D. McCarthy and M. daSilva.

References

[1] *R & S FSVR Data Sheet*. Münich, Germany: Rohde & Schwarz GmbH KG, 2010.

[2] *Implementation of Real-Time Spectrum Analysis*. Munich, Germany: Rohde & Schwarz GmbH KG, 2011.

[3] *Gauss Instruments TDEMI 40G Data Sheet*. Münich, Germany: Gauss Instruments GmbH, 2012.

[4] *Fundamentals of Real-time Spectrum Analysis*. Beaverton, OR: Tektronix Inc., 2008.

[5] F. G. Stremler, *Introduction to Communications Systems*. Boston, MA: Addison-Wesley, 1990.

[6] A. V. Oppenehim and R. W. Schafer, *Discrete Time Signal Processing*. Englewood Cliffs, NJ: Prentice-Hall, 1989.

[7] L. R. Rabiner and B. Gold, *Theory and Application of Digital Signal Processing*. Englewood Cliffs, NJ: Prentice-Hall, 1975.

[8] A. V. Oppenheim, A. S. Willsky, and I. T. Young, *Signals and Systems*. Englewood Cliffs, NJ: Prentice-Hall, 1983.

[9] Y. He, D. McCarthy, and M. daSilva, "Different measurement methods for characterizing and detecting memory effects in nonlinear RF power amplifiers," ARFTG Conference, Phoenix, December 2007.

[10] *Measurement of the nonlinearities of RF amplifiers using signal generators and a spectrum analyzer*. Münich, Germany: Rohde & Schwarz GmbH KG, 2006.

[11] *Fundamentals of Digital Phosphor Technology in Real-time Spectrum Analyzers*, Beaverton, OR: Tektronix Inc., 2008.

[12] *DPX turns a light on in a dark room*, Beaverton, OR: Tektronix Inc., 2006.

[13] K. Bernard and E. Gee, "Real time power mask trigger," US Patent 7 251 577, July 31, 2007.

[14] K. Bernard, "Time-arbitrary signal power statistics measurement device and method," US Patent 7 298 129, November 20, 2007.

[15] K. Bernard and E. Gee, "Real time power mask trigger," US Patent 7 418 357, August 26, 2007.

[16] M. Agoston, W. B. Harrington, and S. L. Halsted, "Method of generating a variable persistence waveform database," US Patent 7 216 046, May 8, 2007.

[17] S. R. Morton and J. C. Demogalla, "Method and apparatus for identifying, saving, and analyzing continuous frequency domain data in a spectrum analyzer," US Patent 5 103 402, April 7, 1992.

[18] J. D. Earls and A. K. Hillman, "Multichannel simultaneous real-time spectrum analysis with offset frequency trigger," US Patent 7 352 827, April 1, 2008.

5 Vector network analyzers

Mohamed Sayed and Jon Martens

5.1 Introduction

The VNA is the instrument that measures the S-parameters (and related quantities) of passive and active devices and components. The phase and magnitude of these S-parameters are displayed in different formats in accordance with the user's application. Scalar network analyzers measure only the magnitude of the device's performance and that is not the focus of this chapter.

VNA measurements can be done using one or many ports, over swept frequency or swept power and with a variety of receiver configurations, depending on the measurement requirement. This chapter explores the history of this instrument, some aspects of its structure and performance, and a brief introduction on how specific measurement applications are affected by the VNA attributes. Many microwave measurement concepts and instruments are based on the VNA and some are discussed later in this book. As such, this chapter serves as something of an introduction to many subtopics.

5.2 History of vector network analyzers

5.2.1 Pre-HP-8510 VNA – 1950–1984

Rohde and Schwarz introduced the first impedance measuring device that could warrant the term "network analyzer" in 1950. Wiltron introduced the 310 VNA in 1965. This was followed by the HP VNA in 1966, 1968, and 1970.

Table 5.1 shows the VNA model numbers and the years for this time period. In [1], Doug Rytting describes in detail these early VNAs.

During this period, sweepers were used as narrow-band sources (2–4 GHz, 4–8 GHz, etc.). The displays shown on the monitor were rectangular or Smith Charts. External computers were used to control the measurements and displays. To cover the 2–20 GHz frequency range, multiple sources were used. Calibration was done manually with external calibration kits.

Four full racks of instruments were needed to perform as an "Automatic Network Analyzer" or ANA. BWOs were used as the source. Multiple plug-ins were used to cover different types of displays and there was one rack for computing and displaying results. All these instruments were located in measurement rooms staffed by engineers and

Table 5.1 Pre-HP 8510 Vector Network Analyzers (VNA)

Date	Company	Country	Model number	Frequency range
1950	Rohde & Schwarz	Germany	DuZ-g Diagraph	30–300 MHz
1952	Rohde & Schwarz	Germany	ZD-9D Diagraph	300–2400 MHz
1965	Wiltron	USA	W-310/311	1-2 / 2-4 / 4-8 GHz
1966	HP	USA	HP-8405	18 GHz
1967	HP	USA	HP-8410	12.4 GHz
1968	HP	USA	HP-8540	18 GHZ
1970	HP	USA	HP-8542	18 GHZ
1972	HP	USA	HP-8409	18 GHZ

technicians. Lab engineers needed to reserve specific times to take their measurements and make calibrations.

Microwave samplers were used for down-converting the input signal to a fixed IF. Harmonics of the low-frequency oscillator were used to mix with the input microwave frequency. Thus, the system performance degraded as frequency increased. The noise floor and stability were not as good as at the present time.

Waveguide frequency bands were used to define different frequency ranges, e.g. L-Band of 1–2 GHz, S-Band of 2–4 GHz, C-Band of 4–8 GHz, X-Band of 8–12 GHz, Ku-Band of 12–18 GHz, K-Band of 18–26.5 GHz, and Ka-Band of 26.5–40 GHz. The Type N connector was the typical one used up to 18 GHz. Precision connectors were developed to go higher in frequency, e.g. APC–7 will go up to 20 GHz and APC–3.5 will go up to 26.5 GHz.

5.2.2 HP-8510 VNA System – 1984–2001

The 26.5 GHz HP-8510 VNA system was introduced in 1984. This VNA included a synthesized source (HP-8340A), error correction, time domain and pulse measurements. The analyzer system bus was used to choose the frequency range and number of points (51, 101, 201, or 401). The IF frequency was 20 MHz and the lowest frequency was 45 MHz. The HP-IB data bus was used for automatic operation of the VNA.

Several test sets were introduced to cover different frequency ranges: 45 MHz–26.5 GHz, 2–18 GHz, 2–20 GHz, and 45 MHz–20 GHz. The HP–8510 rack system consisted of one test set, one source, and an HP–8510 display / user interface. The display was black and white and a magnetic tape was used to collect test results. This system became known as the HP-8510A.

In 1987, Wiltron introduced the 40 GHz 360 VNA system. This system had a color display and extended the lower frequency to 10 MHz. Users enjoyed the competition between HP and Wiltron and benefitted by getting the best performance per dollars or dB/$.

The upper frequencies of these VNAs started to increase to 50 and 65 GHz in coaxial and to 110, 220, and 325 GHz in waveguide.

Table 5.2 HP 8510 Vector Network Analyzers (VNA) systems

Date	Company	Country	Model number	Frequency range
1984	HP	USA	HP-8510	26.5 GHz
1987	Wiltron	USA	360	40 GHz
	HP	USA	HP-8510	50 GHz
	HP	USA	HP-8510	60 GHz
	Wiltron	USA	W-360	65 GHz
	Wiltron	USA	W-360	110 GHz
	HP	USA	HP-8510	110 GHz
	HP	USA	HP-8753	3/6 GHz
	HP	USA	HP-8720	20/40 GHz
1994	Wiltron	USA	37XXX	20/40 GHz
1998	Wiltron	USA	MS462XX	9 GHz
	Rohde & Schwarz	Germany	ZVT8	8 GHz

To extend the HP-8510 system to 40 GHz, a doubler was installed into the test set using a 20 GHz HP-8340 source. The 50 GHz HP-8510 system was introduced along with the APC-2.4 connector. A set of calibration kits, cables, and verification kits was also introduced for each system and each different frequency range.

Many applications in material, measurements, antenna measurements, and radar measurements were shown and used by customers during the period 1984–2001.

A few years later, a one-box VNA was introduced for RF in the HP-8753 and for microwave (HP-8720). Wiltron introduced the Scorpion Network Analyzer which included two sources and noise figure measurements up to 9 GHz. Rohde and Schwarz introduced an 8 GHz ZVT8 which included 8 measurement ports. Table 5.2 shows VNA model numbers and years for this period.

Several technologies were quickly developing: 1) low-cost solutions for low-frequency and production environments; 2) compact sources with fast tuning times and high resolution; 3) calibration routines and kits for higher accuracy; 4) multiple-port applications for production systems and solutions; 5) wideband components such as couplers, cables, and mixers to extend the VNA bandwidth range.

Pulsed measurements for the on-wafer application of high power devices were introduced by HP and Wiltron. Load-pull measurements under pulsed bias and pulsed RF were developed by both companies to test high power devices on wafer. Wafer probes were developed by Cascade Microtech and other vendors.

The customer's need to extend the frequency range and dynamic range of the VNA motivated vendors to develop high-resolution sources, receivers, and calibration routines. During 1984–2000, customer seminars were developed by different vendors and were presented all over the world. Engineers and scientists attended annual International Microwave Symposiums (IMS) to view the latest VNA systems, solutions, and applications.

The automated production of devices using either HP's VNA or Wiltron's VNA made great progress during the period 1984–2001. More details about this history can be found in Rytting's paper [1].

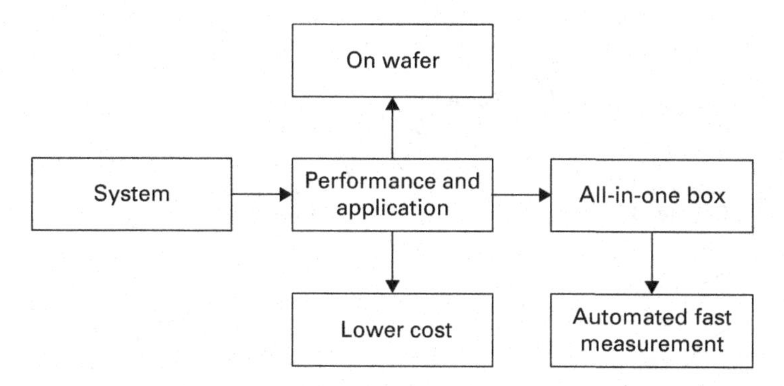

Fig. 5.1 VNA growth and applications.

5.2.3 Evolution of VNA to the Present – 2001–2012

During 2000–2001, all-in-one VNA systems were introduced by Anritsu/Wiltron (Lightning) with a 65 GHz range and HP/Agilent (PNA) with a 50 GHz range.

The need for wider frequency ranges and higher output power has sped up the introduction of the Vector Star from Anritsu and the PNA-X from Agilent during the last few years.

Rohde and Schwarz later introduced the ZVA with four internal sources up to 67 GHz and extended the frequency range up to 220 GHz with external millimeter wave modules.

VNA growth and applications over the last generation are shown in Figure 5.1. The evolution of the VNA is shown in Figure 5.2.

5.3 Authors' remarks and comments

VNA in this chapter refers to Linear VNA. The NVNA and related LSNA are discussed in Chapter 12. In addition, waveform engineering, which is currently being pursued by a number of researchers, is not addressed in this chapter.

There are several factors which will be important in the near future: cost vs. performance, digital designs to 40 GHz and higher, modeling and verification for high power devices, time to market of new technologies, and the role of microwave measurements technology for future devices and mobile technology.

5.4 RF and microwave VNA technology

The most basic objective of the VNA (and many related instruments) is to measure S-parameters or the constituent wave quantities (e.g. $a_1, b_1 \ldots$). In this most basic form, one must acquire incident and reflected waves at each port of interest while providing input signals at the different ports. Carrying this simplistic picture forward, one then requires a signal source, some receivers, and some way of separating the incident and reflected energy. The purpose of this section is to explore some of the elemental blocks

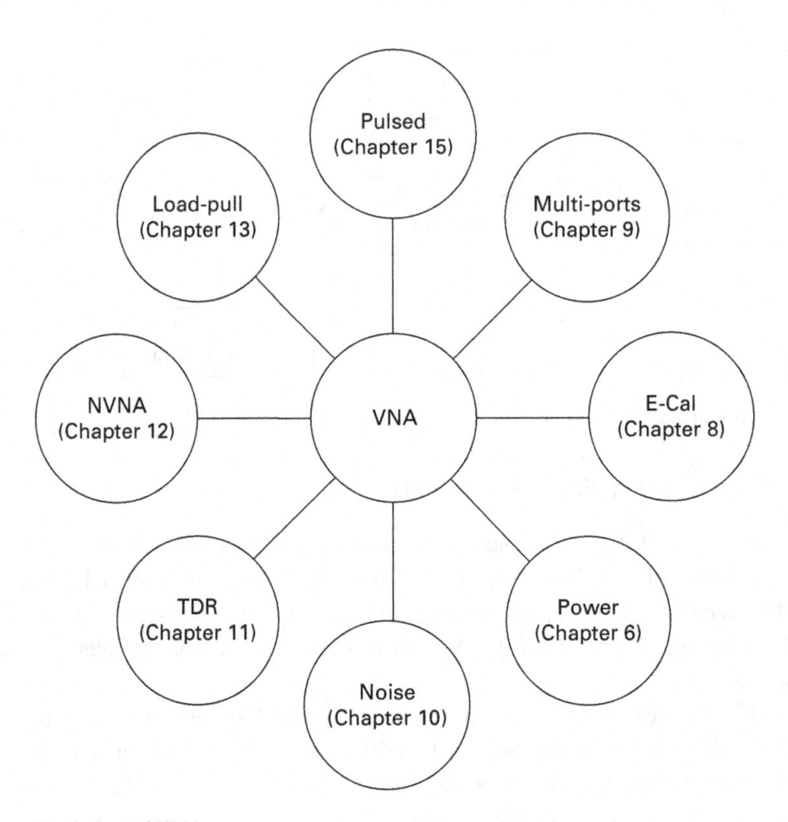

Fig. 5.2 Evolution of VNA.

of a VNA, the performance considerations of relevance, technological evolution over the more recent history of VNAs, and the effect the blocks have on a variety of different measurements. Some of the historical concepts discussed earlier are used by inference and more details can be found elsewhere (e.g. [1–5]).

The concepts above lead to an elemental block diagram (for a 2-port case) like that shown in Figure 5.3, although many variations are possible. Some of the basic elements are:

- One or more signal sources having at least controllable CW/swept frequencies with sufficient spectral purity that measurements can be made. It is also preferred for the power to be controllable.
- Some directional devices (see Chapter 1) for separating incident and reflected waves at the ports. In some cases these devices need not be physically directional, but they could be generalized splitting devices of sufficient stability that they can be computationally directional.
- If there are fewer sources than ports or if there are more or fewer receivers than ports, there must be some means of switching signals.
- One or more receivers, usually incorporating down-converters, to take the incident and reflected waves down to some convenient IF for processing.

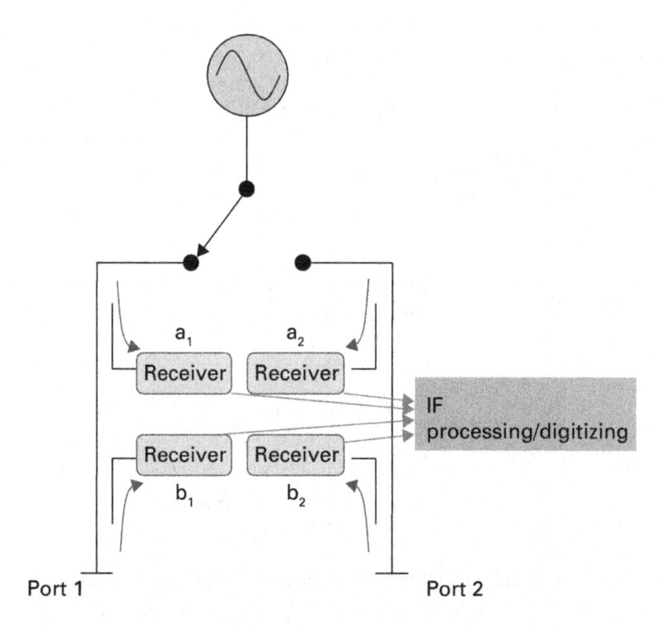

Fig. 5.3 One possible VNA block diagram (for a 2-port case) is shown here that illustrates the key blocks and the flow paths to be discussed.

- An IF section and digitizer to process the converted wave amplitudes into a form useful for computation and display.

Among the possible variations of Figure 5.3, one could use a source per port instead of switching one between two ports. The coupling devices could also be repositioned and, of course, there could be many more ports or just one. The point of the diagram is that the functions listed above are generally present in one form or another.

5.4.1 Sources

Historically, the source in a VNA has taken many forms ranging from simple analog sweepers in the earliest implementations to complex synchronized synthesizers in more modern instruments. Sweepers can be quite fast (and before about 2000, they were generally faster than synthesizers), but the spectral purity is not as good and there are potentially synchronization issues since the LO and IF systems must be semi-coherent with the source system (resulting in sometimes substantial frequency errors). As a result, more of the recent VNAs are synthesizer-based. While there are an infinite number of variations possible, a core block diagram of a synthesized source is shown in Figure 5.4. Considerably more detail on synthesizer structure is presented in Chapter 3 and, increasingly, more of these design concepts are migrating on to VNA platforms. The usual difference is the higher importance that is assigned to point-to-point tuning speed in VNA applications than in classical synthesizer applications, but this line has also been blurring over time.

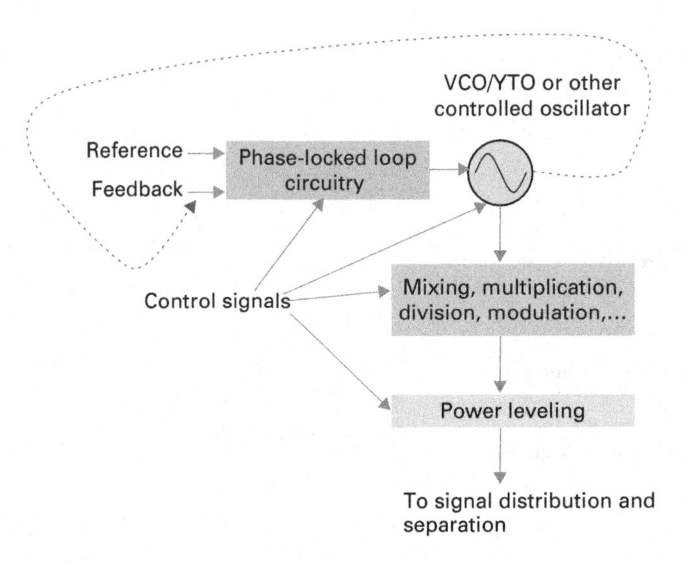

Fig. 5.4 A very generic source block diagram to illustrate some of the choices to be made.

Among the issues to be considered:

- How fast can one get from one frequency to another? While this may not be the dominant part of the measurement speed, it can be relevant.
- How clean is the signal (in terms of phase noise, harmonics, and spurs)? In doing only S-parameter measurements, some of these parameters may not be that important since we know what frequency we are measuring (and there is only one per point) and a narrow IF bandwidth can be used to reduce noise effects. As applications proliferate (IMD, mixer, and nonlinear measurements) and speed needs increase, however, one must pay more attention to the spectral purity.
- How is the frequency plan organized (multipliers, dividers, and mixers operating on a base range)?
- How is power control done? How much power range and accuracy is available? This becomes more important for nonlinear and quasi-linear measurements.

The block diagram in Figure 5.4 is extremely general, since there are many possible structures and large portions of the source may be generated digitally. The reference can come from a crystal oscillator or from some other synthesizer. Feedback for the loop is shown (by the dashed line) to be coming from the VCO, but it could also come from elsewhere in the system (e.g. from a receiver). The oscillator output may be frequency-converted or may be modulated. In some cases, the source may not even be locked (although there are accuracy penalties for that as discussed in the sweeper case).

As discussed before, the source need not actually be locked as Figure 5.4 might suggest. The sweeper type of structure can be quite fast and devoid of some spectral artifacts. The downside is that controlling the timing relationships between source, LO, and acquisition can be challenging. Doing this over temperature and aging can be more difficult and normally requires sophisticated internal calibration structures. Integrating more complicated applications (such as mixer measurements) requiring external sources or otherwise changing the sweep dynamics adds additional challenges.

A fully synthesized approach avoids most of these problems at the expense of some different types of control complexity. In order to make a fully synthesized version fast, but still with good spectral purity, more careful loop design and perhaps more sophisticated control electronics are required. It is not the intent of this chapter to cover details of phase locked loop and synthesizer design (for details, see for example [6]) but there are some key points that may aid the discussion:

- Generally, the wider the loop bandwidth, the faster the settling time. Concomitantly, wider loop bandwidths generally lead to more phase noise in an integrated sense.
- From a measurement point of view, it is the settling of the final receiver IF that is somewhat more important than the independent settling of the source and LO. If the source and LO can settle together, faster net measurement times are possible (there are limits; one must be on frequency to within a certain tolerance).
- Generally the loops will settle faster (to within a fixed tolerance) for smaller frequency steps. For larger steps, dynamic loop response changes can help.
- Increasing levels of (and frequency ranges of) fully digital synthesis can greatly speed the sweep at the cost of some spur control complexity.

At some point in the process, locking is usually required. The next question is where the locking is performed. The simplest approach may be treating both the source and LO as separate synthesizers with their own integrated phase-locked loops with shared reference frequencies at some level. Other possibilities include locking through the receiver, essentially locking the LO to the IF or locking the source to the IF (sometimes termed follower mode and source-locking, respectively). This can reduce the individual loop complexities somewhat and can lead to a very clean received signal since the IF becomes the locking reference. This approach does complicate the application space somewhat since one receiver must be made the locking parent and hence no longer has meaningful phase information. Also the source and LO must have a fixed offset to enable the loop to close which makes mixer, IMD, and other translating measurements quite difficult.

There are many technological decisions to be made on the individual PLLs and most of these are beyond the scope of this chapter, but a few key comments can be made. Historically, YIG oscillators have often been used for the source with a source-locking architecture. While the phase noise of such oscillators is quite good (particularly far from the carrier), they tend to be slow (at least in broadband configurations). More recently, VNAs where all sources are based on varactor-tuned VCOs, which can tune much faster, have been introduced. The trade-off is degraded phase noise at offsets much larger than the loop bandwidth, but even these differences have been shrinking as VCO technology improves.

The fine-tuning structure of these loops has also changed in recent history. Fractional-N structures (e.g. [6]) are very popular and can offer fine-tuning resolution with decent spurious and noise performance. Increasingly wide bandwidth direct digital synthesizers have become more common and have had ever-improving spurious performance. The fine-tuning capabilities of such structures are needed since the VNA tuning resolution

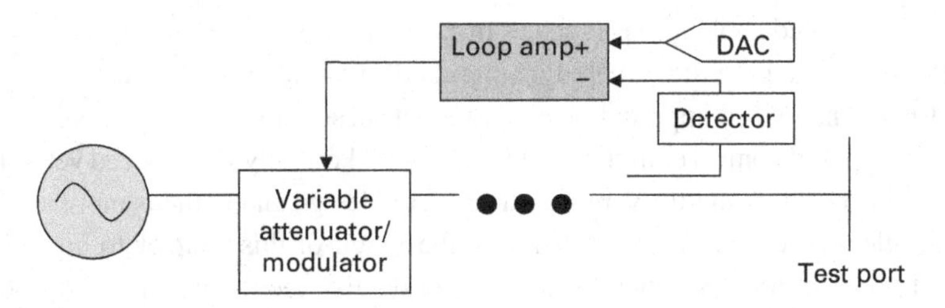

Fig. 5.5 A very simplified ALC loop.

must typically be of the order of 1 Hz (or better if high-order multipliers are part of the system).

Another important aspect of the source side of the system is power control. Aside from having a vague idea of what the DUT is being driven with, swept power measurements are increasingly important to the VNA user for measurements such as gain compression, IMD vs. power, harmonics vs. power, etc. Thus a reasonably accurate and wide range ALC is of importance. Complicating things, like so much else in the system, is that this leveling system must be fast enough to keep up with the measurement.

Leveling subsystems are used in many applications and are conceptually quite simple. They use a power detector of some kind and, in the context of a negative feedback loop, compare the detected output to some desired reference voltage (usually from a DAC) and feed the result to a power modulator of some kind (see Figure 5.5).

For the purposes of illustration, a number of assumptions were built into this diagram that are not mandatory:

- A coupled detector is shown for power detection.
    - Sometimes non-directional splitters are used instead of a coupler. This is much simpler and can lead to a more fixed delivered power instead of incident power at the potential expense of stability.
    - Arrays of detectors are sometimes used for improved control range and even a thermal sensor could be used, although there may be a speed penalty.
- A variable attenuator is shown for power modulation.
    - Amplifier bias is sometimes used for this kind of control. Harmonic generation in that case could be a concern as the requested power is reduced.
    - Cold FET and PIN diode attenuators are both popular for variable attenuators. PIN diode structures often have an advantage in power handling and FET structures often perform better at low frequencies (although there are exceptions to these generalizations and the technologies are constantly evolving). Hybrids are possible.
- A simple loop amp (often an integrator) is shown.
    - Multi-stage and distributed loop amps are often used for more control of loop gain.
    - Variable poles are often used for stability in different operating modes.

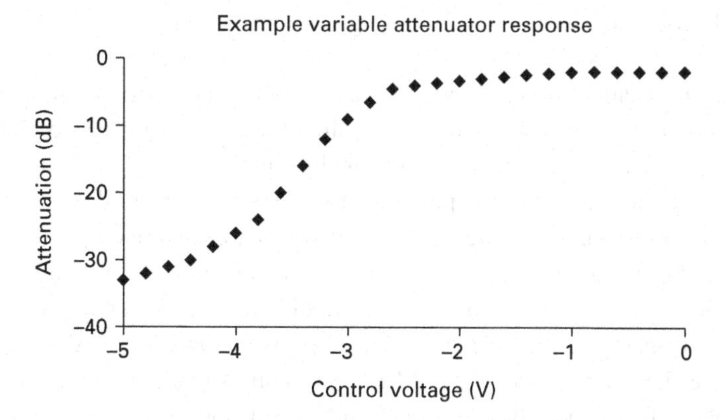

Fig. 5.6 An example response curve of a commercial voltage-variable attenuator.

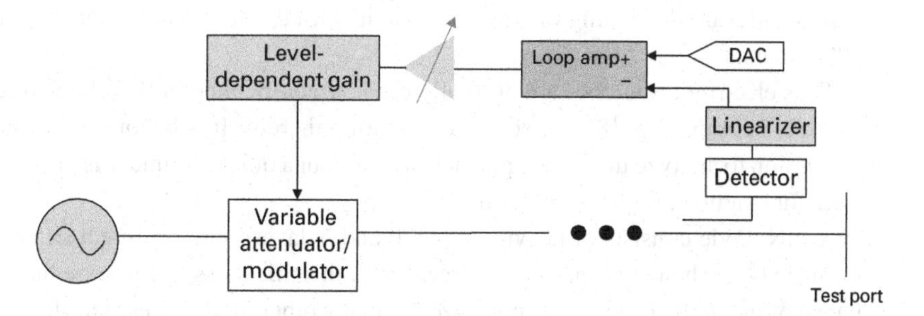

Fig. 5.7 A more complete ALC block diagram.

The issue of loop bandwidth is an important one to consider. Since a VNA has to operate over wide frequency ranges and, often, wide power ranges, the overall loop gain will not be flat. To see this, consider the attenuation curve of a commercially available voltage-variable attenuator (Figure 5.6).

The slope variations in this curve represent changes in loop gain. If this was uncompensated, the bandwidth of the loop could become very small at some states (making the measurement slow at low attenuation levels) and very large at other states (potentially leading to oscillation at higher attenuation levels). In addition to simple level-dependent gain changes, there may be other frequency-dependent gain changes such as when one moves from a fundamental source band to a multiplied source band that may use a different variable attenuator. Since detectors have nonlinear responses over wide power ranges as well, some linearization may be desirable again to keep loop gain relatively flat. From all of these complications, one may end up with a leveling system that looks more like Figure 5.7.

5.4.2 Switches

RF switches are needed in VNAs for a number of reasons including the desire to allow one source to drive two or more ports (thus saving the expense of multiple sources) or to selectively route to multiple receivers (e.g. in a multi-port scenario or to different application-specific receivers).

Very often, the demands on the switches can be extreme in terms of isolation, insertion loss, bandwidth, and, perhaps, power handling/linearity. Using a 2-port VNA as an example, there is usually a main switch (normally called a transfer switch even if it is SPDT) allowing one source to drive port 1 or port 2. The isolation of this switch directly translates to the raw isolation of the VNA. The insertion loss and linearity directly affect what the maximum available port power can be and its bandwidth can limit that of the VNA. For a high-performance microwave VNA, this can be a challenging combination.

In the distant past, electromechanical switches were sometimes used due to their favorable insertion loss/isolation ratio. The repeatability of these switches, typically no better than a few hundredths to a tenth of a dB at microwave frequencies, led to some measurement errors. Also, the lifetime of many mechanical switches does not exceed 10 million cycles. Even at a slow sweep rate of one sweep per second (which may be relevant depending on how the switch is used), this switch would last less than 3000 hours.

Thus electronic switches are normally used, typically either a PIN diode (e.g. [7]) or cold FET circuit (e.g. [8]), or some combination thereof. It is beyond the scope of this discussion to analyze the device physics in detail but a quick summary is provided below (see, for example, [7–9] for more information).

A PIN diode consists of heavily doped P and N layers surrounding a relatively thick intrinsic layer (hence the acronym). Because of this thickness, the reverse biased capacitance of the diode is quite low compared to many other diode types. This leads to better isolation when used in a series construction and less insertion loss in a shunt topology. When forward biased, carriers are injected into the intrinsic layer but do not recombine immediately. This leads to some complications at lower frequencies since the RF frequency can be on the same scale as the recombination rate and distortion occurs.

A typical cold FET switch is just that: a MESFET or similar structure setup with no drain bias. When the gate is biased strongly negative, no carriers are available in the channel and the device provides reasonable isolation in a series sense. Like the PIN diode, the off capacitance (drain to source) is fairly low due to the geometry, so shunt-topology insertion losses can be low as well (although typically worse than with a PIN structure). The elevated capacitance can be mitigated by embedding the switch in a transmission line structure. When the gate is near ground potential, carriers are available in the channel and a relatively low series resistance is available. Unlike the PIN diode, the recombination time remains fast so there are few low-frequency effects. Since one is usually operating against a 0-bias limit, there can be linearity issues, although these have been overcome at least in part with more novel topologies.

Whichever technology, or combination of technologies, is chosen, the issue of switch topology is critical. For simple applications requiring low isolation, a single series-shunt element per arm may be appropriate. In some cases, even a single element can be used but there may be severe match implications on a multi-throw switch. When high levels of isolation are required, more elements are often used per path. There are many choices one can make about the combinations of elements, but a few items can help:

- Series elements generally become less effective at higher microwave frequencies and more shunt elements will be used in that frequency range.
- Sometimes series-shunt pairs are available as a single die or cell and they are often convenient to bias that way.
- Proper allocation for biasing inductors must be made (for PIN switches primarily) and their layout is critical since above 50 GHz or so, bias circuit parasitics may contribute as much to insertion loss as the switch itself.
- Isolation may end up being limited by radiative effects thus making housing design and layout quite important.
- As has been pointed out in the literature, the switch spacing in higher isolation structures is quite important due to the standing waves that will appear between switches.
- Terminating switches are often required, which usually means an additional branch to a load is needed at the output ports although there are other approaches.

As an example of some of these behaviors, an empirical curve of isolation and insertion loss for a SPDT switch using a particular PIN diode technology operating at up to 70 GHz is shown in Figure 5.8. The return loss of the output is roughly constant in on- and off-states due to the presence of a terminating load (with phasing differences and multiple reflection interactions causing differences). The isolation in this measurement is largely noise floor limited but one can see that the insertion loss to isolation ratio supports a reasonable dynamic range.

5.4.3 Directional devices

Key to the concept of reflectometry is the directional device used to collect the incident and reflected wave energy (see also Chapter 1). While there are a number of ways to do this, broader band microwave VNAs tend to rely on directional couplers, while RF VNAs may use a directional bridge (or the above concepts may be combined). It is beyond the scope of this chapter to delve into the theory of the directional devices (see, for example, [9]) but we need the following definitions (see Figure 5.9 for port assignment, reverse coupling assumed for the drawing):

Coupling: S_{24} (sometimes coupling is defined to include the insertion loss, much like S_{41} with a perfect reflect connected)

Insertion loss: S_{21}

Isolation: S_{41}

Directivity is often defined as $|S_{24}/S_{41}|$, although variations on this definition may include insertion loss as part of the numerator (in the sense of reducing the numerator).

Coupling is usually dictated by the signal levels needed by the rest of the system subject to the constraint that directivity usually worsens (for a broadband coupler) if the coupling gets too tight (i.e. $|S_{24}|$ gets larger). With these constraints in mind, coupling factors usually end up in the 10–20 dB range although there are exceptions. Of course, one wants a minimal forward insertion loss (to maximize the available port power) and

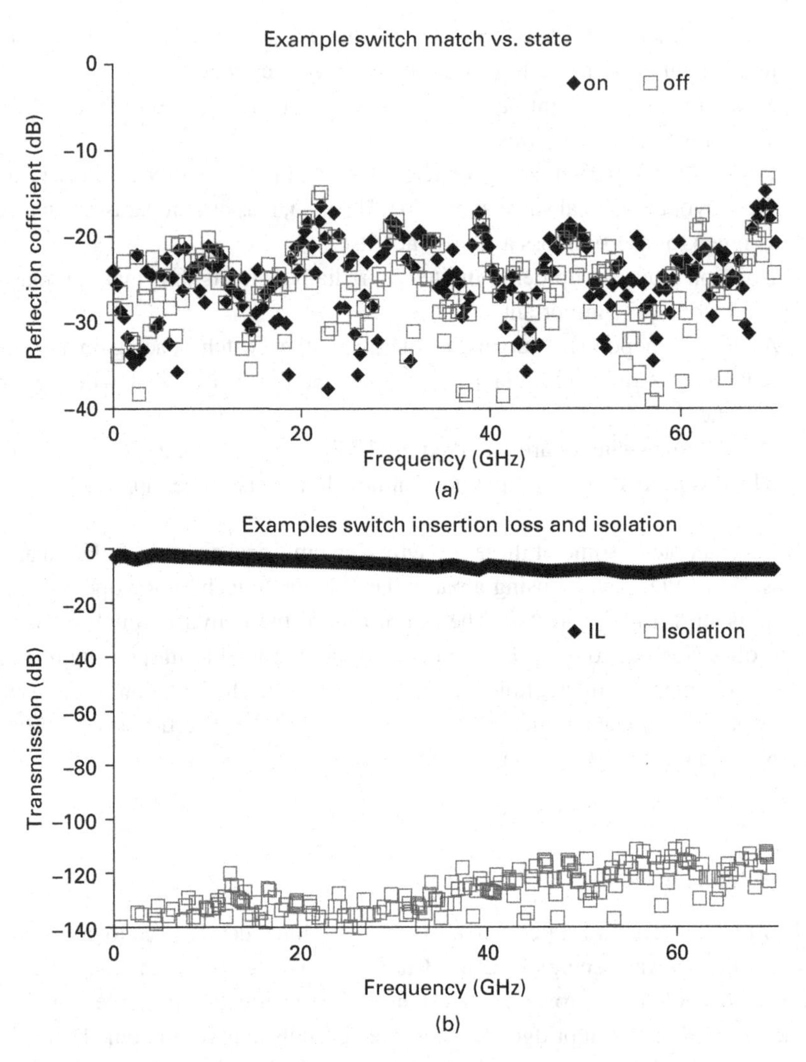

Fig. 5.8 Example insertion loss and isolation of a broadband, high-isolation switch construction.

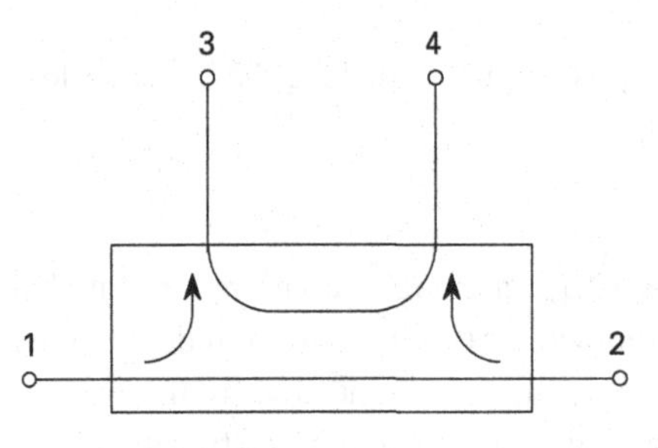

Fig. 5.9 An example coupler block is shown here to help with the definitions. Assume the path 2-4 is the desired coupling direction.

reasonable match (since this may dictate the raw port match and is usually connected to directivity).

The wildcard, which is principally a function of the construction techniques and level of assembly tuning, is directivity or isolation. In view of the power of VNA calibrations, one may wonder how important these raw parameters are to overall system performance. In an instantaneous sense, the answer is usually not significantly. In the longer term (in the sense of calibration stability), it can be considerably more important.

Before exploring these comments, we must revisit the concepts of residual vs. raw parameters (such as directivity and source match). The raw parameters describe the physical performance of the components involved such as the directivity described above for the directional device. The residual directivity is that left after the calibration and also describes the quality of the calibration components, the calibration algorithm, and the calibration process. This concept is discussed in more detail in Chapter 8. It is the residuals, at the time of DUT measurement, that describe the measurement uncertainty to a great degree, not the raw parameters. Now an individual DUT may be sensitive to the raw parameters (e.g. an amplifier may or may not be stable for a given raw port match on the VNA) but the measurement itself can be largely invariant to them.

To see this, consider two calibrations performed on a VNA. The first is with the VNA as it is normally configured, with a raw directivity of about +10 to +15 dB across the band 70 kHz–70 GHz. Also, perform a calibration with a 10 dB pad on the test port so the raw directivity in that case (ignoring pad mismatch so this is an upper bound) is −5 to −10 dB. With the two calibrations, measure the return loss of the same delay line. The results are shown in Figure 5.10 and indicate agreement (to within connector repeatability limits) even in the deepest notches. This indicates the residual directivities are nearly identical.

In a practical sense, however, it is important since the raw parameters have an impact on the stability of the calibration. Consider the directivity correction. In a reflection measurement, the directivity error adds to the DUT's reflected wave to produce the net

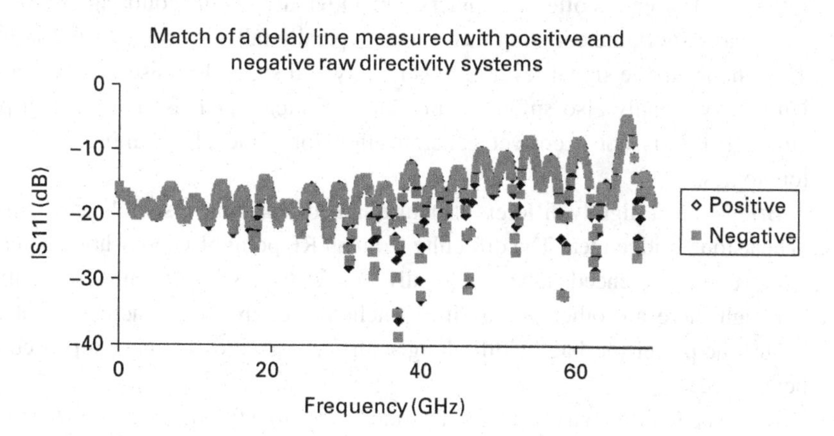

Fig. 5.10 The impact of positive and negative raw directivities on a calibrated measurement. As long as the environment is stable, both calibrations are roughly equivalent.

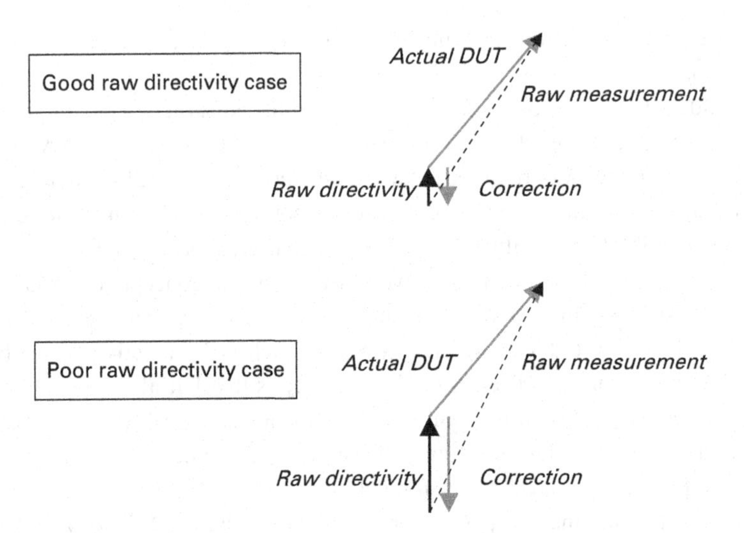

Fig. 5.11 The mathematics of directivity correction for two different raw directivities.

measurement. In the correction, the directivity is subtracted out (as well as other tasks being performed). If that subtraction is small in magnitude, a small drift in the actual amount of directivity does not affect the end result very much. If the subtraction is large, however, a fairly minor drift in that directivity vector can result in a substantial change in the final result (see Figure 5.11).

Thus one often strives for the best directivity possible within the boundaries of the other constraints. In the example of Figure 5.10, both measurements were done shortly after the calibrations. Had the delay line been measured several hours after the calibration in a thermally dynamic environment, the results might have been quite different.

One of the other constraints on the directional element is bandwidth. While the upper end is relatively easy to understand with the collapse of directivity under the wavelength limits, the low end is often misunderstood. Obviously as the coupling section becomes electrically short, the coupling factor must typically fall and often at a 6 dB/octave rate. Thus the available signal level decays rapidly and signal to noise becomes a problem. Directivity usually also suffers at this end but more for reasons of match problems, although this is not a correct generalization for some of the more exotic coupling topologies.

Bridges are a slightly different structure and do bear some resemblance to the classical Wheatstone bridge idea. The difficulty from an RF point of view is how to generate the non-ground referenced nodes. Typically this is done with a transmission line balun, although there are other possibilities (including entirely with active elements). This in turn helps set the bandwidth along with the parasitics of the lumped components being used.

Reasonable directivity can be maintained over large frequency ranges through proper balun design. The example shown in Figure 5.12 could be further optimized by use of a more elaborate balun structure at the expense of some insertion loss.

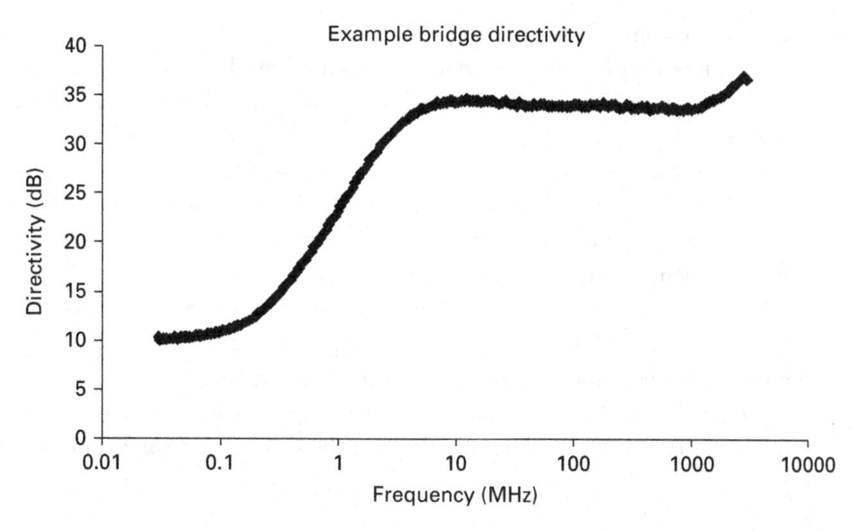

Fig. 5.12 Directivity of an RF bridge structure. Reasonable performance down to very low frequencies is possible with the right balun structure.

Non-directional splitters (including simple three resistor designs) are also sometimes used, particularly for the reference waves. In the reference wave case, the argument can be made that it is far enough from the test port that load-impedance-induced effects are small enough to not impact stability significantly. These non-directional structures have also been used on test ports where the directivity is entirely computational. Stability may be a prime consideration and sometimes extensive thermal stabilization efforts are made in these contexts. Cost, simplicity, and size are the obvious advantages.

5.4.4 Down-converters (RF portion of the receivers)

An entire book could be generated on receiver design, so this discussion is limited to some VNA-specific topics/decisions and some general analyses. In the context of a broadband microwave VNA, one currently has to perform some means of down-conversion. There are many choices to make:

How many down-conversion stages and with what frequency plan?

Classically, measuring receivers have used multiple up and down-conversions to provide better image rejection and to allow a more flexible frequency plan for the purpose of avoiding spurs. In a VNA, the image is usually less of a concern since there is one known signal present (and perhaps its harmonics) that one wants to measure. As the application space changes to include IMD and other more spectrally rich measurements, the image behavior takes on added performance, but there are often ways around the issue. Sources are increasingly clean and converters increasingly linear so that spurs have been a declining problem. Couple that with cost, complexity, and again with the situation of a single known signal and many of the reasons for multiple conversions are mitigated. Depending on how the IF is implemented and for other signal processing reasons, two conversions are sometimes desired but it is less common now to go beyond that. For reasons of stability, homodyne receivers have been avoided in recent years but that may change as the adaptive conversion circuitry used in non-measurement receivers improves.

In an ideal scenario, one would be able to fundamentally mix over the entire frequency range of the instrument. This would have the lowest spurious possibilities and best receiver noise figure (and probably best linearity). For most middle microwave and lower-frequency systems, this is the choice made.

For broadband microwave systems, this can get very expensive since the isolation chains (see Figure 5.13) have to run over this full frequency range, somehow provide enough LO power for the converter (10–20 dBm typically), and provide 100–120 dB of round trip isolation. One could imagine having a separate LO for each of the four or more converters (Figure 5.14) but this gets even more expensive and maintaining phase

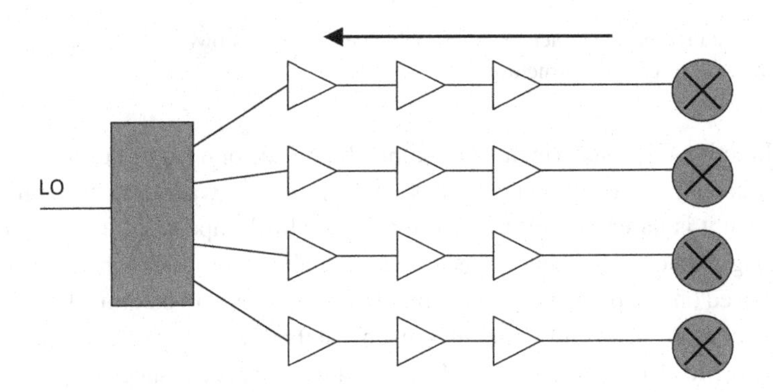

Fig. 5.13 Another four-channel receiver architecture is shown here. Now a single LO is shared between the four down-converters and amplifier chains are used to ensure channel-to-channel isolation. More or fewer amplifiers could be used and, in some circumstances, isolators or filters can be used instead.

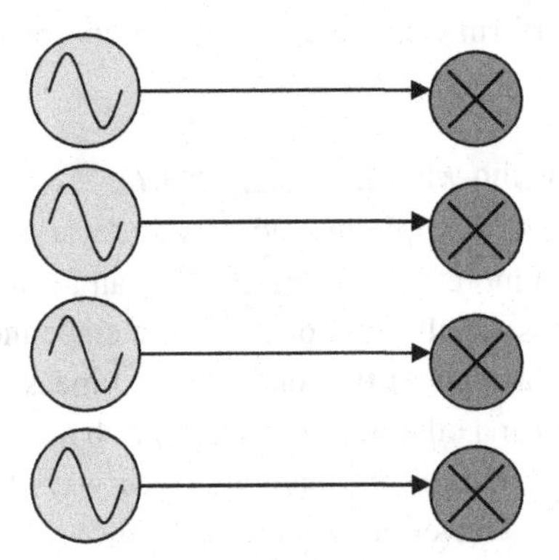

Fig. 5.14 Concept schematic of a four-channel receiver where each down-converter has its own LO. This can be expensive at higher frequencies and it can be challenging to ensure adequate phase synchronization between channels.

synchronization can be challenging. Thus typically, some kind of harmonic conversion process is used to limit the required range of the LO. The harmonic conversion may occur in the final converting device itself or in some pre-multiplier but the important concept is that often, most of the distribution is done at lower frequencies.

Converter type?

There are many possible configurations of converters and the distinctions can be subtle. As indicated above, fundamental mixing (e.g. [10]) is generally used at RF and up into the middle microwave frequency ranges and that usage may continue to march up in frequency. The early VNAs used relatively low LO frequencies (tens to hundreds of MHz) and higher-order harmonic conversion via samplers, in part because of the technologies available (e.g. [11]). The latter structure has many image responses and relatively low conversion efficiency. Depending on how the device is implemented, the linearity can be quite good. The sampler method used involved some form of edge sharpening (originally a step-recovery diode, SRD and more recently using nonlinear transmission lines) followed by a passive differentiator to create a sharp pulse. This pulse turned on the sampling diodes and captured a small window of RF energy at the period of the LO. After filtering, this created an equivalent IF for later processing. The concept is shown in Figure 5.15.

More recently when harmonic conversion has been desired (higher microwave and into the mm-wave range), the trend has been toward harmonic mixers and high LO samplers. Both of these use relatively low harmonic orders and higher LOs (into tens of GHz) and obviously have fewer image responses and tend to have better conversion efficiencies if for no other reason than less energy redistribution but this can vary with implementation. The distinction between a harmonic mixer and sampler of this type can be very subtle (e.g. [12]) and may come down to the degree of LO waveform modification performed prior to the physical converting device. Even this may be a distinction without value as the physical converting device can be engineered to do the waveform modification itself. Instead, we consider some differences with greater or lesser degrees of waveform modification at relatively low orders of harmonic conversion, as is commonly seen in higher frequency VNAs.

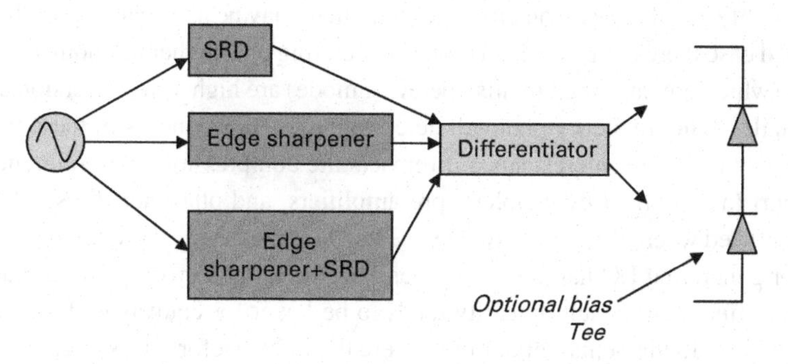

Fig. 5.15 Block diagram of a sampler construction as used in some VNAs.

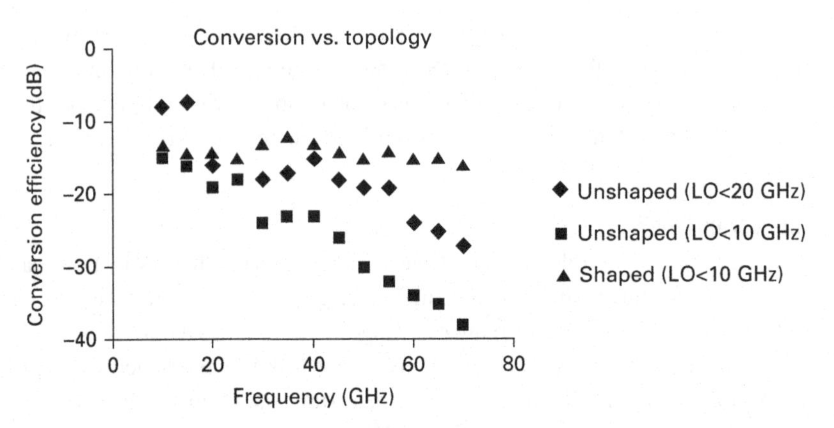

Fig. 5.16 Example plot of conversion efficiency versus frequency and converter LO structure.

Conversion efficiency is of interest since this plays a role in determining the VNA's dynamic range. If the final converting device sees an unmodified sinusoid of some limited frequency range, at higher frequencies less RF energy is captured per cycle and the effective conversion efficiency decreases. The higher the implied harmonic order, the more the decrease. If the LO waveform is highly shaped, the conversion efficiency tends to be flatter but starts off at a lower level, since energy is being somewhat dispersed to the images in all cases. This idea is illustrated in Figure 5.16 where the LO is constrained into two different ranges for the unshaped case and a doubly-balanced structure is assumed that favors odd conversion products. Thus for RF frequencies up to the LO limit, fundamental conversion is used, then 3x the LO is used, then 5x and so on. In the plot, one can see the transitions clearly for the unshaped topology for the first few and then it becomes somewhat muddier as the interaction of many mixing products may occlude the picture. The higher the LO range one can use, the less roll-off one sees in the unshaped cases and the higher the baseline conversion efficiency in the shaped cases. The downside is that the LO distribution becomes increasingly expensive and complicated the higher the LO frequency goes.

The image responses are also of concern, particularly in the non-S-parameter applications. As might be expected from Figure 5.16, the shaped case image responses will all be of roughly equal conversion efficiency and there may be a fair number of them. In the unshaped case there is a roll-off, so if one is operating on a higher harmonic, lower-order images (which are undesired in this operating mode) are higher in conversion efficiency. The details of the frequency plan will determine the relative merits of these two cases.

Linearity is also of interest since it impacts the compression point referenced to the VNA port (as modified by couplers, pre-amplifiers, and other networks). If harmonic mixing is used where the conversion device itself (in the classical diode sense) is responsible for generating LO harmonics, higher drive levels and incomplete saturation are a common effect. As a result, linearity tends to be lower for equivalent device technologies than for a fundamental mix or one where the LO waveform was altered prior to the final conversion device. This concept is shown in Figure 5.17 where the input-referred

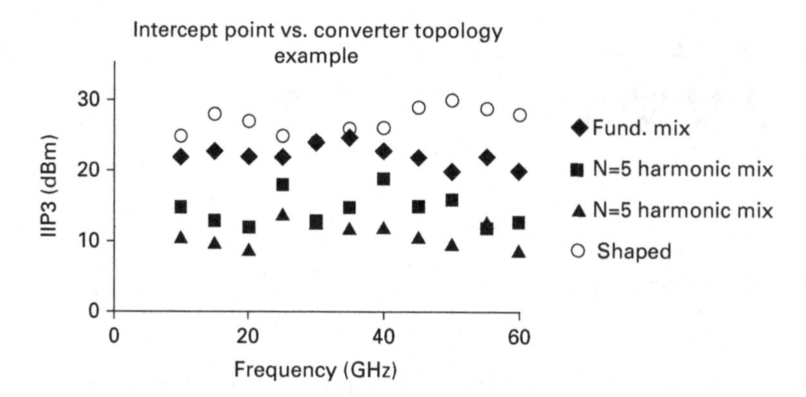

Fig. 5.17 Plot of converter input-referred intercept point vs. frequency and LO structure.

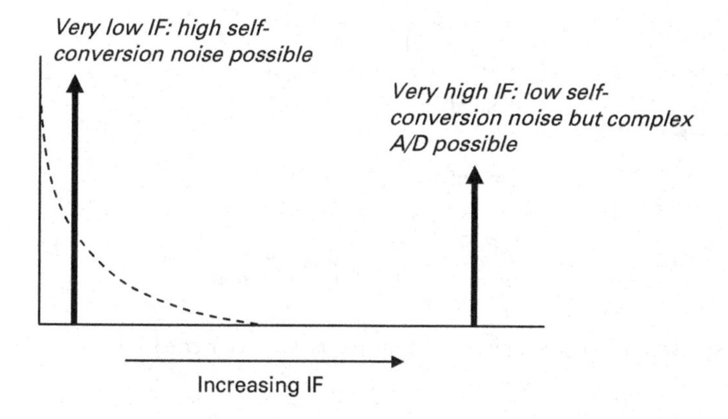

Fig. 5.18 Comparison of the noise effects of high and low IF frequencies.

intercept point for a collection of converters is plotted versus frequency where it was attempted to hold device technology and equivalent drive levels constant. The absolute values of the intercept point and the frequency dependencies will, of course, be a strong function of the technology employed and may not hold for more exotic topologies.

5.4.5 IF sections

The IF section of any receiver often gets less attention from a technological point of view but it is a critical component of instrument performance. The possible floor for speed, dynamic range, and trace noise can be set in this section, although systems are usually designed such that noise performance is not IF-limited. One of the first questions is what IF frequency (or range) should be used. If it is very low, then the A/D circuitry can be simple, but converted LO phase noise becomes more of a problem (depending on the conversion structure) as suggested by Figure 5.18. A very high IF frequency requires a more complex A/D structure and potentially more noise injection at the IF level, but the

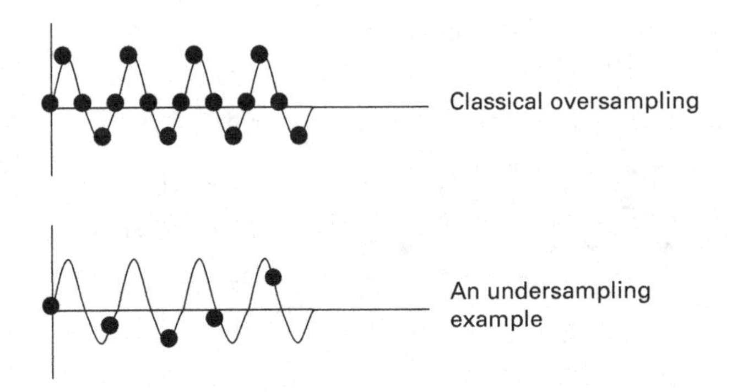

Fig. 5.19 Pictorial diagrams of undersampling and oversampling.

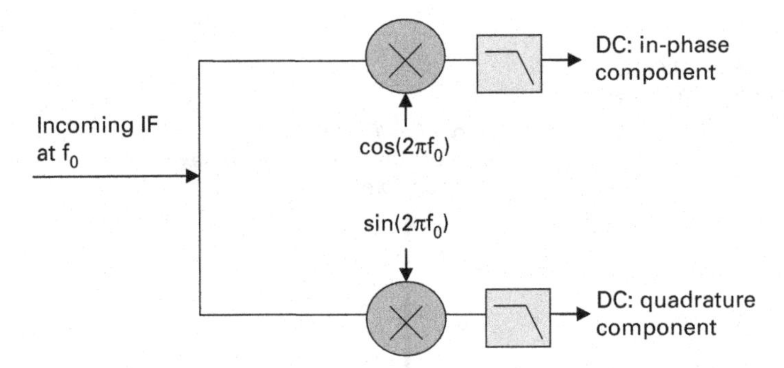

Fig. 5.20 Concept of synchronous detection, where the final IF is down-converted to DC for A/D sampling.

noise and spur contributions from the RF section are usually lower. Certain applications may demand certain ranges of IF frequency (larger bandwidths needed, for example).

Once the IF frequency is selected, the frequency plan for the A/D system usually comes next. Classically, an oversampled structure would be used to allow the extraction of maximal spectral information. This requires a faster A/D clock and places more of a constraint on cleanliness and on the A/D converter. Returning again to the concept of knowing the signal that is being measured, one could move to an undersampled structure which can improve noise and simplicity. The downside is an increase in spurious responses that may require more analog filtering. The classical difference between undersampling and oversampling is illustrated in Figure 5.19.

Another method of detection, termed synchronous detection, works by performing a final down-conversion to DC in an in-phase and in a quadrature sense. Since the A/D converters are operating at DC, the clocking structure can be simpler. Like homodyne systems, however, there are DC defects such as offsets and channel skews that must be minimized and/or corrected for. The concept of synchronous detection is shown in Figure 5.20. This approach was used in many of the earlier VNAs but has been largely supplanted by some of the previously discussed techniques as ADC technology has advanced.

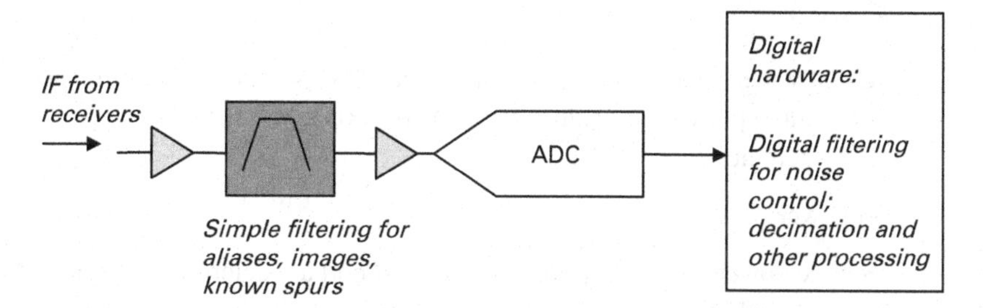

Fig. 5.21 Common IF filtering scheme where some simple filtering is analog but most of the variable filtering (and narrow bandwidth filtering) is done digitally.

Implementation of filtering is another major topic. Usually some analog filtering is required to handle aliases, images, and other known large interferers. This might be simply to avoid overloading any IF amplifiers or the A/D converters. Beyond that, filtering is required for noise reduction (and sometimes reduction of close interferers). This is usually termed an IF bandwidth in VNAs, although an analogous concept of resolution bandwidth applies to spectrum analysis. Historically, this was done with a collection of analog filters but these had issues of stability and measurement accuracy when changing the setting between calibration and measurement. More recent instruments implement all of this filtering digitally. Thus a common filtering implementation may look something like that in Figure 5.21.

5.4.6 System performance considerations

It may be useful to examine how the performance of some of the blocks just discussed **impact the overall system performance parameters.**

- Frequency range
  - Obviously set by source, receiver, and signal separation.
  - Applications requiring only simple S-parameter measurements have clear selection criteria except sometimes extreme out-of-band results are needed for characterization or regulatory reasons. Other applications requiring harmonic measurements have their own demands.
- Dynamic range
  - Two parts to this: noise floor and maximum power. The latter is addressed below (see compression or port power).
  - Noise floor is impacted by front-end loss (couplers and attenuators), conversion loss/gain and initial IF gain stages. An RF pre-amplifier can help at the potential expense of compression and stability. Noise floor and dynamic range are often specified in a 10 Hz bandwidth. Scaling rules help at other bandwidths over at least a limited range.
  - Access loops can be used to improve the reference plane noise floor at the expense of compression limits (skipping coupler loss).

- Trace noise
  - o Usually measured far away from the noise floor so that is not an impact.
  - o LO/source phase noise folds over and converts to the IF.
  - o IF system noise.
- Port power
  - o Source power and loss between source and port are determining factors. Maximum power levels of +10 dBm and higher are increasingly common. Step attenuators are often used to reach −90 dBm or lower.
  - o Compression limits of switches and other test set components may play a role.
- Power accuracy
  - o Generally limited by the structure of the ALC loop, the temperature compensation methods employed, the calibration procedure, and the power ranges involved.
- Harmonics
  - o Usually from the source and related components.
- Compression
  - o Usually the converter linearity sets this limit although any front-end RF amplifiers can sometimes contribute.
  - o RF attenuators can help in some applications.
  - o IF systems can sometimes contribute.
- Raw port parameters
  - o The front-end components (couplers, attenuators, transfer switches, etc.) tend to set these parameters.
- Residual port parameters
  - o Generally the calibration kit and calibration algorithms set these limits. The first instrument parameters to affect the residuals are usually linearity-related.
- Stability
  - o Many factors, some of which are hard to measure (measurement dynamics).
  - o Temperature stability of couplers, switches, and cables.
  - o LO power stability and sensitivity of the converter to its changes.
  - o Linearity of system and stability of port power.
- Measurement speed
  - o A number that is dependent on many variables including frequency step size, number of points, display setups, power levels, and external data transfer setups. It is therefore very hard to compare amongst applications or vendors.
  - o Source tuning speed, receiver acquisition time, digital hardware processing time, and software overhead can all play a significant role.

All of the blocks discussed play a role in how the instrument performs and how these specifications are created. In terms of non-S-parameter measurements, some of the impacts will be discussed in later chapters. While the above are usually still important, the source purity aspects (harmonics, spurs, and phase noise) take on added importance in quasi-linear and nonlinear measurements as does receiver compression. In time domain contexts (when transformed), stability and repeatability play a larger role. Residual port parameters are often dominated by calibration considerations and those are covered in Chapter 8.

5.5 Measurement types in the VNA

5.5.1 Gain, attenuation, and distortion

Basic VNA measurements are gain, attenuation, and distortions. This is similar to measurements with scalar network analyzers. However, the VNA is more accurate, based on correction availability and the greater dynamic range (tuned receiver vs. a broadband receiver). Minor variations in these measurements (in the quasi-linear realm) include DUT harmonic output and intermodulation distortion.

5.5.2 Phase and group delay

The change of phase between the input and output to and from the DUT determines the phase differences going through the DUT. The display of phase on the VNA is usually between $+180^{\circ}$ and -180° although unwrapping of phase (or absolute phase displays) is often available. Group delay is calculated from the derivative of phase with respect to frequency. This derivative is normally calculated numerically, so the interval over which it is calculated can be important (termed the "aperture"): a larger aperture effectively applies smoothing to the data but reduces the resolution (of group delay distortions) in frequency.

5.5.3 Noise figure measurements

More recent VNAs can measure noise figures and noise parameters. Hot/cold and cold only techniques are used for this application. Menus can guide users to connect, calibrate, and measure accurately the DUT noise parameters and figures.

More details are in Chapter 10.

5.5.4 Pulsed RF measurements

Some applications for pulsed RF measurements using VNA are:

- Eliminating thermal effects when testing high power devices on wafer;
- Troubleshooting high power devices to pinpoint trapping effects;

- Testing DUT under similar environments for real-life pulsed applications;
- Optimizing DC pulsed bias to minimize over- and under-shooting;
- Monitoring phase variations through the DUT during RF pulsed signals.

More details are in Chapter 15.

5.5.5 Nonlinear measurements of active and passive devices

Under high-input signals to the DUT, the output signal will contain the fundamentals and harmonics. Thus a nonlinear measurement technique needs to be defined to characterize and test the DUT.

X-parameters are an example of this type of measurement using the Agilent PNA-X [13].

More details are in Chapter 12.

5.5.6 Multi-port and differential measurements

The latest devices include multiple ports and differential ports. Some devices are intrinsically multiport (e.g. couplers) and others are driven differentially. In signal integrity applications (e.g. a backplane) there can be many ports where common measurement needs are insertion loss and crosstalk.

More details are in Chapter 9.

5.5.7 Load-pull and harmonic load-pull

The VNA is used to determine the impedance required to deliver a specific output power. The locus of the output impedance for different powers is drawn on a Smith Chart. Harmonics of the input power are also used to locate the harmonic load pull impedances. This information is used to optimize the output load impedance to cover a specific frequency ranges over a specific power ranges.

More details are in Chapters 13 and 14.

5.5.8 Antenna measurements

A variety of antenna measurements are also coordinated with a VNA including antenna patterns, antenna match, and radar cross section. These measurement systems often include positioners, additional amplifiers (due to remote location of the antennas), and distribution networks.

5.5.9 Materials measurements

The permittivity and permeability of dielectric materials can be measured by the VNA using specific measurement techniques including resonator structures, transmission (free-space or in media), and open reflection probes. This is critical since environmental conditions may be difficult, e.g. very high temperatures or very low temperatures. Material shapes may differ, e.g. biological materials, hot liquids, or large flat surfaces.

5.6 Device types for VNA measurements

5.6.1 Passive devices such as cables, connectors, adaptors, attenuators, and filters

Insertion losses and input and output return losses are the main vector measurements needed for passive devices. Calibration needs may be relatively simple except in cases of very low insertion loss. Dynamic range needs can vary widely, becoming the highest for low crosstalk measurements and filter stopband measurements, for example. Measurement speed, especially for production environments, is essential for this application. Special care needs to taken for millimeter wave measurements of passive devices.

5.6.2 Low power active devices such as low noise amplifiers, linear amplifiers, and buffer amplifiers

Linear amplifiers can also be measured using a VNA. Reverse isolation may be needed in this case, especially for buffer amplifiers. For high-gain amplifiers a dynamic range of more than 100 dB may be needed. A trade-off between speed, accuracy, and the number of points needs to be decided before doing measurements.

Compression can also be measured at 1, 2, or 3 dB, and even the DC current variation may be of interest. Noise figure, intermodulation, distortion, and return loss may also be of interest.

5.6.3 High power active devices such as base station amplifiers and narrow-band amplifiers

High power active devices can be considered as two parts. The linear part is already covered in Section 5.6.2. The nonlinear part is covered in Chapter 12. Attenuators need to be used to protect the VNA from being damaged. Many of these applications are relatively narrowband (10%) and have power higher than watt-scale. The thermal management of devices and device measurement is critical. Pulsing the DC bias can be used to reduce the thermal effects. Within the pulsed bias, the RF measurement can be performed using pulsed RF VNA. This pulsed measurement is covered in Chapter 15.

5.6.4 Frequency translation devices such as mixers, multipliers, up/down-converters and dividers

Newer VNAs are capable of measuring frequency translation devices which have different input and output frequencies. Typical devices are: 1) multipliers where the output frequency is a multiple of the input frequency. The step recovery diode is capable of

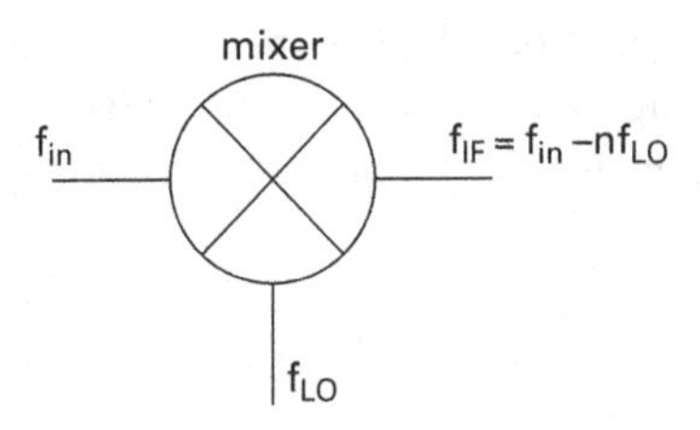

Fig. 5.22 Down converter block diagram.

delivering a comb generator ($n = 1$ to 100) for the input frequency. Common diode-based multipliers are used for lower orders and high-speed digital circuitry is also often used for harmonic generation. 2) Dividers devices where the input frequency is a multiple of the output frequency. 3) Mixer devices are more general where the output frequency is related to the input frequency by the local oscillator frequency as: $f_{IF} = f_{in} + / - f_{LO}$ where the f_{IF} is the intermediate frequency, which is often lower than the input frequency, as shown in Figure 5.22. 4) Harmonic mixers are the same as mixers except that the f_{LO} is a multiple of a lower frequency LO. 5) Up/down-converters are the most general frequency translation devices. They include amplifiers, filters, mixers, or multipliers. Some applications require that the up/down-converter is in the same package and a high isolation switch is used to choose the operation mode. Receivers for consumer electronics often have this or higher levels of integration.

5.6.5 On-wafer measurements of the above devices

On-wafer measurements represent another type of challenge. The probe interface from devices on-wafer to the VNA is an important and challenging aspect of this measurement class as there is additional loss between the couplers and the DUT and the probe contact repeatability can affect uncertainties. A variety of different calibration and de-embedding techniques have been developed for on-wafer measurements that are suitable for the standards involved and are discussed in Chapter 8. For high power on-wafer measurements, a pulsed bias and pulsed RF can be used, as explained in Chapter 15.

At the present time VNAs are used for many applications with the following features:

- High frequency up to 1100 GHz or higher
- Low frequency down to 70 KHz or lower
- New calibration techniques including electronic calibration
- High-speed measurements down to 2.5 microseconds per point
- Multiple ports up to 16 ports or more
- Differential and balanced measurements
- Time domain for signal integrity analysis
- Pulsed RF measurements
- Error-corrected power measurements
- Error-corrected mixer measurements for conversion loss and phase
- Error-corrected noise measurements.

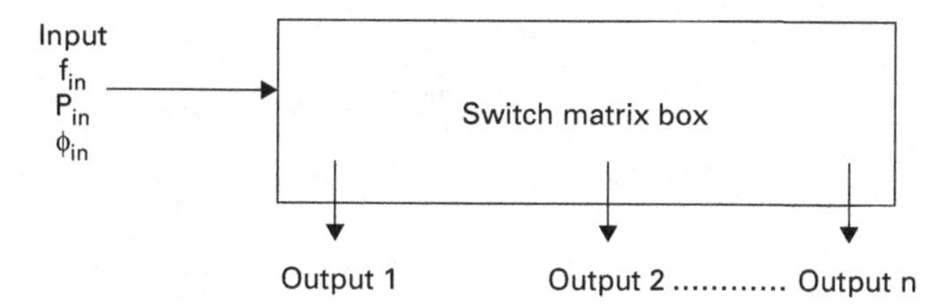

Fig. 5.23 Switch matrix box.

5.7 Improving VNA measurement range

5.7.1 Using a switch matrix box

The basic switch matrix box is used to switch microwave signals between the output of a 2-port VNA and input to one or more input devices as shown in Figure 5.23. Another switch matrix box (or another set of switches in the same box) is used to connect from the output of multiple devices to the input of the 2-port VNA.

Mechanical switches can be used for reasons of linearity and lower insertion losses. This has the disadvantages of slower measurement time, repeatability, and, potentially, switch lifetime (which may be in the tens of millions of cycles). Solid-state switch matrices (using technologies discussed earlier in this chapter) are also used that have excellent repeatability, switching time, and lifetime, but have higher insertion losses and, potentially, worse linearity.

Passive-DUT switch boxes are typically used to test multiple or single devices. However, active-DUT switch boxes are also used to configure the input to devices for specific power, frequency range, noise or multiple signals for intermodulation distortion.

Depending on the software implementation, the switch box may be de-embedded using measured S-parameters or, more commonly, multiport calibrations are applied at the test ports of the switch box.

VNAs used now include many of the switch structures to accomplish the same purpose previously done by the switch matrix box.

5.7.2 Using multiple sources

Testing mixers requires two sources – one for the input signal and the other for the local oscillator. Also testing for intermodulation distortion requires two sources with a frequency offset.

Some VNAs now used include more than one source and a high isolation combiner. The frequency ranges of these sources are important, and there is the need to control each source separately.

The phase noise of these multiple sources can be important for intermodulation measurements where the offset frequency is low.

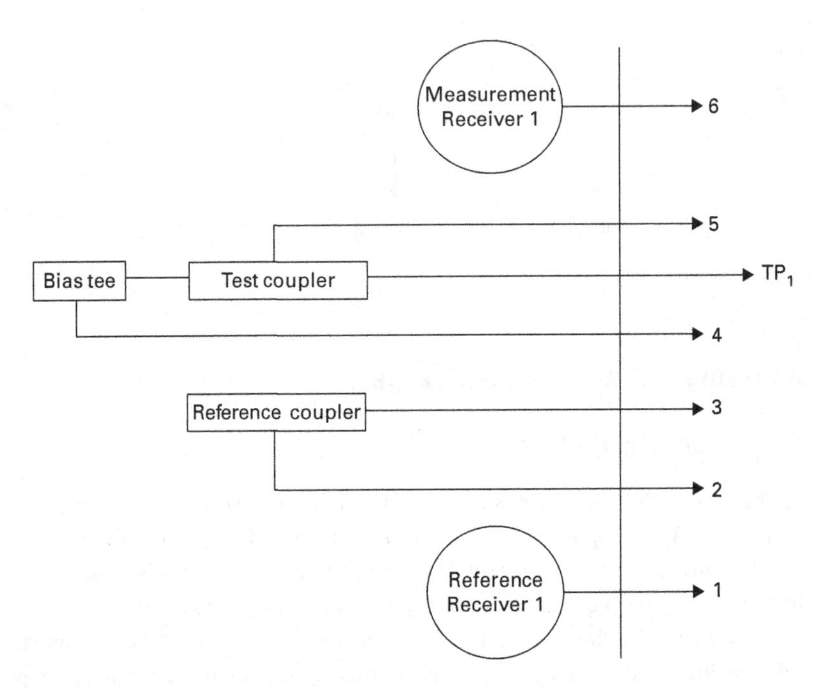

Fig. 5.24 Reverse coupler.

5.7.3 Using reversing couplers

Incident and reflected signals are usually measured using 10–20 dB directional couplers. Thus, reversing the coupler reduces the noise floor by the coupling factor of approximately 10–20 dB. However, this comes at the expense of reducing the maximum input power by 10–20 dB.

Using the switch box to do the signal switching in portions of the measurement routine is typically used for testing low noise devices or systems.

Using reversing couplers does not change the VNA's dynamic range. However, it offsets the dynamic range by 20 dB, as shown in Figure 5.24. Instead of connecting point-1 to point-2, point-3 to point-4, and point-5 to point-6, the reverse coupler technique connects point-1 to point-2, point-4 to point-6, and point-3 to point-5.

5.7.4 Using an external amplifier/attenuator

To characterize a power amplifier for full saturation, higher power than is available from the VNA is needed. Thus, an external amplifier can be connected between point-3 and point-4 as shown in Figure 5.25.

To characterize an amplifier with higher gain which produces higher power than is specified by the VNA input, an external attenuator is needed. It can be connected between point-5 and point-6 as shown in Figure 5.24.

5.7.5 All-in-one VNA box

Adding more couplers, attenuators, receivers, and bias-tees can expand the VNA to a four (or more) port analyzer.

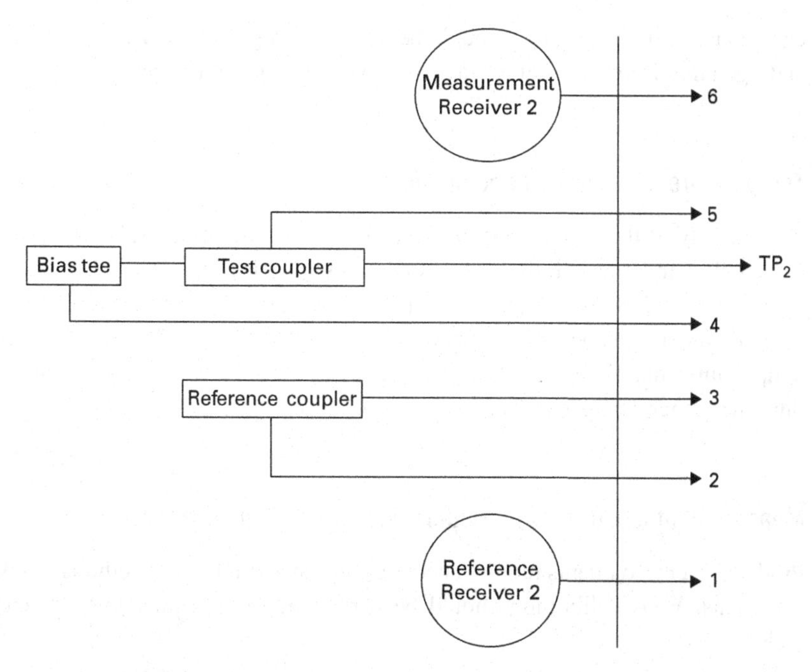

Fig. 5.25 High power amplifier.

Adding more than one source with switches and high isolation combiners can expand the VNA to systems for testing mixers and intermodulation distortion.

Users can then reconfigure the "all-in-one" VNA box into a custom system to meet the test requirements for specific devices or specific applications.

Recently, up to eight sources were introduced by Rohde and Schwarz for the 8 GHz box (ZVT8).

5.8 Practical tips for using VNAs

5.8.1 User training

The user's experience with VNAs can impact measurement results. Thus a basic understanding of operating VNAs and calibration are essential to obtain accurate and repeatable data. The cable type, connector type, operating frequency, and power will influence the levels of user training. When waveguide is used at millimeter wave frequencies, then user training is different from that at the microwave frequencies.

5.8.2 Connector care

Connectors need to be regularly inspected to assure that the connector mating is correct, clean, and precise. At millimeter wave frequencies this is more critical than at RF frequencies. At higher microwave power, connectors need to be checked to see if they can handle the power going through them without degradation. The specific application and operating frequency determine, what type of connector is needed.

5.8.3 Temperature environment and stability

The stability of the temperature of the environment where the VNA is located impacts the stability and repeatability of measurements. This is especially true when comparing devices with mdB variations in their S-parameter values. Therefore, a stable environment is required to ensure accurate measurements. The VNA data is specified for operating in temperatures in a limited range and any deviation from the calibration temperature can introduce other changes in data.

5.8.4 Measurement locations: production, development or research

Production environments require accurate calibration and rapid methods of taking measurements. VNA calibration should be performed on a regular basis to ensure high yields.

Development environments demand a highly reconfigurable VNA system. Measurements need to be performed to prove the development concept. The objective is to increase yield and decrease test time. Automated test systems may be required to measure a large number of devices.

Research environments require much wider frequency range systems and are associated with very different calibration techniques.

5.9 Calibration and calibration kits

Due to imperfections of the VNA and of any networks between the VNA and the DUT, calibrations need to be performed on the VNA to calibrate the measurement system to the DUT reference planes.

Chapter 8 presents different methods of calibration and calibration kits.

5.10 Conclusions

In this chapter, the history and background behind the modern VNA have been discussed as have been many of the fundamental building blocks of that class of instrument. The objective has been to show how the attributes of those building blocks affect measurement performance and how various architectures can optimize or enable certain applications. The core technologies and structures presented here are themselves building blocks for instruments and measurement classes presented in later chapters.

References

[1] D. Rytting, "ARFTG 50 year network analyzer history," *67th ARFTG Conf. Dig.*, pp. 1–8, June 2006.

[2] J. A. C. Kinnear, M.A., A.M.I.E.E. "An automatic swept frequency meter," *British Communications & Electronics*, p. 359, May 1958.

[3] "An advanced new network analyzer for sweep-measuring amplitude and phase from 0.1 to 12.4 GHz," *HP Journal,* Feb. 1967.

[4] R. A. Hackborn, "An automatic network analyzer system," *Microwave Journal,* Vol. 11, pp. 45–52, 1968.

[5] *The essentials of vector network analysis: from α to Z_0*, Anritsu Company, 2008.

[6] J. A. Crawford, *Frequency Synthesizer Design Handbook,* Artech House, 1994.

[7] Microsemi, *The PIN Diode Circuit Designer's Handbook,* 1992.

[8] R. S. Pengelly, *Microwave Field Effect Transistors – Theory, Design and Applications,* Research Studies Press, 1986.

[9] G. D. Vendelin, A. M. Pavio, U. L. Rohde, *Microwave Circuit Design Using Linear and Nonlinear Techniques*, Wiley, 2005, chp. 12.

[10] S. A. Maas, *Microwave Mixers*, Artech House, 1993.

[11] M. Kahrs, "50 years of RF and microwave sampling," *IEEE Trans. Microw. Theory and Tech.*, vol. 51, pp. 1787–1805, June 2003.

[12] J. Martens, "Multiband mm-wave transceiver analysis and modeling," 2012 WAMICON Dig., Apr. 2012.

[13] Company web sites:

- Rohde & Schwarz, http://www2.rohde-schwarz.com/en
- Anritsu Corporation, http://www.anritsu.com
- Agilent Technologies, http://www.agilent.com
- Maury Microwave, http://www.maurymw.com
- Focus Microwaves, http://www.focus-microwaves.com
- NMDG, http://www.nmdg.be

6 Microwave power measurements

Ronald Ginley

6.1 Introduction

In physics, **power** is the rate at which energy is transferred, used, or transformed. For example, the rate at which a light bulb transforms electrical energy into heat and light is measured in watts – the more wattage, the more power, or equivalently the more electrical energy is used per unit time [1]. Energy transfer can be used to do work, so power is also the rate at which this work is performed [2].

For systems or circuits that operate at microwave frequencies, the output power is usually the critical factor in the design and performance of that circuit or system. Measurement of the power (signal level) is critical in understanding everything from the basic circuit element up to the overall system performance. The large number of signal measurements that can be made and their importance to system performance means that the power-measurement equipment and techniques must be accurate, repeatable, traceable, and convenient.

In a system, each component in a signal chain must receive the proper signal level from the previous component and pass the proper signal level on to the succeeding component. If the output signal level becomes too low, the signal becomes obscured in noise. If the signal level becomes too high, though, the performance becomes nonlinear and distortion can result. The uncertainties associated with the measurement of power also play a very important role in the development and application of microwave circuits. For example, a 10 W transmitter costs more than a 5 W transmitter. Twice the power output means twice the geographical area is covered or 40% more radial range for a communication system. Yet, if the overall measurement uncertainty of the final product test is of the order of ± 0.5 dB, the unit actually shipped could have output power as much as 10% higher or lower than the customer expects, with resulting lower operating margins [3].

At low frequencies, the concepts of voltage, impedance, and current can be used to describe how energy is transported through a circuit. At microwave frequencies, voltage and current lose significance and are replaced by "power." The question of how much signal is present is answered by a power measurement. The importance of the measurement of microwave power in microwave circuits is easily seen, as it is the power that "does the work" or in the case of a communication system, it is the power that carries the information [4]. It is at the higher operating power levels that each decibel increase in power level becomes more costly in terms of complexity of design, expense of active devices, skill in manufacture, difficulty of testing, and degree of reliability. "The increased cost per dB of power level is especially true at microwave and higher frequencies, where the high power solid state devices are inherently more costly and the guard-bands designed into the circuits to avoid maximum device stress are also quite costly. Many systems are continuously monitored for output power during ordinary operation" [3].

This chapter covers the fundamentals of making microwave power measurements; beginning with basic concepts, definitions, and terminology. Next, there is a brief discussion on different types of power measurements and how they are applied. To make power measurements, there must be some form of a power detector. The main types of power detectors are covered. The power detectors that are used are not perfect and the results of their measurements must be corrected for different error mechanisms. The first of these, effective efficiency, is discussed in terms of its definition and how it is measured at the highest level of accuracy. The concept of a basic power measurement is explored and the effect of effective efficiency and different error mechanisms and other adaptations are discussed. This leads into a brief discussion of the uncertainty of basic power measurements and a look at the larger contributors to the overall uncertainty of a power measurement. Finally, a few examples of different aspects of power measurements are given.

6.1.1 Why power and not voltage and current?

The first question many ask with regards to measuring microwave power is: why not just use measurements of voltage and current? From a DC and low-frequency perspective, power is defined in terms of the voltage (V), current (I), and resistance (R) as:

$$P = IV = \frac{V^2}{R} = I^2 R. \tag{6.1}$$

Voltage and current measurements are straightforward and easy to make. However, as the frequency approaches 1 GHz, it becomes necessary to directly measure power because voltage and current measurements become impractical. One of the main reasons for this is that voltage and current can vary with position along a lossless transmission line, whereas power maintains a constant value with position. Another example of the decreased usefulness is in waveguide transmission structures, where voltage and current are even more difficult to define due to the structure of the electric and magnetic fields inside the guiding structure. For these reasons, at radio and microwave frequencies, power is more easily measured, easier to understand, and more useful than voltage or current as a fundamental quantity.

6.2 Power basics, definitions, and terminology

Just what do we mean when we talk about microwave power and the measurement of microwave power? First, the unit of power is the Watt. The International System of

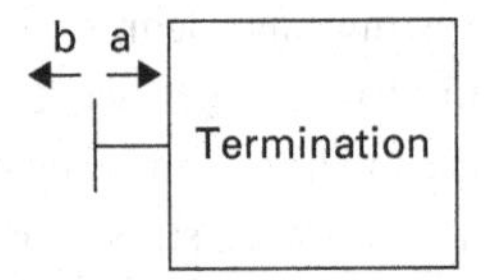

Fig. 6.1 Simple termination to help define power.

Units has established one Watt to be one joule per second. Note that there are no other electrical units used in this definition. We can talk about power in terms of the complex electromagnetic wave amplitudes that are travelling along or incident on a microwave structure. Take, for example, the simple termination shown in Figure 6.1, where "a" is the electromagnetic wave incident on the termination and "b" is the wave that is reflected from the termination. The reflection coefficient is defined as: $S_{11} = \frac{b}{a}$. Different "powers" are defined in terms of the electromagnetic waves as:

incident power: $P_{inc} = a^2$,
reflected power: $P_{ref} = b^2$,
and net power into the termination: $P_{net} = |a|^2 - |b|^2$.

It is difficult to directly measure the complex waveforms "a" and "b". Instead, the measurement of microwave power is performed by transforming the waves into something more easily measured such as a temperature change or rectified energy. There are efforts underway to directly measure the electric or magnetic fields; these techniques will, hopefully, allow us to make much more accurate power measurements [5].

6.2.1 Basic definitions

To be able to understand microwave power measurements, there are a few useful basic definitions:

dB (decibel): the ratio of two powers is often used instead of absolute power. The ratio is dimensionless and is commonly expressed as decibels. The dB is defined as: $dB = 10log_{10}\left(\frac{Power\,Level1}{Power\,Level2}\right)$ where *Power Levels 1* and *2* are arbitrary power levels
dBm: another method of expressing a power level is to reference it to a known level. In the case of dBm, the reference level is 1 mW. Thus, $dBm = 10log_{10}\left(\frac{Power\,Level1}{1mW}\right)$.
dBW: power expressed in dB with a reference level of 1 W.

A comparison of the different terms is given in Table 6.1.

6.2.2 Different types of power measurements

When making a power measurement, what "type" of power measurement must be defined to avoid confusion and incompatible results. There are many different ways to define "power" when looking at sinusoidal or other complex, periodic waveforms. The most

Table 6.1 Comparison of different power definitions

dBm	dBW	Watts
+60	+30	1,000 (1 kilowatt)
+50	+20	
+40	+10	
+30	0	1 (1 Watt)
+20	−10	
+10	−20	
0	−30	1 milliwatt
−10	−40	
−20	−50	
−30	−60	1×10^{-6} (1 microwatt)

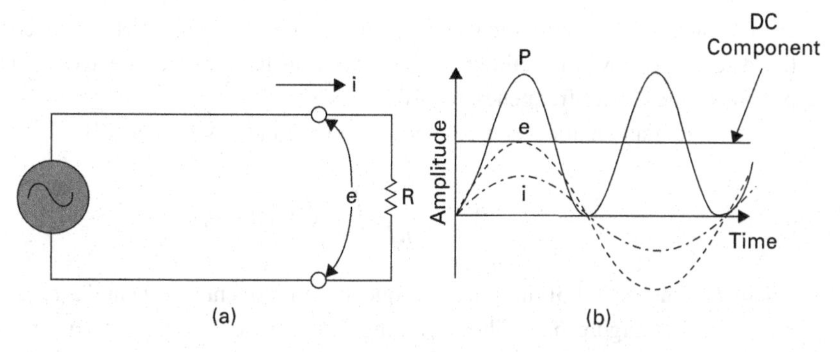

Fig. 6.2 The product P of voltage e and current i varies during the sinusoidal cycle (figure courtesy of Agilent Technologies).

common are average power, pulse power, and peak or peak envelope power. Modern wireless system designs use different complex schemes for combining many channels into broadband complex signal formats. A typical signal, like the EDGE system, requires peak, average, and peak-to-average characterization of power signals.

The term "average power" is very popular and is used in specifying almost all RF and microwave systems. In elementary theory, power is said to be the product of voltage and current. But for an AC voltage cycle, this product V × I varies during the cycle, as shown by curve P in Figure 6.2, according to a 2 × frequency relationship. Using this example, a sinusoidal generator produces a sinusoidal current as expected, but the product of voltage and current has a DC term as well as a component at twice the generator frequency. The word "power" as most commonly used, refers to that DC component of the power product. All the methods of measuring power to be discussed in this chapter use power sensors which, by averaging, respond to the DC component.

The definition of power is energy per unit time. The important question to resolve is over what time is the energy transfer rate to be averaged when measuring or computing power? From Figure 6.2, we clearly see that if too narrow a time interval is used (say close to one cycle) varying answers for energy transfer rate are found. But at microwave

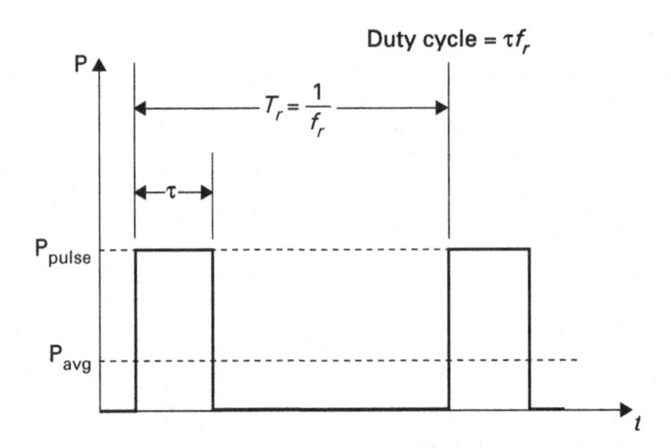

Fig. 6.3 Pulse power P_{pulse} is averaged over the pulse width (figure courtesy of Agilent Technologies).

frequencies, such microscopic views of the voltage-current product are not common. For this discussion, power is defined as the energy transfer per unit time averaged over many periods of the carrier frequency involved.

In a more mathematical sense, average power can be written as [3]:

$$P_{avg} = \frac{1}{nT} \int_0^{nT} e(t)\, i(t)\, dt, \tag{6.2}$$

where T is the period of the lowest frequency component of $e(t)$ and $i(t)$ ($e(t)$ and $i(t)$ are defined in Figure 6.2). The averaging time for average power sensors and meters is typically from several hundredths of a second to several seconds and, therefore, this process obtains the average of most common forms of amplitude modulation [3].

For pulse power, the energy transfer rate is averaged over the pulse width τ (Figure 6.3). Pulse width τ is generally considered to be the time between the 50% rise-time/fall-time amplitude points.

By its very definition, pulse power averages out any aberrations in the pulse envelope such as overshoot or ringing. For this reason it is called pulse power and not peak power or peak pulse power as is done in many radar references. The terms peak power and peak pulse power are not used here for that reason. Peak power refers to the highest power point of the pulse top, usually the risetime overshoot. For certain more sophisticated microwave applications and because of the need for greater accuracy, the concept of pulse power is not totally satisfactory. Difficulties arise when the pulse is intentionally non-rectangular or when aberrations do not allow an accurate determination of pulse width; this is when the peak power method can be used for more accurate measurements [3].

Measurements of modulated signals

Digital modulation provides more information capacity, compatibility with digital data services, higher data security, better-quality communications, and quicker system availability. Developers of communication systems face constraints such as available bandwidth, permissible power, and the inherent noise level of the system. Over the past few years, a major transition has occurred from simple analog amplitude modulation and frequency/phase modulation to new digital modulation techniques. Another layer of complexity in many new systems is multiplexing. Two principal types of multiplexing (or "multiple access") are TDMA and CDMA. These are two different ways to add diversity to signals, allowing different signals to be separated from one another [6].

Although many RF and microwave measurements can be made with CW signals, there are many other signal schemes that require sampling a signal at a certain point in time, or applying non-CW excitation to a circuit under test. Pulses, on-off transitions, power control steps, and some digital modulation schemes are *not* CW signals, and their measurement requires more advanced techniques [7]. Depending on the application, the accuracy of the power meter solution could have a significant impact on the overall performance. For example, the output power transmitted at a cellular base station affects the coverage area. When the base station is installed, the output power is measured and verified. System designers try to optimize the coverage area while balancing trade-offs. More output power leads to a greater coverage area but it can also create interference. If the power output is below a minimum limit, the coverage area is reduced and this could eventually lead to dropped calls and dissatisfied customers.

There are many different ways of looking at a digitally modulated signal. To examine how transmitters turn on and off, a power-versus-time measurement is very useful. In addition, peak and average power levels must be well understood, since asking for excessive power from an amplifier can lead to compression or clipping. These effects distort the modulated signal and usually lead to spectral regrowth as well. The power within one or more cycles of the signal is of interest when developing or troubleshooting mobile radio systems.

When looking at these complex signals, the most appropriate power sensor needs to be selected. Conventional thermal power sensors such as bolometric or thermoelectric detectors (see Section 6.3 for a discussion of these types of sensors) cannot adequately measure complex signal characteristics since they cannot delimit specific areas of power contribution in a timeslot. This is because thermal sensors average the RF power occurring over the entire time period. Sampling the power envelope over time is feasible with diode sensors. However, diode sensors always include signal details such as overshoots, interference pulses, and glitches as well as signal edges of a pulsed RF signal in proportion to their power [8]. Peak power measuring instruments and sensors have time constants in the sub-microsecond region which allow for measurement of pulsed power modulation envelopes [3].

Diode-based power sensors can be used to display power versus time in the same way that an oscilloscope does. This means that you do not miss a single detail of the signal you want to investigate. Furthermore, you can add time-slot and gate structures to your pulsed RF signals and configure them in the manner desired (see Figure 6.4 for an example). By graphically editing the gates added to the "scope" window, you can selectively suppress unwanted components at the beginning and end, which occur, for example, in the transition between two timeslots. Wideband power sensors can quickly and accurately measure peak, average, peak-to-average ratio power measurement, rise/fall time, pulse width, and complementary cumulative distribution function statistical data for wideband signals.

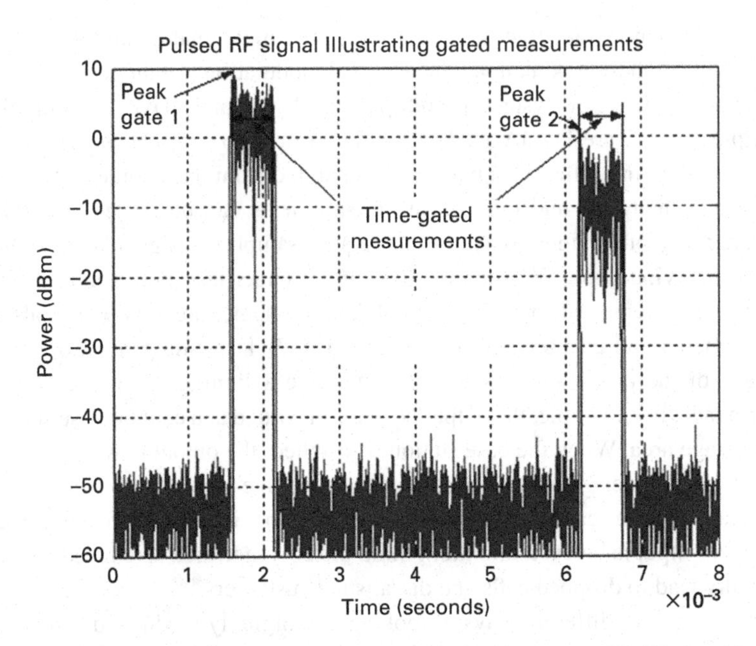

Fig. 6.4 Time domain shot of a wireless signal format, in this case, an EDGE signal in a GSM system. It is an ideal candidate for peak, average, and peak-to-average ratio measurements for time-gated wireless formats (figure courtesy of Agilent Technologies).

The measurement of complex signals is indeed a complex topic. For further information on the topic of modulated signal measurements, see [6, 7, 8, 9, 10].

For the purposes of the discussion in the rest of this chapter, we will focus on average power measurements and CW signals.

6.3 Power detectors and instrumentation

As with power measurements, there are different types of power detectors. It is necessary to have a basic understanding of how power detectors work in order to be able to choose the most appropriate one for the measurement at hand. In addition to learning about the detectors, it is important to understand how the electronic packages associated with the detectors work (the "power meters").

Due to the difficulty of measuring waveforms and power directly at higher frequencies, the techniques used to measure power modify the microwave signal in some manner to allow it to be measured more easily. The three main types of detectors are bolometric, thermoelectric, and diode. Bolometric detectors work by substituting DC power for the RF power; thermoelectric detectors work by substituting a thermally generated voltage for the RF power; and the diode-type sensors work by rectifying the RF signal.

Each type of sensor has its strengths and weaknesses. The bolometric sensors are typically very stable, linear, and have easily modeled behavior: however, they work only in a narrow dynamic range and have limited power capabilities; they also react slowly.

Thermoelectric sensors are also linear, have better sensitivity, good dynamic and power ranges; and for general use, they need a support set of electronics and require a reference point at a known frequency to fix their operation. Diodes are very nonlinear and fast, and the newer generation of diode-based detectors have good dynamic and power ranges; they also need a reference point to fix their operation.

6.3.1 Bolometric detectors

Bolometric detectors use a temperature-sensitive resistor to measure the microwave power. The most common form of bolometric detector is the thermistor detector. Simply put, when microwave power is applied to a thermistor sensor, the resistive element heats up and as a result changes its resistance. By measuring the change in resistance, you can determine the amount of microwave power that was applied. Thus, the microwave power level of the signal being measured is ultimately determined by a DC resistance measurement.

Of course, it isn't really that simple. There are other steps in the process that must be considered. Figure 6.5 shows the basic structure of a generalized bolometric detector and power meter. The DC blocking capacitor, if used, rejects any DC signal coming into the detector from the microwave connector. This is important, as the sensing element reacts to any signal and a DC signal would give a false microwave power level. The sensing element is some form of temperature-sensing resistor. There are many resources that discuss the different types of sensing elements; see [11] for a good analysis of the different types. For this discussion, we consider a thermistor bead sensing element. Thermistor sensors are small beads of metallic oxides with two very small wire leads. The common type of bead is a negative-temperature-coefficient bead which refers to the effect that as the temperature of the bead goes up the resistance of the bead goes down. The resistance of the thermistor is monitored by a power meter circuit. To make sure that there is minimal leakage of the microwave signal beyond the bead, there are filters designed to block any microwave signal, in the operating range of the detector, from leaking out of the DC leads to the power meter circuit.

The power meter circuit in its most basic form is simply a Wheatstone bridge (Figure 6.6). In more complicated schemes, it is a self-balancing bridge circuit with a digital back end. An example of a self-balancing, bridge circuit power meter is the

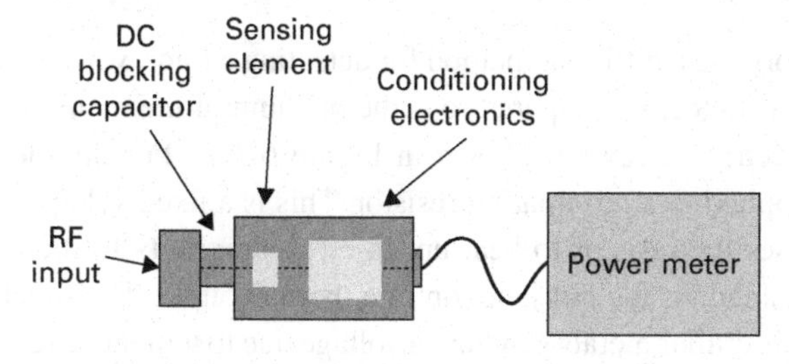

Fig. 6.5 Basic bolometric power sensor design.

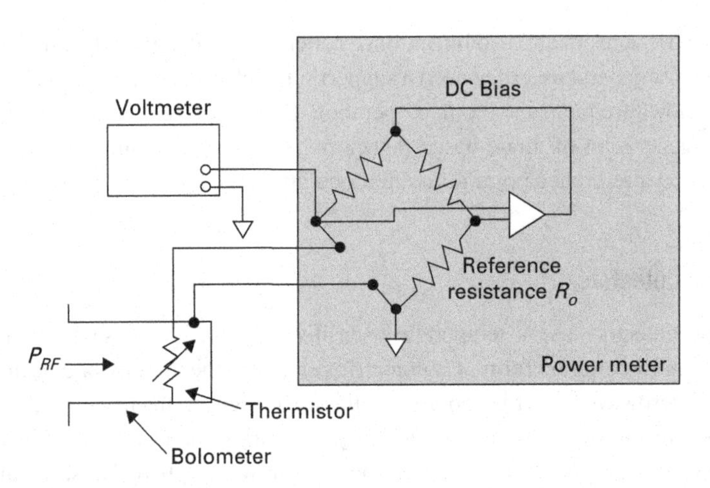

Fig. 6.6 Functional diagram of thermistor and power meter pair.

NIST Type IV power meter [12]. The interaction of the thermistor and the power meter is straightforward. Initially, with no microwave power applied, the power meter supplies a DC current to the thermistor bead. This current heats up the bead and brings its resistance to a point where it is in balance with the internal reference resistor inside the power meter (the balancing arm of the bridge circuit). Once the DC equilibrium is established, the microwave signal is applied to the thermistor. The microwave signal heats up the thermistor bead and, thus, drives the resistance of the bead down. This causes the bridge to go out of balance and it responds by removing enough DC bias from the bead for it to again be balanced against the reference resistor in the power meter. By measuring how much DC power was removed, the amount of microwave power applied is determined. This is termed the DC substituted power. The power for this type of power meter can be calculated as:

$$P_{sub} = \frac{V_{DC-off}^2 - V_{DC-on}^2}{R_0}, \tag{6.3}$$

where V_{DC-off} is the DC bias voltage with no microwave power, V_{DC-on} is the DC bias voltage with microwave power applied and R_0 is the power-meter reference resistance value. Thermistor detectors have a workable dynamic range of -10 to $+10$ dBm.

6.3.2 Thermoelectric detectors

Thermoelectric detectors use a different method for detecting a microwave signal. However, like the bolometric detector, temperature is the medium of the method. The basic circuit for a thermoelectric detector is shown in Figure 6.7(a). For this detector, the microwave signal is applied to a terminating resistor. This is a fixed value resistor. The microwave signal causes the resistor to heat up. In close proximity to the resistor, on a thermally isolated "island" is the hot junction of a thermocouple. Thermocouples are based on the fact that dissimilar metals generate a voltage due to temperature differences at a hot and a cold junction of the two metals, as seen in Figure 6.7(b).

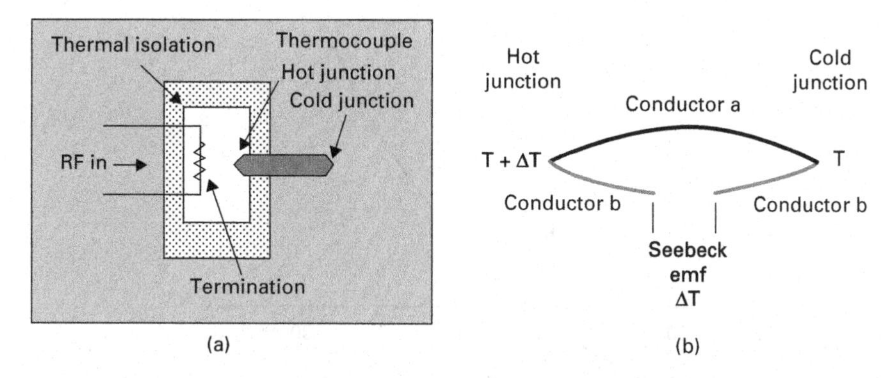

Fig. 6.7 (a) Basic thermocouple detector design and (b) basic thermocouple design (figure courtesy of Agilent Technologies).

When the terminating resistor heats up, it transfers heat energy to the thermocouple, which generates a voltage that can be used to determine the amount of microwave power applied. Since thermocouples and thermistors are heat-based detectors, they are true averaging detectors. In most thermoelectric detectors, the voltage generated by the thermocouple is small and needs to be amplified and conditioned. These electronics are usually housed in the same shell as the terminating resistor and the thermocouple. This overall device is connected to a power meter that converts the thermocouple signal into a power value (a detailed discussion of thermocouples and power meters can be found in [9] and [11]). These power meters have a design that takes advantage of the increased power sensitivity of the thermocouple sensors and is still able to deal with the very low output signals of the detectors. Thermocouple detectors have a dynamic range of approximately -35 to $+20$ dBm.

With thermocouples, where there is no direct power substitution, sensitivity differences between sensors or drift in the sensitivity due to aging or temperature can result in a different DC output voltage for the same RF power. Because there is no feedback to correct for different sensitivities, measurements with thermocouple sensors are said to be open-loop. Thermocouple power meters have solved this need for sensitivity calibration by incorporating a 50 MHz, 1 mW power-reference oscillator whose output power is controlled with great precision (as low as $\pm 0.4\%$) [9]. To verify the accuracy of the system, or adjust for a sensor of different sensitivity, the user connects the thermocouple sensor to the power reference output of the power meter and, by the use of a calibration adjustment, sets the meter to read 1.00 mW. By applying the 1 mW reference oscillator to the sensor's input port, just like when an unknown signal is to be measured, the same capacitors, conductors, and amplifier chain are used for measurement in the same way as for the reference calibration. This feature provides confidence in the power results that the detector/power meter pair is producing.

6.3.3 Diode detectors

Diode detectors use a method for determining power that is very different from the thermistor and thermocouple detectors. Diodes convert AC signals to DC by way of

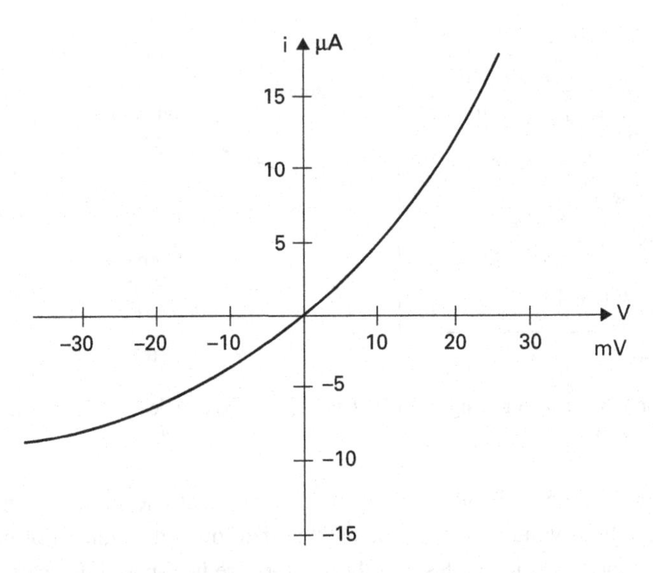

Fig. 6.8 Basic response curve for a semiconductor diode.

their rectification properties. These arise from the nonlinear current-voltage (I-V) characteristics of a semiconductor diode. Metal-semiconductor junctions, exemplified by point-contact technology, exhibit a low potential barrier across their junction, with a forward voltage of about 0.3 V. They have superior RF and microwave performance, and were popular in earlier decades. Low-barrier Schottky diodes, which are metal-semiconductor junctions, succeeded point-contacts and vastly improved the repeatability and reliability [9]. Figure 6.8 shows a typical diode I-V characteristic of a low-barrier Schottky junction, expanded around the zero-axis to show the square-law region.

Mathematically, a detecting diode obeys the diode equation

$$i = I_S\left(e^{\alpha V} - 1\right), \tag{6.4}$$

where $\alpha = q/nKT$, i is the diode current, V is the net voltage across the diode, and I_S is the saturation current and is constant at a given temperature. K is Boltzmann's constant, T is absolute temperature, q is the charge of an electron, and n is a correction constant to fit experimental data (n equals approximately 1.1 for the devices used here for sensing power). The value of α is typically a little under 40 v^{-1} [9].

For a typical diode, the square-law detection region exists for power levels P_{in} from the noise level up to approximately −20 dBm. There is a transition region that ranges from approximately −20 to 0 dBm input power, and there is a linear detection region that extends above approximately 0 dBm. For wide-dynamic-range power sensors, it is crucial to have a well-characterized expression of the transition and linear detection ranges. If you are operating in these regions, then it is necessary to apply some form of correction to the diode curve that is not necessary in the square law region.

The simplified circuit of Figure 6.9(a) represents an unbiased diode device for detecting low-level RF signals. The matching resistor (approximately 50 ohms) is the termination

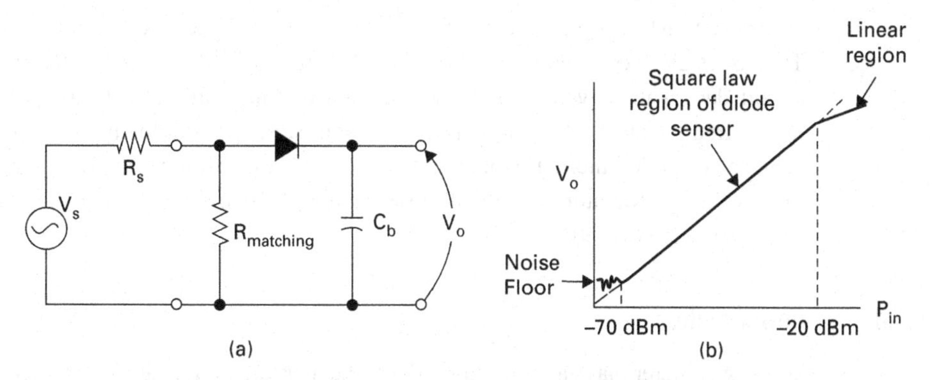

Fig. 6.9 (a) Diagram of a source and a diode detector with matching resistor and (b) power versus voltage curve for diode sensor (figures courtesy of Agilent Technologies).

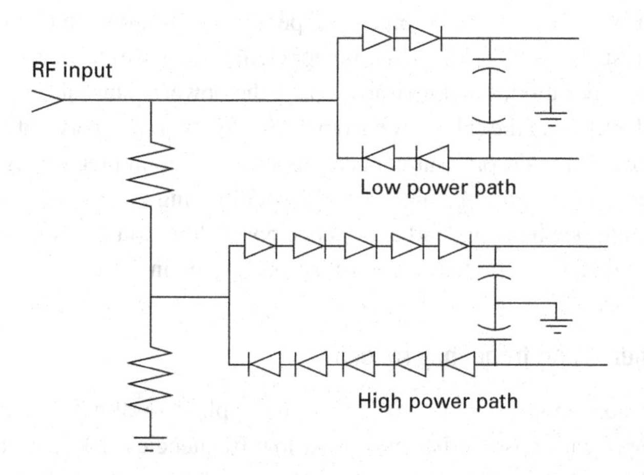

Fig. 6.10 Example of new diode detector schemes (figure courtesy of Agilent Technologies).

for the RF signal. RF voltage is converted to a DC voltage at the diode, and the bypass capacitor is used as a low-pass filter to remove any RF signal that leaks through the diode. This example would work for signals in the square-law region (Figure 6.9(b)) of the diode and would deviate appreciably from that as the power approaches approximately −20 dBm [9].

Digital signal-processing and microwave semiconductor technology have now advanced to the point where dramatically improved performance and capabilities are available for diode power sensing and metering. New diode power sensors are now capable of measuring over a dynamic power range of −70 to +20 dBm (as a reference, thermocouple sensors have a range of approximately −35 to +20 dBm and thermistor sensors have a range of −10 to +10 dBm). This broad range permits the new sensors to be used for CW applications that previously required two separate sensors. A simplified diagram of one of these sensors is shown in Figure 6.10. Here, different diode chains are used for different power levels.

In detecting low power levels of about 100 pW, the diode detector output is about 50 nV. This low-signal level requires a power meter with sophisticated amplifier and chopper circuit design to prevent leakage signals and thermocouple effects from dominating the desired signal. Earlier diode power sensors required additional size and weight to achieve controlled thermal isolation of the diode. A dual-diode configuration balances many of the temperature effects of those highly sensitive circuits and achieves superior drift characteristics [9].

6.3.4 Power meters

The power meter plays an important role in the measurement chain of acquiring a signal, reacting to the signal, and finally producing an output that is understandable. The power meter's function is the last part: producing the results of a measurement in a form that we understand, meaning, in terms of power, a number in dB, dBm, Watts, or something that can be used to signify power. For the bolometric-type sensors, the power meter is the unit that maintains the resistive balance of the thermistor element and uses that process for the calculation of power. For diodes and thermocouples, the power meter takes the voltage from the sensing element and the calibration from the power reference port and calculates power. Note that there are sensors that connect directly to a computer through a USB connector. For these devices, the sensing element, conditioning electronics, and power meter are all built into one housing, and a reference port calibration is not necessary, as the sensor response data is calibrated at the factory and stored in the unit.

6.3.5 Power measurements and frequency ranges

The detector-type concepts described so far can be applied to almost all frequency ranges. Practical application is another matter. At low frequencies, 100 kHz to 1 MHz, bolometric-type detectors have issues with the RF energy leaking through the detector and influencing the power meter and other associated electronics. Thermoelectric-style detectors do not have this problem, as the RF signal terminates in a resistor with no electrical connection to the power meter side. The diode detector circuit also limits the leakage signal. At very high frequencies, above 110 GHz, the waveguide structures become so small that the use of classical thermistor beads and terminating resistors becomes impractical. There are several different techniques that are being used in the 110 GHz to approximately 1 THz range for power measurements. The first technique uses a photo-acoustic method to measure power. "A thin film absorbs incoming THz photons, creating heat. The heat causes expansion of a closed air-cell which is then measured acoustically" [17]. Another technique uses a dry-load, WR-10 (75–110 GHZ) calorimeter operating in an over-moded condition (for more information on dry-load calorimeters, see Section 6.4.2). The detector can be used to measure power in higher-frequency waveguide bands through the use of waveguide tapers or simply attaching the WR-10 port directly to the other waveguide port (if they are compatible) [18]. Finally, there are waveguide-based diode detectors that allow banded power measurements up to 1.7 THz.

While detectors used in the 100 kHz to 50 MHz range are traceable to the voltage and impedance technique, and detectors used in the 50 MHz to 110 GHz range are traceable to calorimetric methods (see Section 6.4), above 110 GHz there are no established methods for tracing power measurements to the SI. There have been efforts to extend the calorimetric methods to higher frequencies. Also, the photo-acoustic and dry-load calorimetric solutions described above try to establish calibration capability through DC-heating substitution-type techniques. These techniques remain unverified, and the problem of establishing traceability in the sub-millimeter ranges still remains.

6.3.6 Power levels and detectors

We have seen that bolometric-style detectors have a limited dynamic range, approximately −10 to +10 dBm, and the modern thermoelectric and diode sensors can have a range of about −70 to +20 dBm. What can you use if you need to measure outside of these levels? For higher power levels there are various forms of microwave wattmeters. These devices measure the higher output directly and generally use rectification, or applying the power to a load and measuring the load's temperature change. They can be either terminating or feed-through devices. Another technique cascades microwave couplers together and after calibration allows standard detectors to measure the coupler-scaled power signal [19]. Due to heating effects caused by the higher levels of power and other error mechanisms, measurements with these detectors tend to have higher uncertainties than those with standard power detectors.

For power levels below −70 dBm, heterodyne detectors/receivers are commonly used. For these devices, the high-frequency signal is mixed with a lower-frequency signal. The resultant signal accurately reproduces the amplitude and phase of the original signal, but at a lower frequency, and simpler low-frequency techniques can be applied to determine the power-signal values providing a larger dynamic range [20].

6.4 Primary power standards

No power detector is perfect. That is to say that power detectors do not indicate in their electronics the exact amount of microwave power being applied to the units. To have an accurate representation of the power incident on the detector, it is necessary to calibrate the detector. While power detectors give readings in terms of the absolute power that is applied to the unit, when we refer to "calibrating" the detector, we do not calibrate it in terms of absolute power; instead, we use the concept of effective efficiency. Figure 6.11 shows a generalized power detector and power meter. Microwave power is applied to the detector. This is represented as P_{inc}, the power incident on the detector. Not all of the incident power reaches the sensing element of the detector. Power is lost through absorption, imperfect conductors, reflections, and other loss mechanisms. P_{net} is the power that is dissipated in the sensing element. Note that while P_{net} takes into account the loss mechanisms of the input section, it does not account for the microwave energy that leaks out of the detector through the DC connections. Finally, after signal processing

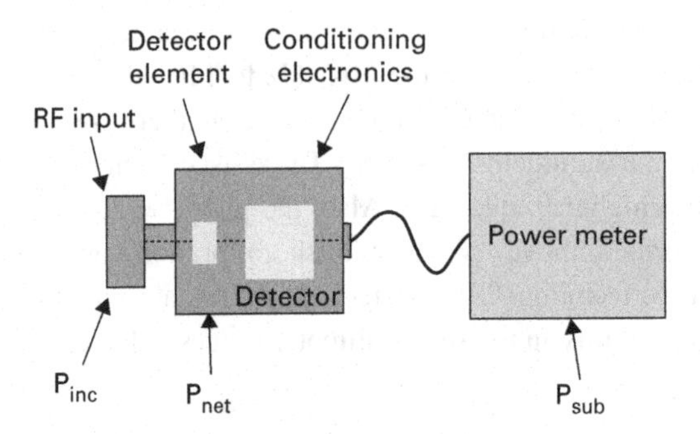

Fig. 6.11 Generalized power detector showing different powers levels that can be determined.

occurs in the power meter electronics, the power meter shows a resultant power level. This is P_{sub}, which is the substituted power (the terms P_{inc}, P_{net}, and P_{sub} were originally defined in terms of bolometric power sensors, but they can be applied to the broader scope of diode and thermoelectric detectors as well).

In an ideal detector, the three powers P_{inc}, P_{net}, and P_{sub} would all be equal. In the real world they are all different. There are two terms that are used to describe the fact that all of the incident power does not reach the sensing element and is not indicated in the substituted power determined by the power meter. The first term is called the effective efficiency (referred to as η or η_e). Effective efficiency is defined as:

$$\eta_e = \frac{P_{sub}}{P_{net}}. \tag{6.5}$$

Thus, effective efficiency η_e for a detector is the ratio of the power determined by the electronics of the power meter P_{sub} to the power dissipated in the sensing element P_{net}. The other term is called the "calibration factor" (also called cal factor or Cf). This is defined as:

$$Cf = \frac{P_{sub}}{P_{inc}} = \eta_e\left(1 - |\Gamma|^2\right). \tag{6.6}$$

The calibration factor is the ratio of the substituted power to the power incident on the detector. This can also be expressed in terms of the effective efficiency and the reflection coefficient of the detector (Γ) as seen in (6.6). Another way of looking at the cal factor is that it relates the substituted power in the detector to the power incident on the detector and takes into account the fact that there is a difference between the reflection coefficient of the detector and the reflection coefficient of the port where the detector is connected. The concept of measurement error due to differences in reflection coefficients is termed the mismatch factor or mismatch correction and is covered in more detail later in this chapter. In the definition above of cal factor, it is assumed that the port to which the detector is connected is non-reflecting.

We see that to make an accurate power measurement, we need to know the effective efficiency (or *CF*) of the detector as well as its reflection coefficient. So how is η_e for a detector determined? In general, η_e is determined through a transfer system that is ultimately traceable to a primary standard for microwave power. The most widely used primary standard is the calorimeter. The calorimeter essentially works by measuring both the bolometric and calorimetric powers (electrical and thermal, respectively) simultaneously and comparing the results. There are several different forms of calorimeters. The "microcalorimeter" and the "dry load calorimeter" are described here, as they account for a large majority of systems used. Good references for information about calorimeters can be found in [11, 13, 14]. An alternative approach for low frequencies, below 100 MHz, using AC voltage and impedance techniques, is also described.

6.4.1 The microcalorimeter

The microcalorimeter (Figure 6.12) measures the temperature rise of the bolometer detector connected to it under different conditions of applied microwave power and DC bias. In the coaxial microcalorimeter, the temperature increase of the detector is measured with a thermopile. The thermopile is connected between the detector being measured and a thermal reference; this is usually a mass with known thermal properties and has roughly

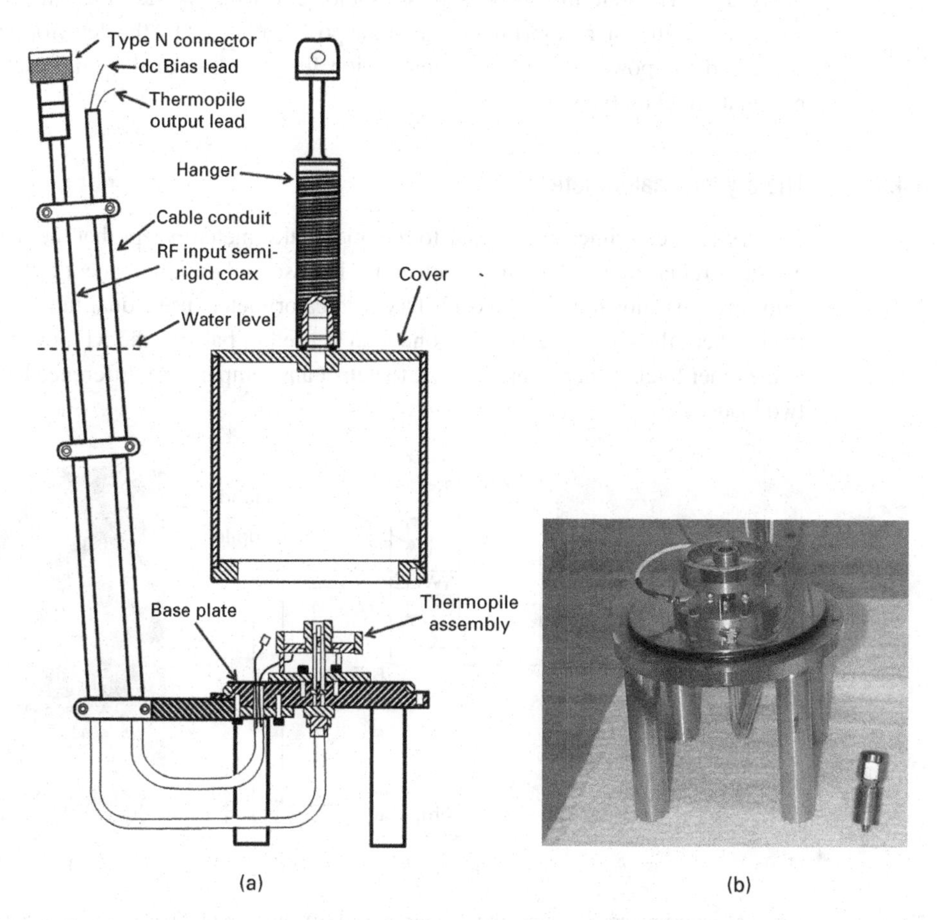

Fig. 6.12 The microcalorimeter. (a) overall descriptive diagram and (b) photograph of Type-N microcalorimeter and a primary transfer standard.

the same thermal mass as the detector. During the measurement, the microcalorimeter is kept in a thermally stable environment to minimize the effect of external temperature changes. A commonly used technique is to immerse the calorimeter in a water-tight housing, in a stable temperature-controlled water bath. To determine the η_e of a detector, measurements are made at each frequency of interest, of the power meter and thermopile output voltages (V_1 and e_1) with only DC applied to the detector, and then again (V_2 and e_2) with both RF and DC applied. The effective efficiency η_e is calculated at each frequency using (6.7) [13]

$$\eta_e = g \frac{1 - \left(\frac{V_2}{V_1}\right)^2}{\frac{e_2}{e_1} - \left(\frac{V_2}{V_1}\right)^2}. \tag{6.7}$$

The term g is a frequency-dependent correction factor for the microcalorimeter-bolometer detector combination, which is also known as the calorimetric equivalence correction. The uncertainty in the measurement is determined primarily by the uncertainty in g. The determination of g is a major effort that is described in [14]. A real advantage of the microcalorimeter is that the power reflected by the detector being measured and the power lost in the transmission lines leading to the calorimeter have a minimal effect on the results.

6.4.2 The dry load calorimeter

The dry load calorimeter is similar to the microcalorimeter except that the thermal reference is replaced by a "dummy" detector. The essential design concept of the dry load calorimeter is shown in Figure 6.13. Dry load calorimeters use a dual input to identical loads where the DC power biasing one load is used to balance the RF power absorbed in the other load. A thermopile is used to detect any temperature difference between the two loads.

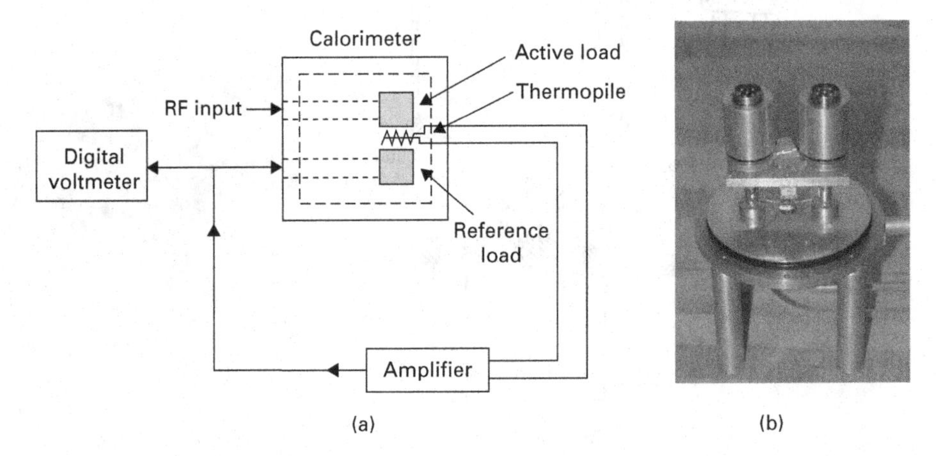

Fig. 6.13 Dry load calorimeters. (a) conceptual diagram and (b) photograph of a system that can be used as a dry load calorimeter with two power detectors attached.

The two loads should see the same external environmental variations and a these should cancel out. Thus, there is no need for as extreme environmental controls as with the microcalorimeter. In microcalorimeters, the effective efficiency of an inserted bolometer mount, which functions as the calorimeter load, is measured. After calibration, this bolometer mount is used as a secondary standard for power measurements. In the case of dry load calorimeters, the calorimeter itself functions as the calorimetric load, and its effective efficiency is determined by measurements and theoretical analysis. Secondary power standards are calibrated using a stable RF generator system by comparing their response with that of the dry load calorimeter.

6.4.3 Voltage and impedance technique

Calorimeter techniques are generally used above 50 MHz. For calibrating power detectors below 50 MHz, the voltage and impedance technique can be used.

If a thermistor detector with an associated power meter and a thermal voltage converter is connected to two sides of a tee, and DC and RF power is supplied through a switch to the third side of the tee (Figure 6.14), the effective efficiency of the thermistor mount can be determined from the electrical parameters of the tee, the thermal voltage converter standard, and the reflection coefficient of the thermistor mount [15, 16]. From before, the effective efficiency of a thermistor mount is defined as

$$\eta_e = \frac{P_{sub}}{P_{RF}}, \tag{6.8}$$

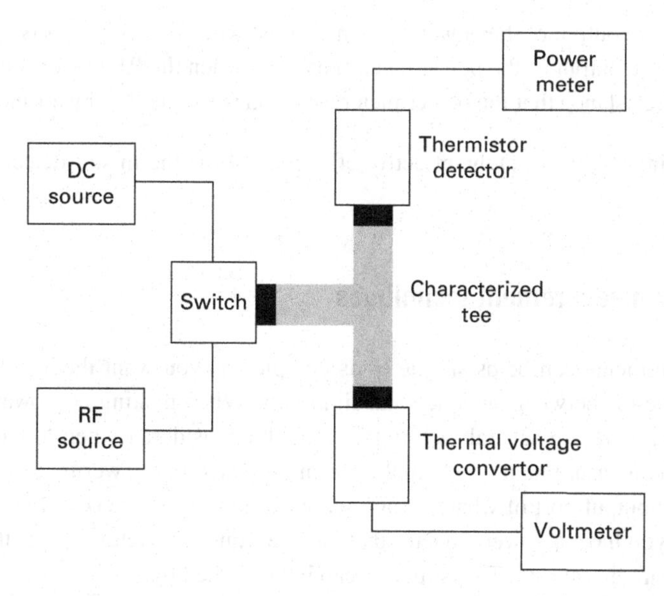

Fig. 6.14 Equipment setup for determining the effective efficiency of a thermistor detector with a thermal voltage convertor.

where P_{sub} is the DC substituted power determined from the electronics of the detector power meter and P_{RF} is the net RF power delivered to the thermistor mount. For the voltage and impedance technique, the RF power delivered to the mount is determined from the following equation:

$$P_{RF} = \left(V_{dc} * C_f * \frac{M_u}{M_S} \right)^2 * Re\left(\frac{1}{Z^*_{par}} \right). \tag{6.9}$$

V_{dc} is the average of the absolute values of the plus and minus DC voltage supplied when the thermal voltage converter output is identical to the output when RF voltage is supplied to the system.
C_f is the correction factor for the thermal voltage converter standard.
M_u and M_s are the mismatch factors for the two sides of the tee.
Z^*_{par} is the complex conjugate of the impedance at the reference plane of the thermistor mount.
Z_{par} can be calculated from the reflection coefficient of the detector using

$$Z_{par} = Z_o \frac{1 + \Gamma_{Det}}{1 - \Gamma_{Det}}, \tag{6.10}$$

where Z_o is the characteristic impedance of the transmission medium (in most cases 50 Ω) and Γ_{Det} is the complex reflection coefficient of the thermistor detector.

The DC substituted power of the thermistor mount is determined from the power meter as

$$P_{sub} = \frac{V_{off}^2 - V_{on}^2}{R_{th}}. \tag{6.11}$$

V_{off} is the voltage output of the power meter measured when the RF power is turned off.
V_{on} is the voltage output of the power meter measured when the RF power is turned on.
R_{th} is the DC resistance that the power meter establishes for the thermistor mount.

By combining (6.8)–(6.11) the effective efficiency of the thermistor detector can be determined.

6.5 Basic power measurement techniques

Power measurements can be as simple or as complex as you want them to be. There is a direct trade-off between accuracy and simplicity. When making microwave power measurements, there are several factors that must be considered when looking at the accuracy of the measurement. The simplest form of measuring power is to connect the detector to the output port of whatever and see what power is there (Figure 6.15). This assumes that you have a power detector that has the same connector type as the output port type you are measuring. This situation can be described by:

$$P_o = P_{sub}, \tag{6.12}$$

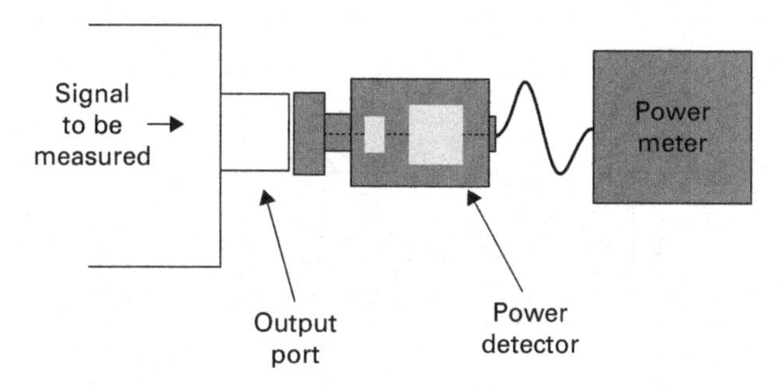

Fig. 6.15 Basic power measurement setup.

where P_o is the power at the output port. If you do this; however, you will have accuracy limitations due to not correcting for the effective efficiency of the detector, not correcting for the mismatch, and not considering the errors in the power meter reference and electronics. There are other additional errors that must be considered. These can include: (a) if the connector of your detector does not match the output port of what you are measuring; (b) uncertainties in the evaluation of the effective efficiency of the power detector; and (c) repeatability.

A more detailed power measurement can be described by:

$$P_o = \frac{P_{sub}}{\eta_{Det} M_{gl}}, \tag{6.13}$$

where P_o is the available power at the port you are measuring that would be delivered to a load, P_{sub} is the substituted power determined from the power meter, M_{gl} is the mismatch factor, and η_{Det} is the effective efficiency of the power detector that is being used for the measurement.

The effective efficiency of the detector can be determined through the use of a calorimetric process or through a transfer process such as a direct comparison power measurement system [21, 22, 23]. The mismatch factor is due to the difference in the reflection coefficient of the output port and the reflection coefficient of the detector.

6.5.1 Mismatch factor

A uniform section of a microwave transmission line is "matched" when it is terminated in such a way that no net reflection of energy occurs. A device that causes a net reflection of energy when connected to this transmission line is termed a "mismatch." The effects of mismatches have long been recognized and evaluated [4, 24, 25, 26, 27]. Mismatch corrections are often neglected when using microwave power detectors. This neglect is often not justified, and large errors result. Returning to the setup to measure power, and now taking reflection coefficients (Γ) into account (Figure 6.16), we can determine the value of the mismatch factor which gives the maximum power

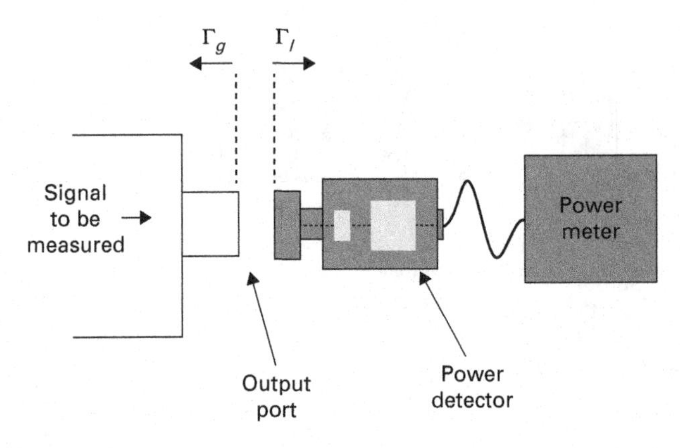

Fig. 6.16 Basic measurement setup with reflection coefficients identified.

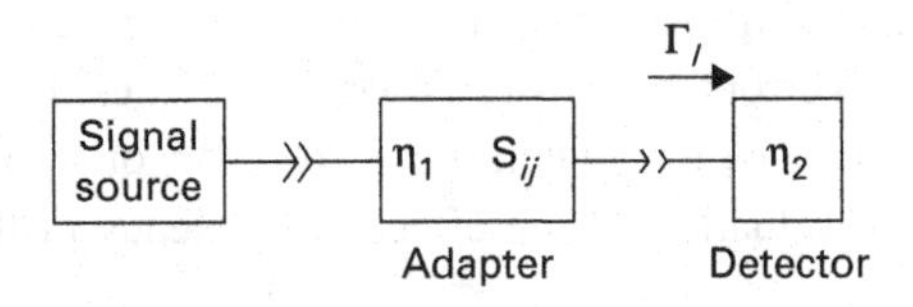

Fig. 6.17 Power measurements using an adapter.

transfer as:

$$M_{gl} = \frac{\left(1 - |\Gamma_g|^2\right)\left(1 - |\Gamma_l|^2\right)}{|1 - \Gamma_g \Gamma_l|^2}; \tag{6.14}$$

Γ_g is the reflection coefficient looking into the port where the detector is connected.

6.5.2 Measuring power through an adapter

Sometimes you end up trying to measure power with a power detector that has a microwave connector type different to the connector type of the port you are trying to measure. The easiest way to make this measurement is simply to place the correct adapter between the output port and the detector. This works well, but the efficiency loss introduced by the adapter must be accounted for to achieve the highest measurement accuracy possible. This situation is seen in Figure 6.17.

The cascaded effective efficiency of the adapter and the detector is

$$\eta = \eta_1 \eta_2, \tag{6.15}$$

where η_1 is the efficiency of the adapter and η_2 is the efficiency of the detector. It can be shown that [28]:

$$\eta_1 = \frac{|S_{12}|^2 \left(1 - |\Gamma_L|^2\right)}{|1 - S_{11}\Gamma_L|^2 - |(S_{12}S_{21} - S_{11}S_{22})\Gamma_L + S_{22}|^2}, \tag{6.16}$$

where S_{ij} are the scattering parameters of the adapter and Γ_L is the reflection coefficient looking into the detector. Both the scattering parameters of the adapter and Γ_L can be determined through measurements. These are most commonly made through the use of a vector network analyzer.

6.5.3 Power meter reference

An inherent characteristic of both diode and thermocouple power measurements is that these measurements are considered open-loop. Thermistor power measurements are inherently more accurate because of their DC-substitution, closed-loop process. The bridge feedback of substituted DC power compensates for differences between thermistor mounts and for drift in the thermistor resistance-power characteristic without recalibration. With diodes and thermocouples, where there is no direct power substitution, sensitivity differences between sensors or drift in the sensitivity due to aging or temperature can result in a different DC output voltage for the same RF power.

All thermocouple and diode power sensors require a power reference to some absolute power level that is traceable to the manufacturer or a national standard to compensate for the open-loop nature of the measurements. Power meters accomplish this power traceability by using a highly stable, internal 50 MHz, 1 mW power reference oscillator. The 1 mW reference power output is near the center of the dynamic range of thermocouple power detectors, but above the range of the sensitive older-style diode detectors. For these detectors, a special 30 dB calibration attenuator, designed for excellent precision at 50 MHz, is generally supplied with the diode detector. When that attenuator is attached to the power reference output on the power meter, the emerging power is 1 μW (−30 dBm) [9].

6.6 Uncertainty considerations

It is very important when making microwave power measurements to understand the accuracy of the measurements that you are making. Also, a detailed look at the uncertainty components for a measurement can help to point out problems in a measurement as well as ways to improve the measurements. Returning to the equation for a general power measurement:

$$P_o = \frac{P_{sub}}{\eta_{Det} M_{gl}}, \tag{6.17}$$

we can use the propagation of uncertainties method and RSS combination of terms to describe the uncertainty in P_o (u_{Po}) [29, 30]. Assuming normal distributions for all uncertainty components (if other than normal distributions are assumed, see [29] and [30] for the proper correction factor to be applied to the component):

$$u_{Po} = \sqrt{\left[\left(\frac{\partial P_o}{\partial \eta_{Det}} d_{\eta Det}\right)^2 + \left(\frac{\partial P_o}{\partial P_{sub}} d_{Psub}\right)^2 + \left(\frac{\partial P_o}{\partial M_{gl}} d_{Mgl}\right)^2\right]} \tag{6.18}$$

where the partial derivatives of (6.17) are determined to be:

$$\frac{\partial P_o}{\partial \eta_{Det}} = \frac{-P_{sub}}{\eta_{Det}^2 M_{gl}},$$
$$\frac{\partial P_o}{\partial P_{sub}} = \frac{1}{\eta_{Det} M_{gl}},$$
$$\frac{\partial P_o}{\partial M_{gl}} = \frac{-P_{sub}}{\eta_{Det} M_{gl}^2}.$$

In (6.18), d_{Psub}, $d\eta_{Det}$, and d_{Mgl} are the uncertainty estimates for the measurements of the substituted power, the effective efficiency of the detector, and the mismatch correction, respectively. These are determined through a thorough examination of the sources of uncertainty for each respective term. Please note that there are other ways of determining the overall uncertainty and the individual components. The method being shown here is primarily for illustration, although it can be used for a complete analysis if desired.

6.6.1 Power meter uncertainty – uncertainty in P_{sub}

Use of thermistors, thermocouples, or diodes requires some form of power meter to be used. The external power meter introduces additional errors that must be accounted for when determining the substituted power. The uncertainty component for the power meter needs to include instrumentation contributions, RF/DC conversion contributions, and contributions from the power meter reference, if it is used. To get the lowest uncertainty using the power reference, the mismatch correction for the reference port/detector pair should be made. There are several good methods for obtaining the reflection coefficient (Γ_g) of the reference port; a good method is described in [31]. Other contributions to the uncertainty, when determining P_{sub}, include general instrumentation errors, power meter zero-setting errors, noise, drift, and power linearity.

6.6.2 η_{Det} uncertainty

The uncertainty in the effective efficiency of the detector comes from the measurement process that was used to determine the value of the effective efficiency. If your detector is a primary transfer standard, then this is the uncertainty that comes from the calorimeter system. If it has been calibrated through a lower-level transfer standard, then this uncertainty component comes from the transfer system used to evaluate the effective efficiency of your detector.

6.6.3 Mismatch uncertainty

References [23] and [26] contain a detailed analysis of the derivation of the uncertainty component due to the mismatch factor. In summation, from the definition of M_{gl}, as shown in (6.14) and repeated here,

$$M_{gl} = \frac{\left(1-|\Gamma_g|^2\right)\left(1-|\Gamma_l|^2\right)}{\left|1-\Gamma_g\Gamma_l\right|^2}. \tag{6.19}$$

The RSS mismatch uncertainty can be found from the separate partial derivatives of (6.19) as:

$$u_{Mgl} = \sqrt{\left[\left(\frac{\partial M_{gl}}{\partial\left|\Gamma_g\right|}d\left|\Gamma_g\right|\right)^2 + \left(\frac{\partial M_{gl}}{\partial\left|\Gamma_l\right|}d\left|\Gamma_e\right|\right)^2 + \left(\frac{\partial M_{gl}}{\partial\phi_{gl}}d\phi_{gl}\right)^2\right]}, \tag{6.20}$$

where $\phi_{gl} = \phi_g + \phi_l$ (ϕ_g and ϕ_l are the phase angles of $\left|\Gamma_g\right|$ and $|\Gamma_l|$, respectively). $d\left|\Gamma_g\right|$, $d|\Gamma_l|$, and $d\phi_{gl}$ are the uncertainty estimates for the magnitude of the output port reflection coefficient, the magnitude of the detector reflection coefficient, and the uncertainty in the reflection coefficient arguments. Explicit expressions for the partial differentials can be determined [23]. The important point here is that the mismatch uncertainty is related to the uncertainties in the determination of the complex reflection coefficients. These uncertainties come from the analysis of the network analyzer system that was used to determine the reflection coefficients.

6.6.4 Adapter uncertainty

If use of an adapter is necessary for a power measurement, then the uncertainty for the cascading of efficiency must be determined. This follows from (6.15) and (6.16). The exact formula for the adapter uncertainty can be determined with the techniques used in the previous sections. It can be seen that the uncertainty is composed of contributions from the determination of the scattering parameters of the adapter. The process of determining these scattering parameters, especially for adapters with different connector types on each end, has fairly high uncertainties and is generally one of the larger contributors to the overall uncertainty of a power measurement.

6.6.5 Device repeatability

The connection repeatability for the detector to the output port must also be considered. The easiest way to estimate this term is to use multiple connections of the detector to the output port. Multiple measurement passes need to be made and the detector should be disconnected and reconnected between each measurement pass. The average of the results for the measurement passes is the final result, and the standard deviation of all of the connections is the uncertainty term as defined by:

$$u_{repeat} = \left[\frac{\sum_{j=1}^{N}\left(\eta_j - \eta_a\right)^2}{N-1}\right]^{1/2}, \tag{6.21}$$

where N is the number of connections/measurements, η_j is the j-th measurement of the effective efficiency, and η_a is the mean of the N measurements.

6.7 Examples

Example 1: The direct comparison power system

While calorimeters are a very good way to evaluate the effective efficiency of a substitution-type detector, they are slow, up to 30 minutes per frequency point, costly, and very hard to evaluate. A much easier approach to determine the effective efficiency of a detector is to use a transfer system. The direct comparison power measurement system is a good example of a simple transfer system [21–23]. The mathematics for the direct comparison system will be developed as an example of the power measurement process.

Figure 6.18 shows a block diagram of a direct comparison system. Overall the system is very simple. A signal generator sends an RF signal into a power splitter/divider that then splits the signal between a monitor detector and either the calibration standard detector or the DUT. The detectors are connected to power meters whose output is connected to a DVM, if needed (depending on the type of power meter), or whose output is read directly by a connected computer through an instrument interface bus.

To calibrate the system, a detector with a known effective efficiency (η_{std}) is connected to the test port of the power splitter. From the known η_{std} of the standard, the reflection coefficient of the standard, the reflection coefficient looking back into the splitter, and the substituted power measured in both the standard and the monitor detector, a value can be determined (K_a) for each measurement frequency that relates the power available at the test port to the power measured in the monitor detector. Mathematically, the calibration is represented by

$$K_a = \frac{P_{dc-std}}{\eta_{std} P_{M-std} M_{gl-std}}, \tag{6.22}$$

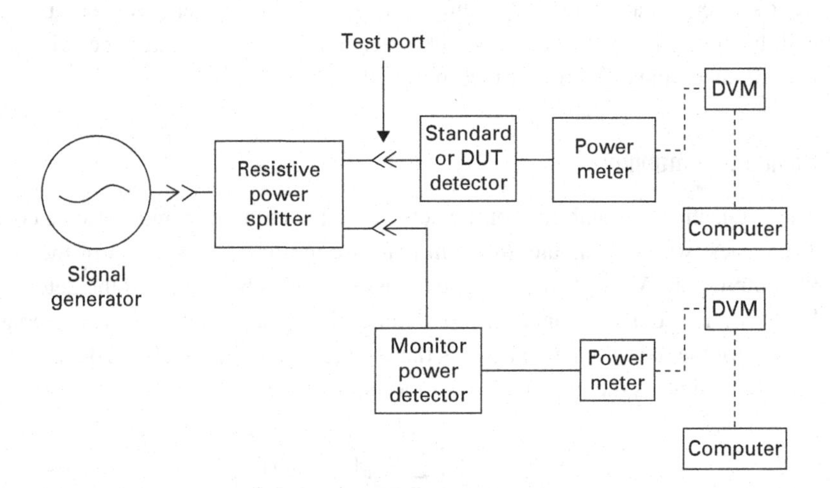

Fig. 6.18 Block diagram of a direct comparison system.

where K_a is the calibration factor for the splitter, P_{dc-std} is the substituted power read from the calibration standard; η_{std} is the known effective efficiency of the calibration standard; P_{M-std} is the substituted power read from the monitor detector during the calibration; and M_{gl-std} is the mismatch factor from the reflection coefficients of the standard and the splitter.

With K_a known, the effective efficiency (η_e) of the DUT can be determined. The process for measuring the η_e of the DUT is the reverse of the calibration process. From the substituted power readings of the monitor detector and the DUT, the reflection coefficient of the DUT, the reflection coefficient looking into the test port, and K_a, the η_e of the DUT can be determined. Note that all of the power readings are used in ratios (monitor detector to the standard or DUT) and are never used as absolute power values. By doing this, any drift of the signal amplitude is negated. The measurement process is represented by

$$\eta_{DUT} = \frac{P_{DC-DUT}}{K_a P_{M-DUT} M_{gl-DUT}}, \tag{6.23}$$

where η_{DUT} is the effective efficiency of the DUT; P_{DC-DUT} is the substituted power read from the DUT; P_{M-DUT} is the substituted power read from the monitor detector during the DUT measurement; and M_{gl-DUT} is the mismatch factor from the reflection coefficients of the DUT and the splitter.

Now, taking (6.23) for η_{DUT} and substituting for K_a from (6.22), we get:

$$\eta_{DUT} = \frac{\eta_{std} P_{DC-DUT} P_{M-std} M_{gl-std}}{P_{DC-std} P_{M-DUT} M_{gl-DUT}}. \tag{6.24}$$

The terms M_{gl-std} and M_{gl-DUT} in (6.24) are expanded as:

$$M_{gl-std} = \frac{\left(1 - |\Gamma_g|^2\right)\left(1 - |\Gamma_{std}|^2\right)}{|1 - \Gamma_g \Gamma_{std}|^2} \tag{6.25}$$

$$M_{gl-DUT} = \frac{\left(1 - |\Gamma_g|^2\right)\left(1 - |\Gamma_{DUT}|^2\right)}{|1 - \Gamma_g \Gamma_{DUT}|^2}, \tag{6.26}$$

with Γ_{std} being the reflection coefficient of the detector with known effective efficiency, Γ_{DUT} the reflection coefficient of the unknown detector, and Γ_g the equivalent source mismatch looking into the splitter.

The reflection coefficient of the standard and the DUT are measured directly with a VNA. There are several techniques for determining the equivalent source mismatch term Γ_g looking into the test port of the splitter. Γ_g is not a true reflection coefficient, as it is determined in such a manner that it is invariant with respect to what is connected to the other ports of the splitter. In terms of the scattering parameters of the splitter (S_{ij}), Γ_g is defined as:

$$\Gamma_g = S_{22} - \frac{S_{12}S_{23}}{S_{13}}. \tag{6.27}$$

Normally the true reflection coefficient of a port of a power splitter is dependent on what is connected to the other ports of the splitter. Good references for the evaluation of Γ_g are [32, 33].

Example 2: Available power calculation

For this example, assume that the power detector you are using for measurement has an effective efficiency of $\eta_{Det} = 0.985$ and this detector is connected to the output of a signal generator. The signal generator's output level is set so that the detector reads 100 mW. The magnitude and angle of the reflection coefficients for the detector and the output port of the generator have been measured to be $\Gamma_{Det} = 0.065$ at 30° and $\Gamma_g = 0.055$ at 67°, respectively.

What is the available power at the output port of the signal generator?

Returning to (6.13) for a general power measurement:

$$P_o = \frac{P_{sub}}{\eta_{Det} M_{gl}}. \tag{6.28}$$

The substituted power reading from the detector needs to be corrected for the effective efficiency of the detector and the mismatch factor, from the signal generator output port and the detector, to determine the actual power at the output port (P_o). Calculating the mismatch factor for maximum power transfer:

$$M_{gl} = \frac{\left(1 - |\Gamma_g|^2\right)\left(1 - |\Gamma_l|^2\right)}{|1 - \Gamma_g \Gamma_l|^2} = \frac{\left(1 - 0.055^2\right)\left(1 - 0.065^2\right)}{\left|1 - \left(0.055\angle 67°\right)\left(0.065\angle 30°\right)\right|^2} = 0.992. \tag{6.29}$$

Thus,

$$P_o = \frac{100\,\text{mW}}{\left(0.985\right)\left(0.992\right)} = 102.34\ \text{mW}. \tag{6.30}$$

Example 3: Power measurement uncertainty calculation

A measurement is made using a Type-N thermoelectric power detector. Calculate both the worst-case uncertainty and the RSS uncertainty for the measurement. For the RSS calculation, use a coverage factor of 2 and assume that all the individual components have a normal distribution.

Assume that the principal error sources are the uncertainty in the calibration of the effective efficiency of the detector, the mismatch uncertainty, the power meter uncertainty (instrumentation and power reference), and the repeatability of the device.

Through evaluation of your system, you have estimated the individual uncertainty terms (with a coverage factor of 2) to be:

$$u_s = \pm 1.8\% \text{ (effective efficiency uncertainty)}$$
$$u_{Mgl} = \pm 2.1\% \text{ (mismatch uncertainty)}$$
$$u_{pm} = \pm 3.1\% \text{ (power meter uncertainty)}$$
$$u_r = \pm 0.8\% \text{ (repeatability uncertainty).}$$

To calculate the worst-case uncertainty, the uncertainty components are linearly summed, giving:

$$U_{WC} = 1.8 + 2.1 + 3.1 + 0.8 = \pm 7.8\%. \tag{6.31}$$

To calculate the RSS uncertainty with a coverage factor of two and normal distributions:

$$U_{RSS} = 2 \times \sqrt{\left(\frac{1.8}{2}\right)^2 + \left(\frac{2.1}{2}\right)^2 + \left(\frac{3.1}{2}\right)^2 + \left(\frac{0.8}{2}\right)^2} = \pm 4.23\%. \tag{6.32}$$

The RSS method is a more realistic method for calculating the uncertainty, as the worst case method assumes that all the components have their maximum value and are in such a direction as to add together constructively.

6.8 Conclusions

Power measurements are very important in describing how microwave circuits work and how information is transferred within and through the circuits. To be able to make a good microwave power measurement, several questions need to be considered: how the data is going to be used, what level of accuracy is necessary, and how the signal is to be measured. The proper choice of microwave power detector type, understanding the way these detectors work, different considerations for making measurements, and the uncertainty components related to the measurements have been described to help answer the above questions. A firm understanding of the basics of power measurements will allow you to more easily make the measurements that support research and product development, and evaluation.

References

[1] D. Halliday and R. Resnick, *Fundamentals of Physics*. Wiley Illustrated, 1974, Chapter 6, Section 7.

[2] R. P. Feynman, R. B. Leighton, and M. Sands, *The Feynman Lectures on Physics*. Pearson Education, Volume I, 1963, pp. 13–23.

[3] "Fundamental of RF and Microwave Power Measurements (Part 1)," Agilent Technologies, Application Note 64-1, 5988-9213EN, April 2003.

[4] G. F. Engen, *Microwave Circuit Theory and Foundations of Microwave Metrology*. London, UK: Peter Peregrinus Ltd., 1992, pp. 103–128.

[5] M. Kinishita, "Atomic microwave power standard based on the Rabi frequency," *IEEE Trans. Instrum. and Meas.*, vol. 60, issue 7, July 2001, pp. 2696–2701.

[6] "Digital Modulation in Communications Systems – An Introduction," Agilent Technologies, Application Note 1298, 5965–7160E, 2001.

[7] G. Breed, "Fundamentals of pulsed and time-gated measurements," *High Frequency Electronics*, Nov. 2010, pp. 52–56.

[8] "Gated Measurements Made Easy," News from Rohde & Schwarz, No. 185, 2005, pp. 15–17.

[9] "Fundamentals of RF and Microwave Power Measurements (Part 2)," Agilent Technologies, Application Note 64-2, 5988-9214EN, July 2006.

[10] "Using Error Vector Magnitude Measurements to Analyze and Troubleshoot Vector-Modulated Signals," Agilent Technologies Product Note 89400-14, 2000.

[11] A. Fantom, *Radio Frequency and Microwave Power Measurement*. London, UK: Peter Peregrinus Ltd., 1990, chapters 2–8.

[12] N. T. Larsen, "A new self-balancing DC-substitution RF power meter," *IEEE Trans. Instrum. and Meas.*, vol. IM-25, no. 4, Dec. 1976, pp. 343–347.

[13] F. R. Clague and P. G. Voris, "Coaxial Reference Standard for Microwave Power," NIST Technical Note 1357, April 1993.

[14] F. R. Clague, "A Calibration Service for Coaxial Reference Standards for Microwave Power," NIST Technical Note 1374, May 1995.

[15] P. S. Filipski, R. F. Clark, and D. C. Paulusse, "Calibration of HF thermal voltage converters using an asymmetrical tee," *IEEE Trans. Instrum. and Meas.*, vol. 50, no. 2, April 2001, pp. 345–348.

[16] M. Halawa and N. Al-Rashid, "Performance of the single junction thermal voltage converter at 1 MHz via equivalent circuit simulation," *Cal Labs Magazine*, Apr./May/Jun. 2009, pp. 40–45.

[17] R. J. Wylde, "Installation and Operating Instructions for the TK TeraHertz Absolute Power Meter System," product manual, Nov. 2002.

[18] "Power Measurement above 110 GHz," VDI Application Note, Oct. 2007.

[19] K. E. Bramall, "Accurate microwave high power measurements using a cascaded coupler method," *Journal of Research of the National Bureau of Standards*, vol. 75c, no. 3 and 4, July–Dec. 1971, pp. 185–192.

[20] "Fundamental of RF and Microwave Power Measurements (Part 4)," Agilent Technologies, Application Note 64-2, 5988-9214EN, Sept. 2008.

[21] R. A. Ginley, "A direct comparison system for measuring radio frequency power (100 kHz to 18 GHz)," *Proceedings of the Measurement Science Conference*, 2006.

[22] J. R. Juroshek, "NIST 0.05–50 GHz direct comparison power calibration system," *Proceedings of the Conference on Precesion Electromagnetic Measurements*, 2000, pp. 166–167.

[23] M. P. Weidman, "Direct comparison transfer of microwave power sensor calibrations," NIST Technical Note 1379, Jan. 1996.

[24] R. W. Beatty and A. C. McPherson, "Mismatch errors in microwave power measurements," *Proceedings I.R.E.*, 1953, vol. 41, Sept. 1953, pp. 1112–1119.

[25] A. Y. Rumsfelt and L. B. Elwell, "Radio frequency power measurements," *Proc. IEEE*, vol. 55, no. 6, June 1967, pp. 837–850.

[26] "Fundamental of RF and Microwave Power Measurements (Part 3), Agilent Technologies, Application Note 64-32, 5988-9215EN, April 2011.

[27] G. F. Engen, "A method of determining the mismatch correction in microwave power measurements," *IEEE Trans. Instrum. Meas.*, vol. IM-17, no. 4, Dec. 1968, pp. 392–395.

[28] D. M. Kerns and R. W. Beatty, *Basic Theory of Waveguide Junctions and Introductory Microwave Network Analysis*. London, UK: Pergamon Press, pp. 42–50.

[29] "Evaluation of Measurement data – Guide to the Expression of Uncertainty in Measurement," BIPM – JCGGM 100:2008, 1995.

[30] B. N. Taylor and C. E. Kuyatt, "Guidelines for Evaluating and Expressing the Uncertainty of NIST Measurement Results", NIST, Technical Note 1297, 1993.

[31] K. Shimaoka, M. Shida, and K. Komiyama, "Source reflection coefficient measurements of the power reference of power meters," *Proceedings of the Conference on Precision Electromagnetic Measurements*, July 2006,

[32] J. R. Juroshek, "A direct calibration method for measuring equivalent source mismatch," *Microwave Journal*, Oct. 1997, pp. 106–118.

[33] H. Jager, "Measurement Method for Determining the Equivalent Reflection Coefficient of Directional Couplers and Power Splitters," Rohde & Schwarz Application Note 08.02-1EZ51_1E.

7 Modular systems for RF and microwave measurements

Jin Bains

7.1 Introduction

One of the major progressions in RF, microwave, and wireless testing is the ability to make fast, flexible, and accurate measurements using software-designed, modular instruments. As RF applications have become increasingly more complex and challenging, legacy test, validation, and design systems, which are generally expensive and rigid, have become increasingly less competitive, and are being replaced by modular, software-designed instruments that are more flexible, extensible, and designed to keep up with the rapid pace of change in the RF and wireless industry. There has been an inflection point in the industry, and the momentum behind software-designed modular instruments is expected to continue accelerating.

Combining Moore's Law with advances in RF technologies and processes has enabled the development of smaller form-factor, lower-cost modular products to match the performance and features of more traditional test products. Modular systems can take full advantage of multi-core processors and make use of the latest FPGA technologies to allow for the greatest measurement flexibility and timing control. These advances have resulted in measurement devices whose core functionality is designed, at least partially, by software written by the system designer(s). Software-designed instruments are mainstream in today's test systems. They allow scientists and engineers to use software to specify pass/fail criteria, test execution flow, signal processing and mathematics, data/logging, and other required elements of test and measurement systems. Software-designed instruments may have vendor-defined elements as part of the system, but unlike purely vendor-defined solutions, software-designed instruments empower engineers with the ability to design their test systems and instruments specifically for their needs.

The expandability of modular systems allows for synchronized, phase-coherent measurements on systems comprising multiple sources or receivers. The increased RF performance of modern modular products has enabled highly accurate measurements with greatly reduced time, space, and cost.

There is an increasing importance in the role of modular measurement systems in radio frequency (RF) and microwave applications. This chapter discusses the fundamentals of modular instruments and reviews some of the salient features of these systems which allow them to be highly effective for many RF and microwave measurement applications.

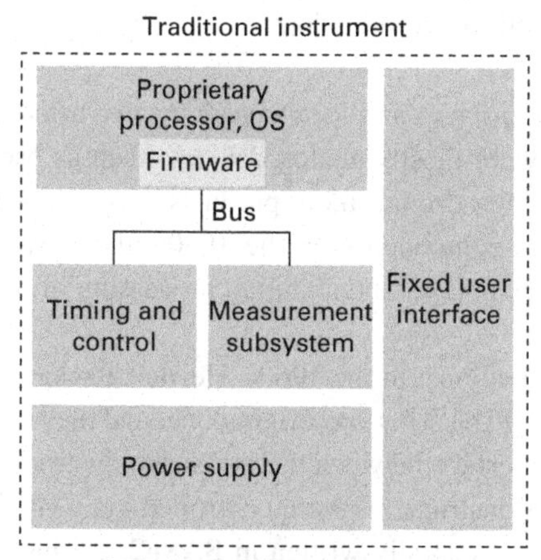

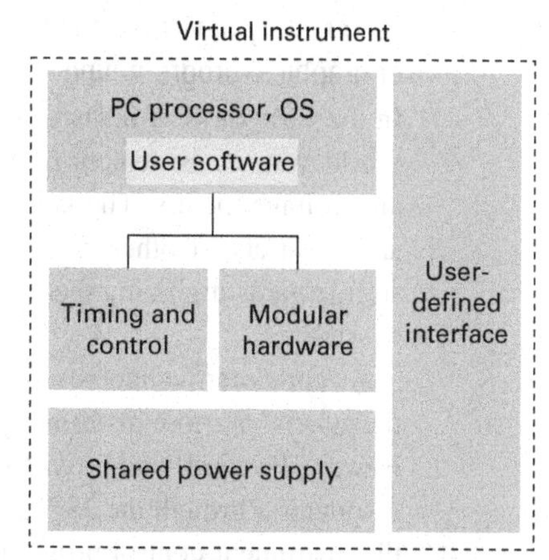

Fig. 7.1 Traditional and virtual instrumentation.

7.1.1 Virtual instrumentation

The rapid adoption of the PC in the last 20 years catalyzed a revolution in instrumentation for test, measurement, and automation. One major development resulting from the omnipresence of the PC is the concept of virtual instrumentation. Virtual instruments offer several benefits to engineers and scientists who require increased productivity, accuracy, and performance [1]. Figure 7.1 illustrates a high-level comparison of a virtual instrument versus a traditional instrument. Note that the virtual instrument is always composed of modular hardware.

"Virtual instruments" is a term that has a long history and it is a broad term used in various industries and many applications. We will briefly explain the various definitions and history of "virtual instruments" or "virtual instrumentation" and then continue the rest of the chapter referring only to "modular instruments," which is a more common term to use in the current era to represent measurement systems composed of modular hardware products. There are several definitions associated with the term virtual instruments [2]. In a general sense, a virtual instrument consists of an industry-standard computer or workstation equipped with powerful application software, cost-effective hardware such as plug-in boards, connected by a high-speed bus, utilizing driver software, which together perform the functions of traditional instruments. Virtual instruments represent a fundamental shift from traditional hardware-centered instrumentation systems to software-centered systems that exploit the computing power, productivity, display, and connectivity capabilities of popular desktop computers and workstations. Although the PC and integrated circuit technology have experienced significant advances in the last two decades, it is software that truly provides the wherewithal to build on this powerful hardware foundation to create virtual instruments, thus providing better ways to innovate and significantly reduce cost. With virtual instruments, engineers and scientists can build measurement and automation systems that suit their needs exactly, instead of being limited by traditional fixed-function instruments.

Brief history of virtual instruments and the introduction of graphical programming

In the early days of instrumentation, prior to and for a decade or two after the Second World War, measurement products were mostly analog devices, such as oscilloscopes and voltage meters. The key technology driving these products was the vacuum tube, and a variety of other purely analog components. In the 1950s, there was a gradual shift towards digitizing the measured signals, so that digital processing of data became possible.

As computer technology became available in the 1960s, Hewlett-Packard developed the general purpose instrument bus (GPIB). This was originally called the HPIB bus (for Hewlett-Packard) and provided an interface between the measuring instrument and the computer. Through the 1970s, virtual instruments were controlled via a GPIB bus, and the instrument control programs were generally written in BASIC. During this period there was a massive expansion in measurement capability provided by test instruments. This expansion was enabled to a large extent by significant progress in integrated circuit technology, which was following a pace of evolution described very accurately by Moore's Law [3].

With further rapid advances in computer technology, and in particular with the advent of the Macintosh, the instrumentation world was prepared for a new paradigm. This came with the emergence of LabVIEW (Laboratory Virtual Instrument Engineering Workbench), developed by National Instruments in 1986 [4]. LabVIEW greatly enhanced the ease of use of instrument products through a graphical user interface, resulting in a friendly, powerful method for control, measurement, and analysis. LabVIEW also allowed customers to extend or add functionality to previously fixed, closed instruments and as such, it started a trend of software being used to design the functions of test systems. In addition to the growth of graphical programming, the performance capabilities of virtual instruments were also enhanced by the rapid evolution of high-speed computer buses, enabling measurement data to be transferred to a computer processor with increasing bandwidth and decreasing latency. All of these factors suggest that the most significant future advances in measurement products are likely to be driven by software technologies.

Virtual instruments versus traditional instruments

Stand-alone traditional instruments, such as oscilloscopes and waveform generators, can be very powerful and expensive, and are generally designed to perform one or more specific tasks defined by the vendor. However, it is generally not possible to extend or customize them. The knobs and buttons on the instrument, the built-in circuitry, and the functions available, are all specific to the nature of the instrument. In addition, special technology and expensive components must be developed to build these instruments, making them very expensive and slow to adapt.

Virtual instruments, by virtue of being PC-based, inherently take advantage of the benefits of the latest technology incorporated into off-the-shelf PCs. For example, virtual instruments are generally significantly faster than boxed instruments because users can always upgrade to the latest desktop PC processors rather than relying on the older processors built inside the boxes. Traditional instruments also frequently lack portability, whereas virtual instruments running on notebooks automatically incorporate their portable nature. Engineers and scientists whose needs, applications, and requirements change very quickly, need flexibility to create their own solutions. You can adapt a virtual instrument to your particular needs without having to replace the entire device because of the application software installed on the PC and the wide range of available plug-in hardware.

7.1.2 Instrumentation standards for modular instruments

Modular instrumentation concepts have evolved over time to deal with increased demand for lower cost and greater lifetime of instruments. Figure 7.2 shows a basic modular instrumentation system, comprising the chassis, embedded controller, high-speed data bus on the chassis backplane, the modular instruments themselves (cards inserted in the chassis), and the graphical user interface.

Various instrumentation standards have been developed to meet the growing requirement for flexible test systems. These instrumentation standards have generally made use of an existing PC bus technology. Some of the modular instrument standards are discussed here.

VXI

VMEbus eXtensions for Instrumentation (VXI) is a standard platform for instrumentation systems based on the VMEbus standard. Besides using the VME bus in the backplane, VXI also implemented timing and synchronization features that were required for instruments. The VXIbus Consortium was formed in 1987 with the intention of defining a multivendor instrument-on-a-card standard. This consortium has defined system-level components required for hardware interoperability. The IEEE officially adopted the VXI specification, IEEE 1155, in March 1993. The VXIplug&play Systems Alliance, founded in September 1993, sought a higher level of system standardization to cover all VXI system components. By focusing on software standardization, the alliance defined standards to make VXI systems easy to integrate and use while maintaining multivendor software interoperability. The success of VXI as an open, multivendor platform is a testament to the value of multivendor standards, and for a period of time, made VXI the platform of choice for open instrumentation systems.

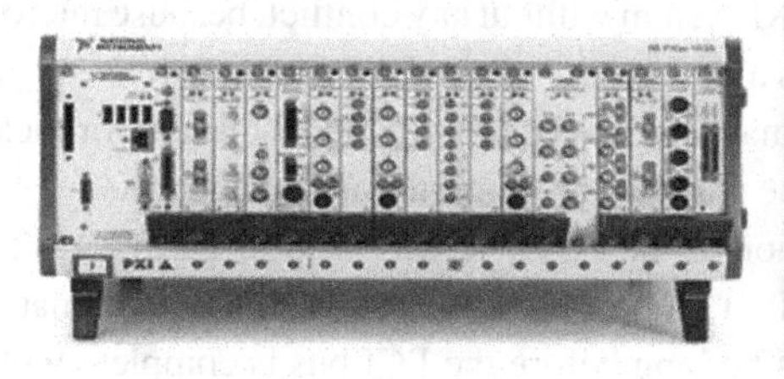

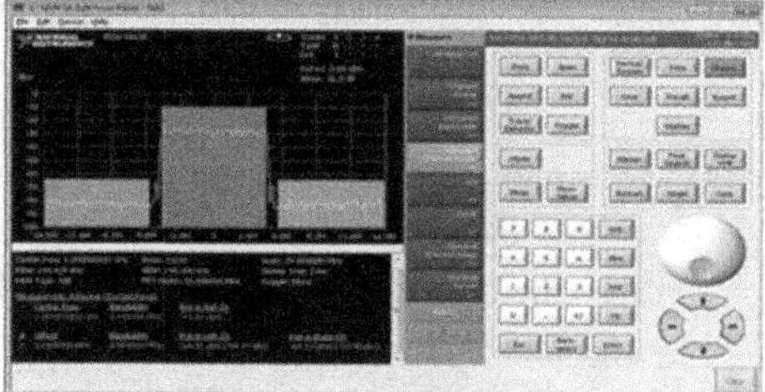

Fig. 7.2 Modular instrumentation system, with graphical user interface.

The demand for an industry-standard instrument-on-a-card architecture was driven by the need for a reduction in the size of rack-and-stack instrumentation systems, tighter timing and synchronization between multiple instruments, and transfer rates faster than the 1 MB/s rate of the 8-bit General Purpose Interface Bus (GPIB). The modular form-factor, high bandwidth, low-latency, and commercial success of the VMEbus made it particularly attractive as an instrumentation platform.

LXI

LAN eXtensions for Instrumentation (LXI) is an instrument control standard based on Local Area Network (LAN) and Ethernet technologies, web interfaces, and IEEE 1588 [5]. LXI offers three levels of synchronization that vendors can choose to implement in their boxes. The LXI Consortium was founded in 2004 and the LXI 1.0 specification was released in September 2005. The LXI Consortium's goals were to increase the interoperability and functionality of Ethernet-based instruments by standardizing common operations and interfaces and to develop, support, and promote the LXI standard.

The need for the LXI standard arose owing to the widespread use of Virtual Private Networks (VPN) and an increase in the number of instruments available on the Internet.

PXI/PXI Express

PCI eXtensions for Instrumentation (PXI) is a rugged, PC-based platform [6]. PXI combines the Peripheral Component Interconnect (PCI) electrical bus with the rugged, modular Eurocard mechanical packaging of CompactPCI and adds specialized synchronization buses and key software features. PXI also adds mechanical, electrical, and software features that define complete systems for test and measurement, data acquisition, and manufacturing applications. These systems are used for applications such as manufacturing test, military and aerospace, machine monitoring, automotive, and industrial test. PXI is currently the most popular and fastest-growing modular instruments form factor.

National Instruments developed and announced the PXI specification in 1997 and launched it in 1998 as an open industry specification to meet the increasing demand for complex instrumentation systems. Currently, PXI is governed by the PXI Systems Alliance (PXISA), a group of more than 70 companies that are chartered to promote the standard, ensure interoperability, and maintain the PXI specification. Because PXI is an open specification, any vendor can build PXI products. CompactPCI, the standard regulated by the PCI Industrial Computer Manufacturers Group (PICMG), and PXI modules can reside in the same PXI system without any conflict, because interoperability between CompactPCI and PXI is a key feature of the PXI specification.

The demand for a high-performance, low-cost deployment solution for measurement and automation systems paved the way for developing this specification.

PCI Express, the next-generation of the PCI bus, was introduced in 2004 to increase the measurement throughput of PXI. Today, most PCs ship with a combination of PCI and PCI Express slots. It will not be long before the PCI bus is completely phased out. The integration of PCI Express signaling into the PXI standard increases the backplane bandwidth from 132 MB/s to 6 GB/s, an improvement of 45 times. The PXIe specification also enhances PXI timing and synchronization features by incorporating a 100 MHz differential reference clock and differential trigger lines. The PXI Express specification adds these features to PXI while maintaining backwards compatibility.

7.1.3 PXI architecture

There is a growing emphasis on increasing measurement speed and improving the flexibility and performance of RF instruments. For these reasons, PXI and PXIe (used interchangeably here, since the basic architecture is identical) are considered the solution for next generation RF and microwave test systems. The following section discusses the PXI hardware and software architectures.

Hardware architecture

PXI systems are composed of three basic components – chassis, system controller, and peripheral modules, as shown in Figure 7.3.

PXI chassis

The chassis provides rugged and modular packaging for the system. Chassis are generally available in 4-, 6-, 8-, 14-, and 18-slot 3U and 6U sizes. Some chassis include AC and DC power supplies and integrated signal conditioning.

PXI controllers

Most PXI chassis contain a system controller slot as the leftmost slot of the chassis. There are a few options when determining the best system controller for an application, including remote controllers from a desktop, workstation, server, or laptop computer and high-performance embedded controllers with either a general purpose (OS) such as Windows or a real-time OS.

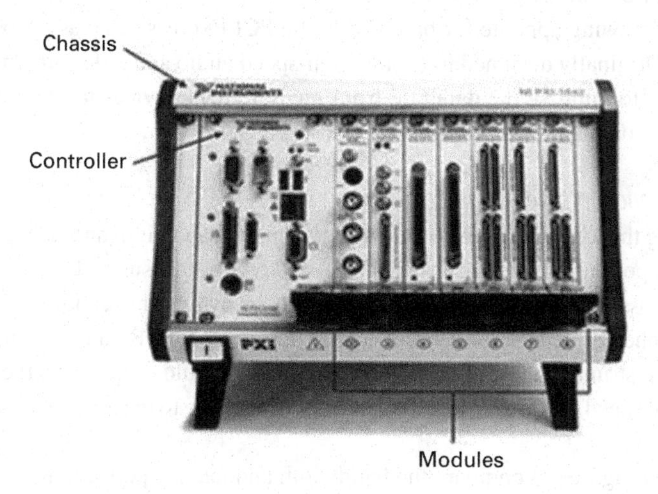

Fig. 7.3 Components of a PXI system.

PXI peripheral modules

Currently, PXI provides the industry's highest-bandwidth and lowest-latency bus with modular I/O for applications ranging from high-resolution DC to 26.5 GHz RF. While the smaller, 3U height instruments are by far the most widely used, because PXI is compatible with CompactPCI, there is an option to use 3U or 6U CompactPCI modules in a PXI or PXI Express system. Additionally, CardBus/PCMCIA and PCI Mezzanine Card (PMC) cards, among others, can be installed in PXI systems using carrier modules.

Software architecture

The development and operation of Windows-based PXI systems is no different from that of a standard Windows-based PC. Additionally, because the PXI backplane uses the industry-standard PCI bus, writing software to communicate with PXI modules is, in most cases, identical to that of PCI boards. Therefore, there is no need to rewrite existing application software, example code, and programming techniques when moving software between PC-based and PXI-based systems.

PXI Express systems also provide software compatibility to help preserve the investment in existing software. Because PCI Express uses the same driver and OS model as PCI, the specification guarantees that there is complete software compatibility among PCI-based systems. As a result, neither vendors nor customers need to change driver or application software for PCI Express-based systems.

As an alternative to Windows-based systems, a real-time software architecture can be used for time-critical applications requiring deterministic loop rates and headless operation (no keyboard, mouse, or monitor). Real-time operating systems help prioritize tasks so that the most critical task always takes control of the processor when needed. With this feature, an application can be programmed with predictable results and reduced jitter. The PXI specification presents software frameworks for PXI systems based on Microsoft Windows operating systems.

Understanding PC technologies

To address the growing appetite for bandwidth, the PCI Express bus was introduced by Intel in 2004. Originally designed to enable high-speed audio and video streaming, PCI Express is used to improve the data rate from measurement devices to PC memory by up to 30 times more than the traditional PCI bus.

PCI and PCIe bandwidth versus latency

For considering the technical merits of alternative buses, bandwidth and latency are two of the most important bus characteristics [8]. Bandwidth measures the rate at which data is sent across the bus, often represented in bits or bytes per second, while latency measures the inherent delay in data transmission across the bus. By analogy, if we were to compare an instrumentation bus to a road, bandwidth would correlate to the width of the road and the speed of travel, while latency would correlate to the number of stoplights in the road.

Table 7.1 and Figure 7.4 compare the bandwidth and latency performance of various instrument buses. An ideal instrument would have a very high bandwidth with very low

Table 7.1 Comparison of instrument control buses

	Maximum Bandwidth (MB/s)	Dedicated Bandwidth	Latency (μs)	Distance (m) (with no repeaters)
GPIB	1.8 (488.1) 8(HS488)	No	30	20
USB	60 (USB HS)	No	1000 (USB 1.1) 125 (USB 2.0)	5
LAN/LXI	12.5 (Fast) 125 (Gigabit)	No	1000 (Fast) 1000 (Gigabit)	100
PXI/PXIe	132 (PXI) 4,000 (PXIe)	No (PXI) Yes (PXIe)	0.7	Internal PC bus

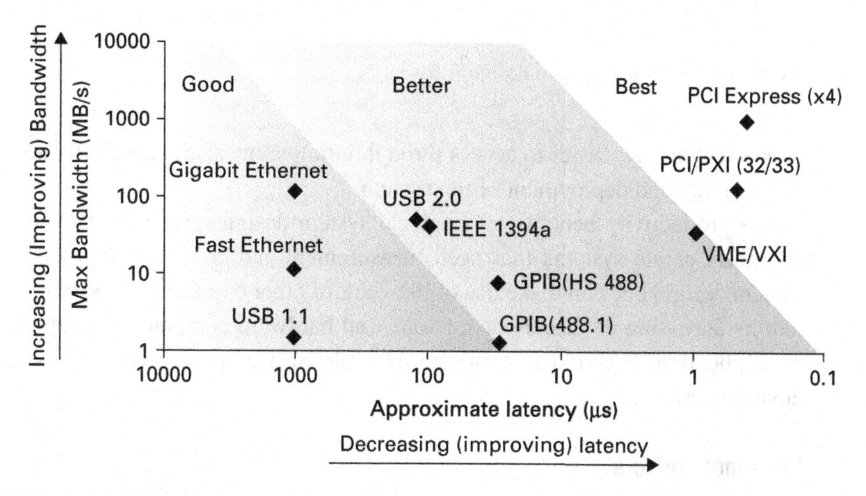

Fig. 7.4 Bandwidth versus latency.

latency. A bus with high bandwidth can transmit more data in a given period than a bus with low bandwidth [7]. A bus with low latency introduces less of a delay between the time data was transmitted from one end and processed at the other end. Bandwidth is important because it determines whether data can be sent as fast as it is acquired and how much onboard memory instruments will need. Latency, while less observable, has a direct impact on applications such as Digital Multimeter (DMM) measurements, switching, and instrument configuration, because it affects how quickly a command sent from one node on the bus, such as the PC controller, arrives at and is processed at another node, such as the instrument.

7.1.4 The role of graphical system design software

Graphical system design, using an open platform of productive software and reconfigurable hardware, shortens the integration cycle for new technology and functionality. It allows engineers to visualize and implement systems faster because the platform

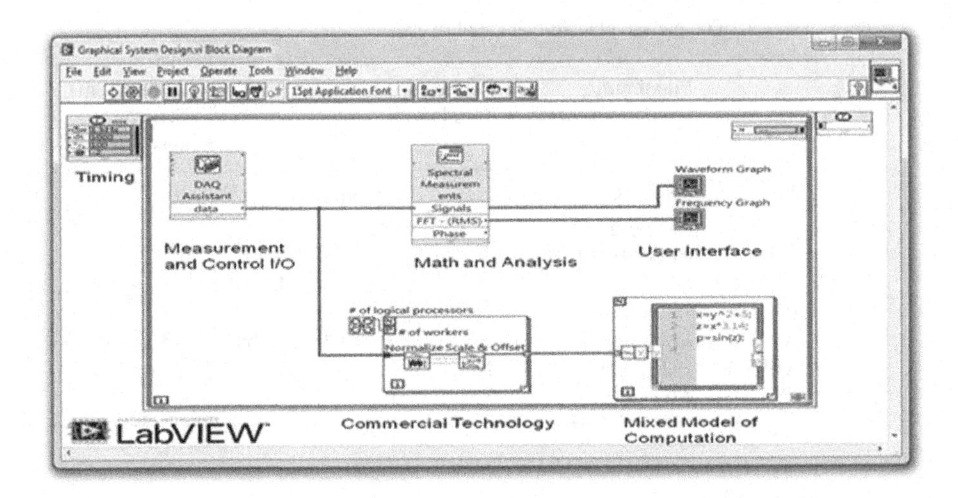

Fig. 7.5 Example of graphical system design software.

makes technology easier to access through intuitive interfaces to accelerate the design, prototyping, and deployment of the system.

The productivity benefits of graphical system design span every industry in which engineers create systems that need measurement and control. When using graphical system design, one can make use of the work of other engineers in the platform ecosystem by accessing thousands of software and hardware components to efficiently build an application. Figure 7.5 shows an example of a graphical system design software environment.

Instrument drivers

Instrument drivers are an integral component in modern automated test systems. They perform the actual communication and control of the instrument hardware in the system, and provide a high-level and easy-to-use programming model that turns complex instrument measurement capabilities into simple software function calls. Instrument drivers are used to simplify instrument control and reduce test program development.

Interchangeable Virtual Instruments (IVI) is a standard for instrument driver software technology. IVI builds on the VXI plug&play specifications and incorporates new features that address issues such as system performance, development flexibility, and instrument interchangeability. IVI drivers also take advantage of the power of the VISA I/O library defined by VXI plug&play to seamlessly communicate with instruments across different I/O buses such as GPIB, VXI, PXI, Serial, Ethernet, and USB.

Analysis routines

Graphical System Design software allows for integrated analysis routines. In general, raw data must be processed before it can be used for collecting information. Signal processing involves analysis, interpretation, and manipulation of signals. Analysis routines give a proper procedure for the analysis process. A very well-tested and correct analysis routine helps to save development time. Analysis routines help to improve efficiency and iterative correlation input variables to process. Therefore, the availability of well-defined and tested analysis routines saves time in creating them.

7.1.5 Architecture of RF modular instruments

RF modular instruments provide all the capabilities of traditional stand-alone instruments, but go well beyond the limitations inherent in the traditional paradigm. The modular instrument breaks down the RF instrument into its key measurement blocks, disaggregating the overall system and thereby opening the way for a lower-cost, more flexible, highly expandable measurement system. As part of the modular instrumentation system, each RF modular product also can take full advantage of the high-speed data buses as well as the triggering and clocking functions available.

To demonstrate the distinction between a traditional instrument and a modular instrument, we will take a closer look at both a vector signal analyzer (VSA) and a vector signal generator (VSG), although we could have just as easily looked at a vector network analyzer (VNA) or any other RF instrument.

Vector signal analyzer

A vector signal analyzer is a measurement device that can measure signals at RF and microwave frequencies. The VSA converts the signal from a time domain representation to a frequency domain representation. The time domain representation gives the amplitudes of the signal at the instants of time during which it was sampled. However, in many cases you need to know the frequency content of a signal rather than the amplitudes of the individual samples. In the frequency domain, you can separate conceptually the sine waves that add to form the complex time domain signal, as shown in Figure 7.6.

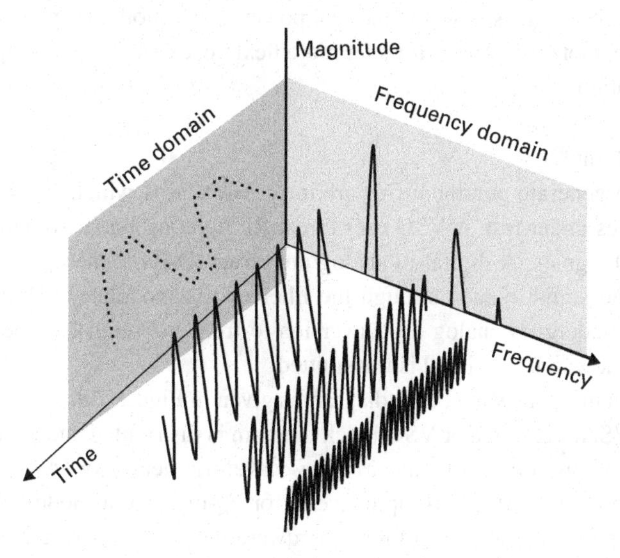

Fig. 7.6 Frequency and time domain measurements.

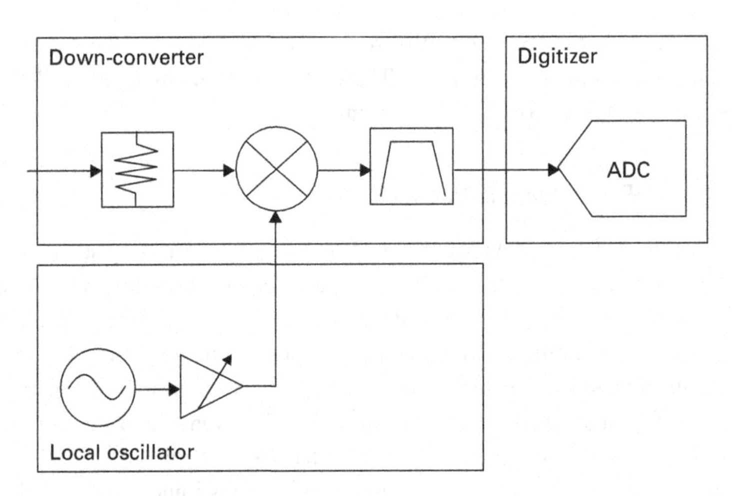

Fig. 7.7 Modular vector signal analyzer block diagram.

An example of a modular VSA block diagram is given in Figure 7.7.

In many ways the block diagram of a modular VSA is identical to that of a traditional stand-alone VSA. The signal to be analyzed passes through the attenuator. The attenuators reduce the gain of the signal. The mixer converts the incoming RF signal to a low-frequency IF signal from the attenuator and local oscillator and passes to the IF amplifier. The signal is amplified by the IF amplifier tuned to the frequency of the down-converted signal, and is then filtered by the IF filter. The IF band pass filter removes the unwanted signals and retains the desired IF signal only. The digitizer then converts the signal to digital form. Finally, the signal data is displayed on the host computer.

However, the key distinction with the modular VSA is that the VSA is implemented in three different modules, one for the down-converter, one for the local oscillator (LO), and one for the digitizer. This is where the real power of the modular approach comes into play, since the door is now open for a far more flexible, expandable, and upgradable measurement solution.

Vector signal generator

VSGs are used to generate continuous or arbitrary waveforms which can be used as an input for devices under test. A VSG uses direct RF up-conversion from differential baseband I and Q signals. A digital-to-analog converter (DAC) generates baseband I and Q signals. The signal passes through the filter and is modulated. The baseband modulated signal undergoes analog up-conversion to frequency-translate the signal to the RF frequency at which the signal is transmitted.

An example of a modular VSG block diagram is given in Figure 7.8.

Similar to the VSA, the modular VSG block diagram is distributed amongst multiple core modules. There are three parts: the arbitrary waveform generator (AWG) module, the local oscillator (LO), and the RF up-converter or IQ-modulator module. The VSG instrument driver software operates all three hardware modules as a single instrument by handling all module programming and interaction. This implementation allows the

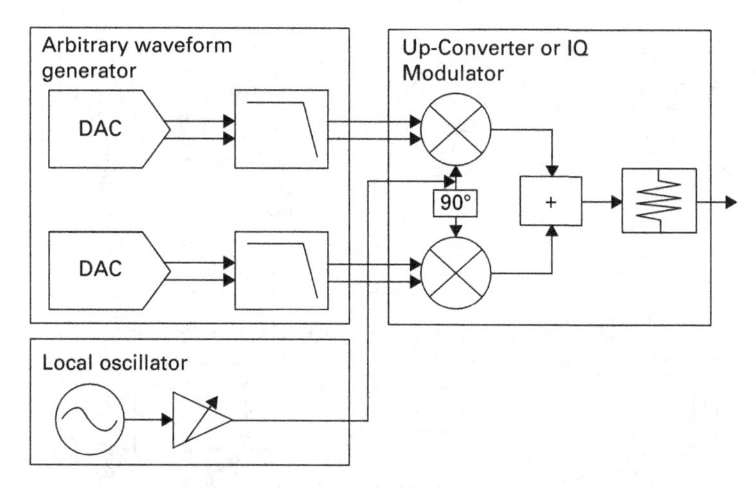

Fig. 7.8 Modular vector signal generator block diagram.

user the flexibility to upgrade any module within the overall VSG system to change or improve the performance of the system.

7.2 Understanding software-designed systems

The term "software-designed systems" means not only using software to measure data, analyze it, and generate results, but also using software to design, prototype, and deploy measurement systems. Using software for measurement systems enables the user to design systems that generate and analyze RF signal measurements four times faster than other modular instrumentation solutions and more than ten times faster than traditional box instruments. Because the solution is software-designed, engineers can easily configure the same measurements used in hardware to fulfill multiple functionalities, which in turn reduces the cost of testing. Another advantage of using software is that the hardware has fixed functionality and vendor defined measurements, while software enables the user to design the measurements required and also has an integrated software GUI.

It is clear that in the modern world of increased abstraction and the need for increasingly flexible systems, a software-designed measurement system has significant advantages over a traditional hardware-centric system.

7.2.1 Measurement speed

To get clear benefits such as productivity and faster time-to-market, measurements have to be performed faster. PXI modular instruments already perform RF measurements significantly faster (around 3 to 10 times) than traditional instruments. This section explains the factors that affect RF measurement time using PXI and how to make the system work faster.

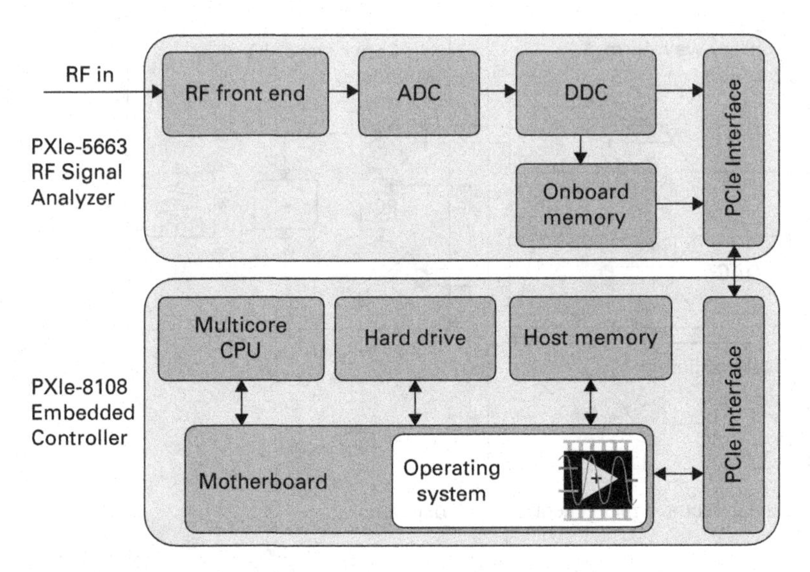

Fig. 7.9 Software-designed PXI instrument using a multicore CPU.

From RF input signal to measurement result

Typically, a software-designed RF measurement system consists of a PXI RF signal analyzer along with a PXI chassis and controller. While the simplest configuration consists of only the analyzer, many PXI RF measurement systems contain additional modules for RF signal generation and mixed signal or DC input and output. Figure 7.9 illustrates the block diagram of a basic PXI RF signal analyzer, such as the NI PXIe-5663.

Measurement begins when an RF vector signal analyzer (VSA) starts to collect an IF signal from the RF front end. The IF signal is processed in the analyzer's digital down-converter (DDC), and IQ samples are generated, which are stored in the onboard memory. A PXI controller then fetches the IQ samples from the VSA's onboard memory through a PCI or PCIe data bus. Once the IQ samples are in the host controller's memory, a software-designed measurement algorithm produces the measurement result. By using different measurement algorithms, PC-based measurement systems can compute a wide range of time and frequency domain measurements including: power, frequency, spectral mask margin, error vector magnitude (EVM), and many others.

The different stages of the measurement process can be evaluated and documented as follows: Step 1: Acquires IQ samples; Step 2: Transfer IQ samples to host PC; Step 3: Apply measurement algorithm.

During each step of the measurement process, various factors affect the overall measurement time. Let us understand and evaluate how each factor affects the measurement time of a typical spectral measurement. Assuming a frequency span of 50 MHz and a resolution bandwidth (RBW) of 30 kHz, we can observe which step takes the longest time. The full analysis is shown in Figure 7.10.

During step 1, factors such as software latency and the signal analyzer's internal acquisition engine are the biggest sources of delay. Because these are minor contributors

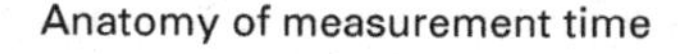

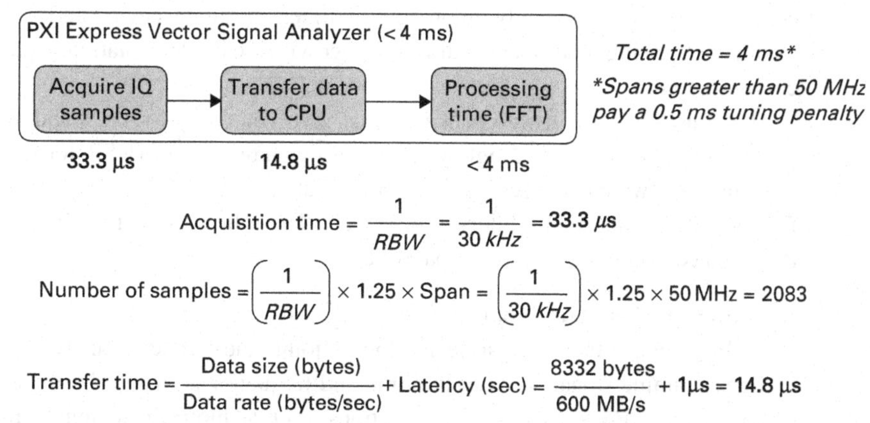

Fig. 7.10 Measurement time for a modular VSA.

to measurement time, a typical RF VSA can produce IQ samples within 30–40 µs of the time that the acquisition was initiated in the software.

In step 2, the data bus bandwidth is the biggest contributor to measurement delays. Step 3, which involves execution of the measurement algorithm, is fundamentally the largest contributor to overall measurement time. In fact, one way to evaluate the influence of signal processing time is to compare the results from steps 1 and 2 to the overall measurement time. Using a PXIe controller, the 50 MHz spectral mask measurement (30 kHz RBW) can typically be performed in 2.8 ms. Given that steps 1 and 2 in the measurement process add up to a maximum of 90 µs, we note that for this example, signal processing accounts for 97% (or more than 2.71 ms) of the total measurement time.

The following section deals with the factors that affect the speed of measurement.

Role of the CPU

By observing the effect of the CPU on the overall measurement time, we can determine that signal processing time is the bottleneck for measurement time. If all the calculations are correct, the measurement time should improve with the use of a more powerful host PC (CPU). Additionally, more intensive measurements take longer to perform.

While many factors, from RBW to number of symbols, affect the RF measurement time, the easiest method to reduce measurement time without affecting measurement quality is to use the fastest CPU available. The availability of a high-performance CPU on a PXI measurement system is the main contributor to the speed of PXI measurements over traditional instrumentation.

Signal processing and parallelism

The need to reduce the cost of wireless handsets continues to increase as wireless devices increase in complexity and volume. Fortunately, multicore processors provide today's software-designed instruments with a high-performance test solution. Multi-core processors significantly improve test times in single device-under test (DUT) testing. Processing time can be reduced further with the implementation of parallel DUT configurations.

There are several software packages that give you ready-to-run, stand-alone signal processing capabilities with high-level digital signal processing (DSP) tools and utilities [9]. These software packages have functions that are designed for performing advanced DSP and designing digital filters interactively. The following section explains some of the advanced signal processing capabilities:

1. **Joint time-frequency analysis**
   Unlike conventional analysis technologies, joint time-frequency analysis (JTFA) routines examine signals in both the time and frequency domains simultaneously. You can apply JTFA in almost all applications, such as biomedical signals, radar image processing, vibration analysis, machine testing, and dynamic signal analysis.
2. **Wavelet analysis**
   Wavelets are a relatively new signal processing method. A wavelet transform is almost always implemented as a bank of filters that decomposes a signal into multiple signal bands. Wavelet transform separates and retains the signal features in one or more of these sub-bands. Thus, you can easily extract signal features. In many cases, a wavelet transform outperforms the conventional FFT in feature extraction and noise reduction. Because the wavelet transform can extract signal features, it is used for data compression, echo detection, pattern recognition, edge detection, cancellation, speech recognition, texture analysis, and image compression.
3. **Super-resolution spectral analysis**
   FFT is the primary tool for spectral analysis. For high-resolution spectra, FFT-based methods need a large number of samples. However, in many cases, the data set is limited because of a genuine lack of data or because users need to ensure that the spectral characteristics of the signal do not change over the duration of the data record. For cases where the number of data samples is limited, a model-based analysis can be used to determine spectral characteristics. By using spectral analysis, a suitable signal model is assumed and the coefficients of the model are determined. Based on this model, the application can predict the missing points in the given finite data set to achieve high-resolution spectra. In addition, model-based methods can be used to estimate the amplitude, phase, damping factor, and frequency of damped sinusoids. Additionally, super-resolution spectral analysis can be used in diverse applications including biomedical research, economics, geophysics, noise, vibration, and speech analysis.

Parallel programming techniques

The parallel execution structure of programming languages and the test executive introduces a variety of potential techniques to reduce overall test time. Some test executive software actually contains built-in functionality to allow engineers to more easily configure parallel device testing. Programming techniques such as data and task parallelism

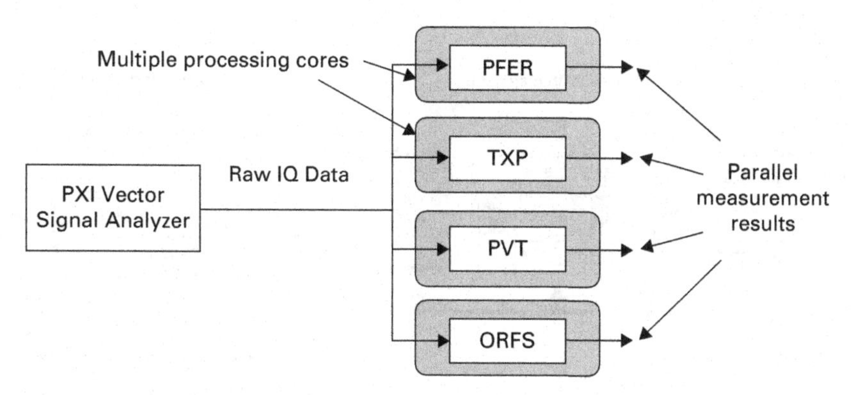

Fig. 7.11 Multicore processors enable parallel measurement algorithms.

allow measurement algorithms to operate more efficiently on multicore processors. When using task parallelism, a programmer will structure the algorithm such that multiple operations are performed simultaneously on a single data set.

In cellular measurement systems, task parallelism enables measurements such as PFER, TxP, PvT, and ORFS to be performed simultaneously using a single set of IQ data. This is illustrated in Figure 7.11.

Some typical GSM measurements are described below:

Phase and Frequency Error (PFER) – Phase and frequency error is a measurement performed on signals that use the Gaussian Minimum Shift Keyed (GMSK) modulation scheme. It is a comprehensive measurement of modulation quality that can identify a wide variety of signal impairments.

Transmit Power (TXP) – The transmit power measurement describes the average power of a GSM burst.

Power Versus Time (PVT) – A PVT measurement is actually a time domain power measurement which compares the power of a GSM burst to a power mask. This measurement is used to ensure that the ramp-up power of the transmitting device does continue for the duration of the burst.

Output RF spectrum (ORFS) – ORFS is used to characterize the power output of the transmit signal at a series of offsets from the carrier. Nonlinear components such as power amplifiers can contribute to poor ORFS performance.

These results indicate that parallel programming techniques enable efficient processor utilization on multicore CPUs. Because measurement speed is processor-limited, virtual instrumentation systems can be upgraded to faster processors to reduce measurement time further.

7.3 Multi-channel measurement systems

Modular instruments have many significant advantages when an application requires multiple channel measurements. Figure 7.12 compares the architecture of a modular

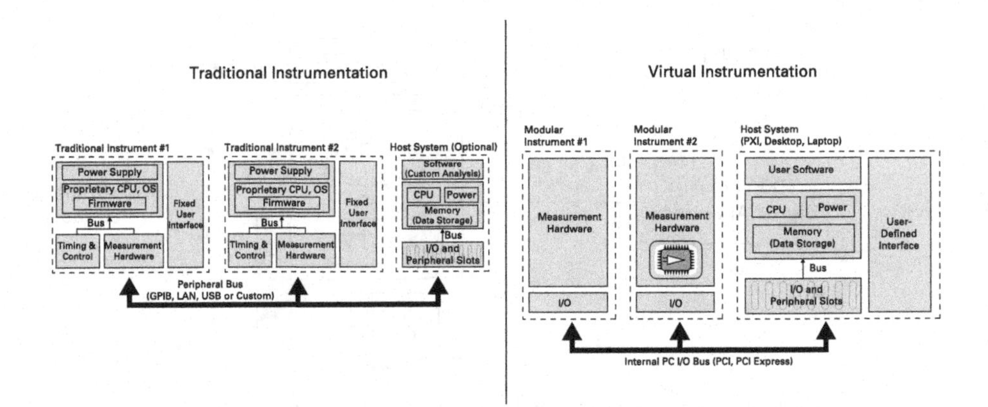

Fig. 7.12 Comparison of virtual instruments and traditional instruments for multi-channel systems.

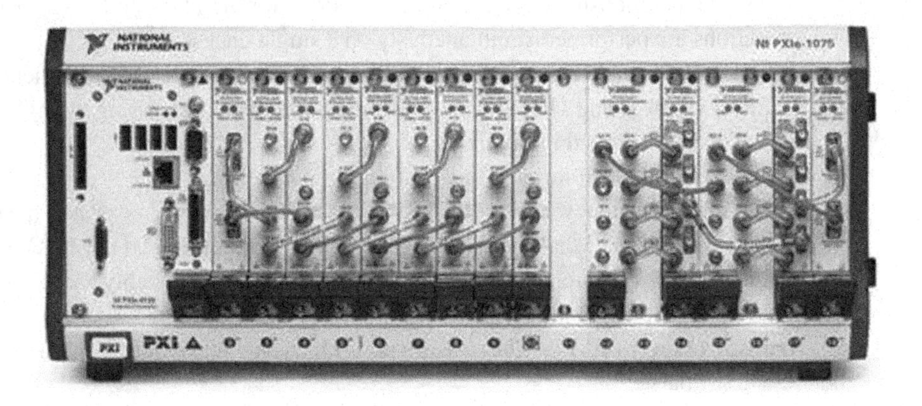

Fig. 7.13 Typical PXI phase-coherent RF measurement system.

system versus a traditional system for multi-channel systems, clearly showing the benefits of the modular system, in addition to the configurability possible through the built-in FPGA.

7.3.1 Phase coherence and synchronization

The modular architectures of PXI RF instruments lend themselves to the phase-coherent RF measurements required for multiple-input multiple-output (MIMO) and beam-forming applications. Figure 7.13 illustrates a modular measurement system with four synchronized RF analyzers (left side of chassis) and two synchronized RF signal generators (right side of chassis).

Phase-coherent RF signal generation

The configuration of any phase-coherent RF system requires synchronization of every clock signal present on the devices. There are different degrees of phase coherency. The

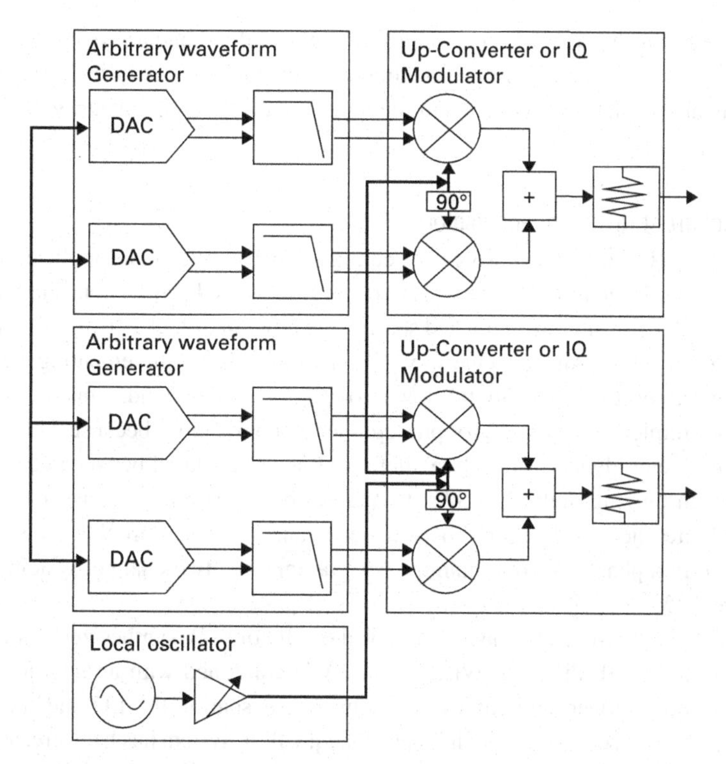

Fig. 7.14 Synchronization of two RF generation channels.

phase in phase coherency refers to the relative phases of the different clock signals used in the instruments. The best situation is when all channels share the same clock signals. The next degree of phase coherency is when the different channels share a common reference clock (10 MHz clock is common). Each channel uses the same reference clock as an input to a PLL that synthesizes an LO for mixing, but each PLL introduces unique phase noise to that channel's LO. Finally, there is no phase coherency when no synchronization or timing signals are shared between channels, making correlation of timing/phase data between those channels difficult or impossible.

Some RF VSGs use direct up-conversion to translate baseband waveforms into RF signals. Figure 7.14 illustrates the basic architecture of a two-channel RF VSG. Note that both baseband sample clocks and the local oscillators (LOs) are shared between both channels.

In Figure 7.14, observe that the VSG again consists of three modules: the local oscillator (CW synthesizer), the arbitrary waveform generator, and the RF up-converter or IQ modulator. These modules are used together as a single-channel RF vector signal generator or they can be combined (as illustrated) with additional arbitrary waveform generators and RF IQ modulators for multichannel signal generation applications. Additional cables are necessary for daisy-chaining the LO signal from the first IQ modulator to the second as well as for daisy-chaining the reference clock from the first baseband AWG to the second.

It is clear that this approach can be used to extend the modular system to more than two channels. The limitation on the number of channels is really only driven by the requirements on phase coherence and the ability to preserve the integrity of the shared LO signal.

Phase-coherent RF signal acquisition

Similarly, an RF/microwave VSA can be configured for multichannel applications. When configuring multiple modules for phase-coherent RF signal acquisition, sharing the LO between the RF down-converters, and sharing the reference clock between the IF digitizers ensures best performance. One way to build a VSA is by implementing signal stage down-conversion to IF and digital down-conversion to baseband. This architecture is one of the simplest to configure for phase-coherent applications because, unlike a three-stage super-heterodyne VSA, only a single LO must be shared between each channel. While synchronizing multiple VSAs, to achieve best performance, distribute a shared baseband reference clock and LO between each analyzer to ensure that each channel is configured in a phase-coherent manner. An example of a two-channel system is shown in Figure 7.15.

The RF/microwave VSA consists of a local oscillator (CW synthesizer), an RF down-converter, and an IF digitizer. When the VSA is combined with an additional down-converter and digitizer, and cables are included for sharing the LO and the digitizer reference clock, a complete two-channel RF acquisition system has been created. Again, this approach allows the system to expand to well beyond two channels. The number of channels is only really limited by the requirements on phase coherence and the ability to preserve the integrity of the shared LO signal.

To understand the method of synchronization between multiple phase-coherent RF VSAs, consider the block diagram of a typical RF signal analyzer as given in Figure 7.16. Observe that even though a single LO is used to down-convert from RF to IF, each analyzer must share three clocks. The analyzers share two signals – the

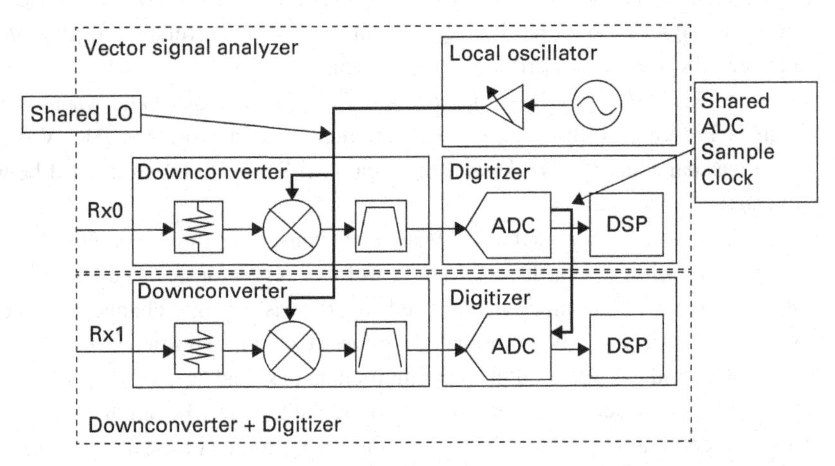

Fig. 7.15 Synchronization of a two-channel phase-coherent RF VSA system.

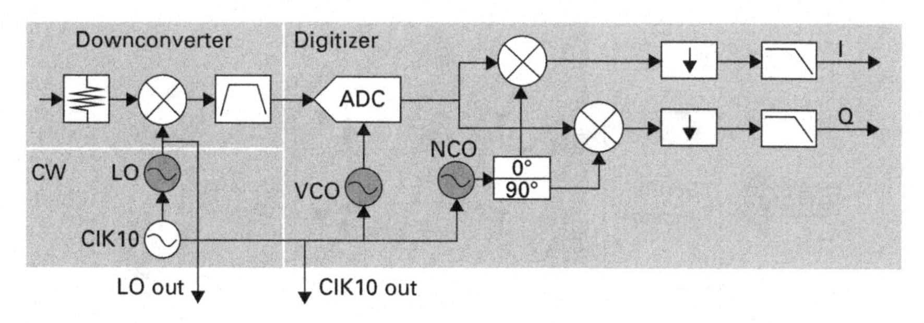

Fig. 7.16 Block diagram of typical VSA, showing sharing of clocks.

LO and the reference clock. The digitizer synthesizes the ADC sample clock and the NCO from the reference clock. An additional channel of phase-coherent acquisition could be added by adding another analyzer that shares the LO and reference clocks.

The local oscillator and 10 MHz digitizer clock are being shared between each RF channel. While sharing a reference clock between each digitizer introduces uncorrelated channel-to-channel phase jitter on the ADC sample clocks, the level of phase noise introduced at IF is negligible compared to the phase noise of the rest of the system.

While emerging technologies such as MIMO and beam-forming produce new challenges for test engineers, modular RF instrumentation provides a cost-effective and high-performance measurement solution to meet these challenges.

7.3.2 MIMO

As the prevalence of wireless communications continues to grow, there is an increasing demand for more effective use of channel bandwidth. One of the most recent innovations that helps achieve this is the development of multiple-input multiple-output (MIMO) technology. MIMO uses multiple transmitters and receivers to increase the effective signal-to-noise ratio (SNR). MIMO exploits a radio-wave phenomenon called multipath: transmitted information bounces off walls, doors, and other objects, reaching the receiving antenna multiple times through different routes and at slightly different times. MIMO harnesses multipath with a technique known as spatial multiplexing. Spatial multiplexing is a process by which a single data stream is multiplexed into multiple data streams, within the same channel. In a physical channel with sufficient multipath reflections, the maximum theoretical improvement in data rates scales linearly with the number of spatial streams. You can refer to Figure 7.17 for a block diagram of spatial multiplexing. MIMO is thus significantly different from the traditional method, whereby the data rate is increased by using more of the limited bandwidth resources. MIMO is the foundational technology used in the WLAN standard 802.11n, which can transmit as high as 140 Mbits per second. The more recent 802.11ac VHT specification allows for use of up to 8x8 MIMO, which allows for double the maximum spatial streams of 4x4 MIMO offered in 802.11n.

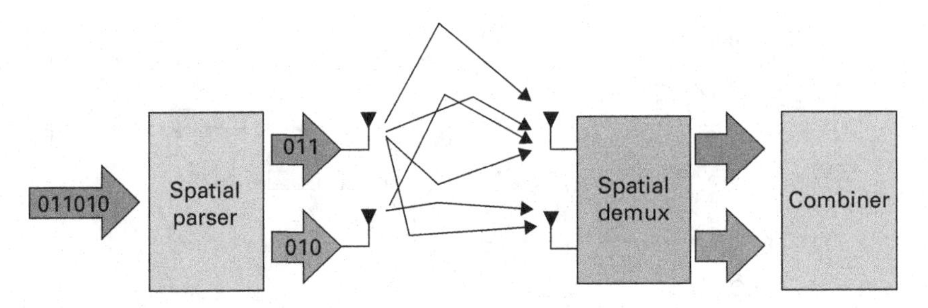

Fig. 7.17 Block diagram of spatial multiplexing.

Spatial multiplexing requires multiple antennas at both the transmitter and receiver. Accurate testing of MIMO transceivers presents significant challenges to existing test instrumentation architectures. New architectures require advanced signal processing algorithms to multiplex and de-multiplex various spatial streams, and tight synchronization between each transmit and receive antenna. Efficient MIMO transceivers cannot be designed by merely imitating or replicating the architecture of the traditional test equipment. A more versatile and efficient solution to the challenge of designing MIMO equipment is the use of software-designed modular instruments.

A MIMO system requires that the phase between each transmitter antenna remain constant. When synchronizing baseband I and Q signals, skew must be minimal to prevent distortion of the RF signal. When synchronizing multiple RF signals, phase skew between each of the RF signals is tolerable but must be minimized as much as possible. One of the challenges that MIMO poses for RF instrumentation is the need for increased levels of phase coherence between instruments.

The existing architecture of traditional RF instruments produces uncorrelated phase noise. Each instrument has a 10 MHz reference input and output, which is used for synthesizing the required local oscillator and baseband clock signals. Each instrument has an independent frequency synthesizer, which produces phase noise. The phase noise acts as an additional source of EVM to the system. Sharing a common clock causes significant phase-shifts, even for relatively small thermal changes and generates a need for frequent calibration. Testing some of the advanced MIMO operating modes requires knowledge about the relative phase difference between channels, which becomes difficult due to the phase noise. One method of overcoming this problem is to distribute the local oscillator from a single RF frequency source to the various RF up-converters and down-converters, as explained in section 7.3.1. This modular architecture with shared LOs leads to a better error vector magnitude (EVM) performance and requires less frequent calibration.

The modular PXI platform can be used to design and deploy MIMO systems quickly, while ensuring both baseband and RF synchronization.

7.3.3 Direction finding

Direction finding is one of many applications that benefits from phase-coherent analysis and generation. With phase-coherent RF vector signal analyzers (VSAs), a phase-comparison direction-finding system can be built relatively easily.

Remember jumping on a trampoline as a child and "stealing the bounce" of a friend? A perfectly timed jump would create the destructive interference necessary to bring the unfortunate jumper to their knees. Sometimes, you would try "giving a bounce," using constructive interference to send your friend much higher than they could have reached on their own. This behavior was observed in waves long before the introduction of the modern trampoline and has found its way into many applications, such as direction finding.

Figure 7.18 illustrates two transmitters and two possible scenarios of a signal source creating constructive and destructive interference at the receiver with the phase shown in the simplified polar plots. In beam-forming applications, a delay in the transmission (phase change) from one of the sources will steer the direction of highest RF intensity, controlling the direction of transmission.

In Figure 7.19, two phase-coherent receivers are used to measure the difference in phase of a signal received by two different paths. Using this relative phase comparison, the direction from which the transmission originated can be determined.

Even though this may not be practical on a trampoline, detecting the direction of an RF transmission in this way is possible using the right tools and techniques.

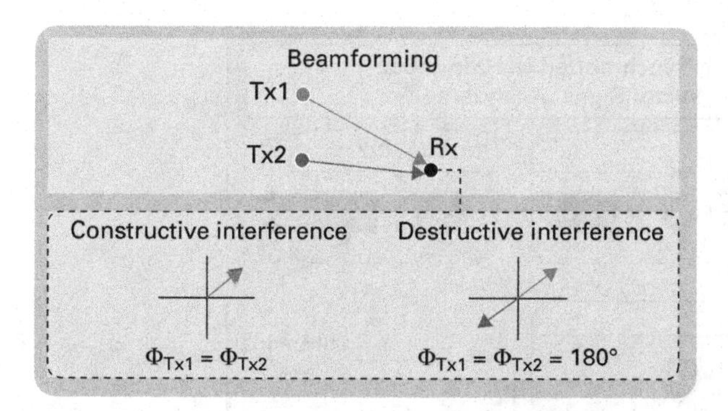

Fig. 7.18 Controlling direction of optimal transmission by adjusting the phase difference between two transmitters.

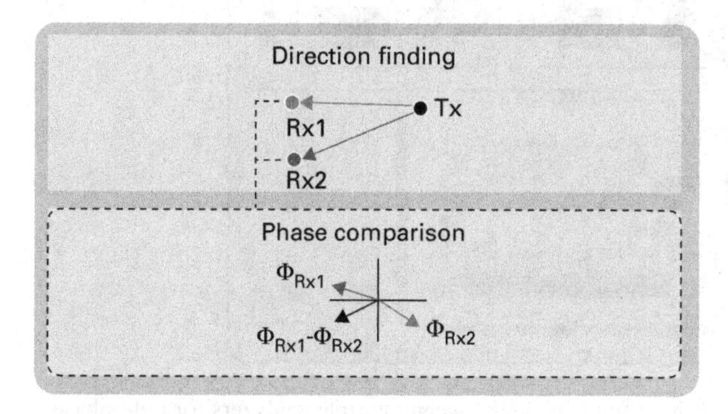

Fig. 7.19 Determining a signal's direction of arrival.

Constructing a direction finder

To construct a basic phase-comparison direction finder, multiple receivers are needed, as well as the ability to measure the phase difference between the received signals, and some math. The first requirement is satisfied by adding more analyzers to the system, but accurately measuring the difference in phase between two signals is more challenging. To compare the phase difference between two measurements, the phase differences between each oscillator used along the down-conversion path from the RF must be known precisely, as well as the time difference between multiple records from the analog-to-digital converters (ADCs).

Figure 7.20 shows a solution using two VSAs sharing a common local oscillator (LO) for down-conversion from RF and a 10 MHz reference clock.

With two phase-coherent signal analyzers, any phase difference between the two RF channels can be easily measured and applied to direction-finding applications. For example, a two-way family radio is used as a transmitter at 462.56 MHz with a pair of general-purpose ultra-high-frequency (UHF) telescoping antennas connected to two VSAs. By positioning the antennas 32.3 cm (one-half wavelength) apart, the expected phase difference is 180 degrees when the antennas share a line of sight to the receivers

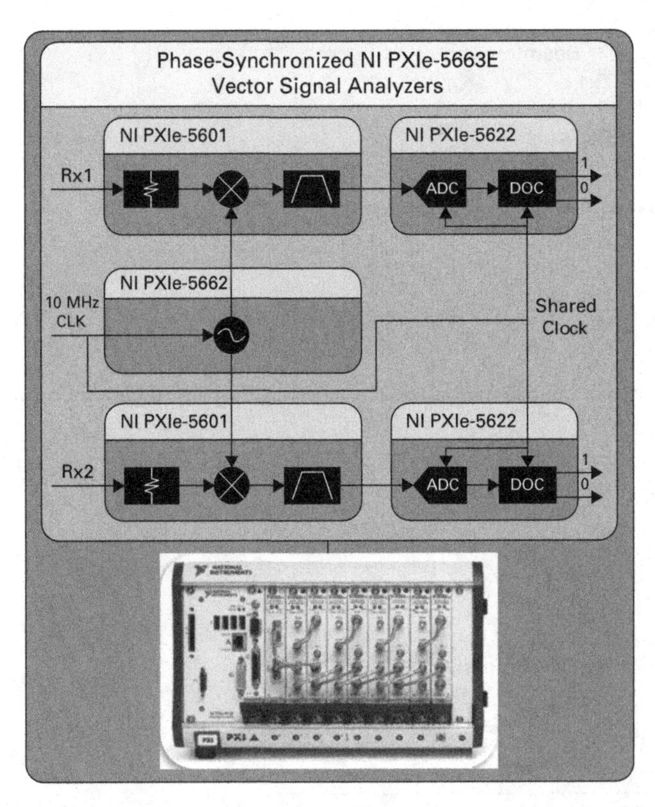

Fig. 7.20 Sharing common LO and sample clock between multiple analyzers for tight phase synchronization.

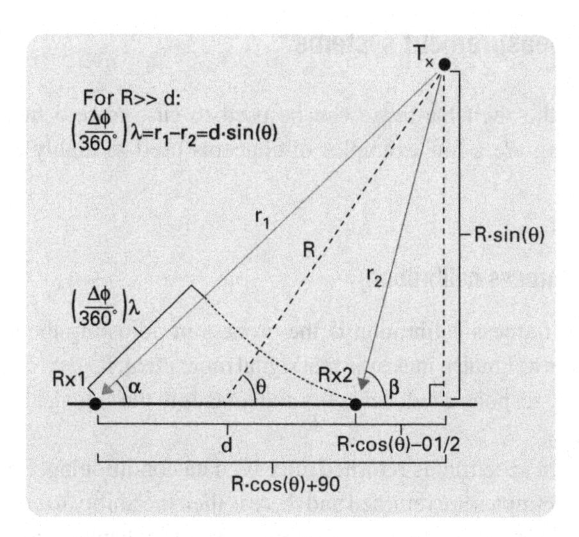

Fig. 7.21 Phase-comparison direction finding using two receivers.

and the expected phase difference is zero degrees when the transmitter is equidistant from both antennas.

By tuning the VSAs to the carrier frequency of 462.56 MHz, I and Q samples are continuously acquired to extract the phase. Verify the zero and 180 degree cases by observing the difference between the phase measurements of the VSAs. The last step is to solve for the intermediate cases.

As shown in Figure 7.21, the goal of a direction finder is to solve for θ. This math is greatly simplified if R is assumed to be much larger than d, which is a valid approximation for most signals of interest.

Knowing the frequency of interest, the distance between the antennas, and the difference in the measured phase, it is possible to solve for the corresponding values for θ. Measuring a phase difference between two analyzers of 58 degrees would translate to a θ of 71.2 degrees, whereas a phase difference of -121 degrees would yield a θ of 132.2 degrees.

7.3.4 Phase array

A phase array is an array of antennas in which the relative phases of the respective signals feeding the antennas are varied in such a way that the effective radiation pattern of the array is reinforced in a desired direction and suppressed in undesired directions. A phased array antenna is composed of multiple radiating elements each with a phase-shifter. Beams are formed by shifting the phase of the signal emitted from each radiating element, to provide constructive or destructive interference and steer the beams in the desired direction. Similar to the examples given above with direction finding and multiple-channel systems, a modular measurement system can give tremendous advantages over traditional measurement systems.

7.4 Highly customized measurement systems

Open software and off-the-shelf hardware can be used to customize a measurement system, and the following are a few examples of concepts used in highly customized measurement systems:

7.4.1 IQ data conditioning (flatness calibration)

IQ data conditioning or flatness calibration is the process of adjusting the strength of certain frequencies within a signal to make the real signal more ideal. IQ data conditioning creates a flatter frequency response, reduces spurs and images in the signal, and increases the linear phase response.

For fixed instruments, a spectrum is returned after IQ data conditioning. For modular instruments, the raw I/Q samples are returned and there is then the ability to apply various manners of correction, before translating it to the typical spectral display, with improved performance.

7.4.2 Streaming

Streaming is the process of transferring data to or from an instrument at a rate high enough to sustain continuous acquisition or generation. Streaming involves direct data transfer to or from memory. This memory can be the onboard memory of the instrument, the RAM of the controller, or the hard drive of the controller. The rate at which data is transferred to these various types of memory is limited by several factors, from the system's bus bandwidth to the read/write speed of the memory media.

The following are the advantages of using streaming for measurement systems:

- Higher and sustained acquisition and generation rates
- Reduction in measurement times due to the elimination of delays associated with downloading or uploading waveforms to or from instruments
- Enables new applications.

A variety of media can be used for streaming applications, including IDE (Integrated Drive Electronics) drives, SATA (Serial Advanced Technology Attachment) drives, and RAID (Redundant Array of Inexpensive Disks) drive systems. For example, a chunk of real-world RF spectrum can be recorded to test a device, and played back in the lab for a virtual field test. If the device doesn't behave as expected, the exact same scenario which caused it to fail is replayed, and allows the problem to be debugged.

Streaming is also very useful in spectrum monitoring and signal intelligence systems, where a large amount of spectral content needs to be analyzed.

Peer-to-peer (P2P) streaming

P2P streaming technology uses PCI Express to enable direct, point-to-point transfers between multiple instruments without sending data through the host processor or memory

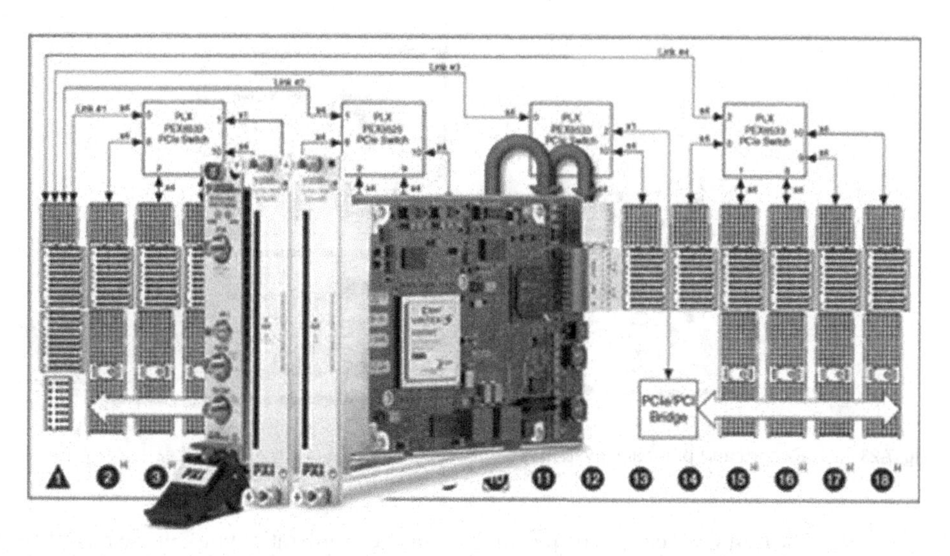

Fig. 7.22 NI FlexRIO peer-to-peer architecture.

[10]. This enables devices in a system to share information without burdening other system resources.

With peer-to-peer technology, data streaming rates of more than 800 MB/s are possible in a single direction. Maximum throughput is dependent on the streaming modules, chassis, and, if the configuration warrants it, the controller. Generally, the lowest of these rates is the maximum possible P2P bandwidth. Peer-to-peer transfers are designed to have a very low latency, but it varies depending on the system configuration. The main advantage of peer-to-peer streaming is that the data need not travel through the host, reducing latency, increasing reliability, and increasing total system bandwidth. Figure 7.22 shows the NI FlexRIO peer-to-peer architecture.

Device-to-host or host-to-device streaming

Device-to-host or host-to-device streaming is the process of streaming data from an instrument (e.g. FlexRIO) to the host controller, and vice versa, for either subsequent processing, or writing to disk.

With this technology, data streaming rates for a PCI Express Gen1 ×4 device of more than 800 MB/s are possible in a single direction. The data transfer rate depends on the bus technology, such as Gen1 or Gen2 PCI Express with a ×4 or ×8 link.

Integrating hard drives for record and playback

Record and playback is the process of using platforms that use high-bandwidth buses (e.g. PCI and PCIe) to enable instruments to stream data to and from RAID 0 (striped) hard drive arrays at high sustained rates [11]. RF record and playback systems combine PXI RF signal analyzers and RF signal generators with RAID arrays for high-speed, long-duration recording and playback. These systems reduce the cost for memory expansion by using PC memory for contiguous acquisitions. You can use record and playback systems to perform host-to-device or device-to-host streaming. For low-rate applications, where

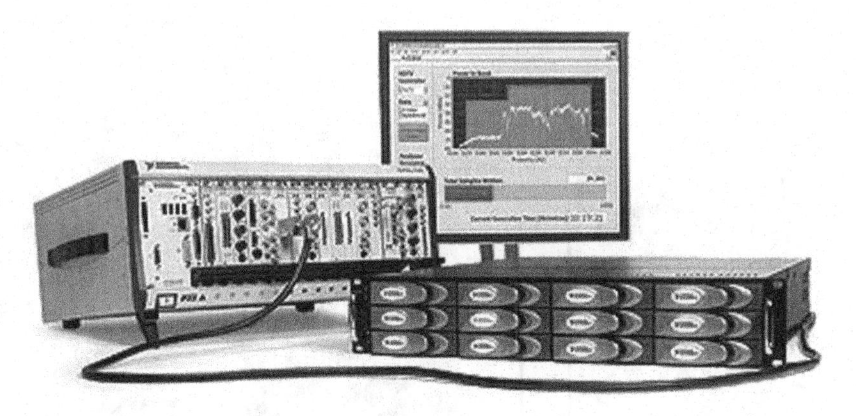

Fig. 7.23 Record and playback system.

the host CPU can keep up with the data streaming rates, you can use record and playback systems for inline or real-time processing.

The ability to generate or acquire terabytes of continuous data can help you implement applications previously possible only with custom hardware, such as the following:

- spectrum monitoring,
- packet sniffing,
- wireless receiver design, validation, and verification,
- digital video broadcasting Bit Error Rate (BER) tests.

Figure 7.23 shows a typical RF record and playback system.

7.4.3 Integrating FPGA technology

Field-programmable gate arrays (FPGAs) are reprogrammable silicon chips. Unlike multi-core PC processors, which run software applications to implement functionalities, programming an FPGA rewires the chip itself to implement the functionalities.

As shown in Figure 7.24, FPGA chip specifications include the amount of configurable logic blocks, number of fixed function logic blocks, such as multipliers, and size of memory resources such as embedded block RAM. There are many other parts to an FPGA chip, but these are typically the most important when selecting and comparing FPGAs for a particular application.

At the lowest level, configurable blocks of logic, such as slices or logic cells, are made up of two basic things: flip-flops and look-up tables (LUTs). This is important to note because the various FPGA families differ in the way flip-flops and LUTs are packaged together. Virtex-II FPGAs for example, have slices with two LUTs and two flip-flops, whereas Virtex-5 FPGAs have slices with four LUTs and four flip-flops. The LUT architecture itself may also differ (4-input versus 6-input).

Every FPGA chip is made up of a finite number of predefined resources with programmable interconnects to implement a reconfigurable digital circuit. For any given

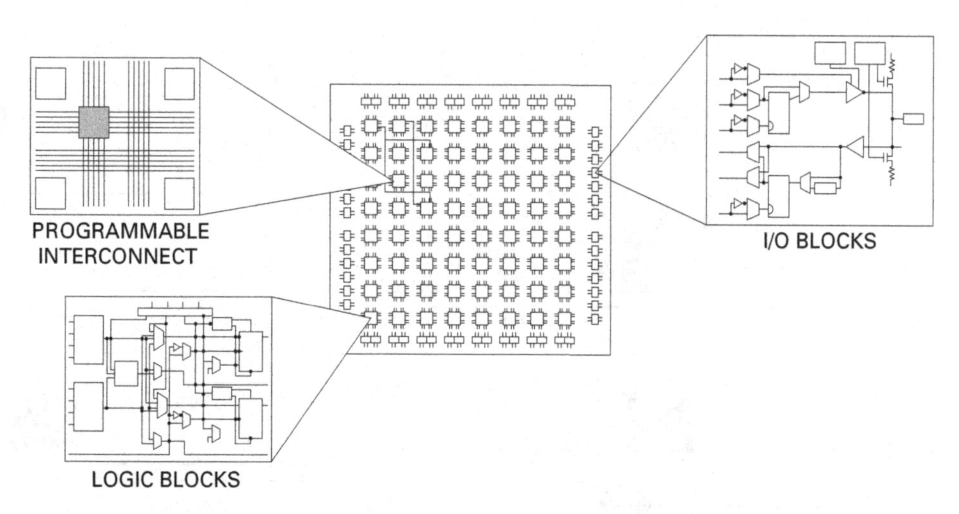

Fig. 7.24 Parts of an FPGA.

piece of synthesizable code, either graphical or textual, there is a corresponding circuit schematic that describes how logic blocks should be wired together. Synthesis is the process of translating high-level programming languages into true hardware implementations.

Beyond being user-programmable, FPGAs offer hardware-timed execution speed as well as high determinism and reliability. They are truly parallel, so different processing operations do not have to compete for the same resources. Each independent processing task has its own dedicated section of the chip, and each task can function autonomously without any influence from other logic blocks. As a result, adding more processing does not affect the performance of another part of the application.

To increase performance even further, FPGAs offer the computational performance to provide real-time measurements that occur faster than the time it takes to acquire the data [12].

Figure 7.25 shows the difference in computational performance between a host-based implementation and a FPGA-based implementation of an adjacent channel leakage ratio (ACLR) calculation.

While the host-based implementation takes advantage of multiple high-performance CPU cores and the high-bandwidth PXI Express data bus, the FPGA implementation reduces measurement time even further by using dedicated, real-time processing and eliminating unnecessary host data transfers. Furthermore, the peer-to-peer FIFO is configured only once regardless of the number of averages, so measurement time scales are based on the amount of time you need to acquire the RF data necessary to perform the measurement.

To be useful in a software-designed instrumentation context, FPGAs must be reprogrammable by the test engineer in software; in other words, they should be used to push software programmability down into the hardware itself. In the past, FPGA technology was available only to engineers with a deep understanding of digital hardware design

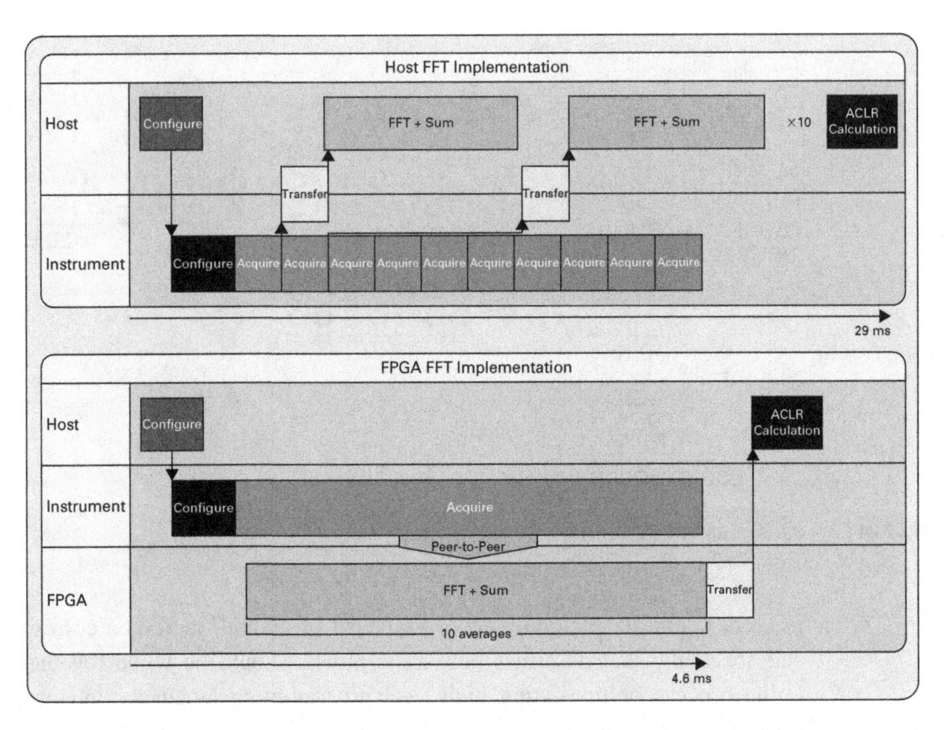

Fig. 7.25 Comparing performance of host and FPGA implementations.

software, such as hardware description languages like Verilog or VHDL, which use low-level syntax to describe hardware behavior. Most test engineers do not have expertise in these tools. However, the rise of high-level design tools is changing the rules of FPGA programming, with new technologies that convert graphical block diagrams or even C code into digital hardware circuitry. These system-level tools that abstract the details of FPGA programming can bridge this gap.

The following are the advantages of using FPGA-based test methods:

- **Real-time, continuous measurements**
  With their processing throughput, FPGAs can perform measurements faster than the I/O hardware can acquire data, so it is now possible to test the DUT continuously. Instead of acquiring, transferring data, then processing, which has a limited duty cycle, we can now acquire and measure continuously.
- **Custom triggering and acquisition**
  With an FPGA always acquiring data, we can define the measurement back-end by adding custom triggering and data recording. We may want to add a complex digital trigger, OR-ing and AND-ing several digital lines to detect a trigger condition.
- **Closed-loop and dynamic test**
  As modern devices are increasingly integrated into the world around them, testing them without incorporating feedback into the test system may not provide adequate test coverage. For instance, modern communication schemes often incorporate acknowledgement packets or bits. If the test system does not correctly interpret these and

respond appropriately and in a timely manner, then it is not obeying the protocol and the DUT may not be accurately tested. Often, only an FPGA can provide these kinds of low latency responses.

- Protocol emulation
  Instead of constantly using software to go back and forth between protocol-level and signal-level information, which can be tedious and slow, we can actually implement protocols on the FPGA, allowing the test system to interact with general test hardware at a protocol level.

7.5 Evolution of graphical system design

Graphical system design is an approach to designing an entire system, using more intuitive graphical software and off-the-shelf (non-custom) hardware devices.

Graphical system design is a valuable technique for creating completely user-designed instrumentation, in which the instrument is completely user-programmable, down to the pin, allowing customization of every aspect of its behavior. A default behavior may be used which resembles that of a "traditional" modular instrument, incorporating record-based data movement paradigms, or a custom protocol may be implemented on the instrument, allowing it to interact with the DUT as if the DUT is in a real-world operating environment, not just a test mode.

The following are a few important milestones in the evolution of graphical system design:

- **Portable measurement algorithms**
  If the microprocessor initiated the virtual instrumentation revolution, then the FPGA is ushering in its next phase. FPGAs have been used in instruments for many years. For instance, today's high-bandwidth oscilloscopes collect so much data that it is impossible for users to quickly analyze all of it. Hardware-defined algorithms on these devices, often implemented on FPGAs, perform data analysis and reduction (averaging, waveform math, and triggering), compute statistics (mean, standard deviation, maximum, and minimum), and process the data for display, all to present the results to the user in a meaningful way. While these capabilities present obvious value, there is lost potential in the closed nature of these FPGAs. In most cases, users cannot deploy their own custom measurement algorithms to this powerful processing hardware. New software tools strive to open these FPGAs to a variety of test and measurement algorithms, effectively and automatically porting them from the CPU to the higher-performance capabilities of the FPGA.
- **Reconfigurable instruments**
  Test systems are reconfigured for endless reasons – from adapting to new test requirements to accommodating instrument substitutions during calibration and repair cycles. Software-designed instrumentation is based on a modular architecture that enables a high degree of reconfigurability. Software-designed instruments consist of modular acquisition/generation hardware whose functionality is characterized through user-defined software running on a host multicore processor.

The new software-designed architecture can meet application challenges that are impossible to solve with traditional methods that require real-time decision making by the host to properly test the device. Instead, engineers can fully deploy the intelligence to the FPGA embedded on the instrument for pass/fail guidance. For some applications, engineers also perform the communication over a protocol – wireless or wired – which requires a significant layer of coding and decoding before making a decision.

- **Heterogeneous computing**
  Automated test systems have always comprised multiple types of instruments; each best suited to different measurement tasks. An oscilloscope, for example, can make a single DC voltage-level measurement, but a DMM provides better accuracy and resolution. It is this mix of different instrumentation that enables tests to be conducted in the most efficient and cost-effective manner possible.
  A heterogeneous computing architecture is a system that distributes data, processing, and program execution among different computing nodes that are each best suited to specific computational tasks. For example, an RF test system that uses heterogeneous computing may have a CPU controlling program execution with an FPGA performing inline demodulation and a graphics processing unit (GPU) performing pattern matching before storing all the results on a remote server. Test engineers need to determine how to best use these computing nodes and architect systems to optimize processing and data transfer.
- **IP to the pin**
  The next phase in integrating design and test is the ability for engineers to deploy design building blocks, known as intellectual property (IP) cores, to both the device under test (DUT) and the reconfigurable instrument. This capability is called "IP to the pin" because it drives user-defined software IP as close to the I/O pins of next-generation reconfigurable instruments as possible. The software IP includes functions/algorithms such as control logic, data acquisition, generation, digital protocols, encryption, math, RF, and signal processing.

Graphical programming languages allow the creation of virtual instruments that can be easily tested before embedding as a subroutine into a larger program. The graphical approach also allows non-programmers to build programs by dragging and dropping virtual representations of lab equipment with which they are already familiar.

7.6 Summary

There has been a rapid increase in the importance of the role of modular measurement systems in RF and microwave applications. This chapter discussed the fundamentals of modular systems and reviewed some of the salient features of these systems which allow them to be highly effective for many RF and microwave measurement applications.

There is no doubt that the advent of modular instruments has been one of the major progressions in RF and microwave testing recently. This approach has resulted in the ability to make fast, flexible, and accurate measurements using SW-designed modular test products. This is a trend that has gained momentum and will continue to accelerate. For instance, it can be very difficult to solve the ever-changing needs of the wireless industry with traditional test products that are often expensive, fairly large, and usually rigid. The expandability of modular systems allows for synchronized, phase-coherent measurements on systems comprising multiple sources or receivers. The increased RF performance of modern modular products has enabled highly accurate measurements in a fraction of the time, space, and cost.

Further advances in RF technologies and processes have enabled the development of smaller form-factor, lower-cost modular products to match the performance and features of more traditional test products. Modular systems, wrapped in graphical design software, can take full advantage of multi-core processors and make use of the latest FPGA technologies to allow for the greatest measurement flexibility and timing control. This capability extends from the early design phase all the way through deployment of a measurement system, and is commonly referred to as Graphical System Design.

References

[1] National Instruments. (2009, May). *Virtual Instrumentation*. [Online]. Available: http://zone.ni.com/devzone/cda/tut/p/id/4752

[2] C. F. Coombs, "Virtual Instruments," in *Electronic Instrument Handbook,* 2nd ed. New York: McGraw-Hill, 1995.

[3] G. E. Moore, "Cramming more components onto integrated circuits," *Electronics*, vol. 38, no. 8, pp. 114–117, Apr. 1965.

[4] M. Santori, "An Instrument that isn't really," *IEEE Spectrum*, vol. 27, no. 8, pp. 36–39, Aug. 1990.

[5] National Instruments. (2011, March). *What is LXI?*. [Online]. Available: http://zone.ni.com/devzone/cda/tut/p/id/7255

[6] National Instruments. (2011, Dec.). *What is PXI?*. [Online]. Available: http://zone.ni.com/devzone/cda/tut/p/id/4811

[7] National Instruments. (2010, Oct.). *PXI Express FAQ*. [Online]. Available: http://zone.ni.com/devzone/cda/tut/p/id/3882

[8] M. Friedman and J. Schwartz, "Techniques for Architecting High-Performance Hybrid Test Systems," in *IEEE Autotestcon 2008*, Salt Lake City, UT: 2008, pp. 282–285.

[9] A. V. Oppenheim and R. W. Schafer, *Discrete-Time Signal Processing*. New Jersey: Prentice-Hall, Inc, 1999.

[10] National Instruments. (2011, Dec.). *An Introduction to Peer to Peer Streaming*. [Online]. Available: http://zone.ni.com/devzone/cda/tut/p/id/10801

[11] National Instruments. (2011, Aug.). *Introduction to Record and Playback*. [Online]. Available: http://zone.ni.com/devzone/cda/tut/p/id/7209

[12] National Instruments. (2012, Feb.). *Make Your Measurements Faster With FPGA Technology*. [Online]. Available: http://zone.ni.com/devzone/cda/pub/p/id/1513

Part III

Linear measurements

8 Two-port network analyzer calibration

Andrea Ferrero

8.1 Introduction

Although VNAs are probably the most advanced microwave systems, with broadband sources, high-speed and high-dynamic-range receivers, the intrinsic property of distribute components makes a calibration procedure mandatory to obtain reasonable results due to an enormous systematic error. To stress this fundamental problem, imagine weighing 300 g of ham with a one-ton plate scale! This is more or less the same influence as systematic phase error introduced by a 1 meter teflon cable in front of a VNA port at 10 GHz if we are trying to measure 1 degree of phase-shifting on a DUT S_{11} parameter. Not only the phase, but also the magnitude as well is affected, due to different attenuation paths in various system sections. Clearly without a proper correction the measurement quality would be unacceptable.

In the early development of VNA, hardware compensation with line stretchers and variable gain amplifiers was attempted, but it's only with the introduction of computer-controlled digital VNAs, that specific signal processing techniques allow a real-time correction of the most important errors. During the last forty years several algorithms have been proposed especially for one- or two-port VNAs; some of them like TRL, SOLT, LRM, and SOLR, became a de-facto standard in all modern VNA firmware; however, many others have been proposed to solve particular problems [1–4]. This chapter presents a review of the error models and the main VNA calibrations, by focusing the attention on their commonalities and by pointing out their different fields of application. The development follows the system approach born in the early 1990s, rather than the traditional one based on the analysis of all possible sources of error [3–5].

8.2 Error model

Let's consider a one-port VNA essential block scheme, as shown in Figure 8.1; from the microwave ports to the digital data we have the following significant parts:

- the source,
- the microwave test set,
- the down-converter,
- the IF Digitizer.

The source provides the required microwave signals, while the microwave test set includes all the microwave components such as couplers, cables, and adapters from the source reference plane to the DUT one. This block is modeled as a 4-port network, where two ports are loaded with the mixers/samplers. The basic hypothesis to develop a general error model is the overall system LINEARITY, i.e. it's assumed that every component from the cables to the A/D is linear. Thus a set of linear equations link the digital outputs of the IF A/D converters with the **a** and **b** waves at the DUT reference plane. The validity of such an approach is mainly constrained by the linear region of the mixer/sampler, while the nonlinearity of the A/D converter is negligible. The system works in the frequency domain, i.e., the source is supposed to be sinusoidal and the receiver is strictly narrowband, typically 100 Hz, for accurate measurements. At each frequency we have a pair of complex numbers at the two IF digitizer outputs usually called the **measured waves,** a_m and b_m. The linearity assumption sets a $\mathbb{C}^2 \Longrightarrow \mathbb{C}^2$ linear application which obviously can be written as:

$$\begin{bmatrix} b_1 \\ a_1 \end{bmatrix} = \begin{bmatrix} d_{11} & d_{12} \\ d_{21} & d_{22} \end{bmatrix} \begin{bmatrix} a_{m1} \\ b_{m1} \end{bmatrix}, \tag{8.1}$$

where the d_{ij} are four error coefficients. However, it's worth deriving (8.1) from classical network theory. Let **S** be the scattering matrix of the four-port test set with the wave convention of Figure 8.1 as:

$$\begin{bmatrix} b_0 \\ a_1 \\ b_3 \\ b_4 \end{bmatrix} = \begin{bmatrix} S_{11} & S_{12} & S_{13} & S_{14} \\ S_{21} & S_{22} & S_{23} & S_{24} \\ S_{31} & S_{32} & S_{33} & S_{34} \\ S_{41} & S_{42} & S_{43} & S_{44} \end{bmatrix} \begin{bmatrix} a_0 \\ b_1 \\ a_3 \\ a_4 \end{bmatrix}. \tag{8.2}$$

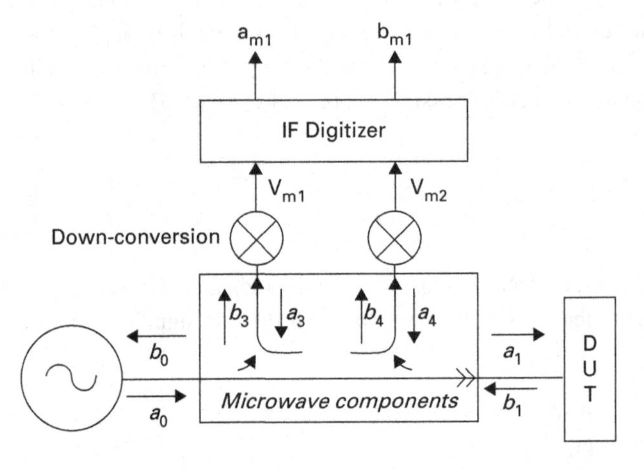

Fig. 8.1 One-port VNA essential block scheme.

The two mixers load ports 3 and 4 with two generic reflection coefficients Γ_3 and Γ_4 as:

$$a_3 = \Gamma_3 b_3 \tag{8.3}$$

$$a_4 = \Gamma_4 b_4. \tag{8.4}$$

The IF output voltages V_{m1} and V_{m2} are the low-frequency images of the total voltage at the RF input ports, i.e.

$$V_{m1} = \alpha_3(1+\Gamma_3)b_3 \tag{8.5}$$

$$V_{m2} = \alpha_4(1+\Gamma_4)b_4, \tag{8.6}$$

where α_i are the proper mixer conversion factors. Finally the readings are linked with the IF voltages by the A/D coefficients:

$$a_{m1} = \beta_1 V_{m1} \tag{8.7}$$

$$b_{m1} = \beta_2 V_{m2} \tag{8.8}$$

We have:

$$\begin{array}{rcl} \beta_1\alpha_3(1+\Gamma_3)b_3 & = & a_{m1} \\ \beta_2\alpha_4(1+\Gamma_4)b_4 & = & b_{m1} \end{array} \Rightarrow \begin{bmatrix} b_3 \\ b_4 \end{bmatrix} = \underbrace{\begin{bmatrix} \xi_1 & 0 \\ 0 & \xi_2 \end{bmatrix}}_{\Xi} \begin{bmatrix} a_{m1} \\ b_{m1} \end{bmatrix} \tag{8.9}$$

while from (8.2)–(8.4):

$$\underbrace{\begin{bmatrix} -S_{11} & 1 & -S_{12} & 0 \\ -S_{21} & 0 & -S_{22} & 1 \\ -S_{31} & 0 & -S_{32} & 0 \\ -S_{41} & 0 & -S_{42} & 0 \end{bmatrix}}_{\mathbf{W}} \begin{bmatrix} a_0 \\ b_0 \\ b_1 \\ a_1 \end{bmatrix} = \underbrace{\begin{bmatrix} S_{13}\Gamma_3 & S_{14}\Gamma_4 \\ S_{23}\Gamma_3 & S_{24}\Gamma_4 \\ (S_{33}\Gamma_3 - 1) & S_{34}\Gamma_4 \\ S_{43}\Gamma_3 & (S_{44}\Gamma_4 - 1) \end{bmatrix}}_{\mathbf{Q}} \begin{bmatrix} b_3 \\ b_4 \end{bmatrix}. \tag{8.10}$$

Finally:

$$\begin{bmatrix} a_0 \\ b_0 \\ b_1 \\ a_1 \end{bmatrix} = \mathbf{W}^{-1}\mathbf{Q}\Xi \begin{bmatrix} a_{m1} \\ b_{m1} \end{bmatrix} = \mathbf{D} \begin{bmatrix} a_{m1} \\ b_{m1} \end{bmatrix}. \tag{8.11}$$

By taking the last two rows of (8.11), the linear system of (8.1) is obtained. It's worth noting that the elements of the matrix **D** are independent of the loading conditions at the source port, i.e. *the error coefficients of a full reflectometer-based VNA are independent of the source*, which means the source can be changed **after** the calibration without affecting its validity. This important result is not obvious and fundamental for many applications. Traditionally the error coefficients are organized as S parameters of a fictitious network, called the **error box**, which is interposed between the DUT and an ideal VNA, as shown

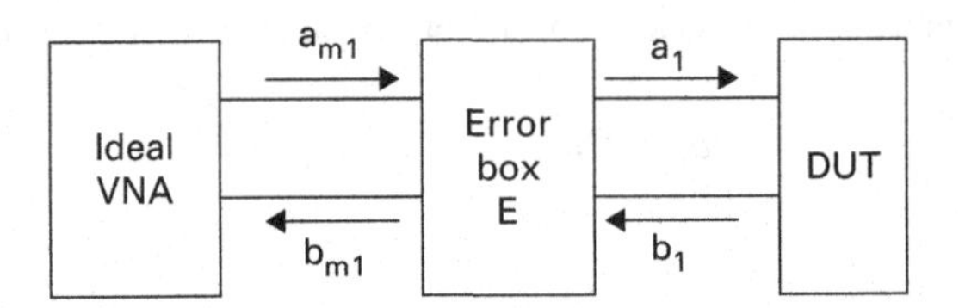

Fig. 8.2 One-port error box.

in Figure 8.2; thus the usual form of the error coefficient becomes:

$$\begin{bmatrix} b_{m1} \\ a_1 \end{bmatrix} = \underbrace{\begin{bmatrix} e_{11} & e_{12} \\ e_{21} & e_{22} \end{bmatrix}}_{\mathbf{E}} \begin{bmatrix} a_{m1} \\ b_1 \end{bmatrix}. \tag{8.12}$$

It's straighforward to obtain the e_{ij} from the d_{ij} as:

$$\begin{bmatrix} -d_{32} & 0 \\ -d_{42} & 1 \end{bmatrix} \begin{bmatrix} b_{m1} \\ a_1 \end{bmatrix} = \begin{bmatrix} d_{31} & -1 \\ d_{41} & 0 \end{bmatrix} \begin{bmatrix} a_{m1} \\ b_1 \end{bmatrix}$$

$$\Downarrow$$

$$\begin{aligned} \begin{bmatrix} e_{11} & e_{12} \\ e_{21} & e_{22} \end{bmatrix} &= \begin{bmatrix} -d_{32} & 0 \\ -d_{42} & 1 \end{bmatrix}^{-1} \begin{bmatrix} d_{31} & -1 \\ d_{41} & 0 \end{bmatrix} \\ &= \frac{1}{d_{32}} \begin{bmatrix} -d_{31} & 1 \\ d_{41}d_{32} - d_{31}d_{42} & d_{42} \end{bmatrix}. \end{aligned} \tag{8.13}$$

However the elements of **E** do not behave in any way as a scattering matrix, i.e. they do not have any particular properties of physical networks, but they are only four complex numbers for each frequency, which include all the systematic, i.e. time-invariant, characteristics of the whole system. Since no assumption has been made about the nature of the error terms, it follows that the calibrated values of the a and b waves are a function of how the error coefficients are computed, or:

the reference impedance of a VNA is set by the calibration and not by the hardware

The VNA readings have no physical meaning until the calibration is performed. To calibrate a VNA means to determine the required set of error coefficients which define the error model.

8.3 One-port calibration

Let's proceed toward the complete solution in the elementary case of one port; from (8.12) we have:

$$\frac{b_{m1}}{a_{m1}} = \Gamma_m = \frac{e_{11} - \overbrace{(e_{11}e_{22} - e_{12}e_{21})}^{\Delta}\Gamma}{1 - e_{22}\Gamma}, \tag{8.14}$$

where $\Gamma = \frac{b_1}{a_1}$ is the desired reflection coefficient while Γ_m is defined as the measured one. From (8.14) we can see that only three error coefficients are needed to compute the corrected reflection coefficient: e_{11}, e_{22}, and Δ, and the de-embedding equation, i.e. the formula which gives the corrected value from the measured one, follows from (8.14) as:

$$\Gamma = \frac{\Gamma_m - e_{11}}{e_{22}\Gamma_m - \Delta}. \tag{8.15}$$

To solve the calibration problem it's useful to write (8.15) as:

$$\begin{gathered} e_{22}\Gamma\Gamma_m - \Delta\Gamma = \Gamma_m - e_{11} \\ \Downarrow \\ e_{11} + e_{22}\Gamma\Gamma_m - \Delta\Gamma = \Gamma_m. \end{gathered} \tag{8.16}$$

This equation shows a simple and effective way to compute the error terms by measuring three different *standards* Γ_i and stacking the corresponding equations as in (8.16) for each Γ_{mi} measurement to form the linear system:

$$\begin{bmatrix} 1 & \Gamma_1\Gamma_{m1} & -\Gamma_1 \\ 1 & \Gamma_2\Gamma_{m2} & -\Gamma_2 \\ 1 & \Gamma_3\Gamma_{m3} & -\Gamma_3 \end{bmatrix} \begin{bmatrix} e_{11} \\ e_{22} \\ \Delta \end{bmatrix} = \begin{bmatrix} \Gamma_{m1} \\ \Gamma_{m2} \\ \Gamma_{m3} \end{bmatrix}. \tag{8.17}$$

The first one-port technique was called **SOL** because the three standards were a **S**hort, an **O**pen, and a **L**oad [5]. The difficulties of making precise microwave standards were immediately obvious, especially for the open and the load ones; furthermore the frequency behaviour of these devices was not ideal nor constant, so a set of electrical models were developed and included in the VNA firmware to describe the response vs. frequency of the standards. Figure 8.3 shows the adopted standard models. They are simple networks where a parameter is obtained as a polynomial fitting of the frequency

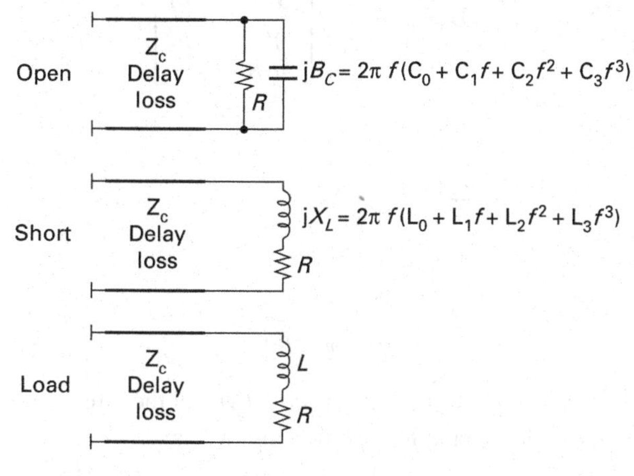

Fig. 8.3 One-port standard models.

response. These models became a de facto standard and every standard manufacturer publishes the parameters of its devices in this way.

Since all the measurements are functions of the calibration standards, to obtain their parameters with high accuracy is a must and cannot be done through experiments. For a coaxial environment the parameters were, in the past, obtained from a scale model of the most critical one (the open) measured at low frequency. Nowadays FEM simulators are used, which poses the metrological question of how accurate they are. Among the three standards, the easiest one to manufacture is the short one; thus a simple solution takes three offset shorts as the three standards. This technique is mathematically identical to the SOL, but the three standards are of the same type and have only a different delay. The problem arises when the line lengths resonate, thus the linear system becomes undeterminate. For this reason the offset short technique is narrowband and has its main application in waveguides. To complete the one-port case, it's worth noting that if we make the following assumptions:

- the dual directional coupler is well matched, balanced, but has finite isolation, i.e. its S-matrix becomes:

$$\mathbf{S} = \begin{bmatrix} 0 & \alpha & \beta & \gamma \\ \alpha & 0 & \gamma & \beta \\ \beta & \gamma & 0 & \alpha \\ \gamma & \beta & \alpha & 0 \end{bmatrix}.$$

- the two mixers are identical and perfectly matched, i.e.:

$$\xi = \xi_1 = \xi_2 = 1$$
$$\Gamma_3 = \Gamma_4 = 0.$$

From (8.10) and (8.11) it follows:

$$\mathbf{D} = \frac{1}{\gamma^2 - \beta^2} \begin{bmatrix} -\beta & \gamma \\ \alpha\gamma & -\alpha\beta \\ \gamma & -\beta \\ -\alpha\beta & \alpha\gamma \end{bmatrix} \tag{8.18}$$

$$\Downarrow$$

$$e_{11} = \frac{\gamma}{\beta} e_{22} = \frac{-\alpha\gamma}{\beta} \Delta = \alpha \tag{8.19}$$

$$\Downarrow$$

$$\Gamma = \frac{\beta\Gamma_m - \gamma}{-\alpha\gamma\Gamma_m - \beta\alpha}. \tag{8.20}$$

Thus in this particular case, e_{11} represents the *directivity* of the directional coupler. In the ideal case of infinite isolation and no insertion loss we have:

$$e_{11} = 0 \quad e_{22} = 0 \quad \Delta = 1 \tag{8.21}$$

Finally let's consider (8.14). If $\Gamma = 0$, i.e. we are connecting a perfect load, we have: $\Gamma_m = e_{11}$. Thus the direct measurement of a standard gives the directivity†. Unfortunately an ideal matched load does not exist, but only an approximation ($\Gamma \approx -40$dB) can be manufactured, which means limiting the VNA accuracy to that level. This was the reason to develop the so called **sliding load** calibration [6]. This calibration uses a sliding load, i.e. a transmission line with a load that can slide along it. From (8.14) note that if $\Gamma \approx 0 \Rightarrow \Gamma_m \approx e_{11} - \Gamma$, and by measuring this device for several load positions, i.e. for different phases, we obtain a small circle on the complex plane whose center is e_{11}. This techinique coupled with a short and an open was widely used on one-port VNAs.

8.4 Two-port VNA error model

Two-port VNAs generally have two different architectures which are modeled by different error models:

- a reflectometer on each port, as shown in Figure 8.4,
- one reference coupler and a single coupler on each port, as shown in Figure 8.5.

The first case has more complex hardware, but, as demostrated above, does not suffer from the switch imperfection or repeatability due to the independence of the error terms from the source termination. The latter fewer components, but it requires a highly repeatable switch and its error model does not allow the use of more modern calibration techniques.

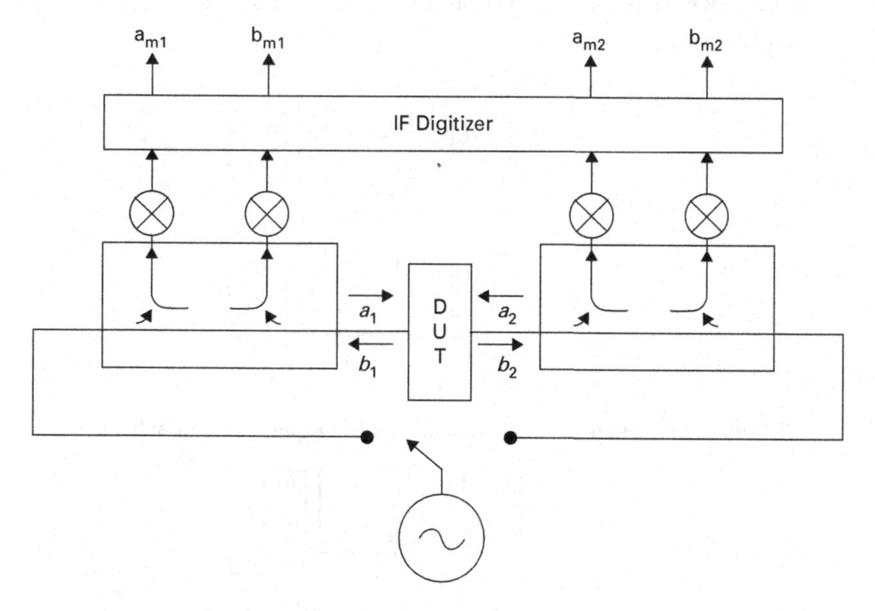

Fig. 8.4 Two-port VNA with a complete reflectometer on each port.

† However, this is true only if the S matrix of the microwave part was referred to the same impedance as the load standard

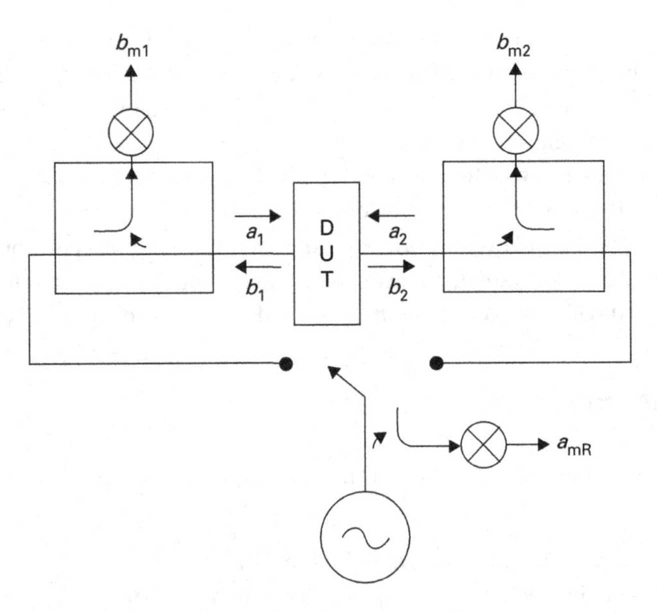

Fig. 8.5 Two-port VNA with a single reference channel.

8.4.1 Eight-term error model

The dual reflectometers VNA of Figure 8.4 is the generalization of the one port. Its error model is straightforward to obtain with the introduction of a second error box as shown in Figure 8.6. Here the four measured quantities a_{m1}, b_{m1}, a_{m2} and b_{m2} are linked to the corresponding waves as:

$$\begin{bmatrix} b_{m1} \\ a_1 \end{bmatrix} = \underbrace{\begin{bmatrix} e_{11}^A & e_{12}^A \\ e_{21}^A & e_{22}^A \end{bmatrix}}_{\mathbf{E_A}} \begin{bmatrix} a_{m1} \\ b_1 \end{bmatrix} \tag{8.22}$$

$$\begin{bmatrix} b_{m2} \\ a_2 \end{bmatrix} = \underbrace{\begin{bmatrix} e_{11}^B & e_{12}^B \\ e_{21}^B & e_{22}^B \end{bmatrix}}_{\mathbf{E_B}} \begin{bmatrix} a_{m2} \\ b_2 \end{bmatrix}. \tag{8.23}$$

It's more convenient to use a cascade matrix representation, where:

$$\begin{bmatrix} b_{m1} \\ a_{m1} \end{bmatrix} = \underbrace{\begin{bmatrix} t_{11}^A & t_{12}^A \\ t_{21}^A & t_{22}^A \end{bmatrix}}_{\mathbf{T_A}} \begin{bmatrix} b_1 \\ a_1 \end{bmatrix} \tag{8.24}$$

$$\begin{bmatrix} b_1 \\ a_1 \end{bmatrix} = \underbrace{\begin{bmatrix} t_{11}^{DUT} & t_{12}^{DUT} \\ t_{21}^{DUT} & t_{22}^{DUT} \end{bmatrix}}_{\mathbf{T_{DUT}}} \begin{bmatrix} a_2 \\ b_2 \end{bmatrix} \tag{8.25}$$

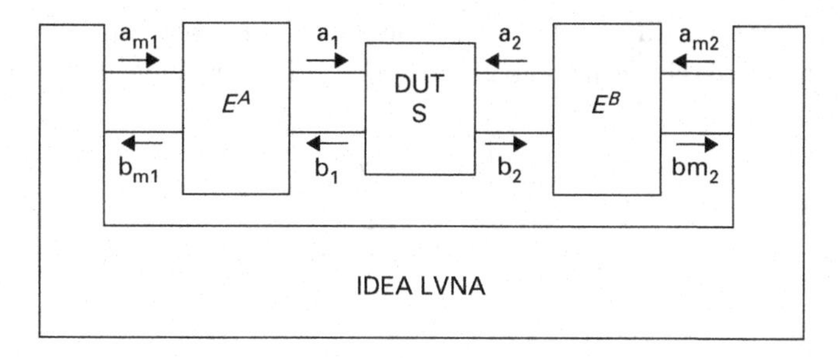

Fig. 8.6 Error model for 2-port VNA with two reflectometers.

$$\underbrace{\begin{bmatrix} t_{11}^B & t_{12}^B \\ t_{21}^B & t_{22}^B \end{bmatrix}}_{\mathbf{T_B}} \begin{bmatrix} a_2 \\ b_2 \end{bmatrix} = \begin{bmatrix} a_{m2} \\ b_{m2} \end{bmatrix} \tag{8.26}$$

and the relationships among the parameters become:

$$\begin{bmatrix} S_{11}^{DUT} & S_{12}^{DUT} \\ S_{21}^{DUT} & S_{22}^{DUT} \end{bmatrix} = \frac{1}{t_{22}^{DUT}} \begin{bmatrix} t_{12}^{DUT} & \Delta_T^{DUT} \\ 1 & -t_{21}^{DUT} \end{bmatrix} \tag{8.27}$$

$$\underbrace{\begin{bmatrix} t_{11}^A & t_{12}^A \\ t_{21}^A & t_{22}^A \end{bmatrix}}_{\mathbf{T_A}} = \frac{1}{e_{21}^A} \underbrace{\begin{bmatrix} -\Delta_E^A & e_{11}^A \\ -e_{22}^A & 1 \end{bmatrix}}_{\mathbf{X_A}} \tag{8.28}$$

$$\underbrace{\begin{bmatrix} e_{11}^A & e_{12}^A \\ e_{21}^A & e_{22}^A \end{bmatrix}}_{\mathbf{E_A}} = \frac{1}{t_{22}^A} \begin{bmatrix} t_{12}^A & \Delta_T^A \\ 1 & -t_{21}^A \end{bmatrix} \tag{8.29}$$

$$\underbrace{\begin{bmatrix} t_{11}^B & t_{12}^B \\ t_{21}^B & t_{22}^B \end{bmatrix}}_{\mathbf{T_B}} = \frac{1}{e_{21}^B} \underbrace{\begin{bmatrix} 1 & -e_{22}^B \\ e_{11}^B & -\Delta_E^B \end{bmatrix}}_{\mathbf{X_B}} \tag{8.30}$$

$$\underbrace{\begin{bmatrix} e_{11}^B & e_{12}^B \\ e_{21}^B & e_{22}^B \end{bmatrix}}_{\mathbf{E_B}} = \frac{1}{t_{11}^B} \begin{bmatrix} t_{21}^B & \Delta_T^B \\ 1 & -t_{12}^B \end{bmatrix}. \tag{8.31}$$

The fundamental calibration equation, i.e. the relationship among the measured and desired quantities, can be now written in terms of the **T** matrix[†] as:

$$\begin{bmatrix} b_{m1} \\ a_{m1} \end{bmatrix} = \underbrace{\mathbf{T_A T_{DUT} T_B}^{-1}}_{\mathbf{T_M}} \begin{bmatrix} a_{m2} \\ b_{m2} \end{bmatrix}$$

$$\Downarrow$$

$$\mathbf{T_M} = \mathbf{T_A T_{DUT} T_B}^{-1} \tag{8.32}$$

[†] For historical reasons we adopt the convention where the T matrix of port 2 is used inverted.

To obtain $\mathbf{T_M}$, two sets of different measurements are required, which are normally given by switching the source between port one and port two. By combining the eight readings we have:

$$\underbrace{\begin{bmatrix} b'_{m1} & b''_{m1} \\ a'_{m1} & a''_{m1} \end{bmatrix}}_{\mathbf{M_1}} = \mathbf{T_M} \underbrace{\begin{bmatrix} a'_{m2} & a''_{m2} \\ b'_{m2} & b''_{m2} \end{bmatrix}}_{\mathbf{M_2}} \qquad (8.33)$$
$$\Downarrow$$
$$\mathbf{T_M} = \mathbf{M_1}\mathbf{M_1}^{-1}$$

where $a'_{m1}, b'_{m1}, a'_{m2}, b'_{m2}$ are with the source at port 1 (*forward* measurements) while $a''_{m1}, b''_{m1}, a''_{m2}, b''_{m2}$ are with the source at port 2 (*reversed* measurements). (8.32) can also be written as:

$$\mathbf{T_M} = \alpha \mathbf{X_A} \mathbf{T_{DUT}} \mathbf{X_B}^{-1}. \qquad (8.34)$$

where $\alpha = \frac{e_{21}^B}{e_{21}^A}$ and

$$\mathbf{T_{DUT}} = \frac{1}{\alpha} \mathbf{X_A}^{-1} \mathbf{T_M} \mathbf{X_B}. \qquad (8.35)$$

which shows that the number of error coefficients required to obtain the corrected **S** matrix in the two-port case is seven and not eight. However the common name for this model is the *eight-term error model* [3].

8.4.2 Forward reverse error model

The second possibility for a two-port VNA is to adopt only one reference coupler and a single coupler on each port, as shown in Figure 8.5. Here the error model discussed above must be changed by assuming that each port has two different states, as shown in Figures 8.7 and 8.8.

1. When the source is connected to the port, the reference and the test coupler form a complete reflectometer and the model is identical to the one discussed above, but here the slightly different notation of (8.36) and (8.37) is introduced.
2. However, when the port is not connected to the source, it is loaded with the internal switch termination and a different model must be adopted where the linearity implies that the two waves will be linked with the single reading $\widehat{b}_{mi}$ as in (8.38) and (8.39).

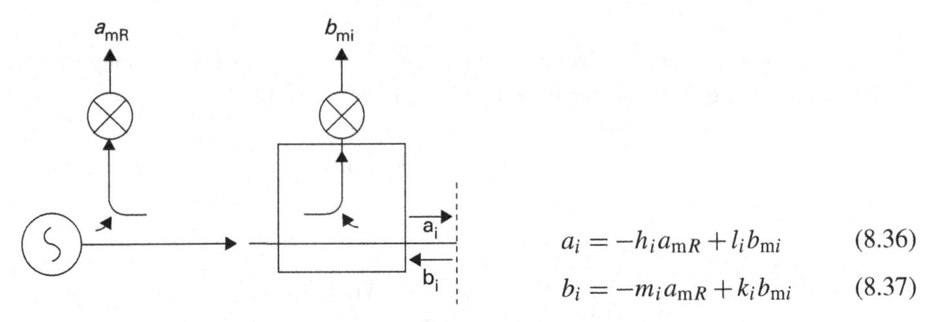

$$a_i = -h_i a_{mR} + l_i b_{mi} \qquad (8.36)$$

$$b_i = -m_i a_{mR} + k_i b_{mi} \qquad (8.37)$$

Fig. 8.7 Two-state error model with the port i connected to the reference channel.

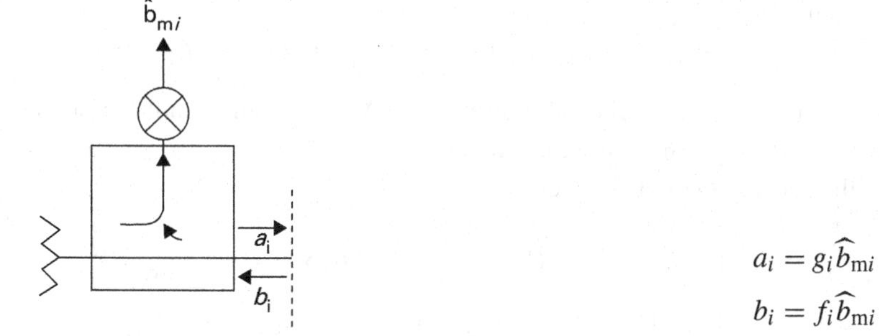

$$a_i = g_i \widehat{b}_{mi} \tag{8.38}$$

$$b_i = f_i \widehat{b}_{mi} \tag{8.39}$$

Fig. 8.8 Two-state error model with the port i teminated

To derive a single equation for calibration and de-embedding, let's note that the total possible measurements obtainable by switching the source on the two ports are six, and let's organize them as follows:

$$\widetilde{\mathbf{A}}_{\mathbf{m}} = \begin{bmatrix} a'_R & 0 \\ 0 & a''_R \end{bmatrix} \widetilde{\mathbf{B}}_{\mathbf{m}} = \begin{bmatrix} b'_{m1} & 0 \\ 0 & b''_{m2} \end{bmatrix} \widehat{\mathbf{B}}_{\mathbf{m}} = \begin{bmatrix} 0 & \widehat{b}''_{m1} \\ \widehat{b}'_{m2} & 0 \end{bmatrix}. \tag{8.40}$$

At the same time let's write the eight DOT waves, two sets of four waves for each source position, as:

$$\mathbf{A} = \begin{bmatrix} a'_1 & a''_1 \\ a'_2 & a''_2 \end{bmatrix} = \widetilde{\mathbf{A}} + \widehat{\mathbf{A}} = \underbrace{\begin{bmatrix} a'_1 & 0 \\ 0 & a''_2 \end{bmatrix}}_{\widetilde{\mathbf{A}}} + \underbrace{\begin{bmatrix} 0 & a''_1 \\ a'_2 & 0 \end{bmatrix}}_{\widehat{\mathbf{A}}} \tag{8.41}$$

$$\mathbf{B} = \begin{bmatrix} b'_1 & b''_1 \\ b'_2 & b''_2 \end{bmatrix} = \widetilde{\mathbf{B}} + \widehat{\mathbf{B}} = \underbrace{\begin{bmatrix} b'_1 & 0 \\ 0 & b''_2 \end{bmatrix}}_{\widetilde{\mathbf{B}}} + \underbrace{\begin{bmatrix} 0 & b''_1 \\ b'_2 & 0 \end{bmatrix}}_{\widehat{\mathbf{B}}} \tag{8.42}$$

and the error coefficients as well as:

$$\begin{aligned} \boldsymbol{H} &= \begin{bmatrix} h_1 & 0 \\ 0 & h_2 \end{bmatrix}, & \boldsymbol{K} &= \begin{bmatrix} k_1 & 0 \\ 0 & k_2 \end{bmatrix}, \\ \boldsymbol{L} &= \begin{bmatrix} l_1 & 0 \\ 0 & l_2 \end{bmatrix}, & \boldsymbol{M} &= \begin{bmatrix} m_1 & 0 \\ 0 & m_2 \end{bmatrix}, \\ \boldsymbol{F} &= \begin{bmatrix} f_1 & 0 \\ 0 & f_2 \end{bmatrix}, & \boldsymbol{G} &= \begin{bmatrix} g_1 & 0 \\ 0 & g_2 \end{bmatrix}. \end{aligned} \tag{8.43}$$

From (8.36)–(8.43) we obtain the calibration equation (8.46) as:

$$\begin{gathered} \widetilde{\mathbf{A}} = \boldsymbol{L}\widetilde{\mathbf{B}}_{\mathrm{m}} - \boldsymbol{H}\widetilde{\mathbf{A}}_{\mathrm{m}} \\ \widetilde{\mathbf{B}} = \boldsymbol{K}\widetilde{\mathbf{B}}_{\mathrm{m}} - \boldsymbol{M}\widetilde{\mathbf{A}}_{\mathrm{m}}, \\ \widehat{\mathbf{A}} = \boldsymbol{G}\widehat{\mathbf{B}}_{\mathrm{m}} \\ \widehat{\mathbf{B}} = \boldsymbol{F}\widehat{\mathbf{B}}_{\mathrm{m}}, \end{gathered} \tag{8.44}$$

$$\begin{gathered} \mathbf{A} = \widetilde{\mathbf{A}} + \widehat{\mathbf{A}} = \boldsymbol{L}\widetilde{\boldsymbol{B}}_{\mathrm{m}} - \boldsymbol{H}\widetilde{\boldsymbol{A}}_{\mathrm{m}} + \boldsymbol{G}\widehat{\boldsymbol{B}}_{\mathrm{m}} \\ \mathbf{B} = \widetilde{\mathbf{B}} + \mathbf{B} = \boldsymbol{K}\widetilde{\boldsymbol{B}}_{\mathrm{m}} - \boldsymbol{M}\widetilde{\boldsymbol{A}}_{\mathrm{m}} + \boldsymbol{F}\widehat{\boldsymbol{B}}_{\mathrm{m}}, \\ \mathbf{S} = \mathbf{B}\mathbf{A}^{-1}, \end{gathered} \tag{8.45}$$

and finally

$$-\boldsymbol{SG}\widehat{\mathbf{B}}_{\mathrm{m}}+\boldsymbol{F}\widehat{\mathbf{B}}_{\mathrm{m}}-\boldsymbol{SL}\widetilde{\mathbf{B}}_{\mathrm{m}}+\boldsymbol{K}\widetilde{\mathbf{B}}_{\mathrm{m}}+\boldsymbol{SH}\widetilde{\mathbf{A}}_{\mathrm{m}}-\boldsymbol{M}\widetilde{\mathbf{A}}_{\mathrm{m}}=\boldsymbol{0}. \tag{8.46}$$

Equation (8.46) is also valid for multiport VNAs as explained in the following chapter; however here we write it in scalar form for 2-port VNAs [7]. In the *forward case*, i.e. with the source at port 1, it gives:

$$\begin{cases} -S_{12}g_2\widehat{b}'_{m2}+(k_1-S_{11}l_1)b'_{m1}+(S_{11}h_1-m_1)a'_R=0 \\ (f_2-S_{22}g_2)\widehat{b}'_{m2}-S_{21}l_1b'_{m1}+S_{21}h_1a'_R=0 \end{cases} \tag{8.47}$$

which can be normalized by k_1 and if we define:

$$\begin{aligned} S_{m11} &= \frac{b'_{m1}}{a'_R} \\ S_{m21} &= \frac{\widehat{b}'_{m2}}{a'_R} \end{aligned} \tag{8.48}$$

as a measured S-parameter we obtain:

$$\begin{cases} -S_{12}\frac{g_2}{k_1}S_{m21}+(1-S_{11}\frac{l_1}{k_1})S_{m11}+(S_{11}\frac{h_1}{k_1}-\frac{m_1}{k_1})=0 \\ (\frac{f_2}{k_1}-S_{22}\frac{g_2}{k_1})S_{m21}-S_{21}\frac{l_1}{k_1}S_{m11}+S_{21}\frac{h_1}{k_1}=0. \end{cases} \tag{8.49}$$

By doing the same procedure for the *reverse* case, i.e. with the source on port 2, we obtain:

$$\begin{cases} -S_{21}g_1\widehat{b}''_{m1}+(k_2-S_{22}l_2)b''_{m2}+(S_{22}h_2-m_2)a''_R=0 \\ (f_1-S_{11}g_1)\widehat{b}''_{m1}-S_{12}l_2b''_{m2}+S_{12}h_2a''_R=0 \end{cases} \tag{8.50}$$

which can be normalized by k_2 and if we define:

$$\begin{aligned} S_{m22} &= \frac{b''_{m2}}{a''_R} \\ S_{m12} &= \frac{\widehat{b}''_{m1}}{a''_R} \end{aligned} \tag{8.51}$$

as a measured S-parameter we obtain:

$$\begin{cases} -S_{21}\frac{g_1}{k_2}S_{m12}+(1-S_{22}\frac{l_2}{k_2})S_{m22}+(S_{22}\frac{h_2}{k_2}-\frac{m_2}{k_2})=0 \\ (\frac{f_1}{k_2}-S_{11}\frac{g_1}{k_2})S_{m12}-S_{12}\frac{l_2}{k_2}S_{m22}+S_{12}\frac{h_2}{k_2}=0. \end{cases} \tag{8.52}$$

Equations (8.49) and (8.52) form a linear system in the four unknown DUT S-parameters that can be easily solved once the error coefficients have been computed. Note that the same system can be applied to calibrate the VNA by using a set of proper standards, i.e. by knowing S_{ij}, and by solving it for the error coefficients. There are ten unknowns so this model is known as the *ten-term error model*. Apparently, this model should have eleven unknowns (six error coefficient times 2 ports minus 1 for the normalization); however the two-port case is the degenerative one and the coefficients reduce to ten, leaving

Table 8.1 Equivalence between different error coefficients notations

VNA	This Book
E_{dF}	m_1/k_1
E_{sF}	l_1/k_1
E_{rF}	$(m_1/k_1)(l_1/k_1) - h_1/k_1$
E_{tF}	f_2/k_1
E_{lF}	$\frac{g_2/k_1}{f_2/k_1}$
E_{dR}	m_2/k_2
E_{sR}	l_2/k_2
E_{rR}	$(m_2/k_2)(l_2/k_2) - h_2/k_2$
E_{tR}	f_1/k_2
E_{lR}	$\frac{g_1/k_2}{f_1/k_2}$

the forward and reverse case equations completely independent. With the formulation here presented the eight-term model can be considered as a subcase of the more general one, but the link between the two is possible only for those VNA architectures with a reflectometer on each port. In this case the eight-term model and the ten-term one are both applicable and interchangeable. However, when there is only one reference coupler this is not possible and only the ten-term model can be used. Table 8.1 reports the formulas for the conversion of the error coefficients notation introduced here with the more common one which is typically included in the VNA firmware.

8.5 Calibration procedures

The two error models can be identified by means of particular procedures which require the measurement of known or even partially known standards. There are many techniques which differ by the kind of standards and the math adopted. Here the more common ones are presented and in particular those called:

- **T**hru-**S**hort-**D**elay,
- **T**hru-**R**eflect-**L**ine,
- **S**hort-**O**pen-**L**oad-**R**eciprocal,
- **L**ine-**R**eflect-**M**atch,
- **S**hort-**O**pen-**L**oad-**T**hru.

All the above methods but the last are usable ONLY with the *eight-term* error model, while the SOLT is usable also on the *forward/reverse* model. This is due to the need for the knowledge of eight readings during the two-port standard measurements, while the SOLT is the only one which does not have this requirement and where the six measurements of the forward/reverse model are enough.

8.5.1 TSD/TRL procedure

This technique uses a Thru which is typically either a direct connection between the ports or a short straight line, a longer **Line** and a known or even unknown reflection standard. Let's consider the two-port calibration equation as (8.32) and let's write it for the Thru, $\mathbf{T}_{mT}$ and Line $\mathbf{T}_{mL}$ measurements:

$$\mathbf{T}_{mT} = \mathbf{T}_A \mathbf{T}_T \mathbf{T}_B^{-1} \tag{8.53}$$

$$\mathbf{T}_{mL} = \mathbf{T}_A \mathbf{T}_L \mathbf{T}_B^{-1}, \tag{8.54}$$

where $\mathbf{T}_T$ and $\mathbf{T}_L$ are the transmission matrix of a fully known standard Thru and of a Line. Let's compute:

$$\begin{aligned} \mathbf{R}_m &= \mathbf{T}_{mL}\mathbf{T}_{mT}^{-1} = \mathbf{T}_A \mathbf{T}_L \mathbf{T}_B^{-1} (\mathbf{T}_A \mathbf{T}_T \mathbf{T}_B^{-1})^{-1} = \\ &= \mathbf{T}_A \mathbf{T}_L \mathbf{T}_T^{-1} \mathbf{T}_A^{-1} = \mathbf{T}_A \Lambda_m \mathbf{T}_A^{-1} \end{aligned} \tag{8.55}$$

$$\begin{aligned} \mathbf{R}_n &= \mathbf{T}_{mT}^{-1}\mathbf{T}_{mL} = \mathbf{T}_B \mathbf{T}_T^{-1} \mathbf{T}_A^{-1} \mathbf{T}_A \mathbf{T}_L \mathbf{T}_B^{-1} = \\ &= \mathbf{T}_B \mathbf{T}_T^{-1} \mathbf{T}_L \mathbf{T}_B^{-1} = \mathbf{T}_B \Lambda_n \mathbf{T}_B^{-1}. \end{aligned} \tag{8.56}$$

If the thru and the line have the same characteristic impedance and their transmission matrices are referenced to an impedance equal to the characteristic one, then:

$$\Lambda_m = \Lambda_n = \Lambda = \mathbf{T}_L \mathbf{T}_T^{-1} = \mathbf{T}_T^{-1} \mathbf{T}_L = \begin{bmatrix} e^{\gamma(l_L - l_T)} & 0 \\ 0 & e^{-\gamma(l_L - l_T)} \end{bmatrix}. \tag{8.57}$$

where l_L and l_T are the line and thru electrical lenghts and γ is the propagation constant. (8.55) and (8.56) are eigenvalue equations where Λ is the eigenvalues matrix. The corresponding eigenvector matrices are the desired error coefficient ones:

$$\mathbf{T}_A = p \begin{bmatrix} \frac{k}{p} a & b \\ \frac{k}{p} & 1 \end{bmatrix} = p \mathbf{X}_A \tag{8.58}$$

$$\mathbf{T}_B = w \begin{bmatrix} 1 & \frac{u}{w} \\ f & \frac{u}{w} g \end{bmatrix} = w \mathbf{X}_B. \tag{8.59}$$

From the solution of the eigenvalues/vector problem f, g, a, and b are known, but $\frac{k}{p}$, $\frac{u}{w}$ and $\alpha = \frac{p}{w}$ (see (8.34)) are still unknown. Let's first consider the measurement of a fully known reflective standard Γ_S^A, as an ideal short at port 1. From (8.14), (8.28), and (8.58), we have:

$$\Gamma_{mS}^A = \frac{e_{11}^A - \Delta_E^A \Gamma_S^A}{1 - e_{22}^A \Gamma_S^A} = \frac{b + \frac{k}{p} a \Gamma_S^A}{1 + \frac{k}{p} \Gamma_S^A} \tag{8.60}$$

$$\Downarrow$$

$$\frac{k}{p} = a \frac{b - \Gamma_{mS}^A}{\Gamma_S^A (\Gamma_{mS}^A - 1)} \tag{8.61}$$

Once $\frac{k}{p}$ is known, $\mathbf{X}_A$ in known. From the thru measurement (8.53), we have:

$$\mathbf{T}_{mT} = \alpha \mathbf{X}_A \mathbf{T}_T \mathbf{X}_B^{-1} \tag{8.62}$$

$$\Downarrow$$

$$\mathbf{Y} = \alpha \mathbf{X}_B^{-1} = \mathbf{T}_T^{-1} \mathbf{X}_A^{-1} \mathbf{T}_{mT} \tag{8.63}$$

Given $\mathbf{Y}$, the calibration problem is solved since the de-embedding equation (8.35) can be now written as:

$$\mathbf{T}_{DUT} = \mathbf{X}_A^{-1} \mathbf{T}_M \mathbf{Y}^{-1}. \tag{8.64}$$

Since the used standards are: a **T**hru, a **S**hort and a **D**elay line the **TSD** acronym was used. The evolution of this technique was the so-called **T**hru-**R**eflect-**L**ine which assumes that the reflective standard Γ_X is **not known**, but is measured at the two ports, so:

$$\Gamma^A_{mX} = \frac{e^A_{11} - \Delta^A_E \Gamma_X}{1 - e^A_{22} \Gamma_X} = \frac{b + \frac{k}{p} a \Gamma_X}{1 + \frac{k}{p} \Gamma_X} \tag{8.65}$$

$$\Gamma^B_{mX} = \frac{e^B_{11} - \Delta^B_E \Gamma_X}{1 - e^B_{22} \Gamma_X} = \frac{f + \frac{u}{w} g \Gamma_X}{1 + \frac{u}{w} \Gamma_X}. \tag{8.66}$$

Let's compute the symbolic form of the thru measurement matrix:

$$\mathbf{T}_{mT} = \alpha \mathbf{X}_A \mathbf{T}_T \mathbf{X}_B^{-1} = \frac{\alpha}{e^{\gamma l_T} \frac{u}{w} (g - f)} \begin{bmatrix} e^{2\gamma l_T} a g \frac{k}{p} \frac{u}{w} - bf & e^{2\gamma l_T} a \frac{k}{p} \frac{u}{w} - b \\ e^{2\gamma l_T} g \frac{k}{p} \frac{u}{w} - f & e^{2\gamma l_T} \frac{k}{p} \frac{u}{w} - 1 \end{bmatrix}. \tag{8.67}$$

If we consider the element $\mathbf{T}_{mT}(1,2)/\mathbf{T}_{mT}(2,2)$ we obtain:

$$\mathbf{S}_{mT}(1,1) = \frac{\mathbf{T}_{mT}(1,2)}{\mathbf{T}_{mT}(2,2)} = \frac{e^{2\gamma l_T} a \frac{k}{p} \frac{u}{w} - b}{e^{2\gamma l_T} \frac{k}{p} \frac{u}{w} - 1} \tag{8.68}$$

and solving the nonlinear system formed by (8.65), (8.66), and (8.68), Γ_X is obtained as:

$$\Gamma_X = \pm e^{\gamma l_T} \sqrt{\frac{(f - \Gamma^B_{mX})(b - \Gamma^A_{mX})(a - \mathbf{S}_{mT}(1,1))}{(g - \Gamma^B_{mX})(a - \Gamma^A_{mX})(b - \mathbf{S}_{mT}(1,1))}}. \tag{8.69}$$

Once Γ_X is known the procedure either follows the **TSD** algorithm or $\frac{k}{p}$ and $\frac{u}{w}$ are given by (8.65) and (8.66), while α can be computed from $\mathbf{T}_{mT}(2,2)$ as:

$$\mathbf{T}_{mT}(2,2) = \alpha \frac{e^{2\gamma l_T} \frac{k}{p} \frac{u}{w} - 1}{e^{\gamma l_T} \frac{u}{w} (g - f)} \Rightarrow \alpha = \mathbf{T}_{mT}(2,2) \frac{e^{\gamma l_T} \frac{u}{w} (g - f)}{e^{2\gamma l_T} \frac{k}{p} \frac{u}{w} - 1}. \tag{8.70}$$

The sign of Γ_X must be known and typically is given by a rough knowledge of the reflection type (a short or an open). Finally note that Γ_X cannot be a match load, i.e

$\Gamma_X = 0$, because we would simply obtain $\Gamma^A_{mX} = b$ and $\Gamma^B_{mX} = f$ which are already known.
The main characteristics of the **TRL** calibration are:

1. The propagation constant of the line is obtained as a by-product of the calibration.
2. The characteristic impedance of the line sets the reference impedance of the VNA.
3. If a zero length thru, i.e. a direct port connection, is used, only the line must be known.
4. The length difference between the line and the thru does not have to be a multiple of the wavelength.

The S_{21} parameter of the line is given by the solution of the eigenvalues problem, while the propagation constant is obtainable if the length is known without the calibration. For this reason the line can be a *partially* known standard. This property of TRL was successfully used to characterize different structures [8]. However the characteristic impedance of the line **automatically** becomes the reference impedance of the VNA since the diagonal property of the eigenvalues matrix Λ is obtainable only by assuming that the reference impedance is equal to the characteristic one. Furthermore its value MUST be known a priori and not from the measurement. The TRL procedure is the only one which sets the reference impedance based on a distributed component, the LINE, while all the other calibration methodologies use a lumped component to set the reference impedance. Thus it's questionable if waves are really measured using calibration procedures other than TRL [9] and all the national metrology labs in the world use a set of lines as their primary microwave coaxial and waveguide standards. If the THRU has zero length, i.e. a unitary transmission matrix, the characteristic impedance of the line remains the ONLY parameter required to obtain a successful calibration and this property makes TRL easily traceable to the mechanical dimension of the standard. The main drawback of the TRL technique is the relatively small bandwidth because of the line resonance. At the frequency where $e^{\gamma(l_L - l_T)} = 1$, the eigenvalue matrix becomes unitary and obviously the problem becomes undeterminate. The calibraton fails and typically, a glitch appears in the measurement date. To avoid this problem, for a broadband calibration several lines are mandatory and a *Multiline TRL* is used [10].

8.5.2 SOLR procedure

The **S**hort-**O**pen-**L**oad-**R**eciprocal calibration was introduced in 1992 and it avoids the use of a fully known two-port device, generally the THRU, which is common in all the other techniques[4]. If a linear system based on (8.65) and (8.66) is formed with three different standards on each port, as done in the one-port case, the matrices $\mathbf{X}_A$ and $\mathbf{X}_B$ are easily obtained, i.e six out of seven error coefficients are given by one-port measurements. Let's now take the measurement of a reciprocal device $\mathbf{T}_{mR}$, i.e with $S_{12} = S_{21}$, from (8.34):

$$\mathbf{T}_{mR} = \alpha \mathbf{X}_A \mathbf{T}_R \mathbf{X}_B^{-1}. \tag{8.71}$$

If the determinant on both sides on (8.71) is taken and by noting that the reciprocity condition implies $\det(\mathbf{T}_R) = 1$, the last term α is easily obtained as:

$$\det(\mathbf{T}_{mR}) = \alpha^2 \frac{\det(\mathbf{X}_A)\overbrace{\det(\mathbf{T}_R)}^{=1}}{\det(\mathbf{X}_B)} \Rightarrow \alpha = \pm\sqrt{\frac{\det(\mathbf{X}_B)\det(\mathbf{T}_{mR})}{\det(\mathbf{X}_A)}}. \qquad (8.72)$$

The characteristics of this technique are

1. Easily applicable with ordinary one-port standards.
2. Any reciprocal device can be used as THRU.
3. Different connection problems can be easily solved.
4. The one-port standards must be fully known.

The SOLR uses the same one-port standard set as the old **SOLT** and this means a straighforward applicability to all the full reflectometer VNAs. However the freedom from the THRU device, the main characteristic of this technique, allows a much easier solution of the calibration problem in many situations. As an example, if the two ports are far apart a fully known THRU may be difficult to obtain while a simple cable, used as reciprocal, is a very easy replacement. Another typical example is on-wafer measurements with right-angle probes, where a bended THRU is far from being an ideal line while the reciprocity condition is easily achieved with non-giromagnetic structures. The main constraint given by the SOLR is the need for perfectly known one-port standards.

8.5.3 LRM procedure

One of the more interesting two-port techniques is the **L**ine-**R**eflect-**M**atch. Here a different notation is used instead of the original one [3], but this one is also useful for the multiport case in the next paragraph. Let's consider (8.36) and (8.37), in the full reflectometer case:

$$a_1 = -h_1 a_{\mathrm{m}1} + l_1 b_{\mathrm{m}1} \qquad (8.73)$$

$$b_1 = -m_1 a_{\mathrm{m}1} + k_1 b_{\mathrm{m}1} \qquad (8.74)$$

$$a_2 = -h_2 a_{\mathrm{m}2} + l_2 b_{\mathrm{m}2} \qquad (8.75)$$

$$b_2 = -m_2 a_{\mathrm{m}2} + k_2 b_{\mathrm{m}2}. \qquad (8.76)$$

As done before for the ten-term model, we can organize the eight readings, four for each source position, as:

$$\mathbf{A_m} = \begin{bmatrix} a'_{m1} & a''_{m1} \\ a'_{m2} & a''_{m2} \end{bmatrix} \quad \mathbf{B_m} = \begin{bmatrix} b'_{m1} & b''_{m1} \\ b'_{m2} & b''_{m2} \end{bmatrix}, \qquad (8.77)$$

the eight DUT waves as:

$$\mathbf{A}=\begin{bmatrix} a_1' & a_1'' \\ a_2' & a_2'' \end{bmatrix} \quad \mathbf{B}=\begin{bmatrix} b_1' & b_1'' \\ b_2' & b_2'' \end{bmatrix} \tag{8.78}$$

and also the error coefficients as (8.43). From (8.77), (8.78), and (8.43) we obtain the calibration equation (8.80) as:

$$\begin{aligned} \mathbf{A} &= \boldsymbol{L}\,\mathbf{B}_{\mathrm{m}} - \boldsymbol{H}\,\mathbf{A}_{\mathrm{m}} \\ \mathbf{B} &= \boldsymbol{K}\,\mathbf{B}_{\mathrm{m}} - \boldsymbol{M}\,\mathbf{A}_{\mathrm{m}} \end{aligned} \tag{8.79}$$

$$-\mathbf{S}\boldsymbol{L}\mathbf{B}_{\mathrm{m}} + \boldsymbol{K}\mathbf{B}_{\mathrm{m}} + \mathbf{S}\boldsymbol{H}\mathbf{A}_{\mathrm{m}} - \boldsymbol{M}\mathbf{A}_{\mathrm{m}} = \mathbf{0}. \tag{8.80}$$

This equation is the scattering version of (8.32). Let's introduce the line measurement matrices

$$\mathbf{A}_{\mathrm{Lm}}=\begin{bmatrix} a_{\mathrm{L}m1}' & a_{\mathrm{L}m1}'' \\ a_{\mathrm{L}m2}' & a_{\mathrm{L}m2}'' \end{bmatrix} \quad \mathbf{B}_{\mathrm{Lm}}=\begin{bmatrix} b_{\mathrm{L}m1}' & b_{\mathrm{L}m1}'' \\ b_{\mathrm{L}m2}' & b_{\mathrm{L}m2}'' \end{bmatrix} \tag{8.81}$$

and the line S-parameter matrix

$$\mathbf{S_L}=\begin{bmatrix} S_{L11} & S_{L12} \\ S_{L11} & S_{L22} \end{bmatrix}. \tag{8.82}$$

Equation (8.80) becomes:

$$-\mathbf{S_L L B_{Lm}} + \mathbf{K B_{Lm}} + \mathbf{S_L H A_{Lm}} - \mathbf{M A_{Lm}} = \mathbf{0}, \tag{8.83}$$

i.e. in scalar form:

$$\begin{aligned} &-\begin{bmatrix} S_{L11} & S_{L12} \\ S_{L11} & S_{L22} \end{bmatrix}\begin{bmatrix} l_1 & 0 \\ 0 & l_2 \end{bmatrix}\begin{bmatrix} b_{\mathrm{L}m1}' & b_{\mathrm{L}m1}'' \\ b_{\mathrm{L}m2}' & b_{\mathrm{L}m2}'' \end{bmatrix}+ \\ &+\begin{bmatrix} k_1 & 0 \\ 0 & k_2 \end{bmatrix}\begin{bmatrix} b_{\mathrm{L}m1}' & b_{\mathrm{L}m1}'' \\ b_{\mathrm{L}m2}' & b_{\mathrm{L}m2}'' \end{bmatrix}+ \\ &+\begin{bmatrix} S_{L11} & S_{L12} \\ S_{L11} & S_{L22} \end{bmatrix}\begin{bmatrix} h_1 & 0 \\ 0 & h_2 \end{bmatrix}\begin{bmatrix} a_{\mathrm{L}m1}' & a_{\mathrm{L}m1}'' \\ a_{\mathrm{L}m2}' & a_{\mathrm{L}m2}'' \end{bmatrix}+ \\ &-\begin{bmatrix} m_1 & 0 \\ 0 & m_2 \end{bmatrix}\begin{bmatrix} a_{\mathrm{L}m1}' & a_{\mathrm{L}m1}'' \\ a_{\mathrm{L}m2}' & a_{\mathrm{L}m2}'' \end{bmatrix}=\begin{bmatrix} 0 & 0 \\ 0 & 0 \end{bmatrix}. \end{aligned} \tag{8.84}$$

Equation (8.84) provides four independent equations, one for each *ij* element. By doing the same procedure for the load measurement matrices:

$$\mathbf{A}_{\Gamma\mathbf{m}}=\begin{bmatrix} a_{\Gamma m1}' & 0 \\ 0 & a_{\Gamma m2}'' \end{bmatrix} \quad \mathbf{B}_{\Gamma\mathbf{m}}=\begin{bmatrix} b_{\Gamma m1}' & 0 \\ 0 & b_{\Gamma m2}'' \end{bmatrix} \tag{8.85}$$

and by remembering that the reflection coefficient for the load is null, we have:

$$\begin{aligned}
&-\begin{bmatrix}0 & 0\\ 0 & 0\end{bmatrix}\begin{bmatrix}l_1 & 0\\ 0 & l_2\end{bmatrix}\begin{bmatrix}b'_{\Gamma m1} & 0\\ 0 & b''_{\Gamma m2}\end{bmatrix}+\\
&+\begin{bmatrix}k_1 & 0\\ 0 & k_2\end{bmatrix}\begin{bmatrix}b'_{\Gamma m1} & 0\\ 0 & b''_{\Gamma m2}\end{bmatrix}+\\
&+\begin{bmatrix}0 & 0\\ 0 & 0\end{bmatrix}\begin{bmatrix}h_1 & 0\\ 0 & h_2\end{bmatrix}\begin{bmatrix}a'_{\Gamma m1} & 0\\ 0 & a''_{\Gamma m2}\end{bmatrix}+\\
&-\begin{bmatrix}m_1 & 0\\ 0 & m_2\end{bmatrix}\begin{bmatrix}a'_{\Gamma m1} & 0\\ 0 & a''_{\Gamma m2}\end{bmatrix}=\begin{bmatrix}0 & 0\\ 0 & 0\end{bmatrix},
\end{aligned} \tag{8.86}$$

i.e. other two independent equations are given by (8.86), because only the *ii* elements are not null. These six equations form a linear calibration system as:

$$\underbrace{\begin{bmatrix}
-a'_{Lm1} & 0 & -b'_{Lm1}\mathbf{S}_{L11} & -b'_{Lm2}\mathbf{S}_{L12} & a'_{Lm1}\mathbf{S}_{L11} & a'_{Lm2}\mathbf{S}_{L12} & b'_{Lm1} & 0\\
-a''_{Lm1} & 0 & -b''_{Lm1}\mathbf{S}_{L11} & -b''_{Lm2}\mathbf{S}_{L12} & a''_{Lm1}\mathbf{S}_{L11} & a''_{Lm2}\mathbf{S}_{L12} & b''_{Lm1} & 0\\
0 & -a'_{Lm2} & -b'_{Lm1}\mathbf{S}_{L21} & -b'_{Lm2}\mathbf{S}_{L22} & a'_{Lm1}\mathbf{S}_{L21} & a'_{Lm2}\mathbf{S}_{L22} & 0 & b'_{Lm2}\\
0 & -a''_{Lm2} & -b''_{Lm1}\mathbf{S}_{L21} & -b''_{Lm2}\mathbf{S}_{L22} & a''_{Lm1}\mathbf{S}_{L21} & a''_{Lm2}\mathbf{S}_{L22} & 0 & b''_{Lm2}\\
-a'_{\Gamma m1} & 0 & 0 & 0 & 0 & 0 & b'_{\Gamma m1} & 0\\
0 & -a''_{\Gamma m2} & 0 & 0 & 0 & 0 & 0 & b''_{\Gamma m2}
\end{bmatrix}}_{\mathbf{N}}
\underbrace{\begin{bmatrix}m_1\\ m_2\\ l_1\\ l_2\\ h_1\\ h_2\\ \hline k_1\\ k_2\end{bmatrix}}_{\mathbf{u}}=\mathbf{0}$$

$$\Downarrow$$

$$\underbrace{\begin{bmatrix}
-a'_{Lm1} & 0 & -b'_{Lm1}\mathbf{S}_{L11} & -b'_{Lm2}\mathbf{S}_{L12} & a'_{Lm1}\mathbf{S}_{L11} & a'_{Lm2}\mathbf{S}_{L12}\\
-a''_{Lm1} & 0 & -b''_{Lm1}\mathbf{S}_{L11} & -b''_{Lm2}\mathbf{S}_{L12} & a''_{Lm1}\mathbf{S}_{L11} & a''_{Lm2}\mathbf{S}_{L12}\\
0 & -a'_{Lm2} & -b'_{Lm1}\mathbf{S}_{L21} & -b'_{Lm2}\mathbf{S}_{L22} & a'_{Lm1}\mathbf{S}_{L21} & a'_{Lm2}\mathbf{S}_{L22}\\
0 & -a''_{Lm2} & -b''_{Lm1}\mathbf{S}_{L21} & -b''_{Lm2}\mathbf{S}_{L22} & a''_{Lm1}\mathbf{S}_{L21} & a''_{Lm2}\mathbf{S}_{L22}\\
-a'_{\Gamma m1} & 0 & 0 & 0 & 0 & 0\\
0 & -a''_{\Gamma m2} & 0 & 0 & 0 & 0
\end{bmatrix}}_{\widetilde{\mathbf{N}}}
\underbrace{\begin{bmatrix}m_1\\ m_2\\ l_1\\ l_2\\ h_1\\ h_2\end{bmatrix}}_{\widetilde{\mathbf{u}}}+$$

$$+\underbrace{\begin{bmatrix}
b'_{Lm1} & 0\\
b''_{Lm1} & 0\\
0 & b'_{Lm2}\\
0 & b''_{Lm2}\\
b'_{\Gamma m1} & 0\\
0 & b''_{\Gamma m2}
\end{bmatrix}}_{\widehat{\mathbf{N}}}
\underbrace{\begin{bmatrix}k_1\\ k_2\end{bmatrix}}_{\widehat{\mathbf{u}}}=\mathbf{0}$$

$$\Downarrow$$

$$\widetilde{\mathbf{N}}\widetilde{\mathbf{u}}+\widehat{\mathbf{N}}\widehat{\mathbf{u}}=\mathbf{0}$$

$$\Downarrow$$

$$\widetilde{\mathbf{u}}=-\widetilde{\mathbf{N}}^{-1}\widehat{\mathbf{N}}\widehat{\mathbf{u}}=\mathbf{W}\widehat{\mathbf{u}}. \tag{8.87}$$

The matrix $\mathbf{W}$ is fully known from the measurements and the definitions of the LINE and MATCH standards. (8.87) defines a subset of the normalized error coefficients $\widetilde{\mathbf{u}}$ as a linear combination of the ratio $\frac{k_2}{k_1}$ as:

$$
\begin{aligned}
\frac{m_1}{k_1} &= w_{11} + w_{12}\frac{k_2}{k_1} \\
\frac{m_2}{k_1} &= w_{21} + w_{22}\frac{k_2}{k_1} \\
\frac{l_1}{k_1} &= w_{31} + w_{32}\frac{k_2}{k_1} \\
\frac{l_2}{k_1} &= w_{41} + w_{42}\frac{k_2}{k_1} \\
\frac{h_1}{k_1} &= w_{51} + w_{52}\frac{k_2}{k_1} \\
\frac{h_2}{k_1} &= w_{61} + w_{62}\frac{k_2}{k_1}.
\end{aligned}
\tag{8.88}
$$

The unknown reflection measured at the two ports gives:

$$
\underbrace{\begin{bmatrix} -a'_{Rm1} & 0 & -b'_{Rm1}\Gamma_R & 0 & a'_{Rm1}\Gamma_R & 0 \\ 0 & -a''_{Rm2} & 0 & -b''_{Rm2}\Gamma_R & 0 & a''_{Rm2}\Gamma_R \end{bmatrix}}_{\widetilde{\mathbf{N}}_R} \underbrace{\begin{bmatrix} m_1 \\ m_2 \\ l_1 \\ l_2 \\ h_1 \\ h_2 \end{bmatrix}}_{\widetilde{\mathbf{u}}} +
$$

$$
+ \underbrace{\begin{bmatrix} b'_{Rm1} & 0 \\ 0 & b''_{Rm2} \end{bmatrix}}_{\widehat{\mathbf{N}}_R} \underbrace{\begin{bmatrix} k_1 \\ k_2 \end{bmatrix}}_{\widehat{\mathbf{u}}} = \mathbf{0}. \tag{8.89}
$$

Let $\Gamma_{Rm1} = \frac{b_{Rm1}}{a_{Rm1}}$ and $\Gamma_{Rm2} = \frac{b_{Rm2}}{a_{Rm2}}$; from (8.87) and (8.89) we have:

$$
\begin{gathered}
\widetilde{\mathbf{N}}_R \mathbf{W} \widehat{\mathbf{u}} + \widehat{\mathbf{N}}_R \widehat{\mathbf{u}} = \mathbf{0} \\
\Downarrow \\
(\widetilde{\mathbf{N}}_R \mathbf{W} + \widehat{\mathbf{N}}_R) \widehat{\mathbf{u}} = \mathbf{0},
\end{gathered}
\tag{8.90}
$$

which has a non-null solution only if the determinant of

$$
\widetilde{\mathbf{N}}_R \mathbf{W} + \widehat{\mathbf{N}}_R =
\begin{bmatrix} -w_{31}\Gamma_{Rm1}\Gamma_R + w_{51} - w_{11} + \Gamma_{Rm1} & -w_{32}\Gamma_{Rm1}\Gamma_R + w_{52} - w_{12} \\ -w_{41}\Gamma_{Rm2}\Gamma_R + w_{61} - w_{21} & -w_{42}\Gamma_{Rm2}\Gamma_R + w_{62} - w_{22} + \Gamma_{Rm2} \end{bmatrix} \tag{8.91}
$$

is null, i.e:

$$
\begin{aligned}
&(-w_{31}\Gamma_{Rm1}\Gamma_R + w_{51} - w_{11} + \Gamma_{Rm1})(-w_{42}\Gamma_{Rm2}\Gamma_R + w_{62} - w_{22} + \Gamma_{Rm2}) \\
&-(-w_{41}\Gamma_{Rm2}\Gamma_R + w_{61} - w_{21})(-w_{32}\Gamma_{Rm1}\Gamma_R + w_{52} - w_{12}) = 0.
\end{aligned}
\tag{8.92}
$$

Equation (8.92) is a second-order equation in Γ_R which can be easily solved. Once Γ_R is known the ratio $\frac{k_2}{k_1}$ is easily computed from (8.90) and all the other error coefficients from (8.88). The main charateristics of **LRM** are:

- Broadband performances.
- Suitable for on-wafer measurement where the use of a load does not require a probe shifting.
- High accuracy as the TRL, if good broadband loads are used.
- The VNA reference impedance is set by the load.

8.5.4 SOLT procedure

The **S**hort-**O**pen-**L**oad-**T**hru is the oldest, most used and widely adopted calibration technique. This is due to the applicability of the technique to both a complete or a partial reflectometer VNA architecture, as shown in the following. The SOLT was first implemented inside the VNA firmware due to its simplicity and applicability on both models; furthermore the commercial availability of coaxial standards made it easy to manage.

Based on the mathematics presented in the previous paragraphs, we use (8.80) to write the corresponding calibration system for the seven measured standards (Opens: Γ_{O1}, Γ_{O2}; Shorts: Γ_{S1}, Γ_{S2}; Loads: $\Gamma_{L1} = \Gamma_{L2} = 0$ and Thru $\mathbf{S}_T$) as [11]:

$$\mathbf{N} = \begin{bmatrix} -a'_{Tm1} & 0 & -b'_{Tm1}S_{T11} & -b'_{Tm2}S_{T12} & a'_{Tm1}S_{T11} & a'_{Tm2}S_{T12} & b'_{Tm1} & 0 \\ -a''_{Tm1} & 0 & -b''_{Tm1}S_{T11} & -b''_{Tm2}S_{T12} & a''_{Tm1}S_{T11} & a''_{Tm2}S_{T12} & b''_{Tm1} & 0 \\ 0 & -a'_{Tm2} & -b'_{Tm1}S_{T21} & -b'_{Tm2}S_{T22} & a'_{Tm1}S_{T21} & a'_{Tm2}S_{T22} & 0 & b'_{Tm2} \\ 0 & -a''_{Tm2} & -b''_{Tm1}S_{T21} & -b''_{Tm2}S_{T22} & a''_{Tm1}S_{T21} & a''_{Tm2}S_{T22} & 0 & b''_{Tm2} \\ -a'_{Lm1} & 0 & 0 & 0 & 0 & 0 & b'_{Lm1} & 0 \\ 0 & -a''_{Lm2} & 0 & 0 & 0 & 0 & 0 & b''_{Lm2} \\ -a'_{Sm1} & 0 & -b'_{Sm1}\Gamma_{S1} & 0 & a'_{Sm1}\Gamma_{S1} & 0 & b'_{Sm1} & 0 \\ 0 & -a''_{Sm2} & 0 & -b''_{Sm2}\Gamma_{S2} & 0 & a''_{Sm2}\Gamma_{S2} & b'_{Sm2} & 0 \\ -a'_{Om1} & 0 & -b'_{Om1}\Gamma_{S1} & 0 & a'_{Om1}\Gamma_{O1} & 0 & b'_{Om1} & 0 \\ 0 & -a''_{Om2} & 0 & -b''_{Om2}\Gamma_{O2} & 0 & a''_{Om2}\Gamma_{O2} & b'_{Om2} & 0 \end{bmatrix}$$

$$\mathbf{u} = \begin{bmatrix} m_1 & m_2 & l_1 & l_2 & h_1 & h_2 & k_1 & k_2 \end{bmatrix}^T$$

$$\mathbf{Nu} = \mathbf{0}. \tag{8.93}$$

The linear system (8.93) is overdetermined in this case† and it can be either solved in the least squared sense or better, three equations are non-necessary. This case was called **QSOLT** (Quick SOLT) because it may avoid the use of one-port standards at port 2 [12, 13]. However for the *forward/reverse* model, since we have more unknowns, all the standard measurements are required. Let's take (8.49) and specialize it for the corresponding standard measurements with the source at port 1, i.e. the *forward case*, as:

$$\begin{bmatrix} 1 & S_{T11}S_{Tm11} & -S_{T11} & 0 & S_{T12}S_{m21} \\ 0 & S_{T21}S_{Tm11} & -S_{T21} & -S_{Tm21} & S_{T22}S_{Tm21} \\ 1 & 0 & 0 & 0 & 0 \\ 1 & \Gamma_{O1}S_{Om11} & -\Gamma_{O1} & 0 & 0 \\ 1 & \Gamma_{S1}S_{Sm11} & -\Gamma_{S1} & 0 & 0 \end{bmatrix} \begin{bmatrix} m_1/k_1 \\ l_1/k_1 \\ h_1/k_1 \\ f_2/k_1 \\ g_2/k_1 \end{bmatrix} = \begin{bmatrix} S_{Tm11} \\ 0 \\ S_{Lm1} \\ S_{Om1} \\ S_{Sm1} \end{bmatrix} \tag{8.94}$$

† We have 7 unknowns and ten equations.

While in the *reverse case*, i.e source at port 2, we use (8.52) and obtain:

$$\begin{bmatrix} 1 & S_{T22}S_{Tm22} & -S_{T22} & 0 & S_{T21}S_{m12} \\ 0 & S_{T12}S_{Tm22} & -S_{T12} & -S_{Tm12} & S_{T11}S_{Tm12} \\ 1 & 0 & 0 & 0 & 0 \\ 1 & \Gamma_{O2}S_{Om22} & -\Gamma_{O2} & 0 & 0 \\ 1 & \Gamma_{S2}S_{Sm22} & -\Gamma_{S2} & 0 & 0 \end{bmatrix} \begin{bmatrix} m_2/k_2 \\ l_2/k_2 \\ h_2/k_2 \\ f_1/k_2 \\ g_1/k_2 \end{bmatrix} = \begin{bmatrix} S_{Tm22} \\ 0 \\ S_{Lm2} \\ S_{Om2} \\ S_{Sm2} \end{bmatrix}. \quad (8.95)$$

The solution of (8.94) and (8.95) directly gives the error coefficients.

8.6 Recent developments

The VNA error models here presented were all based on the *Non Leakage* hypothesis, i.e it's assumed that port 1 and port 2 are isolated and no signal appears on the other side channel unless a 2-port DUT is connected. This assumption limits the applicability of the model where the crosstalk signals are significantly lower than the DUT ones, which is the typical case with coaxial or waveguide devices measurements. However for high attenuation testing or for on-wafer critical applications the crosstalk cannot be negleted. The leakage error models have been introduced since the 1970s, but a successful standard sequence was only invented in the 1990s [14], [15]. Following the linearity principle it's easy to develop a *leakage* model by considering the eight-port network formed by the two sides of the VNA, as shown in Figure 8.9. As done before there is a $\mathbb{C}^4 \Longrightarrow \mathbb{C}^4$ linear

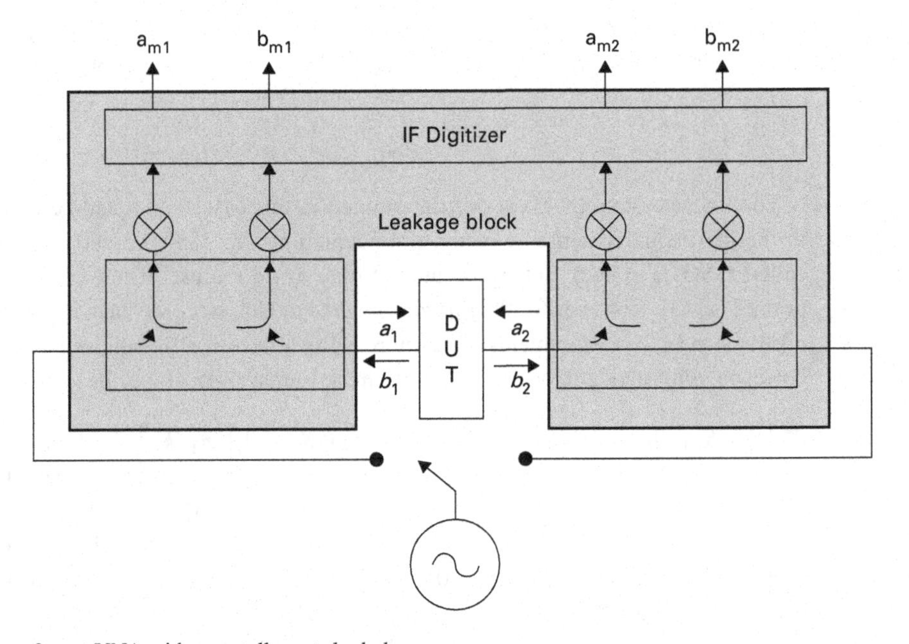

Fig. 8.9 2-port VNA with crosstalk part shaded.

application between the measured and DUT quantities as:

$$\begin{bmatrix} b_1 \\ a_1 \\ b_2 \\ a_2 \end{bmatrix} = \begin{bmatrix} d_{11} & d_{12} & d_{13} & d_{14} \\ d_{21} & d_{22} & d_{23} & d_{24} \\ d_{31} & d_{32} & d_{33} & d_{34} \\ d_{41} & d_{42} & d_{43} & d_{44} \end{bmatrix} \begin{bmatrix} a_{m1} \\ b_{m1} \\ a_{m2} \\ b_{m2} \end{bmatrix}. \tag{8.96}$$

Equation (8.96) contains 16 error terms and a particular solution of the identification problem was given in [14]. Although the leakage calibration has been formally and experimentally tested, the leakage terms are often position-dependent, as in the on-wafer environment where the probes distance dramatically affects the crosstalk. In this case, the correction given by the 16-term calibration may fail because the error terms may change after the calibration. Finally, the author wishes to point out that the simple solution of the leakage problem that can be found in many VNA firmware, which adds two error terms to the *ten-term model*, may lead to incorrect measurements.

Since the two-port VNAs are the most widely used and the calibration is a must to obtain reliable measurements, in the 1990s automatic calibrator devices were introduced and called *Electronic Calibrators*. They are typically PIN diode-based networks that contain different loads and are precharacterized with a metrological grade VNA, calibrated with TRL. The broadband measurement of the electronic calibrator for all the possible states is stored in a file shipped with each unit and used by the VNA firmware to solve the calibration, typically with an SOLT or SOLR algorithm. Since these devices substitute the traditional mechanical standards they are called *Transfer Standards* because their electrical behavior is measured and not computed from mechanical dimensions and EM theory[16]. By using an *Electronic Calibrator* the VNA calibration is greatly simplified to a single connection, but it's always better to verify the obtained accuracy by measuring at least one traditional mechanical device on both ports.

8.7 Conclusion

The calibration of two-port VNAs has greatly enhanced the measurement accuracy of microwave devices. During the last 40 years many different algorithms have been proposed, but the error models and the techniques shown here are now established and implemented in the majority of modern VNAs.

References

[1] B. Donecker, *Determining the measurement accuracy of the HP8510 microwave network analyzer*. Santa Rosa, CA: HP, 1985.

[2] R. A. Franzen, N. R. Speciale, "New procedure for system calibration and error removal in automated s-parameter measurements," in *5th European Microwave Conference*, Sept. 1975.

[3] H. Eul and B. Schieck, "A generalized theory and new calibration procedures for network analyzer self-calibration," *IEEE Trans. Microw. Theory Tech.*, vol. MTT-39, pp. 724–731, Apr. 1991.

[4] A. Ferrero and U. Pisani, "Two-port network analyzer calibration using an unknown thru," *IEEE Microw. Guid. Wave Lett.*, vol. 2, pp. 505–507, Dec. 1992.

[5] R. A. Hackborn, "An automatic network analyzer system," *Microwave J.*, vol. 11, pp. 45–52, May 1968.

[6] I. Kasa, "A circle fitting procedure and its error analysis," *IEEE Trans. Instrum. Meas.*, vol. IM-25, p. 8, Mar. 176.

[7] A. Ferrero, V. Teppati, M. Garelli, and A. Neri, "A novel calibration algorithm for a special class of multiport vector network analyzers," *IEEE Trans. Microw. Theory Tech.*, vol. 56, no. 3, pp. 693–699, Mar. 2008.

[8] D. Williams and R. Marks, "Accurate transmission line characterization," *IEEE Microw. and Guid. Wave Lett.*, vol. 3, pp. 247–249, Aug. 1993.

[9] R. Marks and D. Williams, "A general waveguide circuit theory," *J. Res. NIST*, vol. 97, pp. 533–561, Sept. 1992.

[10] R. Marks, "A multiline method of network analyzer calibration," *IEEE Trans. Microw. Theory Tech.*, vol. 39, no. 7, pp. 1205–1215, July 1991.

[11] K. Silvonen, "A general approach to network analyzer calibration," *IEEE Trans. Microw. Theory Tech.*, vol. 40, no. 4, pp. 754–759, Apr. 1992.

[12] A. Ferrero and U. Pisani, "Qsolt: a new fast calibration algorithm for two-port S-parameter measurements," in *38th ARFTG Conf. Dig.*, San Diego, CA, Dec. 1991, pp. 15–24.

[13] H. Eul and B. Schieck, "Reducing the number of calibration standards for network analyzer calibration," *IEEE Trans. Instrum. Meas.*, vol. IM-40, pp. 732–735, Aug. 1991.

[14] J. V. Butler et al., "16-term error model and calibration procedure for on-wafer network analysis measurements," *IEEE Trans. Microw. Theory Tech.*, vol. 39, no. 12, pp. 2211–2217, Dec. 1991.

[15] K. Silvonen, "A 16-term error model based on linear equations of voltage and current variables," *IEEE Trans. Microw. Theory Tech.*, vol. 54, no. 4, pp. 1464–1469, June 2006.

[16] V. Adamian, "Simplified automatic calibration of a vector network analyzer," in *ARFTG Conference*, Nov. 1994, pp. 1–9.

9 Multiport and differential S-parameter measurements

Valeria Teppati and Andrea Ferrero

9.1 Introduction

The last ten years have witnessed an increasing interest in multiport S-parameter measurements, i.e. S-parameter measurements of devices with more than two ports, for two main reasons: the first one is the increasing complexity of modern microwave devices and circuits and the use of more complex MMICs.

But the main reason is definitely the shift toward microwave frequencies of the personal computer's processors speed, which implies that such digital applications must now face typical microwave challenges. These topics have recently been addressed in [1]. Preserving the signal integrity of a microwave signal through the packages, sockets, connectors, and PCB traces, commonly found in today's computer systems, is one of the main issues. System architectures with hundreds of parallel channels, operating at higher and higher data rates, involve microwave multiport measurements for the characterization, design, and analysis of the structures and their effects on the signals. Microwave designers and engineers are thus facing new challenges in multiport measurement hardware and calibrations.

The first challenge comes from the typical media of digital interconnections: the PCB. It can include both planar and three-dimensional (3-D) DUTs, as found, for example, in memory modules. So, on one hand many data lines must be connected and measured simultaneously, and they do not necessarily lie on a single plane. On the other hand, these connections from the boards to the typically coaxial test ports of the VNA must have good performances at microwaves, i.e. be "transparent" for the measurements.

If the structure is not three-dimensional but planar, the best choice for contacts is connecting directly to the PCB surface with high-performance microwave probes. The alternative for 3-D structures, when probing is not possible, is to use coaxial to PCB launchers, but this solution is not the best in terms of insertion losses. Microwave probes provide better high-frequency transitions to the boards, compared with coaxial launchers, both in terms of connection repeatability and of electrical transparency of the transition. Thus, probing typically offers improved calibration and better measurement accuracy.

For multiport devices, as the number of ports increases, traditional single-port probes are unsuitable, since it could be very difficult to mechanically put in place all the probes, even if specific probe stations could be designed for the purpose. The typical solution is to use *multiport probes*, such as the ones shown in Figure 9.1. With these probes it is

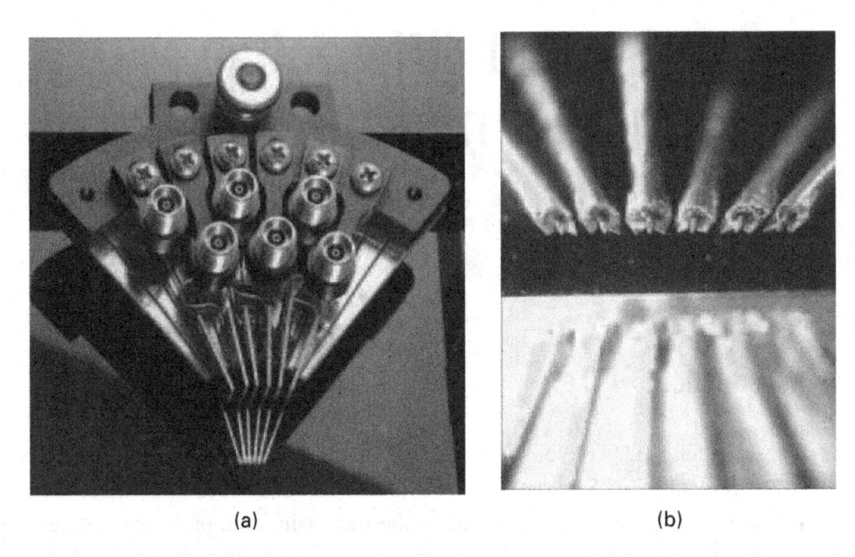

Fig. 9.1 Multiport GSG probe tips on the same probe head [3]. Courtesy of GGB Industries.

possible to measure various channels simultaneously, but the probe pitch and the patterns on the board must be designed to match.

Probe manufacturers provide different types of multiport probes, depending on the customer needs. In particular, one of the aspects to be taken into account for accurate measurements is the presence of crosstalk between probe fingers. Two multiport probe configurations are possible; the first one is obtained simply by tiling standard ground-signal-ground (GSG) probes (GSGGSG), as shown in Figure 9.1. The second configuration allows narrower patterns to be tested as it does not have ground fingers between the various signal lines (GSSG). In this case, the crosstalk between two adjacent fingers might not be negligible, and to achieve accurate calibrations and measurements it should be taken into account. A possible solution to this problem is revised in this chapter [2].

In any case, for both coaxial launchers and probes, the VNA calibration for on-board testing can be performed following two approaches. Either the VNA test ports are calibrated, so that the transition performances are included as part of the DUT, and then a separate de-embedding of the transition is performed. Or, to achieve more accurate results, the reference planes are moved on-board through an on-board calibration [3–5]. In the next sections the problems with the measurement and calibration of multiport systems are described in detail.

9.2 Multiport S-parameters measurement methods

Multiport measurements find the error-corrected S-parameters of a DUT having more than two ports.

There are two approaches to this problem. The first one, available since the early 1980s, consists of performing multiple (*round robin*) measurements, with two-port VNAs and matched loads on the unused ports [6–9]. This method is still used nowadays, by taking into account the non-idealities of the matched terminations [10, 11]. The procedure is quite cumbersome, since it requires $n(n-1)/2$ different two-port measurements, for each n-port measurement. Besides, the overall accuracy is affected by the multiple connections required and by the accuracy of the terminations.

The alternative and more modern approach is the multiport VNA, i.e. a measurement system able to perform straightforward calibrated multiport measurements, with a single DUT connection. The calibration and the measurement problem are the two main aspects to be considered when dealing with a multiport measurement architecture.

Various system architectures are currently available with different numbers of sources and measurement receivers per port, from a maximum of one source and two receivers for each port, as shown in Figure 9.2(a), to a minimum of a single source and two receivers with a proper switch matrix; see Figure 9.1(b). Of course, the solution of Figure 9.2(a) is very expensive, but fast, while the solution of Figure 9.1(b), at the cost of speed, can be used to extend any two-port VNA to multiport.

A number of possible intermediate solutions lies between the two configurations depicted in Figure 9.2, e.g. one source for each couple of ports, etc.

To more clearly delineate why certain choices in the architecture may be made, the following constraints need to be taken into account [1].

(i) Extra sources and receivers are more expensive than extra couplers.
(ii) Cable and connector losses are very high due to the frequencies involved. It is particularly important to minimize losses after the test couplers as this affects the raw directivity.
(iii) Switch isolation becomes worse at higher frequencies.
(iv) At higher frequencies, single-pole double-throw (SPDT) switches perform much better (in terms of isolation) than single-pole triple-throw (SP3T) or single-pole quadruple-throw (SP4T) switches. But of course, substituting SP4T switches with STDT ones complicates the switching matrix.
(v) For error model simplicity, it helps if the load match presented by a port is independent of the driving port.

As the frequency rises, the number of sources, receivers, and directional couplers should be reduced. A fairly good compromise is the *partial reflectometer* architecture (see Figure 9.7), which is analyzed in Section 9.2.2.

9.2.1 Calibration of a complete reflectometer multiport VNA

We now present the generalized and simple error model formulation, that was provided by [12], for a complete reflectometer VNA architecture, i.e. based on the use of two directional couplers for each port. In this case, it is always possible to measure the incident and the reflected waves at each port, wherever the source excitation is. The error

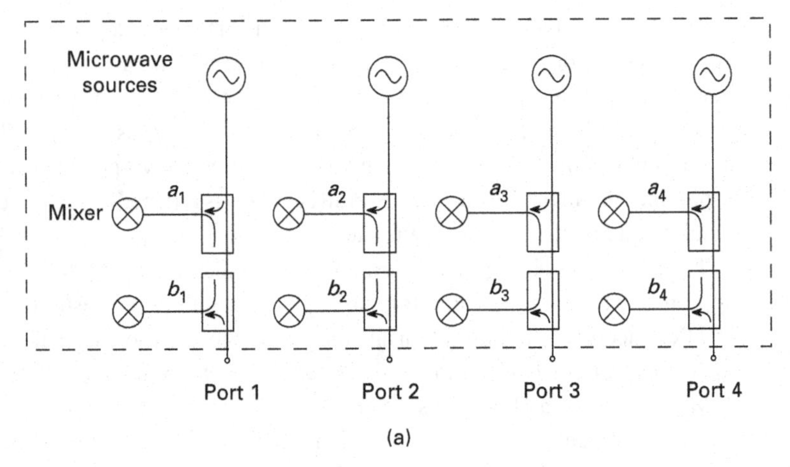

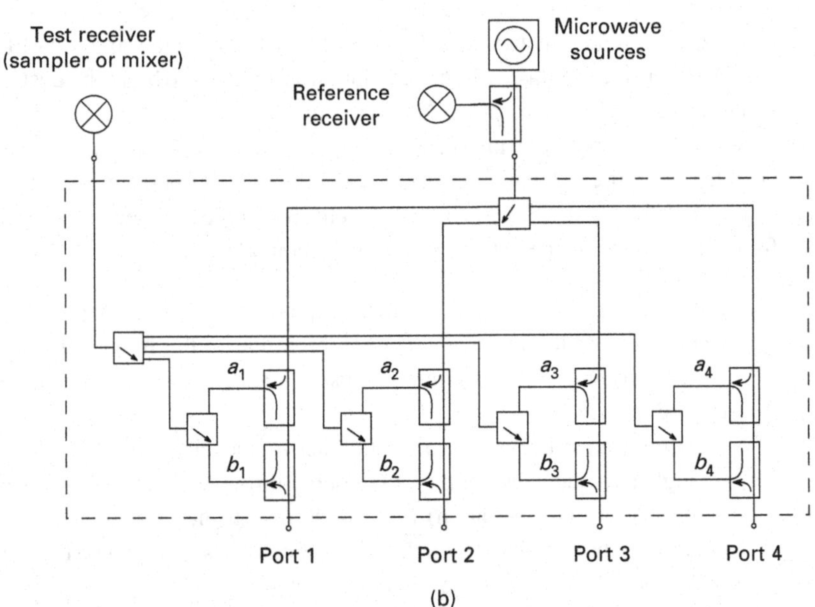

Fig. 9.2 The most expensive (a) and minimum (b) multiport architecture, for a four-port example with complete reflectometers.

model, in the most simple case of no leakage between ports, has $4n - 1$ unknowns, where n is the number of ports.

The incident and reflected waves at each port can be organized in the following matrices:

$$\boldsymbol{A}_m = \begin{bmatrix} a_{m11} & a_{m12} & \cdots & a_{m1n} \\ a_{m21} & a_{m22} & \cdots & a_{m2n} \\ \vdots & \vdots & \ddots & \vdots \\ a_{mn1} & a_{mn2} & \cdots & a_{mnn} \end{bmatrix}, \quad \boldsymbol{B}_m = \begin{bmatrix} b_{m11} & b_{m12} & \cdots & b_{m1n} \\ b_{m21} & b_{m22} & \cdots & b_{m2n} \\ \vdots & \vdots & \ddots & \vdots \\ b_{mn1} & b_{mn2} & \cdots & b_{mnn} \end{bmatrix}$$

$$A = \begin{bmatrix} a_{11} & a_{12} & \cdots & a_{1n} \\ a_{21} & a_{22} & \cdots & a_{2n} \\ \vdots & \vdots & \ddots & \vdots \\ a_{n1} & a_{n2} & \cdots & a_{nn} \end{bmatrix}, \quad B = \begin{bmatrix} b_{11} & b_{12} & \cdots & b_{1n} \\ b_{21} & b_{22} & \cdots & b_{2n} \\ \vdots & \vdots & \ddots & \vdots \\ b_{n1} & b_{n2} & \cdots & b_{nn} \end{bmatrix},$$

where a_{mij} and b_{mij} represent, respectively, the measured incident and reflected waves at port i, when the source excitation is at port j, while a_{ij} and b_{ij} are the actual incident and reflected waves at the port i reference plane, when the source excitation is at port j. Calling $\boldsymbol{S}$ the scattering matrix of the multiport DUT, we can write

$$\boldsymbol{B} = \boldsymbol{S}\boldsymbol{A}. \tag{9.1}$$

The error model is defined by the following:

$$\begin{aligned} \boldsymbol{A} &= \boldsymbol{L}\boldsymbol{B}_\mathrm{m} - \boldsymbol{H}\boldsymbol{A}_\mathrm{m} \\ \boldsymbol{B} &= \boldsymbol{K}\boldsymbol{B}_\mathrm{m} - \boldsymbol{M}\boldsymbol{A}_\mathrm{m}, \end{aligned} \tag{9.2}$$

where $\boldsymbol{L}$, $\boldsymbol{M}$, $\boldsymbol{H}$, and $\boldsymbol{K}$ contain the error coefficients, and can be full matrices (in this case the error model is full leaky), diagonal matrix (for a non-leaky model), or block diagonal (for a partially leaky model [2]).

By combining (9.1) and (9.2), the equation for the error coefficient computation is

$$\boldsymbol{S}\boldsymbol{L}\boldsymbol{B}_\mathrm{m} - \boldsymbol{S}\boldsymbol{H}\boldsymbol{A}_\mathrm{m} - \boldsymbol{K}\boldsymbol{B}_\mathrm{m} + \boldsymbol{M}\boldsymbol{A}_\mathrm{m} = \boldsymbol{0} \tag{9.3}$$

thus for de-embedding we have:

$$\boldsymbol{S} = (\boldsymbol{K}\boldsymbol{B}_\mathrm{m} - \boldsymbol{M}\boldsymbol{A}_\mathrm{m})(\boldsymbol{L}\boldsymbol{B}_\mathrm{m} - \boldsymbol{H}\boldsymbol{A}_\mathrm{m})^{-1}. \tag{9.4}$$

Note that (9.3) is written in terms of measured waves rather than measured S-parameters, as in [13]. In other words, no switch correction technique has been applied here to obtain the measured scattering matrix. The calibration equations are written directly in terms of the measured quantities, with computational advantages.

It is useful to express (9.3) in the following iterative form (written here for simplicity for a non-leaky error model),

$$\sum_{p=1}^{n} S_{ip} L_{pp} b_{\mathrm{m}pj} - \sum_{p=1}^{n} S_{ip} H_{pp} a_{\mathrm{m}pj} - K_{ii} b_{\mathrm{m}ij} + M_{ii} a_{\mathrm{m}ij} = 0 \qquad (i = 1, \ldots, n) \quad (j = 1, \ldots, n) \tag{9.5}$$

because in this form it can be easily used for one- or two-port standards, by simply eliminating the proper rows.

For example, a one-port standard connected at port 1 ($i = 1$, $j = 1$) gives:

$$\Gamma L_{11} \Gamma_\mathrm{m} - \Gamma H_{11} + M_{11} = K_{11} \Gamma_\mathrm{m}, \tag{9.6}$$

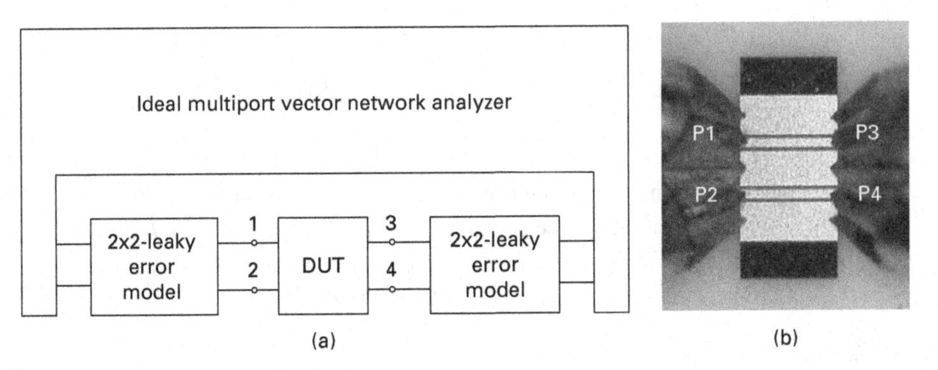

Fig. 9.3 Four-port partially leaky error model (a) and coupled lines, measured with two GSGGSG probes (b). © 2005 IEEE. Reprinted, with permission, from [2].

where Γ_m and Γ are the measured and the defined reflection coefficient of the standard, respectively.

A typical calibration consists of finding the error coefficients by solving a system in the form (9.5), obtained by measuring a proper sequence of one- and two-port standards. Since the coefficients are $4n - 1$, there must be $4n - 1$ linearly independent equations in order to find all the unknowns. The rules to grant the independence of the multiport calibration equations are given in [13].

Calibration of a partially leaky multiport VNA

For some measurement problems, the non-leaky model could be not accurate enough. An example is the multi-finger probes, of Figure 9.1, where crosstalk between fingers belonging to the same probe cannot be neglected. In these cases a more complete error model is required, and is shown in Figure 9.3(a). The measurement system is split in two halves, and the leakage error terms are present in each of the halves, but not between the two parts of the model. This error model is mathematically described by (9.2), where $\boldsymbol{L}$, $\boldsymbol{M}$, $\boldsymbol{H}$, and $\boldsymbol{K}$ are block diagonal [2].

Considering the case of a four-port VNA, with two double signal probes, the partially leaky model takes into account the leakage between the fingers on the same probe, while the side-by-side crosstalk is neglected. It has been proved that, in such cases, this model has better accuracy than both the non-leaky model and the fully leaky one. The crosstalk terms of the half-leaky case are fixed and constant, due to the fixed position of each probe finger, while the side-by-side crosstalk, which is usually minimal, variable, and difficult to model, is neglected. This is better than trying a correction with wrong error terms, as the full-leaky calibration would do [2].

An optimized standard sequence, especially useful for on-wafer measurements, is similar to a classical LRM calibration, where the number of probe touchdowns is minimized: only three on-wafer probe placements are required, as shown in Figure 9.4. The three standard combinations are: 1) thru ports 13 and shorts ports 2 and 4; 2) thru ports 24 and shorts ports 1 and 3; and 3) thru ports 1–4 and loads ports 2 and 3.

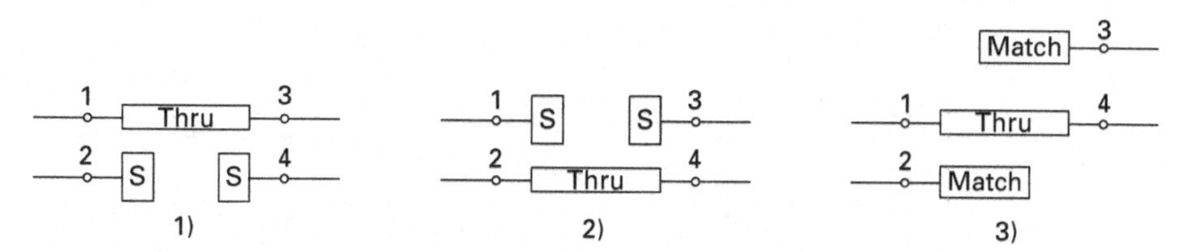

Fig. 9.4 Calibration sequence for a partially leaky four-port VNA, optimized for on-wafer touchdowns. © 2005 IEEE. Reprinted, with permission, from [2].

Figures 9.5 and 9.6 report measurements and simulations of loosely coupled coplanar lines, similar to the ones depicted in Figure 9.3(b). All the simulations were performed with a commercial simulator, implementing a simple circuital model. Figure 9.5(a) refers to 0.58 mm long coupled lines. Scattering parameter S_{12}, i.e. the near-end coupling between the two structures, is clearly overestimated by the non-leaky model, since it does not include the correction for leakage between ports 1 and 2, while the proposed half-leaky and full-leaky calibrations demonstrate a very good agreement with simulations. But if we consider longer (6.6 mm) coupled lines, as shown in Figure 9.5(b), the 10 GHz resonance predicted by simulations is found only with the half-leaky model. The effect of the wrong correction of the full-leaky model is evident since this device is much longer than the thrus used during calibration. Also the non-leaky model does not provide the right value of the resonance frequency.

Finally, in Figure 9.6, the far-end crosstalk of the same (6.6 mm) coupled lines is shown. Also in this case, the half-leaky model agrees better with the simulation than the other two models.

9.2.2 Calibration of a partial reflectometer multiport VNA

Available multiport VNA architectures do not always have complete reflectometers at each port. Some ports might have a partial reflectometer, as the one shown in the example of Figure 9.7. This architecture is typically used, especially for a high number of ports (e.g. twelve or sixteen) to reduce the total number of directional couplers. This has a clear cost advantage, and also improves measurement speed. In principle, the number of directional couplers can be reduced from $2n$ to $n+1$, thus measuring only one incident wave at a time. This, anyway, reduces the calibration possibilities, so it is preferable to have the multiport system split at least in two halves, as shown in Figure 9.7, and the number of directional couplers equal to $n+2$.

For the partial reflectometer multiport VNA, a different error model must be introduced. Due to the presence of switches in the measurement paths, each port is modeled in two different states:

- **state A**: complete reflectometer; see Figure 9.8 (a);
- **state B**: partial reflectometer, i.e. only the reflected wave can be measured; see Figure 9.8 (b).

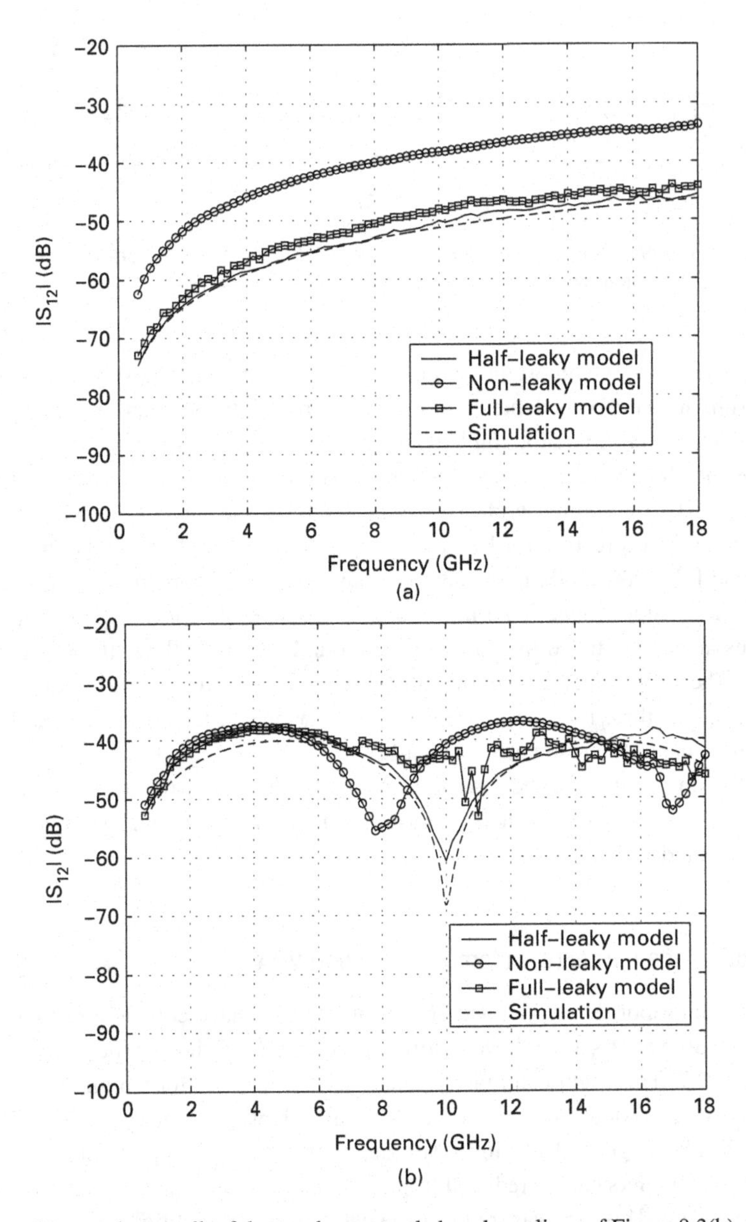

Fig. 9.5 Near-end crosstalk of the two loose coupled coplanar lines of Figure 9.3(b), compared with simulations. (a): 0.58 mm lines, as are the thrus used during calibration. (b) 6.6 mm lines, i.e. more than ten times the thrus used during calibration. © 2005 IEEE. Reprinted, with permission, from [2].

In state A, a full reflectometer is present, thus the state is described by (9.2), that we rewrite here for a non-leaky case:

$$
\begin{aligned}
a_{ii} &= l_i b_{\mathrm{m}ii} - h_i a_{\mathrm{m}ii} \\
b_{ii} &= k_i b_{\mathrm{m}ii} - m_i a_{\mathrm{m}ii}.
\end{aligned}
\tag{9.7}
$$

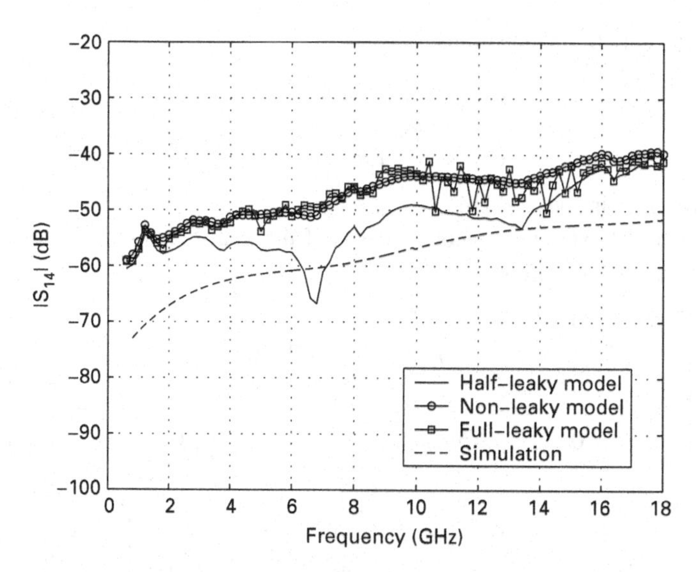

Fig. 9.6 Far-end crosstalk of the two loose coupled coplanar lines of Figure 9.3(b), compared with simulations (6.6 mm long lines). © 2005 IEEE. Reprinted, with permission, from [2].

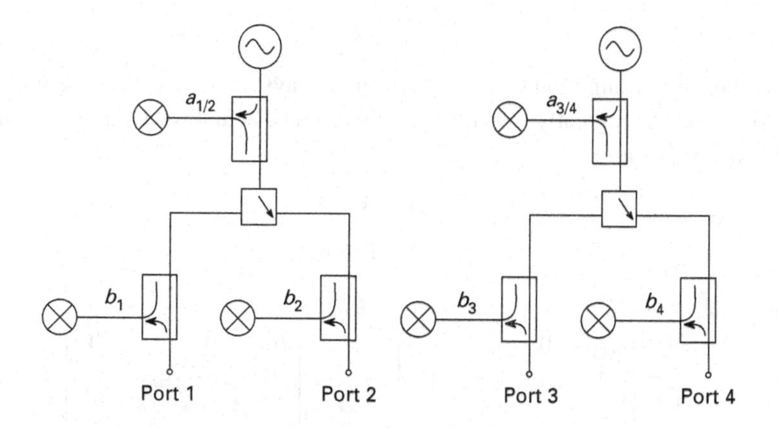

Fig. 9.7 An example of partial reflectometer architecture for a four-port VNA.

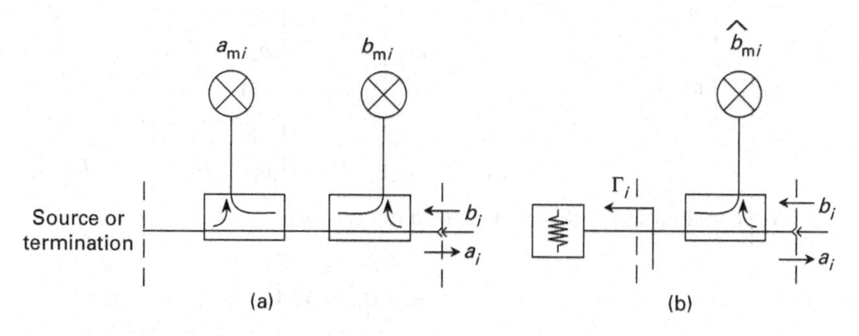

Fig. 9.8 State A and B configurations, (a) and (b), respectively. © 2008 IEEE. Reprinted, with permission, from [14].

Instead, for state B, the equations:

$$\begin{aligned} a_{ij} &= g_i \widehat{b}_{\mathrm{m}ij} \\ b_{ij} &= f_i \widehat{b}_{\mathrm{m}ij} \\ i &\neq j \end{aligned} \tag{9.8}$$

have been recently introduced [14].

The error-model extraction becomes easier if the matrices containing the measurements are organized as follows:

$$\widetilde{\boldsymbol{B}}_{\mathrm{m}} \equiv \begin{bmatrix} b_{\mathrm{m}11} & 0 & \cdots & 0 \\ 0 & b_{\mathrm{m}22} & \cdots & 0 \\ \vdots & \vdots & \ddots & \vdots \\ 0 & 0 & \cdots & b_{\mathrm{m}nn} \end{bmatrix}, \quad \widetilde{\boldsymbol{A}}_{\mathrm{m}} \equiv \begin{bmatrix} a_{\mathrm{m}11} & 0 & \cdots & 0 \\ 0 & a_{\mathrm{m}22} & \cdots & 0 \\ \vdots & \vdots & \ddots & \vdots \\ 0 & 0 & \cdots & a_{\mathrm{m}nn} \end{bmatrix} \tag{9.9}$$

and

$$\widehat{\boldsymbol{B}}_{\mathrm{m}} \equiv \begin{bmatrix} 0 & \widehat{b}_{\mathrm{m}12} & \widehat{b}_{\mathrm{m}13} & \cdots & \widehat{b}_{\mathrm{m}1n} \\ \widehat{b}_{\mathrm{m}21} & 0 & \widehat{b}_{\mathrm{m}23} & \cdots & \widehat{b}_{\mathrm{m}2n} \\ \widehat{b}_{\mathrm{m}31} & \widehat{b}_{\mathrm{m}32} & 0 & \cdots & \widehat{b}_{\mathrm{m}3n} \\ \vdots & \vdots & \ddots & \ddots & \vdots \\ \widehat{b}_{\mathrm{m}n1} & \widehat{b}_{\mathrm{m}n2} & \cdots & \widehat{b}_{\mathrm{m}nn-1} & 0 \end{bmatrix}. \tag{9.10}$$

We make the assumption that all i ports are always in the B state while the source is at port j, because it greatly simplifies the theory. The actual wave matrices at the DUT ports can be seen as

$$\boldsymbol{A} = \widetilde{\boldsymbol{A}} + \widehat{\boldsymbol{A}} \tag{9.11}$$

$$\boldsymbol{B} = \widetilde{\boldsymbol{B}} + \widehat{\boldsymbol{B}}. \tag{9.12}$$

where

$$\widetilde{\boldsymbol{A}} \equiv \begin{bmatrix} a_{11} & 0 & \cdots & 0 \\ 0 & a_{22} & \cdots & 0 \\ \vdots & \vdots & \ddots & \vdots \\ 0 & 0 & \cdots & a_{nn} \end{bmatrix}, \quad \widetilde{\boldsymbol{B}} \equiv \begin{bmatrix} b_{11} & 0 & \cdots & 0 \\ 0 & b_{22} & \cdots & 0 \\ \vdots & \vdots & \ddots & \vdots \\ 0 & 0 & \cdots & b_{nn} \end{bmatrix},$$

$$\widehat{\boldsymbol{A}} \equiv \begin{bmatrix} 0 & a_{12} & a_{13} & \cdots & a_{1n} \\ a_{21} & 0 & a_{23} & \cdots & a_{2n} \\ a_{31} & a_{32} & 0 & \cdots & a_{3n} \\ \vdots & \vdots & \ddots & \ddots & \vdots \\ a_{n1} & a_{n2} & \cdots & a_{nn-1} & 0 \end{bmatrix}, \quad \widehat{\boldsymbol{B}} \equiv \begin{bmatrix} 0 & b_{12} & b_{13} & \cdots & b_{1n} \\ b_{21} & 0 & b_{23} & \cdots & b_{2n} \\ b_{31} & b_{32} & 0 & \cdots & b_{3n} \\ \vdots & \vdots & \ddots & \ddots & \vdots \\ b_{n1} & b_{n2} & \cdots & b_{nn-1} & 0 \end{bmatrix}.$$

We can then write (9.7) and (9.8) in matrix form

$$\begin{aligned} \widetilde{\boldsymbol{A}} &= \boldsymbol{L}\widetilde{\boldsymbol{B}}_{\mathrm{m}} - \boldsymbol{H}\widetilde{\boldsymbol{A}}_{\mathrm{m}} \\ \widetilde{\boldsymbol{B}} &= \boldsymbol{K}\widetilde{\boldsymbol{B}}_{\mathrm{m}} - \boldsymbol{M}\widetilde{\boldsymbol{A}}_{\mathrm{m}} \\ \widehat{\boldsymbol{A}} &= \boldsymbol{G}\widehat{\boldsymbol{B}}_{\mathrm{m}} \\ \widehat{\boldsymbol{B}} &= \boldsymbol{F}\widehat{\boldsymbol{B}}_{\mathrm{m}}, \end{aligned} \tag{9.13}$$

In this formulation leakage is neglected, thus $\boldsymbol{L}, \boldsymbol{M}, \boldsymbol{H}, \boldsymbol{K}, \boldsymbol{F}$, and $\boldsymbol{G}$ matrices are all diagonal.

By substituting (9.13) in (9.11) and (9.12) we have

$$\begin{aligned} \boldsymbol{A} &= \widetilde{\boldsymbol{A}} + \widehat{\boldsymbol{A}} = \boldsymbol{L}\widetilde{\boldsymbol{B}}_{\mathrm{m}} - \boldsymbol{H}\widetilde{\boldsymbol{A}}_{\mathrm{m}} + \boldsymbol{G}\widehat{\boldsymbol{B}}_{\mathrm{m}} \\ \boldsymbol{B} &= \widetilde{\boldsymbol{B}} + \widehat{\boldsymbol{B}} = \boldsymbol{K}\widetilde{\boldsymbol{B}}_{\mathrm{m}} - \boldsymbol{M}\widetilde{\boldsymbol{A}}_{\mathrm{m}} + \boldsymbol{F}\widehat{\boldsymbol{B}}_{\mathrm{m}}, \end{aligned} \tag{9.14}$$

By substituting (9.14) in (9.1), we find the new matrix equation for the error coefficient computation

$$-\boldsymbol{SG}\widehat{\boldsymbol{B}}_{\mathrm{m}} + \boldsymbol{F}\widehat{\boldsymbol{B}}_{\mathrm{m}} - \boldsymbol{SL}\widetilde{\boldsymbol{B}}_{\mathrm{m}} + \boldsymbol{K}\widetilde{\boldsymbol{B}}_{\mathrm{m}} + \boldsymbol{SH}\widetilde{\boldsymbol{A}}_{\mathrm{m}} - \boldsymbol{M}\widetilde{\boldsymbol{A}}_{\mathrm{m}} = \boldsymbol{0}. \tag{9.15}$$

Like (9.3), (9.15) is written in terms of measured waves rather than measured S-parameters.

The generalized system (9.15) can be used to compute the error coefficients from the standard measurements and definitions. As before, it is useful to write the n^2 equations as follows:

$$\begin{aligned} -\sum_{p=1}^{n}(1-\delta_{pj})S_{ip}g_p\widehat{b}_{\mathrm{m}pj} + (1-\delta_{ij})f_i\widehat{b}_{\mathrm{m}ij} - S_{ij}l_j b_{\mathrm{m}jj} + \\ +\delta_{ij}k_i b_{\mathrm{m}ij} + S_{ij}h_j a_{\mathrm{m}jj} - \delta_{ij}m_i a_{\mathrm{m}ij} = 0 \\ (i = 1, \ldots, n) \\ (j = 1, \ldots, n) \end{aligned} \tag{9.16}$$

where δ_{ij} is the Kronecker delta. These equations can be easily used for one- and two-port standard connections. A system of equations is obtained by putting together all the equations coming from the different standard measurements. The solutions of this system are the error coefficients.

Since this system is homogeneous, in order to avoid the trivial zero solution it is normalized to one of the unknown coefficients; thus, the total number of unknown error coefficients is $6n-1$, instead of $4n-1$ as in the complete reflectometer model [15].

From (9.15), the de-embedding equation is the following:

$$\boldsymbol{S} = \left[\boldsymbol{K}\widetilde{\boldsymbol{B}}_{\mathrm{m}} - \boldsymbol{M}\widetilde{\boldsymbol{A}}_{\mathrm{m}} + \boldsymbol{F}\widehat{\boldsymbol{B}}_{\mathrm{m}}\right]\left[\boldsymbol{L}\widetilde{\boldsymbol{B}}_{\mathrm{m}} - \boldsymbol{H}\widetilde{\boldsymbol{A}}_{\mathrm{m}} + \boldsymbol{G}\widehat{\boldsymbol{B}}_{\mathrm{m}}\right]^{-1}. \tag{9.17}$$

9.2.3 Multiport measurement example

Figure 9.9 shows a typical critical DUT. Port 1 and port 2 are connectorized through APC7, and have no gender. Instead, ports 3 and 4 have SMA female connectors. This means that a direct connection, i.e. a “thru” standard, can be inserted only between ports 1 and 2.

Fig. 9.9 An example of multiport DUT: a directional coupler [16]. © 2008 IEEE. Reprinted, with permission, from [17].

One way to solve the calibration problem in this case, without resorting to adapter removal, is to split it into simpler sub-problems. For example, it is possible to perform a classical TRL between ports 1 and 2, while an "unknown thru" calibration can be performed between ports 3 and 4 [18]. At this point, these two sets of error coefficients must be merged into one, by means of another "unknown thru", between ports 1 and 4, for example.

The standard sequence is then:

- APC7 thru ports 1 and 2,
- APC7 line ports 1 and 2,
- APC7 reflect (e.g. a short) at ports 1 and 2,
- SMA female short, open and load at ports 3 and 4,
- SMA female–female adapter at ports 3 and 4,
- SMA female–APC7 adapter at ports 1 and 4.

The measurement results, after this calibration, are shown in Figure 9.10. This multi-port calibration involving a TRL between ports 1 and 2 is capable of resolving very low values of insertion loss.

9.3 Mixed-mode S-parameter measurements

Mixed-mode S-parameters have been introduced for the analysis of lines, circuits and systems in differential configuration at microwaves. In the following we revise the original definition and provide a generalized method to compute the mixed-mode S-parameters from single-ended ones.

The original definition of differential and common mode S-parameters is due to Bockelman and Eisenstadt in 1995 [19]. They introduced the so called *mixed-mode* scattering matrix, a linear transformation from the single-ended S-matrix to this new matrix,

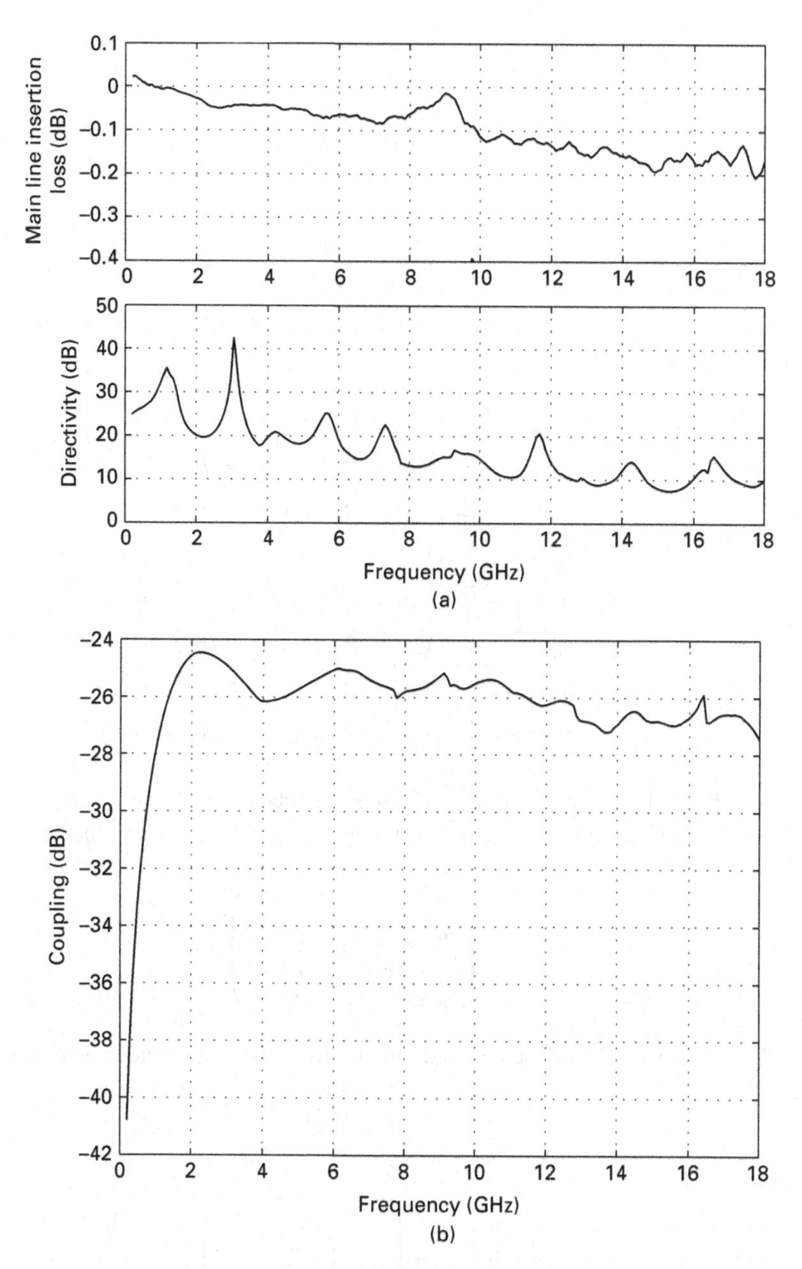

Fig. 9.10 Calibrated measurements of coupler isolation and directivity (a) and coupling factor (b) © 2008 IEEE. Reprinted, with permission, from [17].

and designed and implemented an instrument able to measure directly the mixed-mode S-matrix, i.e. the *pure-mode* VNA [20]. In [19] Bockelman and Eisenstadt showed, for a four-port case, that if the differential and common mode voltages and currents are defined as

$$
\begin{aligned}
V_{d12} &\equiv V_1 - V_2 \\
I_{d12} &\equiv (I_1 - I_2)/2 \\
V_{c12} &\equiv (V_1 + V_2)/2 \\
I_{c12} &\equiv I_1 + I_2 \\
V_{d34} &\equiv V_3 - V_4 \\
I_{d34} &\equiv (I_3 - I_4)/2 \\
V_{c34} &\equiv (V_3 + V_4)/2 \\
I_{c34} &\equiv I_3 + I_4
\end{aligned}
\tag{9.18}
$$

then it is possible to define the differential and common mode waves similarly to the single-ended ones

$$
\begin{aligned}
a_{d12} &\equiv \frac{1}{2\sqrt{R_d}}(V_{d12} + R_d I_{d12}) = \frac{1}{\sqrt{2}}(a_1 - a_2) \\
b_{d12} &\equiv \frac{1}{2\sqrt{R_d}}(V_{d12} - R_d I_{d12}) = \frac{1}{\sqrt{2}}(b_1 - b_2) \\
a_{c12} &\equiv \frac{1}{2\sqrt{R_c}}(V_{c12} + R_c I_{c12}) = \frac{1}{\sqrt{2}}(a_1 + a_2) \\
b_{c12} &\equiv \frac{1}{2\sqrt{R_c}}(V_{c12} - R_c I_{c12}) = \frac{1}{\sqrt{2}}(b_1 + b_2) \\
a_{d34} &\equiv \frac{1}{2\sqrt{R_d}}(V_{d34} + R_d I_{d34}) = \frac{1}{\sqrt{2}}(a_3 - a_4) \\
b_{d34} &\equiv \frac{1}{2\sqrt{R_d}}(V_{d34} - R_d I_{d34}) = \frac{1}{\sqrt{2}}(b_3 - b_4) \\
a_{c34} &\equiv \frac{1}{2\sqrt{R_c}}(V_{c34} + R_c I_{c34}) = \frac{1}{\sqrt{2}}(a_3 + a_4) \\
b_{c34} &\equiv \frac{1}{2\sqrt{R_c}}(V_{c34} - R_c I_{c34}) = \frac{1}{\sqrt{2}}(b_3 + b_4)
\end{aligned}
\tag{9.19}
$$

where R_d and R_c are purely real reference impedances (typically $R_d = 100\ \Omega$ and $R_c = 25\ \Omega$). Consequently, the mixed-mode S-matrix $\boldsymbol{S}_{\mathrm{MM}}$ was defined as,

$$
\begin{pmatrix} b_{d12} \\ b_{d34} \\ b_{c12} \\ b_{c34} \end{pmatrix} \equiv \boldsymbol{S}_{\mathbf{MM}} \begin{pmatrix} a_{d12} \\ a_{d34} \\ a_{c12} \\ a_{c34} \end{pmatrix}.
\tag{9.20}
$$

This matrix can also be computed directly from the single-ended S-matrix as:

$$
\boldsymbol{S}_{\mathrm{MM}} = \boldsymbol{M}\boldsymbol{S}\boldsymbol{M}^{-1}.
\tag{9.21}
$$

where

$$
\boldsymbol{M} = \frac{1}{\sqrt{2}} \begin{pmatrix} 1 & -1 & 0 & 0 \\ 0 & 0 & 1 & -1 \\ 1 & 1 & 0 & 0 \\ 0 & 0 & 1 & 1 \end{pmatrix}.
\tag{9.22}
$$

The measurement of the mixed-mode S-matrix $\boldsymbol{S}_{\mathrm{MM}}$ can then be performed by measuring a single-ended S-matrix with a single-ended multiport VNA, and then applying (9.21). This is the easiest and more common approach. An alternative is a modified VNA, the pure-mode VNA, which is able to separately excite the differential and the common mode, by using a 180° hybrid coupler. Two possible implementations are shown in Figure 9.11.

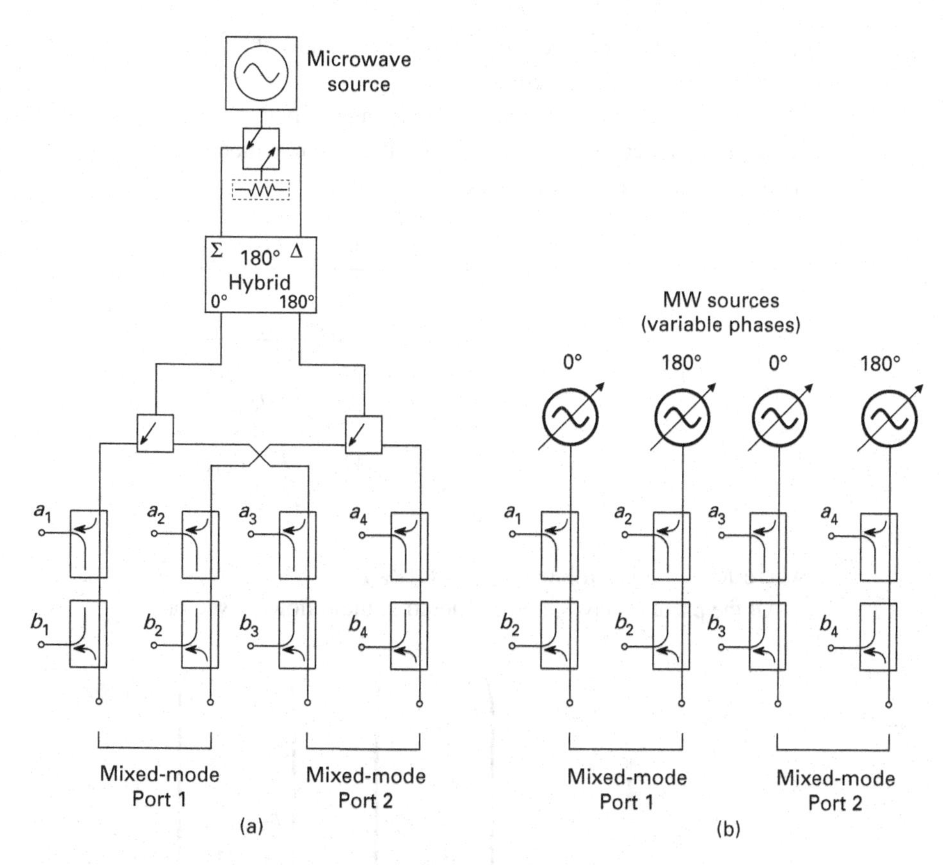

Fig. 9.11 Possible implementations of a pure mode VNA, with hybrid couplers (a) [19], or independent sources (b). © 2008 IEEE. Reprinted, with permission, from [21].

The Bockelman and Eisenstadt formulation is very simple and intuitive, but has the following drawbacks:

- the waves are referred to a real reference impedance, so if a TRL-like calibration is applied, the notation does not take into account the complex characteristic impedance of the reference line, and will be inaccurate [22]
- it cannot be applied to a DUT having one (or more than one) single-ended ports.

These drawbacks have been overcome by a recently introduced formulation [23]. It applies to the measurement of an n-port DUT, with p mixed-mode ports and $n-p$ single-ended ports. For the latter $n-p$ ports, the single-ended pseudo-waves are defined, according to [22]

$$a_j \equiv \alpha\sqrt{R_j}\,\frac{V_j + I_j Z_j}{2|Z_j|}$$

$$b_j \equiv \alpha\sqrt{R_j}\,\frac{V_j - I_j Z_j}{2|Z_j|},$$

where Z_j is the complex reference impedances, $R_j = \Re\{Z_j\}$, and α is a unitary magnitude complex normalization factor.

For the mixed-mode p ports, considering the port-pair j, k of the mixed-mode port set, with reference impedances $Z_{\mathrm{c}jk}$ for the common mode, and $Z_{\mathrm{d}jk}$ for the differential mode, the mixed-mode waves are

$$\begin{aligned} a_{\mathrm{d}jk} &\equiv \sqrt{R_{\mathrm{d}jk}}\,\frac{V_{\mathrm{d}jk} + I_{\mathrm{d}jk} Z_{\mathrm{d}jk}}{2|Z_{\mathrm{d}jk}|} \\ b_{\mathrm{d}jk} &\equiv \sqrt{R_{\mathrm{d}jk}}\,\frac{V_{\mathrm{d}jk} - I_{\mathrm{d}jk} Z_{\mathrm{d}jk}}{2|Z_{\mathrm{d}jk}|} \\ a_{\mathrm{c}jk} &\equiv \sqrt{R_{\mathrm{c}jk}}\,\frac{V_{\mathrm{c}jk} + I_{\mathrm{c}jk} Z_{\mathrm{c}jk}}{2|Z_{\mathrm{c}jk}|} \\ a_{\mathrm{c}jk} &\equiv \sqrt{R_{\mathrm{c}jk}}\,\frac{V_{\mathrm{c}jk} - I_{\mathrm{c}jk} Z_{\mathrm{c}jk}}{2|Z_{\mathrm{c}jk}|}, \end{aligned}$$

where $R_{\mathrm{d}jk} = \Re(Z_{\mathrm{d}jk})$ and $R_{\mathrm{c}jk} = \Re(Z_{\mathrm{c}jk})$.

All the pseudo-waves are re-ordered in the following vectors:

$$\mathring{\boldsymbol{a}} \equiv \begin{pmatrix} a_{\mathrm{d}12} \\ a_{\mathrm{d}34} \\ \cdot \\ \cdot \\ a_{\mathrm{d}(p-1)p} \\ a_{\mathrm{c}12} \\ a_{\mathrm{c}34} \\ \cdot \\ \cdot \\ a_{\mathrm{c}(p-1)p} \\ a_{p+1} \\ \cdot \\ \cdot \\ a_{n-1} \\ a_n \end{pmatrix} \qquad \mathring{\boldsymbol{b}} \equiv \begin{pmatrix} b_{\mathrm{d}12} \\ b_{\mathrm{d}34} \\ \cdot \\ \cdot \\ b_{\mathrm{d}(p-1)p} \\ b_{\mathrm{c}12} \\ b_{\mathrm{c}34} \\ \cdot \\ \cdot \\ b_{\mathrm{c}(p-1)p} \\ b_{p+1} \\ \cdot \\ \cdot \\ b_{n-1} \\ b_n \end{pmatrix}$$

and a *generalized* mixed-mode matrix can now be defined

$$\mathring{\boldsymbol{b}} \equiv \mathring{\boldsymbol{S}}\mathring{\boldsymbol{a}}. \tag{9.23}$$

Starting from these definitions, it is possible to find that the relationship between $\mathring{\boldsymbol{S}}$ and $\boldsymbol{S}$ is a bilinear transformation,

$$\mathring{\boldsymbol{S}} = (\tilde{\Xi}_{21} + \tilde{\Xi}_{22}\boldsymbol{S})(\tilde{\Xi}_{11} + \tilde{\Xi}_{12}\boldsymbol{S})^{-1} \tag{9.24}$$

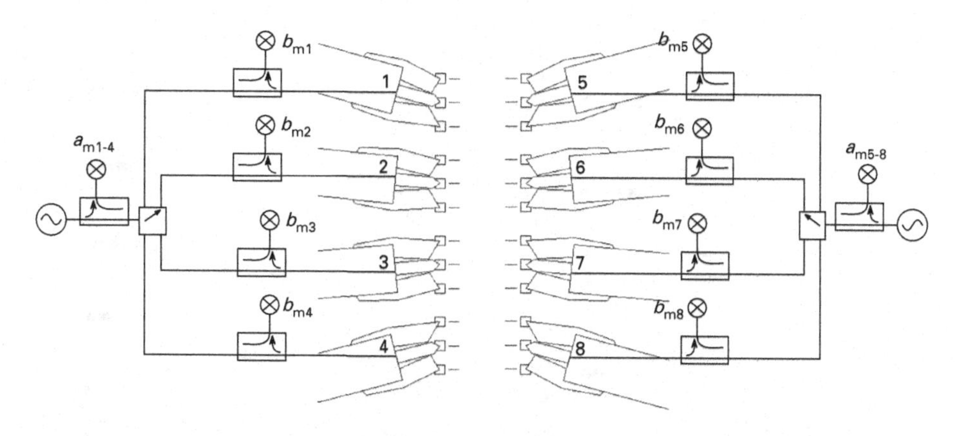

Fig. 9.12 Probe setup and measurement architecture. © 2008 IEEE. Reprinted, with permission, from [21].

where $\tilde{\Xi}_{ij}$ are transformation matrices containing all the single-ended and mixed-mode reference impedances, computed in [23].

9.3.1 Mixed-mode multiport measurement example

The example presented here, courtesy of Intel Corporation, appeared in [24]. The problem is the on-wafer measurements of transmission lines in differential configuration, with an eight-port VNA system.

The probe setup is shown in Figure 9.12, where the measurement architecture is also shown. The multiport system was a partial reflectometer system, divided in two halves (ports 1–4 and 5–8), so that a two-port standard, connected for example between ports 1 and 5, can be measured as if the reflectometer architecture was complete. Thus, any two-port calibration algorithm can be applied to these ports. In particular, a multiline TRL was chosen for ports 1 and 5, while the rest of the ports were calibrated with the thru connections summarized in Table 9.1, using QSOLT-type algorithms [25]. The probe touchdowns for this calibration are shown in Figure 9.13. The total number of

Table 9.1 Calibration matrix

	Port 1	Port 2	Port 3	Port 4	Port 5	Port 6	Port 7	Port 8
Port 1	X	X	X	X	MTRL	X	X	X
Port 2	X	X	X	X	thru	thru	X	X
Port 3	X	X	X	X	X	thru	thru	X
Port 4	X	X	X	X	X	X	thru	thru
Port 5	MTRL	thru	X	X	X	X	X	X
Port 6	X	thru	thru	X	X	X	X	X
Port 7	X	X	thru	thru	X	X	X	X
Port 8	X	X	X	thru	X	X	X	X

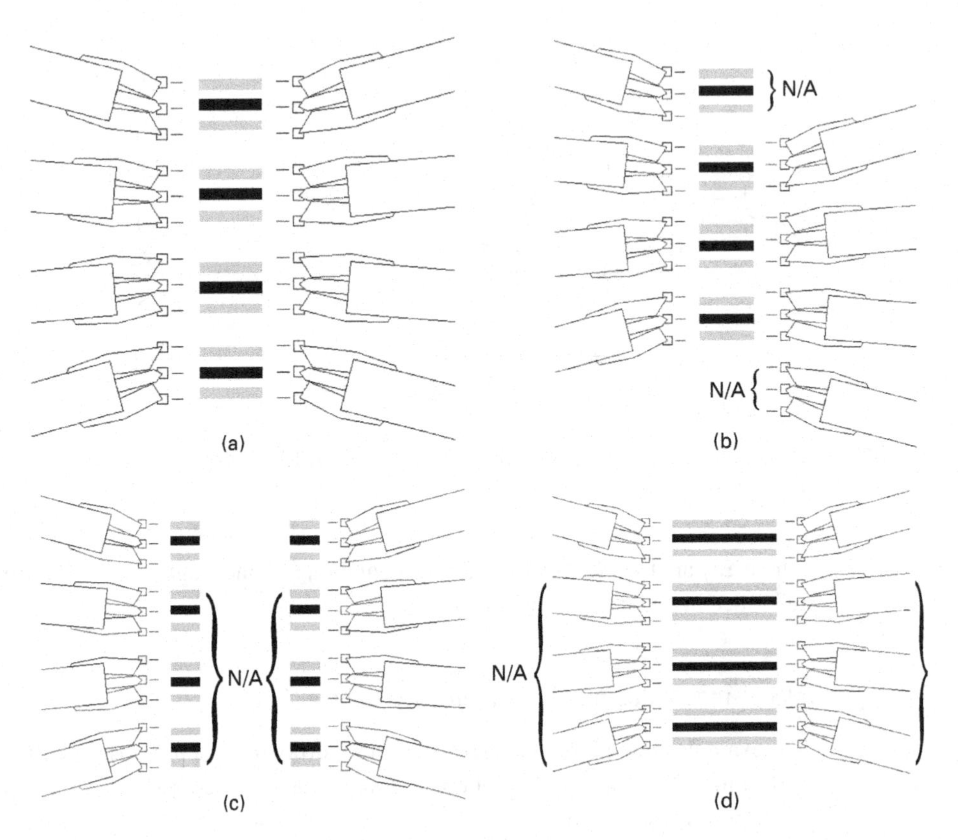

Fig. 9.13 Calibration touchdowns: straight thrus (a), shifted thrus (b), are obtained with the same calibration standard. Offset opens (c) are measured only at ports 1 and 5, as the set of three different length lines (d). © 2008 IEEE. Reprinted, with permission, from [21].

touchdowns for the calibration of this eight-port system is only six, which is rather low, considering that a multiline TRL with three lines would require five touchdowns for a two-port VNA.

The measurement results of a pair of transmission lines in differential configuration are shown in Figure 9.14. The differential parameters of interest, the insertion loss (IL) and the return loss (RL) were computed from the single-ended S-parameter, applying the transformations (9.24). It is interesting to note that the analysis of the mixed-mode performances leads to different conclusions than the analysis of the single-ended ones. If we consider a fixed (target) level of RL, e.g. -12 dB, this performance is achieved for higher frequency if the differential RL is considered instead of the single-ended RL. The same happens for the insertion loss, as shown in Figure 9.14(b).

In conclusion, this example has shown that choosing the proper error model and calibration algorithm can considerably reduce the measurement time and costs. Moreover, the mixed-mode S-parameter matrix can give important information for the design and performance analysis of differential devices and circuits.

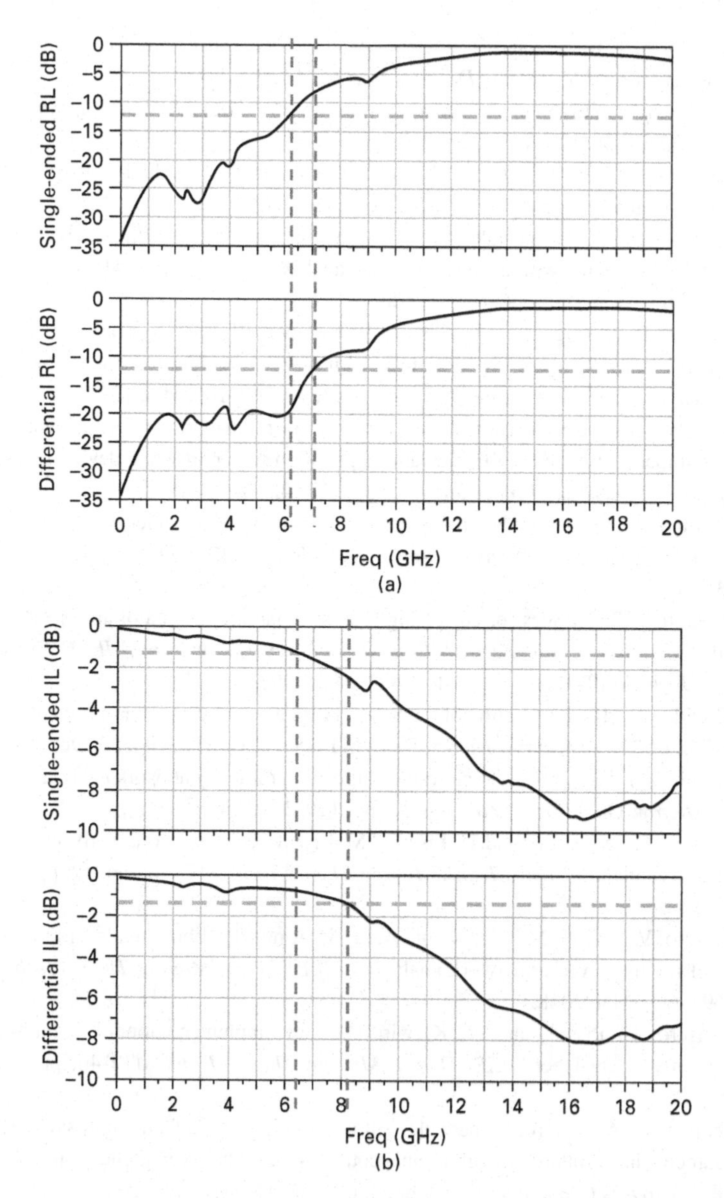

Fig. 9.14 Single-ended and differential return loss (a) and insertion loss (b) of two lines in differential configuration. © 2008 IEEE. Reprinted, with permission, from [21].

References

[1] T. G. Ruttan, B. Grossman, A. Ferrero, V. Teppati, and J. Martens, "Multiport VNA measurements," *IEEE Microwave*, pp. 56–69, June 2008.

[2] V. Teppati and A. Ferrero, "On-wafer calibration algorithm for partially leaky multiport vector network analyzers," *IEEE Trans. Microw. Theory Tech.*, MTT-53, no. 11, pp. 3665–3671, Nov. 2005.

[3] B. Grossman, T. Ruttan, and E. Fledell, "Architectural considerations for multiport vector network analyzers," *Proc. IEC DesignCon 2007*, 13, Feb. 2007.

[4] B. Grossman and T. Ruttan, "Why multi-port VNAs?" *70th ARFTG Conf. Signal Integrity Workshop Dig.*, pp. 120–139, Dec. 2007.

[5] B. Grossman, T. Ruttan, and E. Fledell, "Comparison of multiport VNA architectures measured results," *Proc. 66th ARFTG Conf.*, Dec. 2005.

[6] J.-C. Tippet and R.-A. Speciale, "A rigorous technique for measuring the scattering matrix of a multiport device with a 2-port network analyzer," *IEEE Trans. Microw. Theory Tech.*, MTT-30, no. 5, pp. 661–666, May 1982.

[7] U. Lott, W. Baumberger, and U. Gisiger, "Three-port RF characterization of foundry dual gate FETs using two-port test structures with on-chip loading resistors," *Proc. IEEE Int. Conference on Microelectronics Test Structures*, pp. 167–180, Mar. 1995.

[8] C. S. Hartmann and R. T. Hartmann, "Software for multi-port RF network analysis with a large number of frequency samples and application to 5-port SAW device measurement," *Ultrasonics Symposium Proceedings*, 1, pp. 117–122, Dec. 1990.

[9] H.-C. Lu and T.-H. Chu, "Multiport scattering matrix measurement using a reduced-port network analyzer," *IEEE Trans. Microw. Theory Tech.*, MTT-51, no. 5, pp. 1525–1533, May 2003.

[10] J.-C. Rautio, "Techniques for correcting scattering parameter data of an imperfectly terminated multiport when measured with a two-port network analyzer," *IEEE Trans. Microw. Theory Tech.*, MTT-31, no. 5, pp. 407–412, May 1983.

[11] M. Davidovits, "Reconstruction of the S-matrix for a 3-port using measurements at only two ports," *IEEE Trans. Microw. Theory Tech.*, MTT-5, no. 10, pp. 349–350, Oct. 1995.

[12] A. Ferrero and F. Sanpietro, "A simplified algorithm for leaky network analyzer calibration," *IEEE Microw. Guid. Wave Lett.*, 5, no. 4, pp. 119–121, Apr. 1995.

[13] A. Ferrero, F. Sampietro, and U. Pisani, "Multiport vector network analyzer calibration: a general formulation," *IEEE Trans. Microw. Theory Tech.*, MTT-42, no. 12, pp. 2455–2461, Dec. 1994.

[14] A. Ferrero, V. Teppati, M. Garelli, and A. Neri, "A novel calibration algorithm for a special class of multiport vector network analyzers," *IEEE Trans. Microw. Theory Tech.*, MTT-56, pp. 693–699, Mar. 2008.

[15] A. Ferrero, U. Pisani, and K. Kerwin, "A new implementation of a multiport automatic network analyzer," *IEEE Trans. Microw. Theory Tech.*, MTT-40, pp. 2078–2085, Nov. 1992.

[16] V. Teppati and A. Ferrero, "A new class of non-uniform, broadband, non-symmetrical rectangular coaxial-to-microstrip directional couplers for high power applications," *IEEE Trans. Microw. Wireless Compon. Lett.*, 13, no. 4, pp. 152–154, Apr. 2003.

[17] V. Teppati, A. Ferrero, and U. Pisani, "Recent advances in real-time load-pull systems," *IEEE Trans. Instrum. Meas.*, 57, no. 11, pp. 2640–2646, Nov. 2008.

[18] A. Ferrero and U. Pisani, "Two-port network analyzer calibration using an unknown 'thru'," *IEEE Microw. Guid. Wave Lett.*, MGWL-2, pp. 505–507, Dec. 1992.

[19] D. Bockelman and W. Eisenstadt, "Combined differential and common-mode scattering parameters: theory and simulation," *IEEE Trans. Microw. Theory Tech.*, MTT-43, no. 7, pp. 1530–1539, July 1995.

[20] D. Bockelman and W. Eisenstadt, "Pure-mode network analyzer for on-wafer measurements of mixed-mode S-parameters of differential circuits," *IEEE Trans. Microw. Theory Tech.*, MTT-45, no. 7, pp. 1071–1077, July 1997.

[21] A. Ferrero and V. Teppati, "Multiport and mixed mixed-mode measurements," in *Proc. 72nd ARFTG Conf., Signal Integrity Workshop*, Portland, OR, Dec. 2008.

[22] R. Marks and D. Williams, "A general waveguide circuit theory," *J. Res. NIST*, 97, pp. 533–561, Sept. 1992.

[23] A. Ferrero and M. Pirola, "Generalized mixed-mode S-parameters," *IEEE Trans. Microw. Theory Tech.*, MTT-54, pp. 458–463, Jan. 2006.

[24] E. Fledell and T. Ruttan, "Digital backplane interconnections and bus multi-port differential characterization," in *IEEE MTT-S Intl. Microwave Symp. Dig.*, June 2006.

[25] A. Ferrero and U. Pisani, "Qsolt: a new fast calibration algorithm for two-port S-parameter measurements," in *38th ARFTG Conf. Dig.*, San Diego, CA, Dec. 1991, pp. 15–24.

10 Noise figure characterization

Nerea Otegi, Juan-Mari Collantes, and Mohamed Sayed

10.1 Introduction

Noise is one of the most critical issues in wireless systems because it is a fundamental limiting factor for the performance of microwave receivers. Industry requirements for increasingly higher performing communication systems require tighter noise specifications that make the noise figure measurement a critical step in the characterization of modern microwave circuits and systems.

Noise figure measurements of circuits and sub-systems have been traditionally performed with noise figure meters specifically developed for that purpose. A paradigmatic example is the HP8970 (and associated family) that was considered for years as the reference meter for noise figure characterization. This instrument, as well as other modern equipment, uses the popular Y-factor technique to compute the noise figure from the ratio of two power measurements ("cold" and "hot"). The scalar nature of the measurements allows an easy and straightforward characterization process. This simplicity is undoubtedly part of its large success. However, its accuracy is limited by the match properties of the device under test and measurement setup.

There are two factors that have been driving an evolution in the noise figure characterization schemes. One factor is a growing tendency in microwave instrumentation to integrate different types of measurements into a single instrument box. As a result, noise-figure characterization is now available as an option in modern vector network analyzers (VNA) from different manufacturers. The other factor is that the accuracy requirements in environments that are not perfectly matched (millimeter wave and beyond, on-wafer setups, etc.) demand a noise figure characterization that takes advantage of vector measurements to improve scalar results.

In this context, solutions have been proposed to enhance the original scalar Y-factor technique with vector correction terms that account for systematic errors such as mismatch. Moreover, techniques other than Y-factor are also proposed in modern equipment. This is the case for some new VNAs that use the cold-source technique. Here, the device is measured at a single "cold" state. The cold-source technique was mainly used in the past to characterize the noise parameters of single transistors, which very often present poor match characteristics. That is why, in its classical form, cold-source includes corrections for mismatch errors and requires vector measurements. It is indeed a more complex characterization approach than the scalar Y-factor technique.

Noise figure characterization approaches with the ability to correct for a variety of systematic errors are not exclusive to new VNAs. They can also be applied to other microwave instruments with noise figure capabilities as most modern spectrum analyzers (SA), although additional equipment for vector measurements is required in this case.

In this chapter we provide a detailed description of both Y-factor and cold-source techniques for noise figure characterization. As starting point, the fundamentals of noise figure are briefly summarized in Section 10.2. Section 10.3 is devoted to the classical Y-factor technique, while cold-source is treated in Section 10.4. In Section 10.5 the main sources of systematic errors in a noise figure measurement are analyzed (mismatch, receiver bandwidth and linearity, etc.). Whenever it is relevant, their impact on each technique is comparatively discussed and measurement examples are provided. Finally, Section 10.6 is dedicated to the noise figure characterization of mixers. This is usually a challenging measurement because, in addition to the frequency translation, nearly every drawback affecting ordinary two-ports is magnified in mixers.

10.2 Noise figure fundamentals

10.2.1 Basic definitions and concepts

Noise

Noise is a random process associated with several sources. Thermal noise (Johnson [1] or Nyquist [2] noise) is one of the principal noise mechanisms in RF and microwave systems and it is caused by the random motion of thermally excited charge carriers in any passive circuit element that contains losses. There are several other noise sources such as shot noise, generation-recombination (G-R) noise, flicker noise, quantum noise etc. The analysis of the properties of noise as a random process or the particular nature of these several noise sources is beyond the scope of this chapter. Fundamental readings on noise can be found, amongst others, in [3–6].

Any resistive element at a temperature T different from zero Kelvin generates thermal noise that can be expressed as the available thermal noise power [2]:

$$N = kTB. \tag{10.1}$$

where k is the Boltzmann constant (1.38e-23 Joule/Kelvin) and B is the considered bandwidth (Hertz). Therefore, the available noise power generated by a passive element is proportional to temperature and bandwidth, but does not depend on resistance or frequency. A noise mechanism that fulfils this last property is referred to as white noise. Actually, the spectral density of the available thermal noise power has a slight frequency dependence that can be neglected up to TeraHertz frequencies [4]. Other types of noise, such as shot and G-R noise, also behave as white noise in the RF and microwave frequency ranges, so in a noise characterization the overall effect of these sources is usually treated as equivalent to thermal noise [7].

Noise figure

Any two-port device, in addition to amplifying or attenuating both the signal and the noise present at its input, adds extra noise generated by its own components, thus degrading the signal-to-noise ratio (SNR). The noise figure is a figure of merit that characterizes this degradation and it is defined as the ratio of the SNR at the input and the SNR at the output when the input noise is thermal noise generated by a passive load at a reference temperature of $T_0 = 290$ K [8]:

$$F = \left.\frac{S_i/N_i}{S_o/N_o}\right|_{T=T_0}. \tag{10.2}$$

S_i and S_o are, respectively, the signal powers available at the input and output of the two-port, while N_i and N_o are the available noise powers. This definition of noise figure can be extended to multiport devices [9]. It is worthwhile to note that, because of its definition, the relevance of the noise figure is limited to low input signals and low noise levels. This is why the noise figure has little significance for a power amplifier, where the added noise has a negligible contribution to the degradation of the SNR because of the large levels involved.

The most basic concepts associated with noise figure are graphically shown in Figure 10.1. In Figure 10.1(a), a generic block diagram of a two-port device with a noise source connected at its input can be seen. The noise power available at the output of the device as a function of the noise source temperature, the "noise line," is plotted in Figure 10.1(b).

According to the definition of noise figure, the available noise at the input of the two-port device is thermal noise generated by a passive load:

$$N_i = kBT_0. \tag{10.3}$$

In addition, the noise power available at the output port can be expressed as

$$N_o = G_{av}N_i + N_{add} = kBG_{av}T_0 + N_{add} \tag{10.4}$$

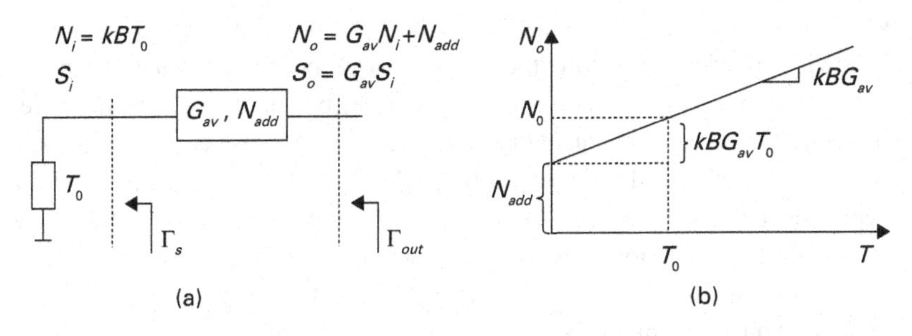

Fig. 10.1 (a) Generic two-port device with noise source connected at input. (b) Graphical representation of the noise at the output of the device as a function of the noise source temperature.

where, N_{add} is the noise added by the two-port device and G_{av} is the two-port available gain [7], defined as

$$G_{av} = \frac{1-|\Gamma_s|^2}{|1-s_{11}\Gamma_s|^2}|s_{21}|^2\frac{1}{1-|\Gamma_{out}|^2}, \tag{10.5}$$

where s_{ij} are the S-parameters of the two-port device, Γ_{out} is its output reflection coefficient, and Γ_s is the reflection coefficient of the passive load connected at the input of the two-port device.

The noise figure definition in (10.2) can be rewritten as:

$$F = \frac{kBG_{av}T_0 + N_{add}}{kBG_{av}T_0}, \tag{10.6}$$

which is the ratio of the noise power available at the output port to the contribution to the output of the input termination, when this termination is at the reference temperature of 290 K. (10.6) represents the formal definition of noise figure adopted by the IRE [10, 11].

It is seen from (10.6) that the noise figure characterizes the noise added by the device and, thus, this added noise can be expressed as a function of the noise figure as

$$N_{add} = kBG_{av}T_0(F-1). \tag{10.7}$$

The IRE introduced the equivalent denomination noise factor for the noise figure (10.6), [12], sometimes called noise figure in linear terms. It is nowadays broadly accepted to use noise figure *NF* for the quantity (10.8), expressed in dB, while noise factor is used for the linear quantity F. From now on, this convention is followed in this chapter.

$$NF = 10\log_{10}(F). \tag{10.8}$$

Noise temperature

Sometimes, especially for low noise devices, the effective input noise temperature, T_e, is used instead of the noise factor to characterize the noise generated by a device. According to [11] the effective input noise temperature is the temperature at which a source termination connected to a noise-free equivalent of the two-port device would lead to the same output noise power of the real two-port device with a noise-free source termination. Figure 10.2 illustrates the meaning of effective input noise temperature.

According to its definition, the noise at the output of the two-port device can be written in terms of this effective input noise temperature as:

$$N_o = kBG_{av}(T_0 + T_e). \tag{10.9}$$

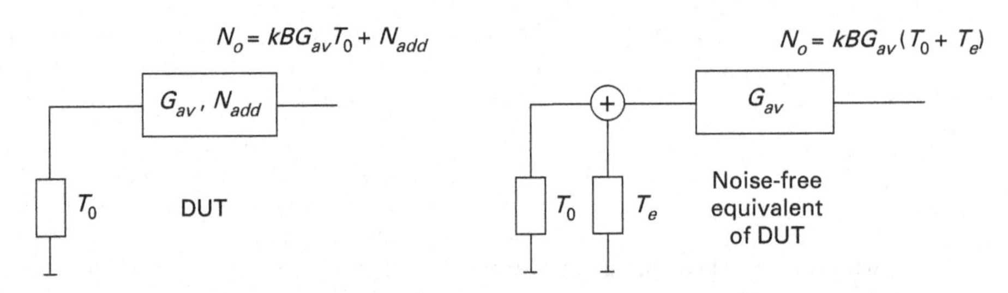

Fig. 10.2 Graphical interpretation of effective input noise temperature.

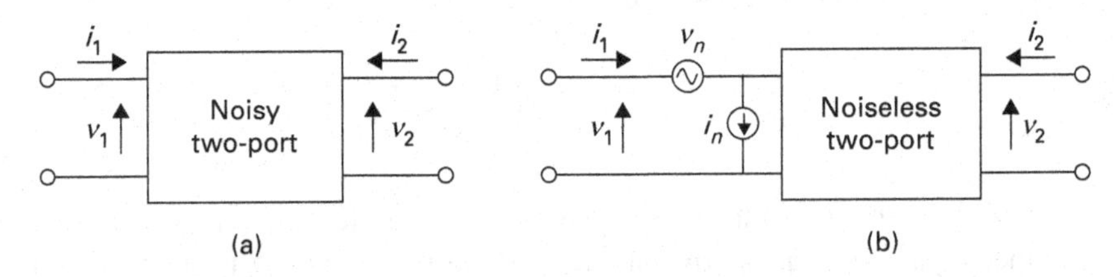

Fig. 10.3 (a) Two-port device with internal noise sources; (b) equivalent circuit with noise voltage source v_n and noise current source i_n at noise-free equivalent input.

It is seen from its definition that the effective input noise temperature is a translation to the input port of the noise added by the device.

$$T_e = \frac{N_{add}}{kBG_{av}}. \tag{10.10}$$

If (10.9) is brought into the definition of noise factor, a simple relationship between the noise factor and the effective input temperature is obtained:

$$F = \frac{T_e}{T_0} + 1. \tag{10.11}$$

This relationship between the effective input noise temperature and the noise factor is limited to two-port transducers with a single input frequency and a single output frequency, as explicitly stated in [12]. This is a non-trivial assessment that is further analyzed in Section 10.6, that is devoted to mixer noise figure characterization.

Noise parameters

The noise behavior of a linear two-port device can be fully modeled by two noise sources added to a noise-free equivalent of the original two-port device [13], shown in the classical representation of Figure 10.3. As there are several internal noise processes that are complex, these equivalent noise sources are generally not independent. Several representations in terms of different equivalent noise sources [13, 14], or parameterizations based on noise-waves [14] can be found in the literature. A compilation of diverse representations is given in [16]. Whatever the noise representation, the noise sources associated with the two-port device are correlated in a general case. Thus, four independent parameters, noise parameters, are required to fully characterize the internal noise of a linear two-port device in terms of its source impedance (leading to the so-called noise correlation matrix): two real parameters, one for each of the sources, and the real and imaginary parts of a complex parameter that takes into account the correlation between the sources. In the model of Figure 10.3, these four parameters are the mean square fluctuations of the noise sources, $\overline{v_n^2}$, and $\overline{i_n^2}$, and a complex correlation parameter $\overline{v_n i_n^*}$.

Accordingly, the noise factor of a two-port device depends on the source termination through a set of four independent noise parameters [14]. $\{F_{min}, R_n, Re(\Gamma_{opt}), Im(\Gamma_{opt})\}$ is the most common set of noise parameters for microwave two-port devices. These noise parameters are derived from the noise fluctuations ($\overline{v_n^2}, \overline{i_n^2}, \overline{v_n i_n^*}$) and completely characterize the noise response of the two-port device [14]. F_{min} is the minimum noise factor of the device and Γ_{opt} is the optimum source reflection coefficient, which provides the minimum noise factor. The "noise resistance" R_n is a parameter that characterizes how rapidly the noise factor diverges from F_{min} as the source reflection coefficient varies from the optimum case. In this representation, the noise factor is given as a function of the reflection coefficient Γ as:

$$F = F_{min} + 4\frac{R_n}{Z_0}\frac{|\Gamma - \Gamma_{opt}|^2}{|1+\Gamma_{opt}|^2(1-|\Gamma|^2)}, \tag{10.12}$$

where Z_0 is the characteristic impedance.

A three-dimensional representation of (10.12) shows a paraboloid with its minimum (F_{min}) at Γ_{opt} (see Figure 10.4). If constant noise factor values are depicted as a function of Γ on the Smith-Chart, the well-known noise circles [17] are obtained. These circles represent the projection of the paraboloid on the Smith-Chart.

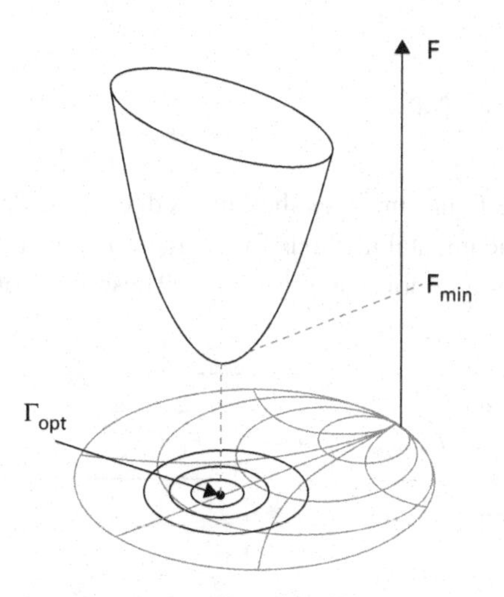

Fig. 10.4 Three-dimensional representation of noise factor versus source reflection coefficient and noise circles on a Smith-Chart.

Obtaining the noise parameters of a two-port device is not an easy task [18]. The first methods for characterizing the noise parameters of a two-port device were based on the actual experimental searching for the minimum noise factor and its corresponding optimum source impedance [10]. However, computer-aided data fitting techniques were soon proposed to extract the noise parameters from measured data [19], leading to faster and more accurate noise parameter characterization techniques (source pull techniques), as [19–23]. For that at least four noise figures corresponding to four source reflection coefficients are required, although more than four terminations are normally used to obtain the parameters from an overdetermined system and minimize errors. The noise parameter extraction was further simplified on the basis of directly measuring noise power values instead of noise figures [24–26]. Impedance tuners are used to synthesize the required source reflection coefficients, which have to be adequately distributed on the Smith-Chart in order to avoid ill-conditioning problems [27–30]. To facilitate the data fitting, linearized versions of the classical noise parameter representation, sometimes referred to as noise pseudoparameters, are used, as in [19]. Linearized parameterizations based on noise-wave descriptions can also be found in [31, 32].

Noise figure of cascaded devices

Let us consider a system formed by several cascaded two-port devices, each of them represented by its noise factor F_j and available gain G_{avj}, like the one shown in Figure 10.5.

As shown by Friis [8], the overall noise factor of a cascaded system is given by

$$F = F_1(\Gamma_s) + \frac{F_2(\Gamma_{out1}) - 1}{G_{av1}} + \frac{F_3(\Gamma_{out2}) - 1}{G_{av1}G_{av2}} + \cdots \tag{10.13}$$

This expression reflects a well-known fact: the importance of the first stage in a system receiver, since the contribution to the SNR degradation of later stages is reduced by the product of gains of the preceding ones.

10.2.2 Two noise figure characterization concepts: Y-factor and cold-source

In essence, characterizing the noise figure involves the knowledge of the "noise line" in Figure 10.1(b). There are two fundamental methods for characterizing a line: obtaining two points of the line or obtaining a point and the slope. These two concepts are

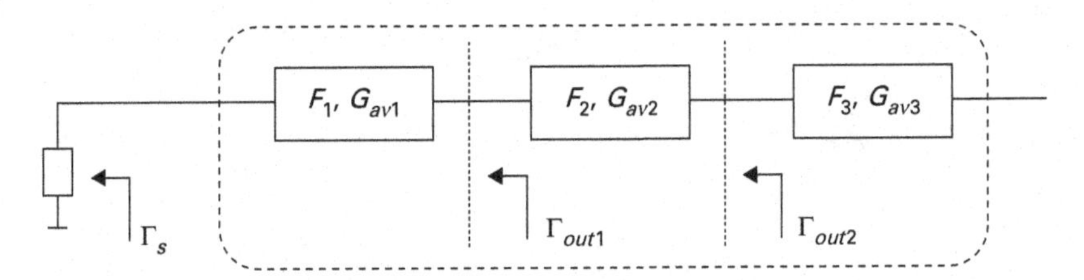

Fig. 10.5 Block diagram of a cascaded system formed by several two-port devices.

directly related to the two basic noise figure measurement methodologies that are further discussed in this chapter: Y-factor and cold-source.

The Y-factor technique obtains the noise factor from two noise powers (N_c, N_h) corresponding to two different input temperatures (T_c, T_h), known as cold and hot temperatures, respectively [11]. The ratio of these two quantities is called Y-factor, $Y = N_h/N_c$, and thus the name of the technique. (10.14) and (10.15) represent the noise powers available at the output of the two-port device for the two input temperatures. The noise factor can be expressed as a function of the Y-factor and both temperatures as shown in (10.16).

$$N_c = kBG_{av}\left[T_c + T_0\left(F-1\right)\right] \tag{10.14}$$

$$N_h = kBG_{av}\left[T_h + T_0\left(F-1\right)\right] \tag{10.15}$$

$$F = \frac{\left(T_h/T_0 - 1\right) - Y\left(T_c/T_0 - 1\right)}{Y-1}. \tag{10.16}$$

Assuming that the cold temperature T_c is very close to the reference temperature T_0, the noise factor can be approximated to:

$$F \approx \frac{\left(T_h/T_0 - 1\right)}{Y-1}. \tag{10.17}$$

The cold-source approach obtains the noise factor from a single noise power corresponding to the cold temperature [24]. The noise factor is directly derived from (10.14) and does not require a hot noise power measurement. The "noise slope" kBG_{av} has to be previously determined for that. The noise factor is given by (10.18), which can be approximated to (10.19) considering again that the cold temperature T_c is close to the reference temperature T_0.

$$F = \frac{N_c}{kBG_{av}T_0} - \left(\frac{T_c}{T_0} - 1\right) \tag{10.18}$$

$$F \approx \frac{N_c}{kBG_{av}T_0}. \tag{10.19}$$

10.3 Y-factor technique

The Y-factor technique is the most popular noise figure measurement methodology, used by the majority of the commercially available noise figure meters from the classical HP-8970 [33], to recent versions of the NFA series N897X-A [34]. Modern noise figure measurement implementations included in spectrum analyzers are also usually based on the Y-factor technique [35, 36]. A very detailed description of this technique can be found in [37]. The basic diagram of a Y-factor measurement is shown in Figure 10.6.

As schematically shown in the previous section, in the Y-factor technique the noise factor is obtained through two noise power measurements for two different input temperatures. In order to physically generate the cold and hot input "temperatures,"

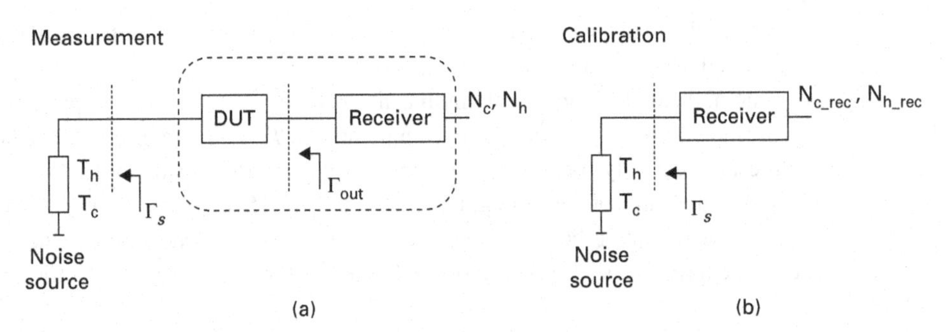

Fig. 10.6 Basic block diagrams for Y-factor characterization technique: (a) measurement step; (b) calibration step.

i.e. available noise powers corresponding to these temperatures, commercial noise sources are normally used. These noise sources are avalanche diodes that provide two different broadband noise levels, depending on whether they are biased or not [38]. When the diode is not biased ("off" or "cold" state), it is basically a resistor that generates a noise power proportional to the ambient temperature. However, when the diode is biased in avalanche ("on" or "hot" state) it provides extra noise, significantly higher than the thermal noise corresponding to the off state, which can be related to a hot temperature. This hot temperature is quantified by the Excess Noise Ratio (ENR) of the noise source: $ENR = 10\log_{10}\left((T_h - T_c)/T_0\right)$.

With the noise source connected to the DUT, two noise powers are measured (Figure 10.6(a)) so that the noise factor can be calculated as shown in (10.16). However, the noise factor obtained in this way is not the device noise factor, F_{DUT}, but the noise factor F_{sys} of the cascaded system composed by the DUT and the noise receiver. Thus, from the measured noise powers (N_c, N_h) the system noise factor is calculated as:

$$F_{sys} = \frac{(T_h/T_0 - 1) - Y\,(T_c/T_0 - 1)}{Y - 1}, \quad Y = \frac{N_h}{N_c}. \tag{10.20}$$

A second stage correction is required to eliminate the noise contribution of the receiver. To this end, a calibration step is needed to characterize the receiver noise factor. With the noise source directly connected to the receiver (Figure 10.6(b)) two noise powers (N_{c_rec} and N_{h_rec}) are measured and the receiver noise factor F_{rec} is again calculated as:

$$F_{rec} = \frac{(T_h/T_0 - 1) - Y_{rec}\,(T_c/T_0 - 1)}{Y_{rec} - 1}, \quad Y_{rec} = \frac{N_{h\_rec}}{N_{c\_rec}}. \tag{10.21}$$

Modern noise figure measurement equipment includes the possibility of correcting the differences between the cold temperature T_c and the reference one T_0 [39]. If the cold temperature can be approximated to the reference, (10.20) and (10.21) can be simplified according to (10.17) as:

$$F_{sys} \approx \frac{(T_h/T_0 - 1)}{Y - 1}, \quad Y = \frac{N_h}{N_c} \tag{10.22}$$

$$F_{rec} \approx \frac{(T_h/T_0 - 1)}{Y_{rec} - 1}, \quad Y_{rec} = \frac{N_{h\_rec}}{N_{c\_rec}}. \tag{10.23}$$

The calibration step is also used to obtain the DUT gain, necessary for the second-stage correction. From the four measured scalar noise powers the insertion gain of the device is computed by means of (10.24).

$$G_{ins} \equiv \frac{N_h - N_c}{N_{h\_rec} - N_{c\_rec}}. \tag{10.24}$$

Finally, the device noise figure is calculated as:

$$NF_{YF} \equiv 10\log_{10}\left(F_{sys} - \frac{F_{rec}(\Gamma_s) - 1}{G_{ins}}\right). \tag{10.25}$$

It is important to remark that all the measurements involved in the determination of (10.25) are exclusively scalar measurements.

10.4 Cold-source technique

Following [24], the cold-source technique is a usual noise-measurement methodology for noise parameter extraction. A basic diagram of the cold-source procedure is given in Figure 10.7. In the cold-source technique the noise figure is characterized from a single noise-power measurement of the device, (N_c), with a source termination at ambient

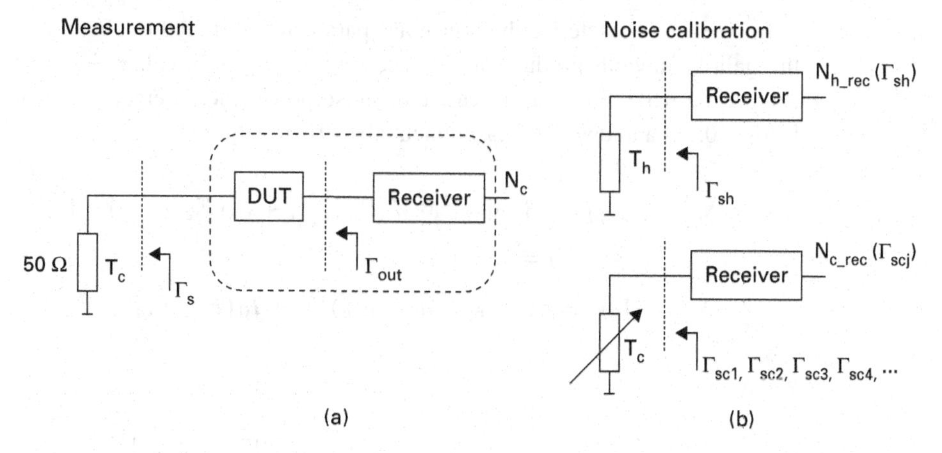

Fig. 10.7 Basic block diagrams for cold-source characterization technique: (a) measurement step; (b) calibration step.

temperature connected at its input (see Figure 10.7(a)). To this end, the device available gain G_{av} and the gain-bandwidth product of the receiver $kB|s_{21rec}|^2$ have to be previously determined. To obtain an accurate measurement of the $kB|s_{21rec}|^2$ term requires the use of a noise source in its cold and hot states, but it should be noted that the hot noise source is only necessary in the calibration step. Cold-source implementations that characterize the receiver gain-bandwidth product without a noise source can also be found [40, 41]. The $kB|s_{21rec}|^2$ term can be estimated by obtaining the gain and bandwidth responses of the receiver. For that, a narrowband frequency sweep is performed and the effective noise bandwidth is computed by integrating this response [10]. A second stage correction is again needed to properly characterize the device noise figure and, thus, the noise contribution of the receiver has to be characterized in the calibration step, as well. Generally devoted to noise parameter extraction, treated in Section 10.2.1, the cold-source technique is normally a fully corrected procedure, including a complete receiver noise calibration to get its four noise parameters. The DUT noise figure is obtained from

$$NF_{CS} \equiv 10\log_{10}\left(\frac{N_c}{kB|s_{21rec}|^2 G_{av} MM(\Gamma_{out}) T_0} - \frac{F_{rec}(\Gamma_{out}) - 1}{G_{av}} - \left(\frac{T_c}{T_0} - 1\right)\right), \tag{10.26}$$

where $F_{rec}(\Gamma_{out})$ is the receiver noise factor corresponding to the output reflection coefficient of the DUT, G_{av} is the device available gain given in (10.5), and $MM(\Gamma_{out})$ is a term accounting for the mismatch between the device and the receiver (with s_{11rec} the input reflection coefficient of the receiver).

$$MM(\Gamma_{out}) = \frac{1 - |\Gamma_{out}|^2}{|1 - s_{11rec}\Gamma_{out}|^2}, \tag{10.27}$$

In order to characterize the four noise parameters required to obtain $F_{rec}(\Gamma_{out})$ and the gain-bandwidth product of the receiver $kB|s_{21rec}|^2$ a calibration step is needed (Figure 10.7(b)). For that, at least four noise power measurements with four passive loads (10.28) and one hot measurement (10.29) are required.

$$N_{c\_rec}(\Gamma_{scj}) = kB|s_{21rec}|^2 MM(\Gamma_{scj})\left[T_c + T_0\left(F_{rec}(\Gamma_{scj}) - 1\right)\right], \quad j = 1, 2, 3, 4\ldots \tag{10.28}$$

$$N_{h\_rec}(\Gamma_{sh}) = kB|s_{21rec}|^2 MM(\Gamma_{sh})[T_h + T_0(F_{rec}(\Gamma_{sh}) - 1)], \tag{10.29}$$

where

$$F_{rec}(\Gamma) = F_{min\_rec} + 4\frac{R_{n\_rec}}{Z_0}\frac{|\Gamma - \Gamma_{opt\_rec}|^2}{|1 + \Gamma_{opt\_rec}|^2(1 - |\Gamma|^2)}. \tag{10.30}$$

The five unknowns ($kB|s_{21rec}|^2$ and the receiver noise parameters $\{F_{min\_rec}, R_{n\_rec}, Re(\Gamma_{opt\_rec}), \mathrm{Im}(\Gamma_{opt\_rec})\}$, or an equivalent set) are extracted from these measured noise powers. For that, linearized versions of (10.28) (10.29), in terms of equivalent noise pseudoparameters, are generally used. Diverse approaches to the noise description, based on different parameterizations, can be found, as for instance in [25, 32, 42].

A simplified estimation of $kB|s_{21rec}|^2$ can be obtained from (10.31) assuming that the receiver noise factor does not vary between one of the cold measurements and the hot one. This estimation is sometimes used as the starting point of an iteration process [24].

$$kB|s_{21rec}|^2 \approx \frac{N_{h\_rec}(\Gamma_{sh})/MM(\Gamma_{sh}) - N_{c\_rec}(\Gamma_{sc1})/MM(\Gamma_{sc1})}{T_h - T_c}, \tag{10.31}$$

where $N_{c\_rec}(\Gamma_{sc1})$ and $N_{h\_rec}(\Gamma_{sh})$ are, respectively, the cold and hot noise powers measured by the receiver with the noise source connected to it (typically $\Gamma_{sc1} \approx 0$, $\Gamma_{sh} \approx 0$). Note however, that the actual $kB|s_{21rec}|^2$ would be given by:

$$kB|s_{21rec}|^2 = \frac{N_{h\_rec}(\Gamma_{sh})/MM(\Gamma_{sh}) - N_{c\_rec}(\Gamma_{sc1})/MM(\Gamma_{sc1})}{T_h - T_c + T_0(F_{rec}(\Gamma_{sh}) - F_{rec}(\Gamma_{sc1}))}. \tag{10.32}$$

As explained in Section 10.2.1, to minimize errors, more than four cold terminations are normally used in the noise calibration. To this end, impedance tuners that synthesize the required impedance states are generally used and the noise parameters are then extracted by means of fitting methods [25, 26, 40]. The terminations have to be adequately distributed on the Smith-Chart in order to provide a well-conditioned set of equations that allow the accurate computation of the four noise parameters [27–30].

The described cold-source technique requires a more complex measurement bench than the scalar Y-factor method [43]: vector measurements, impedance tuning, switching circuitry, etc. Thus, the cold-source method was not the usual option for a standard noise figure measurement of circuits or subsystems, but was mainly focused on noise parameter extraction. Only recently, with the appearance of modern VNAs with noise-figure measurement capabilities [40, 41], has attention been brought to this technique in the context of circuit noise figure characterization. A different noise figure characterization approach implemented in a VNA and based on digital data processing techniques can also be found [44].

10.5 Common sources of error

The accuracy of a noise figure measurement depends on a wide variety of factors including the characteristics of the DUT, the measurement setup and, obviously, the degree of approximations included in the methodology. For instance, when coming to analyze a scalar noise figure measurement technique as the classical Y-factor, mismatches in the measurement path will impact the accuracy of the final result. There are other effects that can have an influence on noise figure accuracy, such as measurement temperature, receiver linearity, and bandwidth. There is also uncertainty associated with the limited accuracy of the measurement instruments; the imperfect knowledge of the noise temperatures; the incomplete knowledge of the correction terms used to remove systematic errors; and random effects such as connector variability, jitter, etc. Evaluating the overall uncertainty of a noise figure measurement is not an easy task because of the complex formulae involved, especially in those methodologies that include a variety of corrections, and it is not treated in this chapter. The fundamentals on measurement uncertainty and guidance on numerical methods for its evaluation are given in [45] and [46], respectively, while works particularly focused on noise measurement uncertainty can be found, for example, in [47–50]. Accuracy and uncertainty issues have been increasingly treated by noise figure measurement equipment manufacturers, as for example in [39, 40, 51–53]. Nowadays, specific uncertainty calculators for noise figure measurements are offered by different manufacturers [54–56].

In this section, some basic sources of error in noise figure characterization are discussed. The analysis is divided into three main categories (mismatch, temperature, and measurement setup). Each systematic effect is treated separately to extract unambiguous conclusions. Where it is pertinent, the accuracy implications for Y-factor and cold-source are compared. Numerical examples are used to help visualize the main results of the analyses. In addition, measurement examples are provided to confirm and illustrate the basic conclusions. Only ordinary two-port devices are considered in this section, since frequency translating devices are specifically addressed in Section 10.6.

10.5.1 Mismatch

In this section, only errors coming from mismatch are considered. Any other source of error in the measurement is neglected. Obviously, mismatch has a significant impact on a scalar methodology such as the classical implementation of Y-factor. Therefore this part is mainly focused on the Y-factor technique. The effect of including correction terms for mismatch systematic errors in Y-factor is also studied.

Y-factor and the second stage correction

The Y-factor technique is based on a scalar approximation (10.25) of the actual Friis formula [8] for the DUT-receiver system, given in (10.33).

$$NF_{DUT} = 10\log_{10}\left(F_{sys} - \frac{F_{rec}(\Gamma_{out}) - 1}{G_{av}}\right). \tag{10.33}$$

It should be noted that (10.25) neglects any mismatch effect in the measurement path and includes two main approximations to the Friis formula (10.33).

The first approximation concerns the available gain. The device available gain, whose accurate characterization requires vector measurements, is substituted by the insertion gain, directly obtained through scalar noise power measurements. The insertion gain is defined as:

$$G_{ins} = \frac{|1 - s_{11rec}\Gamma_s|^2}{|1 - s_{11}\Gamma_s|^2}|s_{21}|^2\frac{1}{|1 - s_{11rec}\Gamma_{out}|^2}. \tag{10.34}$$

If the noise source and the receiver are properly matched, which is usually the case, the insertion gain tends to $|s_{21}|^2$. If, in addition, the device is properly matched the available gain also tends to $|s_{21}|^2$. Hence, if the device is adequately matched, the insertion gain will be a good approximation of the available gain. When this is not the case, both gains can diverge significantly.

The second approximation concerns the receiver noise factor. During the calibration step, the source termination presented to the receiver corresponds to the reflection coefficient of the noise source, Γ_s. Thus, the measured receiver noise factor is $F_{rec}(\Gamma_s)$ instead of $F_{rec}(\Gamma_{out})$, as required by the Friis formula. The noise source can be considered to be a fairly matched device. As a consequence, if the DUT has a poor output match, the receiver noise factor measured in the calibration step, $F_{rec}(\Gamma_s)$, may not be a good approximation of the noise factor that the receiver actually has during the DUT measurement step, $F_{rec}(\Gamma_{out})$.

The expression resultant from the classical scalar Y-factor implementation (10.25) is given by (10.35). This expression shows the error associated with the calculation of the noise figure with respect to the true noise figure of the DUT, $NF_{DUT} = 10\log_{10}(F_{DUT})$. It can be observed how (10.35) converges to the true value, NF_{DUT}, for perfect match conditions ($\Gamma_{out} = \Gamma_s = 0$).

$$NF_{YF} = 10\log_{10}\left(F_{DUT}(\Gamma_s) + \frac{F_{rec}(\Gamma_{out}) - 1}{G_{av}} - \frac{F_{rec}(\Gamma_s) - 1}{G_{ins}}\right). \tag{10.35}$$

Therefore, although the Y-factor technique is a simple technique from the implementation point of view, the lack of vector measurements and the neglecting of the receiver noise factor dependence on the source termination can significantly degrade the accuracy of the measured noise figure, especially when measuring poorly matched, low-gain devices.

In [57], a Y-factor technique complemented with vector measurements is proposed. From these additional vector measurements, the available gain of the device is computed and included in the second stage correction. It is important to note that the receiver noise factor required for this second stage correction is still obtained as in the classical scalar Y-factor technique. Therefore, any dependence of the receiver noise on the source termination is still neglected. This implementation can be formulated as

$$NF_{YF\_Gav\_CORR} \equiv 10\log_{10}\left(F_{sys} - \frac{F_{rec}(\Gamma_s) - 1}{G_{av}}\right), \tag{10.36}$$

in contrast to the fully scalar technique given in (10.25), where the device insertion gain is used in the second stage correction. The expression that results from this implementation, under the assumptions made, is:

$$NF_{YF\_Gav\_CORR} = 10\log_{10}\left(F_{DUT}(\Gamma_s) + \frac{F_{rec}(\Gamma_{out})}{G_{av}} - \frac{F_{rec}(\Gamma_s)}{G_{av}}\right). \tag{10.37}$$

The possible benefits of the partial correction included in (10.36) are not guaranteed and have to be carefully analyzed. As shown in [58], for low gain mismatched devices, the second stage correction requires the knowledge of receiver noise parameters to be efficient and rigorous. Applying vector corrections without an accurate knowledge of the receiver noise factor may end up in poor accuracy, even worse than the basic scalar approach.

Let us illustrate this analysis with the help of a numerical example.

Numerical example 1. The following setup is considered:
Noise source: ENR = 15 dB, $\Gamma_{sc} = \Gamma_{sh} = 0.05\angle{-20^{\circ}}$.
DUT: $s_{11} = 0.08\angle 45^{\circ}$, $s_{22} = variable \angle 180^{\circ}$, $s_{21} = 5$ dB, $s_{12} = -50$ dB; $NF = 3$ dB.
Receiver: $s_{11rec} = 0.06\angle 170^{\circ}$; $NF_{min\_rec} = 6$ dB, $R_{n\_rec} = 50\ \Omega$, $\Gamma_{opt\_rec} = 0.07\angle 60^{\circ}$.

Let us define the error associated with each methodology as the difference between the resultant noise figure and the true noise figure ($e_{YF} = NF_{YF} - NF_{DUT}$, $e_{YF\_Gav\_CORR} = NF_{YF\_Gav\_CORR} - NF_{DUT}$). In Figure 10.8(a) the errors associated with both techniques are plotted as a function of device output match. As can be seen, both methodologies become considerably inaccurate as this output match worsens. The error given by the corrected Y-factor version is slightly lower than the error of the classical approach for good DUT output match. However, as the output match degrades, $e_{YF\_Gav\_CORR}$ increases more rapidly than e_{YF}. This rapid increase is due to the combination of an available gain that tends to infinity with a noise factor obtained at the calibration step that remains constant.

It is clear from the nature of the second stage correction that the effect of any mismatch at the output stage will decrease with increasing gain. A new analysis can be performed as a function of the device gain to see its influence on the resultant errors. For this second analysis a fixed output match of $s_{22} = 0.5$ is taken, while the magnitude of the s_{21} parameter is varied. The errors computed from the classical and corrected Y-factor versions are shown in Figure 10.8(b). As expected, the error provided by both techniques tends to zero as the device gain increases.

Finally, it is important to highlight that the accuracy associated with both techniques depends highly on the characteristics of the DUT and receiver.

Noise source match variations

The above analysis dealt with mismatch effects associated with the second stage correction. However, mismatch effects at the input stage of the DUT are also important in the Y-factor technique, since they affect the very principle of the method itself. Indeed, the formulation of the Y-factor methodology relies on the basis that there is a single device noise factor during the measurement process. However, if the variation of the reflection coefficient of the noise source from the cold to the hot state is not negligible, the device noise factor will actually vary [59], [60].

Let us come back to the basic Y-factor concept presented in Section 10.2.2, where no consideration of the source reflection coefficient has been made. The cold and hot noise powers given in (10.14) and (10.15) depend on the reflection coefficient of the source

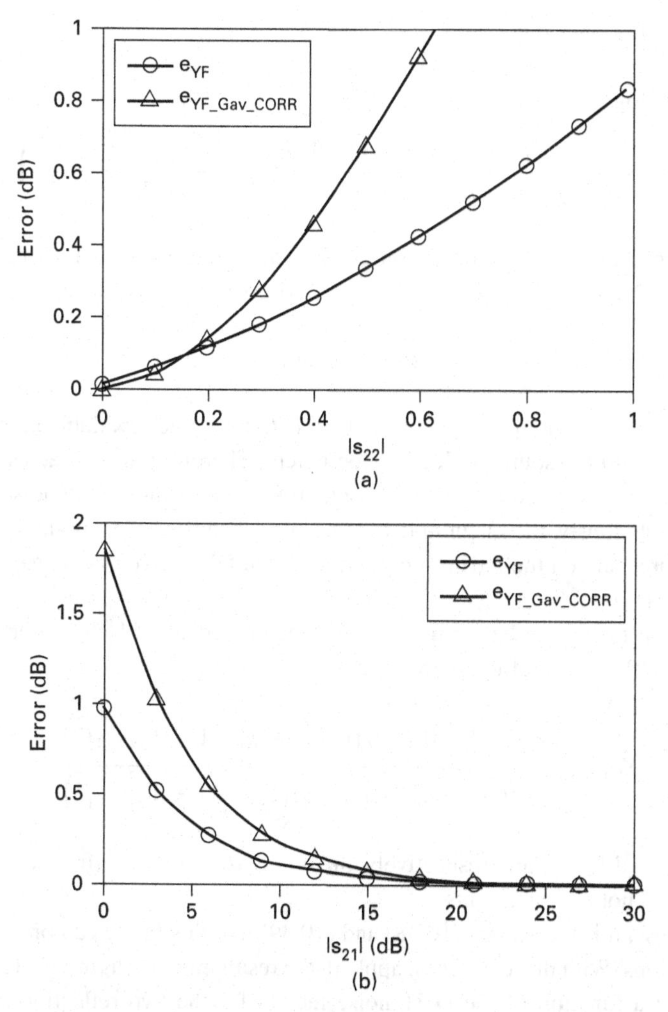

Fig. 10.8 Errors associated with classical and corrected Y-factor methodologies as a function of: (a) DUT output match ($s_{21} = 5$ dB); (b) DUT gain for a given s_{22} of 0.5.

termination at each temperature, as explicitly shown in (10.38) and (10.39)

$$N_c(\Gamma_{sc}) = kBG_{av}(\Gamma_{sc})[T_c + T_0(F(\Gamma_{sc}) - 1)] \tag{10.38}$$

$$N_h(\Gamma_{sh}) = kBG_{av}(\Gamma_{sh})[T_h + T_0(F(\Gamma_{sh}) - 1)], \tag{10.39}$$

where Γ_{sc} and Γ_{sh} are the reflection coefficients of the noise source in its cold and hot states.

Hence, the noise factor computed from these noise powers will accordingly depend on the device noise factors corresponding to both source reflection coefficients. Neglecting any other sources of error, the noise figure calculated from the Y-factor expression (10.17) will be actually given by (10.40), which is a function of the device gains and noise factors at the two noise source states.

$$NF_{YF} = 10\log_{10}$$
$$\left(\frac{G_{av}(\Gamma_{sc})\left(T_h/T_0-1\right)F(\Gamma_{sc})}{\left(G_{av}(\Gamma_{sh})F(\Gamma_{sh})-G_{av}(\Gamma_{sc})F(\Gamma_{sc})\right)+G_{av}(\Gamma_{sh})\left(T_h/T_0-1\right)}\right). \quad (10.40)$$

Only when changes in the reflection coefficient of the noise source are negligible ($\Gamma_{sh} \simeq \Gamma_{sc} = \Gamma_s$), does the noise figure characterized by (10.40) converge to $NF(\Gamma_s)$:

$$NF_{YF}(\Gamma_{sh} \simeq \Gamma_{sc} = \Gamma_s) \simeq 10\log_{10}(F(\Gamma_s)) = NF(\Gamma_s) \quad (10.41)$$

It should be noted that in (10.38) and (10.39) both the device available gain and noise factor change with the source reflection coefficient. Therefore, inaccuracies associated with discrepancies between Γ_{sc} and Γ_{sh} come from both the match and noise variations of the DUT. Obviously, the magnitude of the error depends on the amount of change in the noise source and on the intrinsic properties of the DUT, given by its gain and noise characteristics.

A correction factor can be applied to the Y-factor to deal with the variations in the noise source reflection coefficient [57]:

$$Y_{CORR} = \frac{N_h}{N_c}\frac{\left(1-|\Gamma_{sc}|^2\right)|1-s_{11}\Gamma_{sh}|^2\left(1-|\Gamma_{out\_h}|^2\right)}{\left(1-|\Gamma_{sh}|^2\right)|1-s_{11}\Gamma_{sc}|^2\left(1-|\Gamma_{out\_c}|^2\right)}, \quad (10.42)$$

where Γ_{out_c} and Γ_{out_h} are, respectively, the output reflection coefficients of the DUT in the cold and hot measurements.

Nevertheless, it is clear from (10.38) and (10.39) that this factor can only correct for match variations. With this correction applied, the resultant noise figure leads to (10.43), which is still a function of the DUT noise factors for the two reflection coefficients $F(\Gamma_{sc})$ and $F(\Gamma_{sh})$.

$$NF_{YF\_Y\_CORR} = 10\log_{10}\left(\frac{\left(T_h/T_0-1\right)F(\Gamma_{sc})}{\left(F(\Gamma_{sh})-F(\Gamma_{sc})\right)+\left(T_h/T_0-1\right)}\right). \quad (10.43)$$

In (10.43), match variations from cold to hot measurement have been eliminated and this will in general lead to an improvement in the overall accuracy. Nonetheless, the variations in the device noise factor itself due to source termination changes cannot be taken into account unless the four noise parameters of the device are fully determined.

Numerical example 2. Let us illustrate the above discussion with an example. For the example, the following characteristics have been considered:
Noise source: ENR = 8 dB, $\Gamma_{sc} = 0.22$, $\Gamma_{sh} = 0.19\angle{-30^\circ}$.
DUT: $s_{11} = variable\ \angle{-60^\circ}$, $s_{22} = 0$, $s_{21} = 40$ dB, $s_{12} = -50$ dB; $NF_{min} = 1$ dB, $R_n = 20\ \Omega$, $\Gamma_{opt} = 0.45\angle 200^\circ$.
These match and ENR values could be realistically assigned to a Q347B millimeter wave

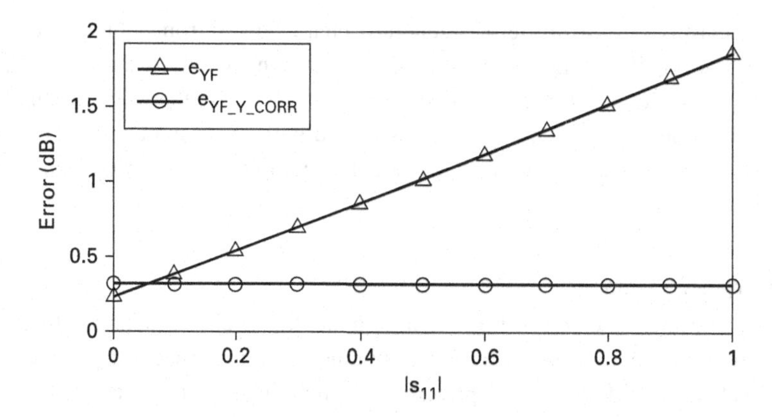

Fig. 10.9 Errors resultant from NF_{YF} and $NF_{YF\_Y\_CORR}$ versus DUT input match.

noise source from Agilent. Second-stage noise is not included. We take $NF\ (\Gamma_{sc})$, the noise figure corresponding to the cold state, as the true value. Then, the errors associated with NF_{YF} and $NF_{YF\_Y\_CORR}$ have been computed as the difference between the values obtained with (10.40) and (10.43), respectively, and the true value. The results are plotted in Figure 10.9 as a function of the DUT input match.

As can be seen in Figure 10.9, NF_{YF} presents an increasing error as the DUT input match worsens. In contrast, $e_{YF\_Y\_CORR}$ does not depend on DUT s_{11}, but a constant non-negligible error remains. This is due to the variation of the device noise figure between cold and hot measurements, as shown by (10.43). For the considered DUT, the errors associated with the typical 346 noise source family from Agilent would be significantly lower (in particular, $e_{YF\_Y\_CORR}$ will be negligible).

A more involved measurement strategy that takes into account noise source variations in the Y-factor technique can be found [61]. For that, a complete noise characterization of the device that obtains its four noise parameters is required.

In order to minimize mismatch-related effects, the use of isolators or attenuator pads is recommended, although this solution has its own drawbacks [39]. The inclusion of isolators limits the frequency range and several isolators may be required to cover the entire band in a wideband measurement. In addition, a rigorous characterization of the influence of the isolator requires a vector characterization. If several isolators are required, this process will accordingly enlarge. In contrast, attenuators have broadband response. However, an attenuator reduces the ENR presented to the DUT by its insertion loss, requiring a vector correction for its accurate characterization.

Cold-source

The cold-source technique presents a significant advantage over the Y-factor when dealing with mismatch-related errors, since in this technique the device noise figure is measured for a single-source impedance state. As a consequence, any inaccuracy related to noise source reflection coefficient variations can be avoided in the cold-source procedure. It should be noted that such variations affect obtaining the gain-bandwidth product of the receiver, because this term is generally characterized from cold/hot measurements [60]. However, if a fully corrected procedure is considered, as the one described by (10.26), these variations can be properly accounted for. Furthermore, in this fully corrected procedure any mismatch error associated with the second stage is eliminated at the cost of a substantial increase in measurement complexity compared with the classical Y-factor technique.

Measurement example

Let us illustrate the effect of mismatch on noise figure characterization by means of a measurement example. To this end, measurements of a mismatched passive device are provided. This kind of device represents a challenging test in this context, because the lack of gain magnifies the mismatch effects in the output stage. Also, note that the true noise factor of a passive device can be calculated analytically from its S-parameters as the inverse of the available gain. The DUT has been built up by combining an attenuator with a mismatch block. Figure 10.10(a) shows the output return loss of the DUT, while the available gain is depicted in Figure 10.10(b). The measurements were carried out in an in-home setup specifically conceived to implement different characterization approaches (Y-factor and cold-source with different levels of corrections). The setup includes a PNA E8358A, a low noise preamplifier, and a commercial 346B noise source.

Figure 10.11 compares the noise figures measured in a 1–2 GHz frequency range through four different characterization approaches. NF_{YF} is the classical scalar Y-factor technique (10.25). NF_{CS} is the cold-source method (10.26), with vector corrections and a full noise receiver calibration. $NF_{YF\_Gav\_CORR}$ represents the partially corrected Y-factor approach (10.36), which makes use of the DUT available gain instead of the insertion gain. Finally, NF_{YF_CORR} is a fully corrected version of the Y-factor technique, including a correction for variations in the noise source match (10.42) and a noise calibration of the receiver. The true DUT noise figure NF_{DUT} computed from S-parameters is also depicted in Figure 10.11. As can be seen, the scalar Y-factor technique NF_{YF} cannot equal the accuracy provided by the cold-source NF_{CS} for this demanding DUT, because of the lack of vector corrections and receiver noise calibration. In contrast, the fully corrected Y-factor NF_{YF_CORR} presents accuracy comparable to the cold-source technique, as expected from the comparable level of corrections included in the methodology. The residual effect of noise source reflection coefficient variations is negligible with the 346B noise source in this case. Finally, it should be noted that the highest error corresponds to the partially corrected Y-factor version $NF_{YF\_Gav\_CORR}$.

10.5.2 Temperature effects

A usual approximation when making noise figure measurements is that of considering the cold temperature T_c to be equal to the reference temperature T_0 (290 K). However, T_c is normally the ambient temperature and does not in general agree with the reference T_0. As previously mentioned, a correction can be included in modern noise figure measurement instruments to account for this difference. Moreover, current noise sources include

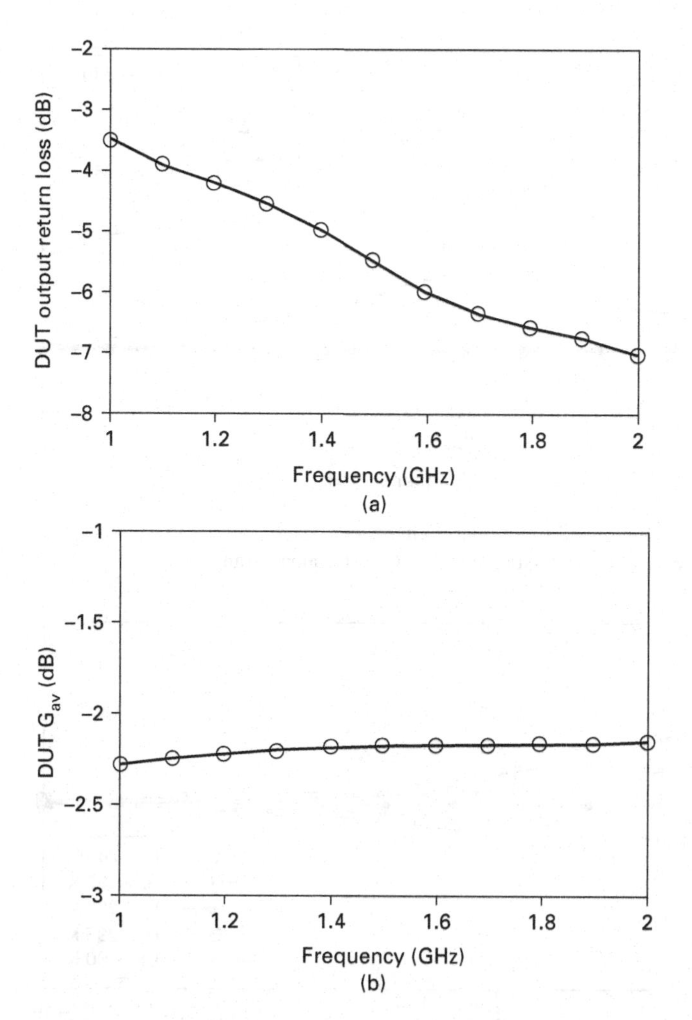

Fig. 10.10 (a) DUT output return loss; (b) DUT available gain.

temperature sensors that can measure their own temperature [62] and automatically perform the temperature compensation. If differences between T_c and T_0 exist and are not corrected, an error is introduced in the measurement performed with both the Y-factor and cold-source techniques. If no other systematic effect is present and the second stage contribution is negligible, the noise figure computed by both techniques is

$$NF_{YF} = NF_{CS} = 10\log_{10}\left(F + \left(\frac{T_c}{T_0} - 1\right)\right) \tag{10.44}$$

where F is the true noise factor.

Numerical example 3. Figure 10.12 shows the error given by (10.44) for a range of T_c between 280 K and 300 K as a function of the noise figure. It is clear from Figure 10.12 that approximating T_c to T_0 is acceptable for DUTs with $NF > 5$ dB.

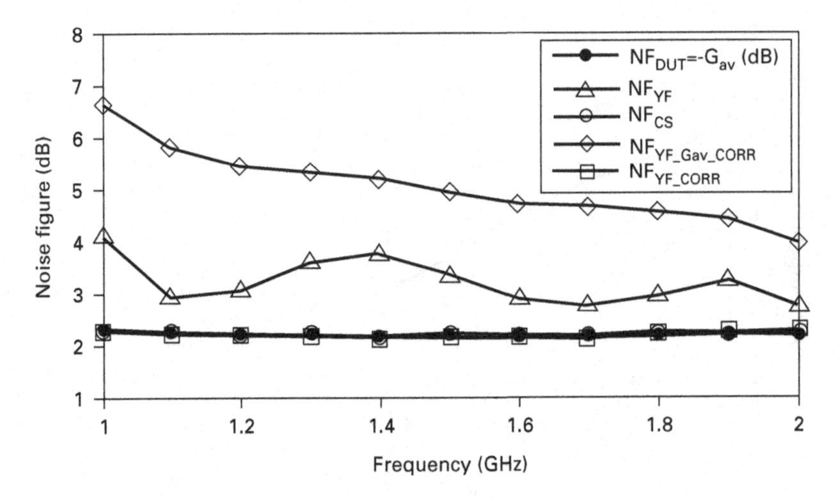

Fig. 10.11 True DUT noise figure and noise figures characterized with Y-factor (scalar, partially corrected, and fully corrected) and cold-source in 1–2 GHz frequency range.

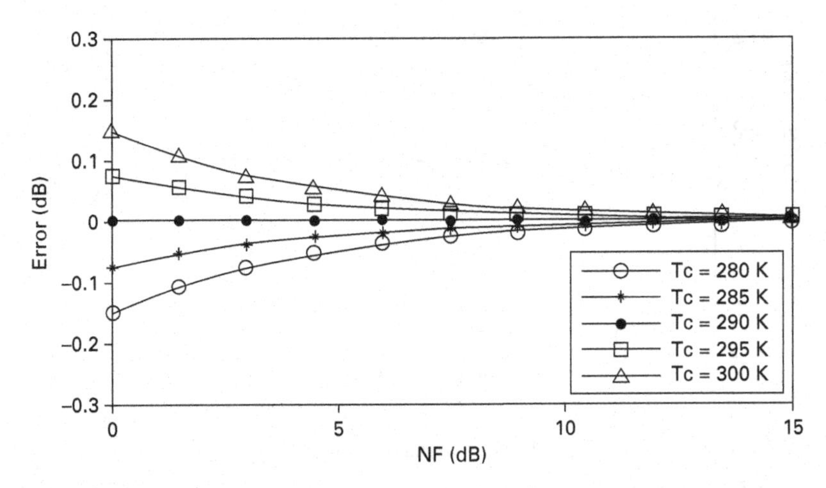

Fig. 10.12 Errors due to $T_C \neq T_0$ as a function of actual noise figure for Y-factor, according to [39], and cold-source.

Measurement example

Figure 10.13 shows the measurement results obtained from Y-factor and cold-source techniques with and without cold temperature correction. The same measurement setup as in Section 10.5.1 is used. The cold temperature T_c is the ambient temperature, 298 K (25°C) in this case. The DUT is a low noise amplifier with an approximately 30–25 dB gain in the 1–2 GHz frequency range. As expected from the analysis, the non-corrected approaches are approximately 0.1 dB over the corrected ones.

10.5.3 Measurement setup

Until this point mismatch and temperature-related errors have been analyzed. We discuss here three important points related to the measurement setup characteristics that

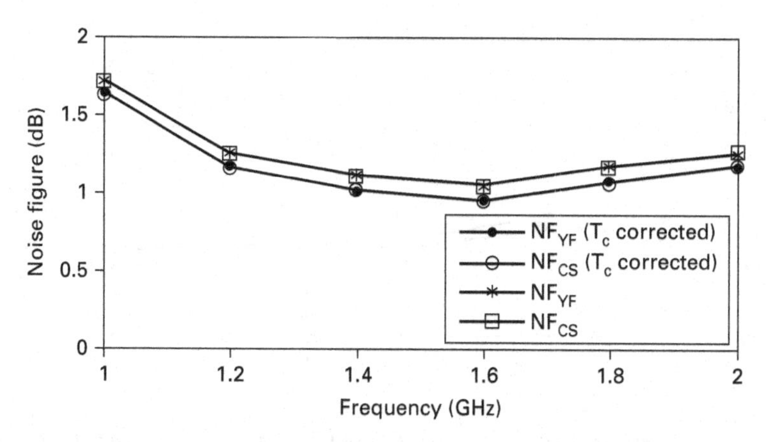

Fig. 10.13 Y-factor and cold-source noise figure measurements of an LNA with and without T_c correction ($T_c = 298$ K).

can have a non-negligible influence on the final accuracy of the measurement results: selection of the ENR, receiver bandwidth, and receiver linearity. Moreover, their impact on measurement accuracy is different depending on the technique we use for the noise figure calculation: Y-factor or cold-source.

Noise source ENR selection

Selecting an adequate ENR can be important for an accurate noise figure measurement. This selection depends on the measurement technique. For the Y-factor technique, several reasons recommend the use of a low ENR [39], unless high noise figures are to be measured. First, a low ENR reduces the possibility of driving the receiver beyond its linear region. Also, even if the receiver maintains its linear response, higher ENR values may require internal attenuation that increases the receiver noise factor, thus reducing the measurement accuracy. In addition, reflection coefficient variations from the cold to the hot state are in general smaller in a lower ENR noise source, because of built-in attenuators. Therefore, inaccuracies due to source reflection coefficient variations are reduced, recalling (10.40) and (10.43).

In contrast, higher ENR values are better suited to the cold-source technique. The use of the noise source, necessary for accurately obtaining the $kB|s_{21rec}|^2$ term, is limited to the calibration step in this technique; thus, receiver compression should not be a problem. When measuring the DUT, the noise power involved is generally much higher than that of a calibration performed with a low ENR noise source. The $kB|s_{21rec}|^2$ characterized in that case magnifies the overall uncertainty of the calculated noise figure. However, a higher ENR allows a calibration in a larger dynamic range, closer to the DUT measurement level and thus providing lower uncertainty. In addition, greater reflection coefficient variations associated with a high ENR are not as critical as in the Y-factor technique since they only affect the characterization of the noise receiver and, thus, can be corrected as mentioned earlier.

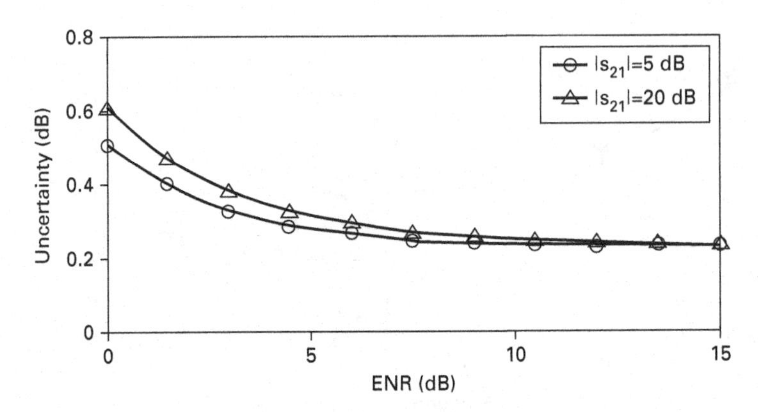

Fig. 10.14 Uncertainty given by cold-source as a function of ENR for a 2% uncertainty in measured noise powers and 0.2 dB uncertainty in ENR for two different DUT gains.

Numerical example 4. The uncertainty increase associated with a low ENR is illustrated in this example. Let us consider the following setup:
Noise source: ENR = *variable* dB, $\Gamma_s = 0$.
DUT: $s_{11} = 0$, $s_{22} = 0$, $s_{21} = 5/20$ dB, $s_{12} = -50$ dB; $NF = 3$ dB.
Receiver: $s_{11rec} = 0$; $NF_{rec} = 6$ dB.
For the analysis a 2% uncertainty due to jitter is assigned to the measured noise powers and a typical 0.2 dB uncertainty is assigned to the noise source ENR. Figure 10.14 shows the uncertainty associated with the noise factor characterized by means of the cold-source technique as a function of the ENR of the noise source for the two DUT gains considered. As can be seen, the resultant uncertainty (standard deviation of the result) increases for decreasing ENR values. This uncertainty increase is slightly magnified by the DUT gain.

Receiver bandwidth

The internal bandwidth of a classical noise figure instrument such as the HP8970 is about 4 MHz. Current noise figure analyzers have variable bandwidths that can be reduced much further [34]. Selecting an adequate receiver bandwidth is fundamental for measurement accuracy. During the calibration step the total noise power within the bandwidth of the receiver is measured. In contrast, if the DUT (or the combination DUT plus noise receiver) has a bandwidth narrower than the receiver itself, the noise bandwidth will be restricted by the presence of the DUT during the measurement and thus, errors can arise (see Figure 10.15). This is a situation that can typically happen when measuring at the passband edge of a very frequency-selective DUT. For the sake of clarity, let us call B_{cal} and B_{meas}, respectively, the noise bandwidths during calibration and during DUT measurement.

Neglecting any systematic error other than the bandwidth variation, the noise figure obtained from a Y-factor technique can be approximated to (10.45), where F_{DUT} is the true DUT noise factor. In this technique, the error becomes insignificant if the gain of the DUT is significantly larger than the ratio B_{cal}/B_{meas} [39].

$$NF_{YF} \approx 10\log_{10}\left(F_{DUT} + \left(\frac{B_{cal}}{B_{meas}} - 1\right)\frac{1}{G_{av}}\right). \tag{10.45}$$

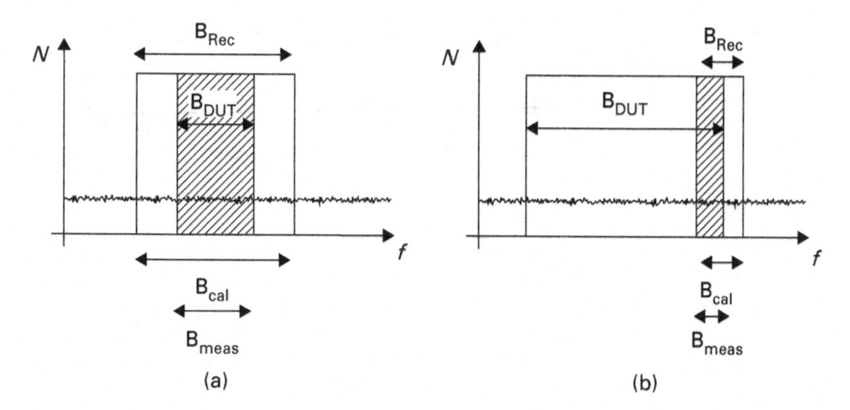

Fig. 10.15 Two possible error sources: (a) DUT bandwidth narrower than receiver bandwidth; (b) measurement in DUT passband edge.

As in the Y-factor case, the fact of having a $B_{meas} < B_{cal}$ is also a source of error in the cold-source technique. Nevertheless, there is a significant difference in the impact of this error from one technique to the other. In the cold-source case, any bandwidth difference between calibration and measurement affects the resultant noise figure, no matter the value of the DUT gain. The noise figure measured with cold-source technique, neglecting again any other source of systematic error, can be approximated to (10.46).

$$NF_{CS} \approx 10\log_{10}\left(F_{DUT}\frac{B_{meas}}{B_{cal}} + \left(1 - \frac{B_{meas}}{B_{cal}}\right)\frac{1}{G_{av}}\right). \tag{10.46}$$

Numerical example 5. This example serves to visualize the differences between the errors associated with both techniques. Characteristics of the setup are:
Noise source: ENR = 15 dB, $\Gamma_s = 0$.
DUT: $s_{11} = 0$, $s_{22} = 0$, $s_{21} = 10$ dB, $s_{12} =$ -50 dB; $NF = 3$ dB.
Receiver: $s_{11rec} = 0$; $NF_{rec} = 6$ dB. Bandwidths: $B_{meas}/B_{cal} = variable/0.5$.
For each technique (Y-factor and cold-source) an error function is calculated as the difference between the computed noise figure and the true one. The error is calculated as a function of the bandwidth ratio with a fixed s_{21} of 10 dB. The errors obtained are plotted in Figure 10.16. As shown in this figure, the error associated with the cold-source technique is significantly larger. This is because in the Y-factor case the error is attenuated by the DUT gain, while in the cold-source case the error tends to the bandwidth ratio B_{meas}/B_{cal} (in dB) as the DUT gain increases. Obviously, both errors disappear if $B_{meas} = B_{cal}$.

Receiver linearity

A noise figure measurement relies on the linearity of the whole measurement system, as is clear from Figure 10.1(b). If the noise powers involved are high enough to drive the receiver into compression, the computed noise figure will not be accurate. As previously stated, the use of a low ENR is good practice to avoid the nonlinear behavior of the receiver in the Y-factor technique. However, if the DUT gain is high, in-line attenuation after the DUT may also be required. If this is the case, a correction has to be applied to

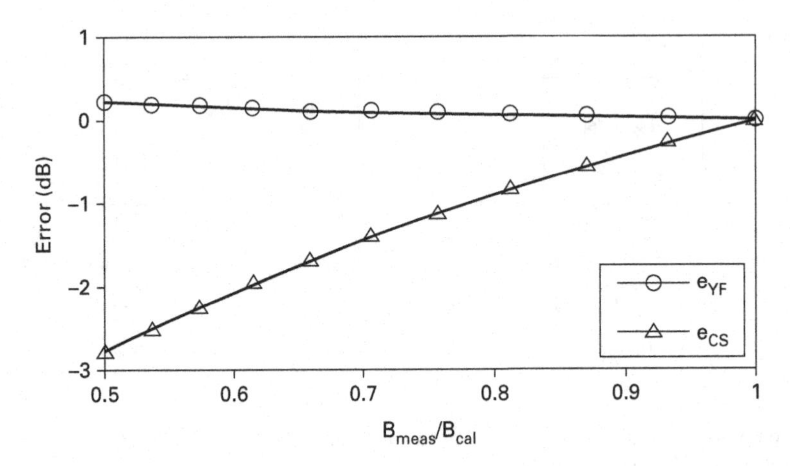

Fig. 10.16 Errors arising from a measurement bandwidth narrower than calibration bandwidth as a function of bandwidth ratio B_{meas}/B_{cal}.

the measurement to eliminate the contribution of the attenuator. It should be noted that, as long as it is not included in the calibration step, an accurate characterization of the attenuator requires vector measurements [39].

Deviations from linear behavior affect the noise figure calculation for the Y-factor as well as for the cold-source technique. Again, errors vary depending on the technique. These errors depend on the DUT characteristics, the ENR value, and the compression curve of the receiver. In contrast to the bandwidth discussion, here no general conclusion can be easily extracted about which technique becomes less accurate when a linearity deviation is taking place during the measurement process. However, the cold-source technique is less susceptible to driving the receiver into its nonlinear range because of the lack of a hot noise power measurement.

(10.47) and (10.48) are, respectively, the approximated noise figures computed from the Y-factor and cold-source techniques when there is a linearity deviation in the receiver. In these expressions C_c and C_h are compression factors (typically $0 < C_h \leq C_c \leq 1$) so that the actual measured cold and hot noise powers are C_cN_c and C_hN_h, instead of the ideals N_c and N_h (see Figure 10.17). Any systematic effect that is different from the compression of the receiver has been neglected in these expressions. In the Y-factor case, the noise figure will be overestimated because of the reduction in the denominator of (10.47) due to $C_h < C_c$. Note that if both measurements presented equal compression factors ($C_h = C_c$) there would be no error in the Y-factor case. However, in the cold-source technique, the noise figure will be underestimated if the measured noise power is compressed.

$$NF_{YF} \approx 10\log_{10}\left(\frac{(T_h/T_0 - 1)}{(C_hN_h/C_cN_c - 1)}\right), \tag{10.47}$$

$$NF_{CS} \approx 10\log_{10}\left(\frac{C_cN_c}{kB|s_{21rec}|^2G_{av}T_0}\right). \tag{10.48}$$

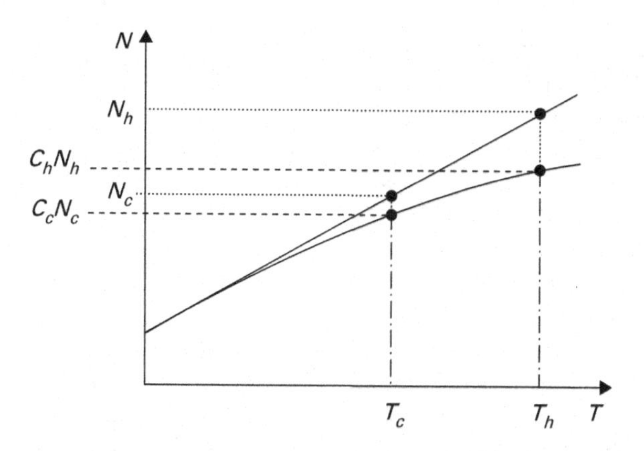

Fig. 10.17 Basic diagram of compression in receiver: ideal noise powers (N_c, N_h) and compressed noise powers (C_cN_c, C_hN_h) with $C_h < C_c$.

Measurement example

This example illustrates the effect of compression in the noise receiver. Again, the same measurement setup described in Section 10.5.1 is used. Figure 10.18(a) shows the noise response of this receiver as a function of its input noise. Compression for high noise-power levels is clearly noticeable. Y-factor and cold-source measurements of a variable-gain amplifier are performed. High gain values of the DUT bring the receiver into compression. The measurement results are plotted in Figure 10.18(b). To compare with a valid reference, the noise figure of the DUT, NF_{DUT}, was measured with a linear receiver that avoids compression through the use of input attenuators (superimposed in Figure 10.18(b)). As previously analyzed, the Y-factor technique tends to overestimate the noise figure because of a larger compression in the hot measurement. Indeed, the computed noise figure is 5 dB over the reference one for a 36 dB DUT gain. In contrast, for such gain the noise figure provided by the cold-source technique is approximately 0.2 dB below the reference NF_{DUT}, due to a compressed cold measurement.

10.6 Noise figure characterization of mixers

Mixers have some particular characteristics that complicate obtaining accurate noise figure measurements. They often present a poor output match (generally worse than amplifiers) and can have losses instead of gain (diode-based and cold-FET mixers). In addition, other effects specific to frequency translation appear. Besides, although mixer noise theory was developed early [63]–[71], some degree of confusion has accompanied mixer noise figure formulation from the very beginning, as was already pointed out in [68]. In this section the noise figure definition specifically provided by the IEEE for frequency translating devices [11] is analyzed. The definition and significance of the single-sideband (SSB) noise figure of a mixer are revisited. Obtaining the SSB noise figure through the Y-factor and cold-source techniques is comparatively discussed [72].

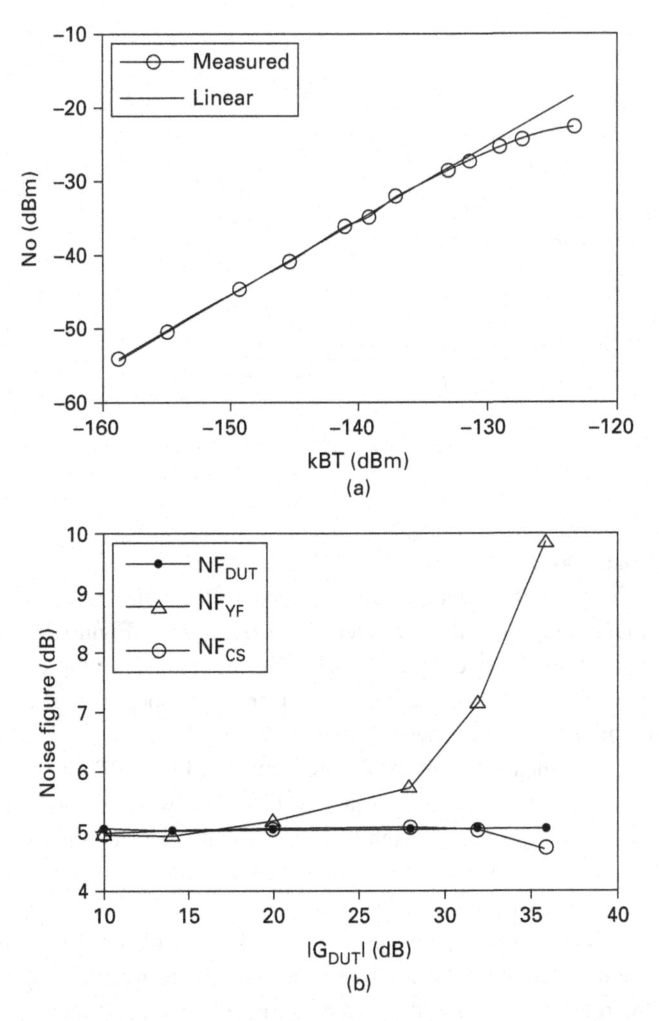

Fig. 10.18 (a) Compression curve of noise receiver. (b) Y-factor, cold-source, and DUT noise figures versus gain.

10.6.1 Noise figure definitions for frequency translating devices

According to [11], the noise factor of a frequency translating device can be expressed mathematically as:

$$F = \frac{N_o}{kBT_0G_{av}}. \tag{10.49}$$

where, N_o is the total noise power available at the output port at the output frequency when the noise temperature of its input termination is $T_0 = 290$ K at all frequencies. kBT_0G_{av} is the portion of N_o that is engendered by the input termination at temperature T_0 at the input frequency/frequencies. It is important to note that in the denominator of (10.49) only the contribution via signal-frequency transformation(s) is included. All

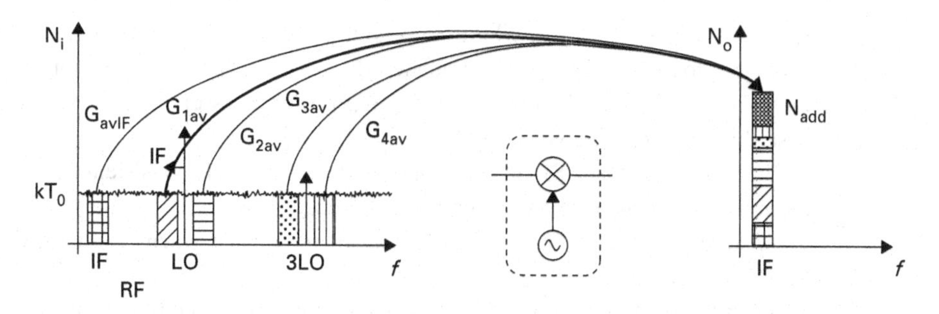

Fig. 10.19 Basic diagram of noise behavior in a mixer. N_o includes noise converted from principal, image, and idler frequencies, as well as noise added by the mixer.

other contributions, i.e. contributions from frequency conversions where the signal is not present in operating conditions, are excluded [11].

Single-sideband, double-sideband, and all-sideband noise figure

Considering that, in the normal operation of a heterodyne system, the signal is only present at a single frequency, G_{av} in (10.49) is simply $G_{av} = G_{1av}$, with G_{1av} being the available conversion gain relating this single input frequency to the output one. Therefore, according to the IEEE definition, the SSB noise factor of a mixer is given by (10.50).

$$F_{SSB} = \frac{N_o}{kBG_{1av}T_0}. \tag{10.50}$$

It should be noted that this definition does not exclude from N_o any noise generated at image or idler frequencies. On the contrary, it excludes from the denominator any gain different from G_{1av}, i.e. any gain not corresponding to a signal-frequency transformation.

Thus, N_o, given in (10.51), includes contributions from every possible conversion from input to output, as schematically shown in Figure 10.19.

$$N_o = kB\left(G_{1av} + G_{2av} + \cdots + G_{nav}\right)T_0 + N_{add}. \tag{10.51}$$

In (10.51) the G_{jav} terms represent each possible available conversion gain from input to output: G_{1av}, as already defined, is the principal available conversion gain, which relates the input RF frequency to the output one; G_{2av} is the image available conversion gain; $G_{3av}, \ldots, G_{nav}$ represent the available gains associated with idler conversions. Finally, N_{add} is the noise added by the mixer (including white noise coming from LO port). Analogous to the available gain of an ordinary two-port device (10.5), the available conversion gains G_{jav} can be defined as:

$$G_{jav} = \frac{1 - \left|\Gamma_s\left(f_j\right)\right|^2}{\left|1 - \Gamma_s\left(f_j\right)s_{11}\left(f_j\right)\right|^2}\left|c_{21}\left(f_j, f_{IF}\right)\right|^2 \frac{1}{1 - \left|\Gamma_{out}\left(f_{IF}\right)\right|^2}, \tag{10.52}$$

where $\Gamma_s\left(f_j\right)$ and $s_{11}\left(f_j\right)$ are, respectively, the source and input reflection coefficients at input frequency f_j. $\Gamma_{out}\left(f_{IF}\right)$ is the output reflection coefficient of the mixer at IF frequency. Finally, $c_{21}\left(f_j, f_{IF}\right)$ is a conversion parameter from input to output frequency, analogous to the standard s_{21} S-parameter. Both $s_{11}\left(f_j\right)$ and $\Gamma_{out}\left(f_{IF}\right)$ must be obtained under operating conditions, i.e. with the LO power at its operating level, to correctly describe the mixer behavior.

The definition of the SSB noise factor (10.50) is completely consistent with considering the noise factor as a figure of merit that characterizes the degradation of the signal-to-noise ratio from the input to the output of the device when operating in the SSB heterodyne mode.

When the denominator of (10.49) includes noise contributions from every possible transformation, an *all-sideband* (ASB) [73] noise factor is obtained, given in (10.53).

$$F_{ASB} = \frac{N_o}{kB\left(G_{1av} + G_{2av} + \cdots + G_{nav}\right)T_0}. \tag{10.53}$$

It is usually assumed that idler contributions are negligible compared to the principal and image contributions. If this is the case, the all-sideband noise factor equals the double-sideband (DSB) noise factor, as defined in (10.54), where only the available gains corresponding to RF and image conversions are considered in the denominator. Obviously, when the system operates in DSB (as in receivers for radiometry applications or in zero-IF receivers), the figure of merit that characterizes the degradation of the SNR is F_{DSB}.

$$F_{DSB} = \frac{N_o}{kB\left(G_{1av} + G_{2av}\right)T_0}. \tag{10.54}$$

It can be directly deduced from (10.50) and (10.53) that the SSB noise factor is equal to the ASB noise factor magnified by the quotient of the sum of all available conversion gains contributing to the output over the principal available conversion gain, as shown by

$$F_{SSB} = F_{ASB}\frac{\left(G_{1av} + G_{2av} + \cdots + G_{nav}\right)}{G_{1av}}. \tag{10.55}$$

Let us now consider a noise figure measurement N_c at a temperature T_c different from the reference temperature T_0. Then, the SSB noise factor is given by:

$$F_{SSB} = \frac{N_c}{kBG_{1av}T_0} - \frac{\left(T_c - T_0\right)}{T_0}\frac{\left(G_{1av} + G_{2av} + \cdots + G_{nav}\right)}{G_{1av}}. \tag{10.56}$$

It is clear from (10.56) that if $T_c \neq T_0$ a correction factor that includes all the available gains, is required. Note that in this case the temperature difference is magnified by the gain ratio, in contrast to ordinary two-ports devices.

Noise temperature in mixers

Sometimes the noise temperature is used instead of the noise figure for characterizing the noise behavior of a mixer [6]. Let us consider a mixer from a system point of view. For the sake of simplicity, only the principal and image conversions are included. Figure 10.20

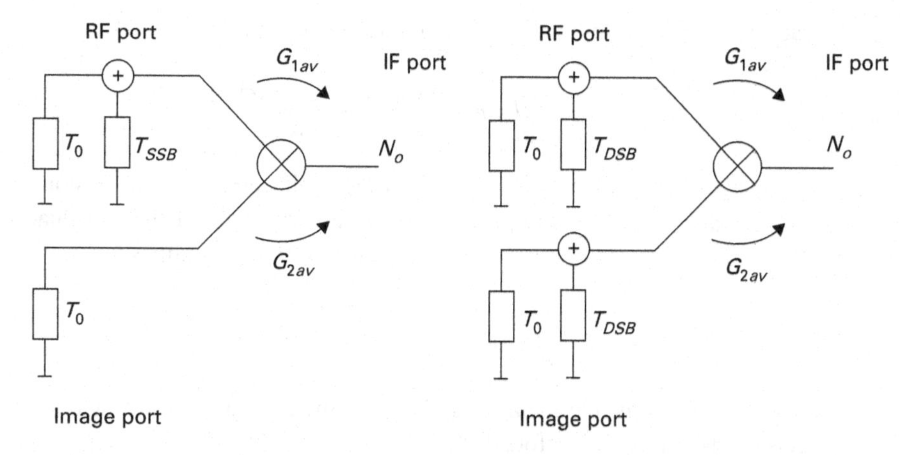

Fig. 10.20 Basic diagram of SSB and DSB noise temperature concepts.

shows a typical representation of SSB and DSB input noise temperature concepts. In such a diagram, the input RF frequency and the image frequency are treated as separate input ports.

In the SSB noise temperature, T_{SSB}, all the noise generated by the mixer is translated to the RF input port. In doing so, the noise available at the output can be given as a function of T_{SSB} by means of (10.57). In contrast, when referring to the DSB noise temperature, T_{DSB}, the total noise generated in the mixer is translated to both input ports, i.e. to RF and image ports. In this case, the output noise can be written as in (10.58).

$$N_o = kB\left[G_{1av}\left(T_0 + T_{SSB}\right) + G_{2av}T_0\right] = kB\left[G_{1av}T_{SSB} + \left(G_{1av} + G_{2av}\right)T_0\right] \quad (10.57)$$

$$\begin{aligned} N_o &= kB\left[G_{1av}\left(T_0 + T_{DSB}\right) + G_{2av}\left(T_0 + T_{DSB}\right)\right] \\ &= kB\left[\left(G_{1av} + G_{2av}\right)T_{DSB} + \left(G_{1av} + G_{2av}\right)T_0\right]. \end{aligned} \quad (10.58)$$

Both concepts are simply a translation to the input (to RF port for SSB and to RF and image ports for DSB) of the noise added by the mixer:

$$T_{SSB} = \frac{N_{add}}{kBG_{1av}} \quad (10.59)$$

$$T_{DSB} = \frac{N_{add}}{kB\left(G_{1av} + G_{2av}\right)}. \quad (10.60)$$

As a consequence of (10.59) and (10.60), the SSB input noise temperature is equal to the DSB input noise temperature multiplied by the ratio of the sum of principal and image conversion gains over the principal conversion gain, as in the noise factor case.

$$T_{SSB} = T_{DSB}\frac{\left(G_{1av} + G_{2av}\right)}{G_{1av}}. \quad (10.61)$$

We can also express the noise factor in terms of these input noise temperatures. If the mixer added noise is rewritten in terms of T_{SSB}, the relationship between the SSB noise factor and the SSB input noise temperature can be achieved:

$$F_{SSB} = \frac{T_{SSB}}{T_0} + \frac{(G_{1av} + G_{2av})}{G_{1av}}. \tag{10.62}$$

Equation (10.62) is the direct result of the application of the IEEE noise factor definition, although other interpretations can be found [74]. If the principal and image conversion gains can be considered to be equal, (10.62) simplifies to:

$$F_{SSB} \approx \frac{T_{SSB}}{T_0} + 2. \tag{10.63}$$

In a similar way, the relationship between the DSB noise factor and the DSB input noise temperature can be found.

$$F_{DSB} = \frac{T_{DSB}}{T_0} + 1. \tag{10.64}$$

Finally, let us generalize the previous analysis to include all mixer responses and let us consider an effective input noise temperature T_e common to all these responses [11]. In doing so, the noise available at the output is:

$$N_o = kB\left[G_{1av}(T_0 + T_e) + G_{2av}(T_0 + T_e) + \cdots + G_{nav}(T_0 + T_e)\right]. \tag{10.65}$$

Then the SSB noise factor can be written in terms of T_e as (10.66), which simplifies to (10.67) when idler conversions are negligible.

$$F_{SSB} = \left(\frac{T_e}{T_0} + 1\right)\frac{(G_{1av} + G_{2av} + \cdots + G_{nav})}{G_{1av}} \tag{10.66}$$

$$F_{SSB} \approx \left(\frac{T_e}{T_0} + 1\right)\frac{(G_{1av} + G_{2av})}{G_{1av}}. \tag{10.67}$$

Equation (10.68) relates the ASB noise factor to the effective input noise temperature T_e. It should be noted that this relationship applies to the DSB noise factor of a mixer with negligible idler conversions, as shown by (10.64).

$$F_{ASB} = \frac{T_e}{T_0} + 1. \tag{10.68}$$

10.6.2 Obtaining the SSB noise figure from Y-factor and cold-source

As previously mentioned, typical noise figure meters such as the classical HP8970 and derived implementations (including spectrum analyzers with noise measurement capabilities) use the Y-factor technique to characterize the noise figure of circuits, including mixers. In the Y-factor technique, the noise figure is characterized from two noise power measurements. If the DUT is a mixer, both noise powers include noise contributions from the image and idler frequencies, because the noise source is a broadband device that provides extra noise in a wide frequency range.

Considering $T_c = T_0$ for simplicity, and analogous to (10.14) and (10.15), the cold and hot noise powers corresponding to a frequency converter are, in that order, (10.69) and (10.70).

$$N_c = kB\,(G_{1av} + G_{2av} + \cdots + G_{nav})\,T_0 + N_{add} \tag{10.69}$$

$$N_h = kB\,(G_{1av} + G_{2av} + \cdots + G_{nav})\,T_h + N_{add}. \tag{10.70}$$

Then, applying the Y-factor expression (10.17), the obtained noise factor is an ASB noise factor:

$$F = \frac{T_h/T_0 - 1}{N_h/N_c - 1} = \frac{(T_h - T_0)\,N_c}{T_0\,(N_h - N_c)} = \frac{N_c}{kB\,(G_{1av} + G_{2av} + \cdots + G_{nav})\,T_0} = F_{ASB}. \tag{10.71}$$

In (10.69) and (10.70) no noise contribution of the receiver has been considered. In fact, in an actual measurement this contribution has to be eliminated, as usual, applying the second-stage correction:

$$NF_{YF} = 10\log_{10}\left(\frac{T_h/T_0 - 1}{N_h/N_c - 1} - \frac{F_{rec} - 1}{G_{ins}}\right), \tag{10.72}$$

where F_{rec} and G_{ins} are computed from the standard calibration step, (10.23) and (10.24), respectively.

In order to obtain the SSB noise figure from a Y-factor technique, the following assumptions are often made [6]: image conversion is equal to principal conversion ($G_{1av} = G_{2av}$) and all idler conversions are negligible ($G_{3av} = \ldots = G_{nav} = 0$). Then, the SSB noise figure is considered to be simply 3 dB higher than the measured one:

$$NF_{SSB\_YF} = NF_{YF} + 3\,\text{dB}. \tag{10.73}$$

Obviously, the above assumptions are not always satisfied. A common approach for obtaining a "true" SSB noise figure measurement through a Y-factor technique includes a filter at the input of the device that filters out image and idler frequencies. However, impedance terminations of the mixer input port at the image and idler frequencies can have a non-negligible influence on the device noise performances [73]. Therefore, if the filter is not required for regular operation of the device, some amount of error should be expected in the noise figure characterization.

In [75], which is a noise figure measurement implementation on a spectrum or signal analyzer, the possibility of including a correction factor to the noise figure measurement as a function of image rejection, instead of a fixed value of 3 dB, is provided. However, this correction factor is only an estimate, because the actual gain ratio of the mixer, i.e.

the ratio of the sum of all available conversion gains over the principal conversion gain, is not characterized.

Inaccuracies are not only related to the ASB to SSB translation. As in ordinary two-port devices, there may be other sources of systematic error in the measurement. Indeed, mixers can present worse measurement conditions for noise characterization than two-port devices. On the one hand, most mixers are passive devices with usually poorer matching conditions than two-port devices [76]. On the other hand, since conversions from many frequencies might be involved, restrictions to usual approximations ($T_c \approx T_0$, $\Gamma_c \approx \Gamma_h$, etc.) are tighter than for the standard two-port devices.

In contrast to the Y-factor technique, cold-source is a straight implementation of the SSB formulation. For that, a cold noise power measurement with a matched load connected at the input of the mixer is performed, in the standard manner of the cold-source approach. In addition, and according to (10.50), the principal available conversion gain G_{1av} has to be characterized. To obtain it, conversion characteristics as well as the input and output matches of the device must be characterized (recall (10.52)). The contribution of the noise receiver must be eliminated by applying the second-stage correction.

The total noise power measured by the noise receiver is:

$$\begin{aligned} N_c = {} & kB|s_{21rec}|^2 MM(\Gamma_{out}) \\ & \times [T_0 G_{1av} F_{SSB} + (T_c - T_0)(G_{1av} + G_{2av} + \cdots + G_{nav})] \\ & + kB|s_{21rec}|^2 MM(\Gamma_{out})\, T_0\, (F_{rec}(\Gamma_{out}) - 1), \end{aligned} \tag{10.74}$$

where Γ_{out} is the output reflection coefficient of the mixer and $MM(\Gamma_{out})$ is the mismatch between the mixer and the receiver, as given by (10.27). Note that these quantities are obtained at f_{IF}.

Therefore, the SSB noise figure of the mixer can be obtained from

$$\begin{aligned} NF_{CS} \equiv 10\log_{10} & \left(\frac{N_c}{kB|s_{21rec}|^2 MM(\Gamma_{out}) G_{1av} T_0} - \frac{F_{rec}(\Gamma_{out}) - 1}{G_{1av}} \right. \\ & \left. - \frac{(T_c - T_0)}{T_0} \frac{(G_{1av} + \cdots + G_{nav})}{G_{1av}} \right), \end{aligned} \tag{10.75}$$

where the $kB|s_{21rec}|^2$ term is characterized in the calibration step at f_{IF}.

Finally, it is important to note that the correction term for $T_c \neq T_0$ requires the knowledge of the overall available gain, i.e. the sum of the available gains corresponding to all significant conversions ($G_{1av} + \cdots + G_{nav}$).

Measurement example

Y-factor and cold-source SSB noise figure measurement results of three diode-based mixers are compared in this section. The three mixers have different gain and match characteristics but the same IF frequency. The measurement setup includes a spectrum analyzer (PSA E4440), a vector network analyzer (PNA E8358A), a 346B commercial noise source, and two signal generators to measure the necessary gain, match, and noise powers. The characterization is performed versus LO power since this power can affect the noise generated by a diode-based mixer [77]. In the calculations, T_c has been realistically approximated to T_0.

Let us first consider the cold-source procedure defined by (10.75). When the difference between T_c and T_0 is negligible (10.75) tends to (10.76).

$$NF_{CS} \equiv 10\log_{10}\left(\frac{N_c}{kB\,|s_{21rec}|^2\,MM(\Gamma_{out})\,G_{1av}T_0} - \frac{F_{rec}(\Gamma_{out})-1}{G_{1av}}\right). \quad (10.76)$$

For the SSB noise figure measurement with the Y-factor technique (labeled as NF_{YF+3}), (10.73) is used, where the SSB noise figure is calculated by simply adding 3 dB to the scalar Y-factor result.

In addition to that, a scalar version of the cold-source technique is also considered for comparison. NF_{CS_SCALAR} is given

$$NF_{CS\_SCALAR} \equiv 10\log_{10}\left(\frac{N_c}{kB\,|s_{21rec}|^2\,G_{1ins}T_0} - \frac{F_{rec}-1}{G_{1ins}}\right), \quad (10.77)$$

where G_{1ins} is the principal insertion gain measured in a spectrum analyzer and no noise calibration of the receiver is considered. This scalar approach is a simple and fast solution to implement in a spectrum analyzer when good match conditions are satisfied.

The results obtained for the three mixers are plotted in Figure 10.21. The first mixer under test, *Mixer 1*, (RF = 0.3 GHz, LO = 1.3 GHz, IF = 1 GHz) presents comparable conversion losses for principal and image frequencies ($G_{1av} \approx G_{2av}$) and conversions from idler frequencies are negligible ($G_{3av} + \cdots + G_{nav} \approx 0$). In addition, output return losses are better than -10 dB in the measurement range. According to the properties of the mixer, the three noise figure calculations (NF_{CS}, NF_{YF+3} and NF_{CS_SCALAR}) lead to similar results (Figure 10.21(a)). For this mixer, NF_{YF+3} provides a good approximation of the SSB noise figure, due to the favorable match and conversion characteristics of the device. In addition, no vector corrections or receiver noise calibration are necessary because of its good match.

Let us now analyze *Mixer 2* (RF = 2 GHz, LO = 3 GHz, IF = 1 GHz). In this case, the sum of gains corresponding to image and idler conversions ($G_{2av} + G_{3av} + \cdots + G_{nav}$) is larger than the principal conversion gain G_{1av}. In addition, the output return losses are again better than -10 dB in the entire measurement band. The measurement results, plotted in Figure 10.21(b), are again consistent with the characteristics of *Mixer 2*. As can be seen, NF_{YF+3} underestimates the SSB noise figure, as expected from its conversion losses. However, the two cold-source approaches, NF_{CS_SCALAR} and NF_{CS} provide identical results due to the good output match of *Mixer 2*. This result shows that NF_{CS_SCALAR} can provide a good estimation of the SSB noise figure of mixers with fair match characteristics.

Finally, the results obtained for the third mixer, *Mixer 3* (RF = 2 GHz, LO = 3 GHz, IF = 1 GHz), are given in Figure 10.21(c). The mixer presents different principal and image conversion losses and it is poorly matched at the output port (worse than -5 dB in the whole measurement range). As shown in Figure 10.21(c), three different responses

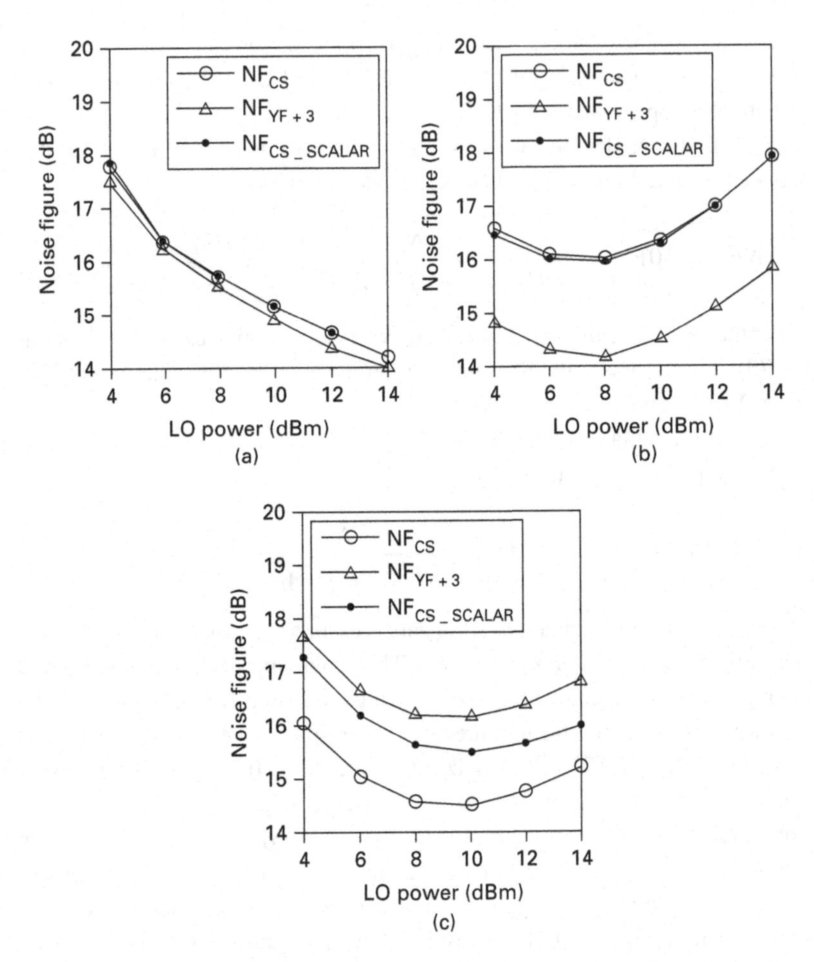

Fig. 10.21 SSB noise figure as a function of LO power [72]: (a) *Mixer 1*; (b) *Mixer 2;* (c) *Mixer 3.* Reprinted with permission of the IEEE.

have been obtained. In this case, the error associated with NF_{YF+3} comes from the non-ideal gain characteristics and from the poor match properties of the mixer. On the one hand, the gain ratio, i.e. the ratio of the sum of all significant available gains over the principal one, does not have a 3 dB value. In addition, the mixer is not adequately matched at the output. For this last reason, NF_{CS_SCALAR} cannot provide an accurate result. As a conclusion, in this challenging case a procedure that includes mismatch corrections and receiver noise calibration is now needed for accurate noise figure characterization.

10.7 Conclusion

In this chapter we have explained two popular methodologies for noise figure characterization: Y-factor and cold-source. These two methodologies have different implications in terms of measurement complexity, calibration, error correction, accuracy, etc. We have tried to provide a clear picture of the principles behind each technique, including their basic equations, calibration processes, and measurement requirements. The most common sources of error affecting both methodologies have been reviewed. These sources have been separated into three categories: mismatch, temperature, and measurement setup. In most cases they affect each technique differently. Comparative analyses have been performed whenever it was relevant. Noise figure of mixers has also been addressed in a final section because of the particularities of this kind of characterization in the context of a frequency conversion. The procedures for obtaining a single-sideband noise figure from Y-factor and cold-source methodologies have been detailed and compared.

References

[1] J. B. Johnson, "Thermal agitation of electricity in conductors," *Phys. Rev.*, vol. 32, pp. 97–109, July 1928.

[2] H. Nyquist, "Thermal agitation of electric charge in conductors," *Phys. Rev.*, vol. 32, pp. 110–113, July 1928.

[3] W. B. Davenport and W. L. Root, *An Introduction to the Theory of Random Signals and Noise*. New York: McGraw-Hill, 1958.

[4] A. Van der Ziel, *Noise in Solid State Devices and Circuits*. New York: Wiley-Interscience, 1986.

[5] A. B. Carlson, *Communication Systems. An Introduction to Signals and Noise in Electrical Communication*, 3rd ed. New York: McGraw Hill, 1986.

[6] S. A. Maas, *Noise in Linear and Nonlinear Circuits*. Norwood: Artech House Inc., 2005.

[7] D. M. Pozar, *Microwave Engineering*. New York: John Wiley and Sons, 1998.

[8] H. T. Friis, "Noise figure of radio receivers," *Proc. IRE*, vol. 32, no. 7, pp. 419–422, July 1944.

[9] H. A. Haus and R. B. Adler, *Circuit Theory of Linear Noisy Networks*. New York: John Wiley and Sons, 1959.

[10] "IRE Standards on methods of measuring noise in Linear twoports, 1959," *Proc. IRE*, vol. 48, no. 1, pp. 60–68, January 1960.

[11] "Description of the noise performance of amplifiers and receiving systems," *Proc. IEEE*, vol. 51, no. 3, pp. 436–442, March 1963.

[12] IRE 7.S2, "IRE standards on electron tubes: Definition of terms, 1957," *Proc. IRE*, vol. 45, no. 7, pp. 983–1010, July 1957.

[13] H. Rothe and W. Dahlke, "Theory of noisy fourpoles," *Proc. IRE*, vol. 44, no. 6, pp. 811–818, June 1956.

[14] H. A. Haus et al. "Representation of noise in linear twoports," *Proc. IRE*, vol. 48, no. 1, pp. 69–74, January 1960.

[15] P. Penfield, "Wave representation of amplifier noise," *IRE Trans. Circuit Theory*, vol. 9, no.1, pp. 84–86, March 1962.

[16] K. Hartmann, "Noise characterization of linear circuits," *IEEE Trans. Circuits Syst.*, vol. 23, no. 10, October 1976.

[17] H. Fukui, "Available power gain, noise figure, and noise measure of two-ports and their graphical representations," *IEEE Trans. Circuit Theory*, vol. 13, no. 2, pp. 137–142, June 1966.

[18] M. W. Pospieszalski, "Interpreting transistor noise," *IEEE Microwave Magazine*, vol. 11, no. 6, pp. 61–69, October 2010.
[19] R. Q. Lane, "The determination of device noise parameters," *Proc. IEEE*, vol. 57, no. 8, pp. 1461–1462, August 1969.
[20] M. Mitama and H. Katoh, "An improved computational method for noise parameter measurement," *IEEE Trans. Microw. Theory Tech.*, vol. 27, no. 6, pp. 612–615, June 1979.
[21] G. Vasilescu, G. Alquie, and M. Krim, "Exact computation of two-port noise parameters," *Electron. Lett.*, vol. 25, no. 4, pp. 292–293, February 1989.
[22] A. Boudiaf and M. Laporte, "An accurate and repeatable technique for noise parameter measurements," *IEEE Trans. Instrum. Meas.*, vol. 42, no. 2, pp. 532–537, April 1993.
[23] L. Escotte, R. Plana, and J. Graffeuil, "Evaluation of noise parameter extraction methods," *IEEE Trans. Microw. Theory Tech.*, vol. 41, no. 3, pp. 382–387, March 1993.
[24] V. Adamian and A. Uhlir, "A novel procedure for receiver noise characterization," *IEEE Trans. Instrum. Meas.*, vol. 22, no. 2, pp. 181–182, June 1973.
[25] A. C. Davidson, B. W. Leake, and E. Strid, "Accuracy improvements in microwave noise parameter measurements," *IEEE Trans. Microw. Theory Tech.*, vol. 37, no. 12, pp. 1973–1978, December 1989.
[26] R. Meierer and C. Tsironis, "An on-wafer noise parameter measurement technique with automatic receiver calibration," *Microwave Journal*, vol. 38, pp. 22–37, March 1995.
[27] G. Caruso and M. Sannino, "Computer-aided determination of two-port noise parameters," *IEEE Trans. Microw. Theory Tech.*, vol. 26, no. 9, pp. 639–642, September 1978.
[28] J. M. O´Callaghan and J. P. Mondal, "A vector approach for noise parameter fitting and selection of source admittances," *IEEE Trans. Microw. Theory Tech.*, vol. 39, no. 8, pp. 1376–1382, August 1991.
[29] J. M. O´Callaghan, A. Alegret, L. Pradell, and I. Corbella, "Ill conditioning loci in noise parameter determination," *Electron. Lett.*, vol. 32, no. 18, pp. 1680–1681, August 1996.
[30] S. Van den Bosch and L. Martens, "Improved impedance pattern generation for automatic noise parameter determination," *IEEE Trans. Microw. Theory Tech.*, vol. 46, no. 11, pp. 1673–1678, November 1998.
[31] S. W. Wedge and D. B. Rutledge, "Wave techniques for noise modeling and measurement," *IEEE Trans. Microw. Theory Tech.*, vol. 40, no. 11, pp. 2004–2012, November 1992.
[32] J. Randa and D. K. Walker, "On-wafer measurement of transistor noise parameters at NIST," *IEEE Trans. Instrum. Meas.*, vol. 56, no. 2, pp. 551–554, April 2007.
[33] "Applications and operation of the 8970A noise figure meter," Hewlett-Packard Product Note 8970A-1 (Agilent Manual 08970–99000), November 1981.
[34] "Agilent N8973A, N8974A, N8975A NFA series noise figure analyzers," Agilent Data Sheet 5980–0164E, November 2007.
[35] "Agilent PSA series spectrum analyzers. Noise figure measurements personality," Agilent Technical Overview 5988–7884EN, August 2005.
[36] "Application firmware for noise figure and gain measurements R&S®FS-K30 for R&S®FSP/FSU/FSQ," Rohde & Schwarz FS-K30 Data Sheet, November 2003.
[37] "Fundamentals of RF and microwave noise figure measurements," Agilent Application Note 57–1, October 2000.
[38] "Agilent 346A/B/C noise source," Agilent Operating and Service Manual 00346–90139, July 2001.
[39] "Noise figure measurement accuracy – the Y-factor method," Agilent Application Note 57–2, March 2004.

[40] "High-accuracy noise figure measurements using the PNA-X series network analyzer," Agilent Application Note 1408–20, September 2010.

[41] "Noise figure measurements, VectorStar, making successful, confident NF measurements on amplifiers," Anritsu Application Note 11410–00637, June 2012.

[42] N. Otegi, J. M. Collantes, and M. Sayed," Receiver noise calibration for a vector network analyzer," *76th Microwave Measurement Symposium (ARFTG)*, December 2010.

[43] "A new noise parameter measurement method results in more than 100x speed improvement and enhanced measurement accuracy," Maury Microwave Application Note 5A–0.42, March 2009.

[44] "Noise figure measurement without a noise source on a vector network analyzer," Rohde & Schwarz Application Note 1EZ61_2E, October 2010.

[45] BIPM, IEC, IFCC, ISO, IUPAC, IUPAP, OIML, "Guide to the expression of uncertainty in measurement," ISO, 1993, corrected and reprinted 1995.

[46] BIPM, IEC, IFCC, ISO, IUPAC, IUPAP, OIML, "Guide to the expression of uncertainty in measurement. Supplement 1. Numerical methods for the propagation of distributions," ISO, 2004.

[47] D. Boyd, "Calculate the uncertainty of NF measurements," *Microwaves & RF*, pp. 93–102, October 1999.

[48] J. Randa, "Uncertainty analysis for noise-parameter measurements at NIST," *IEEE Trans. Instrum. Meas.*, vol. 58, no. 4, pp. 1146–1151, April 2009.

[49] A. Collado, J. M. Collantes, L. De la Fuente, N. Otegi, L. Perea, and M.Sayed, "Combined analysis of systematic and random uncertainties for different noise-figure characterization methodologies," *IEEE MTT-S Int. Microwave Symp. Dig.*, pp. 1419–1422, June 2003.

[50] N. Otegi, J. M. Collantes, and M. Sayed, "Uncertainty estimation in noise figure measurements at microwave frequencies," *AMUEM 2005, International Workshop on Advanced Methods for Uncertainty Estimation in Measurement*, May 2005.

[51] "Noise figure corrections," Anritsu Application Note 11410–00256, November 2000.

[52] "NF accuracy," Anritsu Application Note 11410–00227, November 2003.

[53] "The Y factor technique for noise figure measurements," Rohde & Schwarz Application Note 1MA178_0E, May 2011.

[54] Agilent Technologies, *Noise Figure Uncertainty Calculator* [Online], Available: http://www.agilent.com.

[55] Rohde & Schwarz, *Noise Figure Error Estimation Tool* [Online], Available: http://www2.rohde-schwarz.com.

[56] Anritsu, *Noise Figure Uncertainty Calculator* [Online], Available: http://www.anritsu.com.

[57] D. Vondran, "Noise figure measurement: Corrections related to match and gain," *Microwave Journal,* vol. 42, pp. 22–38, March 1999.

[58] J. M. Collantes, R. D. Pollard, and M. Sayed, "Effects of DUT mismatch on the noise figure characterization: A comparative analysis of two Y-factor techniques," *IEEE Trans. Instrum. Meas.*, vol. 51, no. 6, pp. 1150–1156, December 2002.

[59] G. F. Engen, "Mismatch considerations in evaluating amplifier noise performance," *IEEE Trans. Instrum. Meas.*, vol. 22, no. 3, pp. 274–278, September 1973.

[60] N. J. Khun, "Curing a subtle but significant cause of noise figure error," *Microwave Journal*, vol. 27, no. 6, pp. 85–98, June 1984.

[61] L. F. Tiemeijer, R. J. Havens, R. de Kort, and A. J. Scholten, "Improved Y-factor method for wide-band on-wafer noise-parameter measurements," *IEEE Trans. Microw. Theory Tech.*, vol. 53, no. 9, pp. 2917–2925, September 2005.

[62] "Agilent N4000A, N4001A, N4002A SNS series noise sources 10 MHz to 26.5 GHz," Agilent Product Overview 5988–0081EN, July 2005.

[63] E. W. Herold, "The operation of frequency converters and mixers for superheterodyne reception," *Proc. IRE*, vol. 30, no. 2, pp. 84–103, February 1942.

[64] M. J. O. Strutt, "Noise-figure reduction in mixer stages," *Proc. IRE*, vol. 34, no. 12, pp. 942–950, December 1946.

[65] H. C. Torrey and C. A. Whitmer, *Crystal Rectifiers*. McGraw-Hill: New York, 1948.

[66] W. L. Pritchard, "Notes on a crystal mixer performance," *IRE Trans. Microw. Theory Tech.*, vol. 3, no. 1, pp. 37–39, December 1955.

[67] G. C. Messenger and C. T. McCoy, "Theory and operation of crystal diodes as mixers," *Proc. IRE*, vol. 45, no. 9, pp. 1269–1283, September 1957.

[68] R. D. Haun, "Summary of measurement techniques of parametric amplifier and mixer noise figure," *IRE Trans. Microw. Theory Tech.*, vol. 8, no. 4, pp. 410–415, July 1960.

[69] M. R. Barber, "Noise figure and conversion loss of the Schottky barrier mixer diode," *IEEE Trans. Microw. Theory Tech.*, vol. 15, no. 11, pp. 629–635, November 1967.

[70] D. N. Held and A. R. Kerr, "Conversion loss and noise of microwave and millimeter-wave mixers: Part 1 – theory," *IEEE Trans. Microw. Theory Tech.*, vol. 26, no. 2, pp. 49–55, February 1978.

[71] D. N. Held and A. R. Kerr, "Conversion loss and noise of microwave and millimeter-wave mixers: Part 2 – experiment," *IEEE Trans. Microw. Theory Tech.*, vol. 26, no. 2, pp. 55–61, February 1978.

[72] N. Otegi, N. Garmendia, J. M. Collantes, and M. Sayed, "SSB noise figure measurements of frequency translating devices," *IEEE MTT-S Int. Microwave Symp. Dig.*, pp. 1975–1978, June 2006.

[73] R. Poore, "Noise in ring topology mixers," Agilent EEsof EDA, 2000.

[74] S. A. Maas, *Microwave Mixers*, 2nd ed., Norwood: Artech House, 1992.

[75] "Noise measurement software FS-K3. Noise test system with FSE, FSIQ, or FSP analyzers," Rohde & Schwarz Application Note, News from Rohde-Schwarz 167, 2000/II.

[76] J. Dunsmore, "Novel method for vector mixer characterization and mixer test system vector error correction," *IEEE MTT-S Int. Microwave Symp. Dig.*, vol. 3, pp. 1833–1836, June 2002.

[77] G. M. Hegazi, A. Jelenski, and K. S. Yngvesson, "Limitations of microwave and millimeter-wave mixers due to excess noise," *IEEE Trans. Microw. Theory Tech.*, vol. 33, no. 12, pp. 1404–1409, December 1985.

11 TDR-based S-parameters

Peter J. Pupalaikis and Kaviyesh Doshi

11.1 Introduction

Many engineers are familiar with the VNA as an instrument for measuring S-parameters. The VNA's origins lie in microwave systems analysis and its application has been primarily in the frequency domain. Many are also familiar with the use of TDR for making qualitative measurements of time domain reflections and other phenomena. TDR has its origins in signal integrity analysis, as signal integrity is primarily concerned with time domain effects.

It is less well known that TDR and associated TDT is also a highly useful technique for precise quantitative measurements in signal integrity and can be used effectively for S-parameter measurement.

This chapter deals with the measurement of S-parameters using time domain techniques such as found in TDR and TDT. We cover the topic by first describing the hardware architecture of TDR instruments including the sampling system, the pulser, and the timebase. Then we describe how time domain TDR and TDT measurements are converted to raw, uncalibrated, frequency domain S-parameters. We do not deal with calibration techniques as these are the same for the VNA and TDR once raw S-parameters have been determined. Then, we quantitatively discuss the main element that effects the accuracy of time domain measurements: that of noise or SNR. SNR is such a big problem that it is the major source of error in time domain derived S-parameters and it is worthwhile understanding the sources of dynamic range degradation in TDR systems and the key design areas for improvement. We end the chapter with a consideration of how S-parameter measurements are affected by noise and present equations for determining measurement uncertainty when noise is the primary source of error.

11.2 TDR pulser/sampler architecture

The TDR module consists of a step source (pulser) and a sampler that can measure the reflected signal. Figure 11.1 is an idealized schematic of a pulser-sampler. First we describe the working principles of the sampling system by ignoring the pulse generating system and its output and then describe the operation of the pulse generating system.

The sampling strobe shown in the lower right corner of the figure is a clock signal that controls when the signal from the DUT is sampled. The block marked Impulse Generator consists of a unipolar impulse generator, an amplifier, and a nonlinear transmission line (NLTL). Input to the NLTL is a slow rise-time amplified impulse that is converted to a

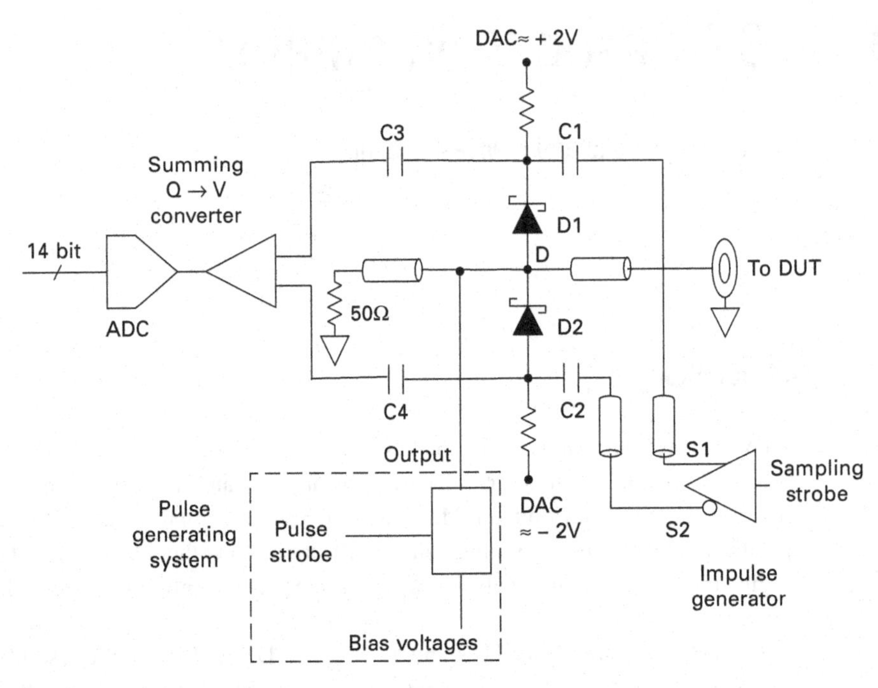

Fig. 11.1 TDR sampler schematic.

faster rise-time impulse. The behavior of an NLTL for generating a faster rise-time output from a slow rise-time step has been studied extensively [1, 2] and the references therein provide more NLTL details. Capacitors C1 and C2 in Figure 11.1 are AC-coupling capacitors, whereas the capacitors C3 and C4 store the charge corresponding to the signal coming from the DUT. Schottky diodes D1 and D2 form a switch that controls the sampling process. The "summing charge to voltage converter" sums the charge on C3 and C4 and converts that to voltage, which is then digitized by the ADC.

To understand the operation of the sampler, note that the polarity of the DAC is such that the two Schottky diodes are reverse biased. Now consider the case when the sampling strobe maintains the diodes in the reversed bias region, and there is no signal from the DUT. When the Schottky diodes are reverse biased, they can be thought of as an open. The DUT in this case sees a matched load of 50 ohms. Capacitor C3 holds the charge due to the 2 V DAC and the capacitor C4 holds the charge due to the −2V DAC. In this state, the output of the "summing charge to voltage converter" is zero. The system under measurement remains undisturbed by the sampler.

Next consider the case when we momentarily forward bias the two Schottky diodes and there is no signal from the DUT. In this state, capacitor C3 is charged to −0.2 V, the forward bias voltage drop across D1. Capacitor C4 is charged to 0.2 V, the forward bias voltage drop across D2. Since there is no signal from the DUT, the node marked D is at 0 V. Now suppose there was some reflected or through signal from the DUT. In this case, the voltage at node D is the voltage due to the signal from the DUT. Let this be x V. Now C3 is charged to $x - 0.2$ V and C4 is charged to $x + 0.2$ V. When the forward

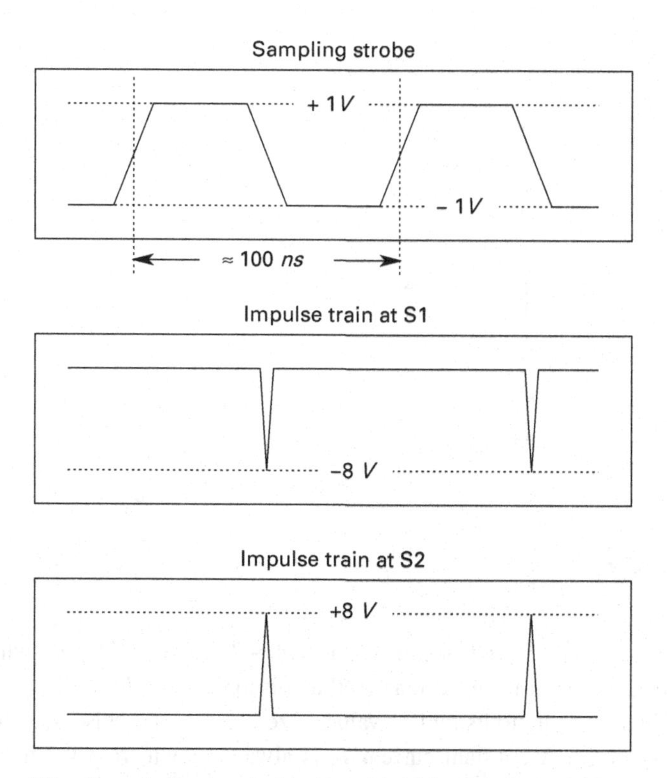

Fig. 11.2 Waveforms indicating the biasing of the Schottky diodes in the sampler.

bias is removed, the two capacitors are discharged and the "summing charge to voltage converter" converts the charge to $2x$ V ($x-0.2+x+0.2$ V).

The appropriate biasing of the Schottky diodes is achieved through the sampling strobe and the block marked Impulse Generator. The sampling strobe is a clock signal of approximately 10 MHz frequency, shown in Figure 11.2. Details of generating the TDR strobe are explained in the next section. This sampling strobe is first converted to an impulse and then amplified. The amplified impulse then passes through the NLTL, the result of which is a faster rise time impulse as shown by the signals marked S1 and S2 in Figure 11.2. As shown, S1 is an impulse with a peak of approximately -8 V whereas S2 is an impulse with a peak of approximately 8 V. This high level of voltage is enough to forward bias the two Schottky diodes. Thus, for the short interval of time corresponding to the impulse width, the diodes are forward biased and the signal from the DUT is recorded by capacitors C3 and C4. For the remaining time, the sampler acts like a matched load for the DUT.

The schematic of the pulser block in Figure 11.1 is shown in Figure 11.3. This is one of the many possibilities as described in [3]. As shown in Figure 11.3 and described in [3], the pulser consists of a constant current source, supplying current I_0; bias voltage $-V$ that drives the current source; bias voltage $+V$ that controls the switching of output diode; fast switch S_1; and resistor R_1, which is usually 50 Ω. Initially when the switch is open, there is a constant current flowing through resistor R_1 and the

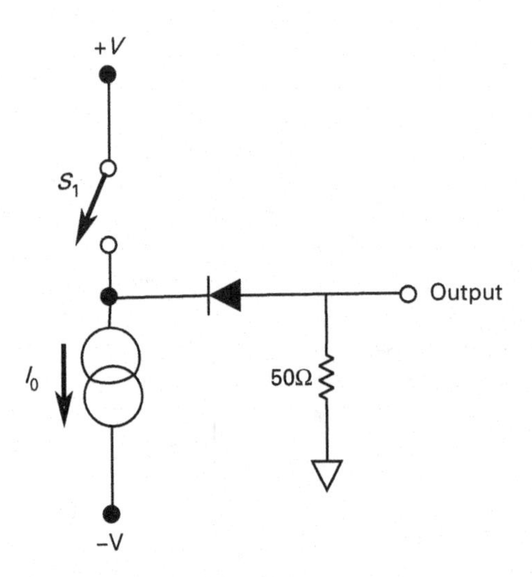

Fig. 11.3 TDR pulser schematic.

diode producing the generator baseline voltage of $-I_0R_1$ volts. When switch S_1 is closed, the diode becomes reversed biased, thus disconnecting I_0 from R_1. The output voltage changes rapidly to its topline value of zero volts. The pulser has the source resistance of R_1 ohms. A constant current I_0 is always drawn from the independent power supply $-V$. The rise time of the step is governed by the switching transients of switch S_1 and the charge storage time in the diode. The opening and the closing of the switch can be controlled by an external pulser strobe signal (not shown here). Note that the schematic described here is to describe the basic operation of a pulse generating system. More advanced pulser designs and details can be found in [3, 5] and references therein.

Referring back to Figure 11.1, when the pulser is active, the sampler will record the voltage due to the pulser as well as any reflections from the DUT. A sampler-only module can be constructed without the pulser generating system. More details about the sampler can be found in [6] and references therein.

11.3 TDR timebase architecture

Since TDR measurements have traditionally been performed on sampling scopes, it is natural to assume that TDR timebases resemble sampling scope timebases. There are two types of sampling scope timebases in use and both are used for TDR with adaptation for the pulser. The most common style of timebase is the *sequential sampling* timebase, which is shown schematically in Figure 11.4. Here we see on the left a reference clock, a pulser strobe, and an arm signal. This is a very simplistic diagram that assumes that the pulser strobe generation time is totally arbitrary, which means it might be free-running or generated manually. If free-running, the arm signal determines when

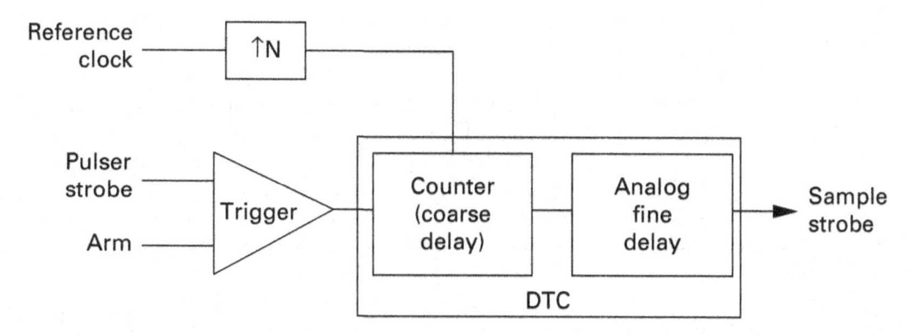

Fig. 11.4 TDR using sequential sampling.

the system is ready to take a sample. Prior to arming, a programmable delay device called a DTC is set to count off a predetermined time following the firing of the pulser which is the trigger event. In this manner, whenever the system is programmed with a delay time and armed, a sampling strobe will be generated a specified time after the pulser strobe. By repeatedly programming different times, arming the system, and generating pulser strobes, an *equivalent time* waveform containing the behavior of the DUT due to the applied step edges can be recorded. We call this waveform equivalent time because it represents a correct, high sample rate waveform despite the fact that the actual samples were taken at dramatically different times. For equivalent time sampling to work in a TDR, the reaction of the DUT to the TDR stimulus must be identical for every pulser strobe. The delay action performed by the DTC is generally performed by two elements: one digital and the other analog. The digital element counts clocks and is the coarse timer. The analog element is fine and depends on some phenomenon like the discharge time of a precharged capacitor. Although the sequential sampling timebase is the most popular, it has many severe disadvantages for a TDR. These are:

- The DTC is nonlinear and requires calibration. Even 1 ps of sampling error creates large inaccuracies in TDR-based S-parameter measurements.
- The system is slow. As we will see later in the chapter, sampling speed is critical for TDR because of the amount of averaging required. TDR measurements using a sequential sampling range in an actual sample rate between 40 and 150 kS/s, so for a 40 Kpoint waveform only one to a few waveforms can be acquired each second.
- Because a trigger system is involved, the jitter of the trigger also adds to measurement error.
- The equivalent time sample rate is dependent on the granularity of the DTC control.

Another type of sampling scope timebase is the CIS timebase. This architecture was originally proposed by LeCroy [7], [8]. This timebase is illustrated in Figure 11.5. The CIS timebase generates continuous pulser and sampler strobes that intentionally beat with each other. The example shown in Figure 11.5 is the timebase arrangement for a LeCroy SPARQ in "normal" mode, meaning that the pulser is pulsing at 5 MHz or with a 200 ns period. The sampler is placed at the seemingly odd sample rate of 9.884647 MHz

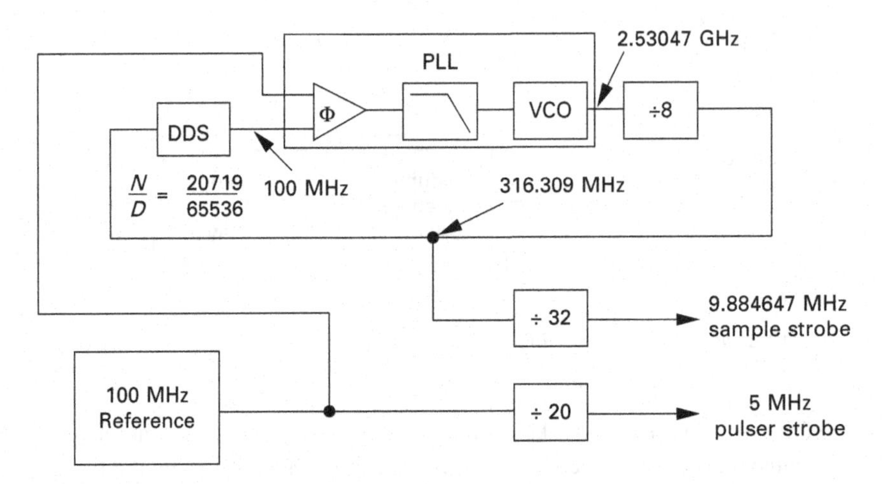

Fig. 11.5 TDR using coherent interleaved sampling.

through the combination of the PLL and the DDS. Here we see that the input to the phase detector is both the 100 MHz reference clock and, in steady-state, a 100 MHz phase locked output from the DDS. The DDS in this example, has been programmed to multiply its input by $20719/65536 \approx 0.316$ so that when the system is locked, the input to the DDS is approximately 316 MHz, requiring the output of the VCO at eight times higher frequency to be approximately 2.53 GHz. When locked, therefore, we have a sample strobe rate of the VCO output divided by 32. If we examine Figure 11.5 we see that this system produces a sample strobe rate of precisely 2048/20719 times the 100 MHz reference. Since the pulser is pulsing at a rate of precisely one-twentieth of the reference clock frequency, we have the following equality:

$$f_{ref}\frac{2048}{20719}\frac{1}{S} = \frac{f_{ref}}{20}\frac{1}{P},$$

where S refers to an integer number of samples and P refers to an integer number of cycles of the pulser. Therefore, we have:

$$\frac{S}{P} = \frac{40960}{20719}.$$

This specific arrangement of frequencies means that exactly 40 960 samples corresponds to exactly 20 719 cycles of the pulser. The Stern-Brocot algorithm described in [9] was used to obtain the rational number equivalent of the ratio of frequencies. Therefore, a back-end memory system can store consecutive samples using a modulo 40 960 counter and even average many results. When read out, the memory is reordered by taking the memory index times 24 079 and again counting modulo 40 960. This reordering produces a 40 960 sample equivalent time waveform of a complete cycle of the pulser at, for this example, an equivalent time sample rate of 204.8 GS/s.

The operation of this system is shown in Figure 11.6. Here we see samples being taken of repeating waveforms representing the reaction of the DUT to repeated step edges. In

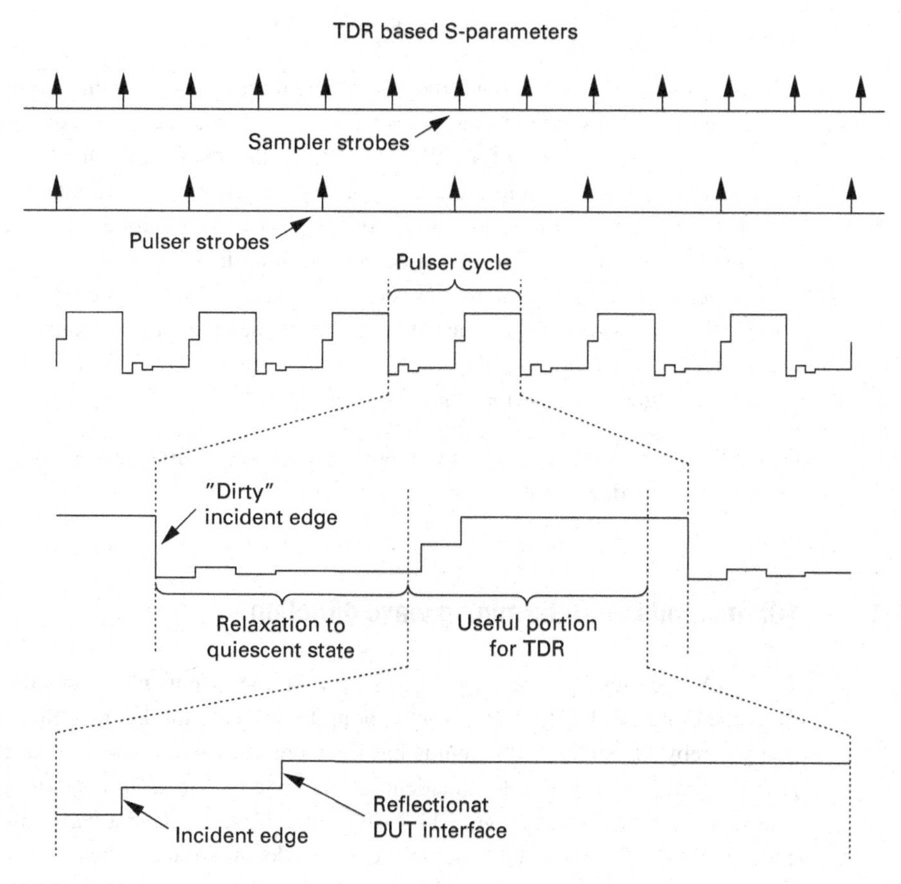

Fig. 11.6 TDR cycles.

the end, an entire waveform is acquired representing one complete cycle. Not all of the cycle is usable. The useful part consists of the portion just before the application of the high-speed edge up to the time that the TDR is turned off. Usually, the edge is really fast in only one direction. Equally important, during the time between the edge application and when it turns off, the TDR presents an ideally 50 Ω load. The reverse going edge (the "dirty" edge in Figure 11.6) is slow and during the off time, the system presents an uncontrolled, non-ideal impedance.

The benefits of the CIS timebase are:

- The sample rate is very high – approximately 10 MS/s in this SPARQ example. The samples are taken consecutively and continuously.
- The system removes the possibility for timebase nonlinearity.
- Presuming the programming capability of the multiplication and division factors in the system, there are no practical limitations on record length or equivalent time sample rate.

There are some drawbacks to this timebase method:

- While the sample rate is high, the system, while not requiring the storage of, does require slipping over unwanted portions of the pulser cycle. In the end, we usually want only a portion of half a cycle of TDR, as illustrated in Figure 11.6 (a slightly smaller portion is shown to account for the potential duty cycle variation).
- While fast, the time between the negative edge and the positive edge must be long enough for the system to reach a quiescent or fully discharged state. This is because TDR depends on the assumption of zero energy storage in the system prior to the step edge. This same assumption must also be met for sequential sampling timebases, but is easier because of the much slower speeds involved. For DUTs with longer electrical lengths, longer pulser periods must be programmed.

Generally speaking, the extreme speed benefits and lack of timebase nonlinearity issues far outweigh the drawbacks of CIS.

11.4 TDR methods for determining wave direction

The VNA operates by sweeping frequencies. At each frequency, a standing-wave is developed in a DUT. The VNA is attempting to simulate the Fourier transform situation whereby the sinusoidal stimulus has been present for all time. Because of this, it is impossible to determine the incident and reflected wave from an acquired voltage waveform as illustrated in Figure 11.7. In Figure 11.7, we have a *buoy man* trying to understand what is going on by the simple up and down motion of the buoy. Because the buoy only moves up and down and because it has been doing it forever, the buoy man cannot determine the direction of the underlying forward and reverse propagating waves. He only senses the sum of the wave effects. Because of this inability to determine wave direction, the use of directional couplers is necessitated in the VNA to perform the separation. Directional couplers are microwave devices that can distinguish the directionality of waves, but they generally suffer from a number of imperfections and limitations. They are specified and rated on two key specifications that are interesting for VNA usage: on directionality and attenuation, especially frequency-dependent effects. Directionality is the ability of the directional coupler to output waveforms going in only one direction. The attenuation of the directional coupler goes up as frequency goes down, meaning that the size of the acquired waveform is small or nonexistent at low frequency giving the VNA poor dynamic range at low frequency and no DC point. Generally, the outputs of the directional-couplers are sampled as voltage waveforms and since the results are sinusoidal waves at a single frequency, a complex amplitude and phase is determined. Since the sinusoid is very narrow-band (ideally a single frequency) the resulting amplitude and phase can be determined very precisely with high dynamic range.

TDR operates by launching an impulsive wavefront in the form of the rising edge of a step into a DUT. The TDR is attempting to simulate the Laplace transform situation whereby the system has never been stimulated prior to the arrival of the step edge and

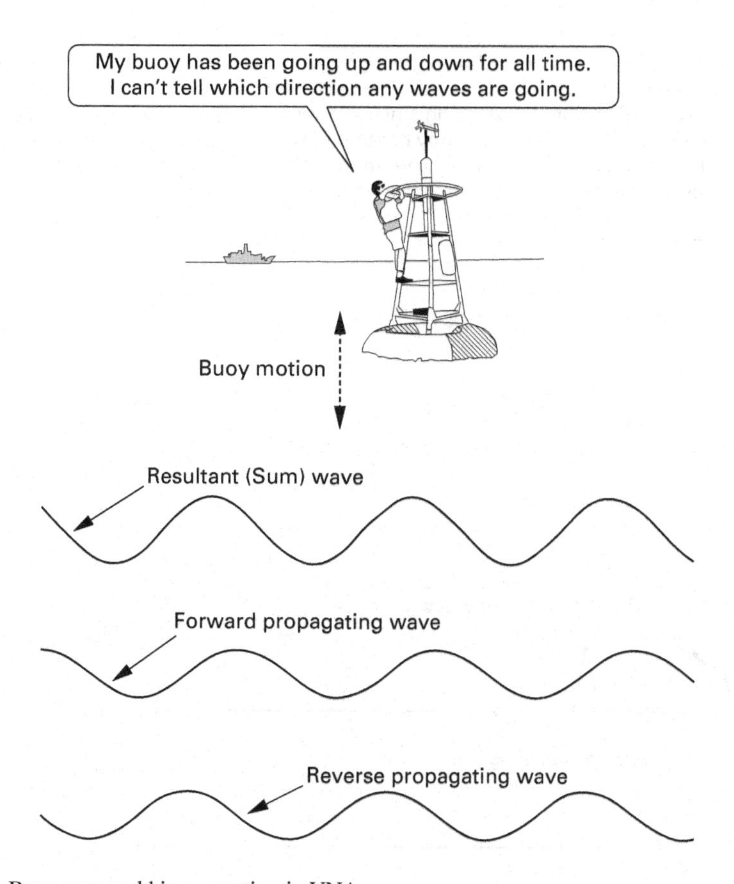

Fig. 11.7 Buoy man and his perception in VNA.

is in a completely quiescent state with no stored energy. In analyzing TDR waveforms, there are two key assumptions made:

(i) Until the TDR edge arrives, there has never been any incident edge from the source.
(ii) After the TDR edge comes and goes, no other edges are generated by the source.

These assumptions are illustrated by buoy man's perceptions in TDR in Figure 11.8. Here, buoy man is waiting and when the first up and down motion of the buoy is detected, he knows that it is the incident wave because of the first assumption. As the buoy keeps going up and down, he knows that these must be reflected waves because of the second assumption. The ability to detect wave direction from a single voltage waveform that is the sum of forward and reverse propagating waves is one of the keys to TDR and means that TDR does not require directional couplers. Of course, by examining Figure 11.6, we see that in order to meet the key assumptions, we must have the repetition rate of the TDR low enough such that on each cycle, the system is totally relaxed and all energy has been removed from the system. In other words, the cycle must be long enough for the effects of the incident edge on the system to die down sufficiently. By the way, this

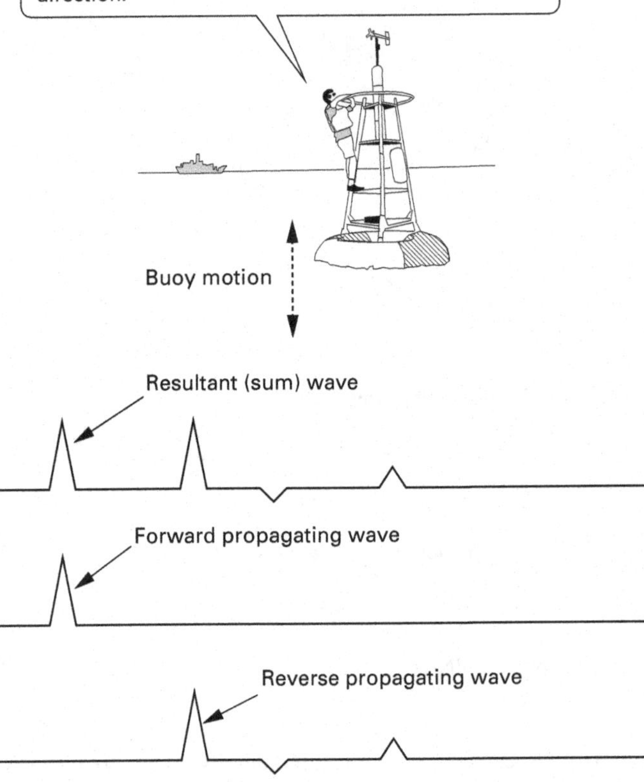

Fig. 11.8 Buoy man and his perception in TDR.

is why TDR is especially bad for handling AC coupled devices because the AC coupling leads to long time constants that make the meeting of these assumptions very difficult.

To summarize, in TDR it is easy to separate the incident from the reflected waveform. The incident wave is the rising edge of the step and the reflected wave is everything later.

A perceptive reader might wonder whether the TDR is actually properly accounting for reflections that occur after the incident waveform has been generated. For example, it is possible (and it occurs in practice) for waves returning from the DUT to be reflected from the source and sent back towards the DUT in the forward direction. This possibility is handled as follows:

(i) The assumption is that at the source side, the sampler and source are sufficiently co-located such that the sampler is seeing only the reflected waves after the incident has been generated. In other words, although forward going waves are retransmitted from reverse going returning waves from the DUT, the system only ever sees the returning waves. This means that the system might see two returning reflections from the DUT due to imperfections of the source termination.

(ii) The waves returning from the DUT due to secondary reflections at the source are either ignored or come out in calibration. In general TDR usage (meaning not for S-parameter determination) the DUT is often assumed to have only one interface, as accounting for all the internal reflections within the DUT is difficult. Therefore, secondary reflections either from the source or within the DUT are moved off the screen when TDR is being used for qualitative measurements. When TDR is being used for S-parameter determination, these secondary reflections must be provided and the waveform must be long enough for all the reflections to die down to essentially zero and be removed through calibration techniques similar to VNA usage. Remember, this same situation exists in the VNA as well and the standing waves generated are a function of similar effects.

The accounting for internal reflections within a DUT in TDR usage is commonly referred to as *peeling* [10], [11]. Peeling accounts for all reflections by remembering all of the reflections in the system and applying this memory to classify each reflection that is seen.

Figure 11.9 shows how TDR is used in principle and how users mentally separate the incident waveform from the reflected waveform. At the top of Figure 11.9 we see three overlaid waveforms for the cases of an open, short, and matched load (the load is the same impedance as the line in which the TDR waveform was propagating). Pay attention to the different labeling of the y-axis for the top drawing portion. All the waveforms begin with a step from 0 to half the source voltage level. Here we presume that the impedance of the source is the same as the impedance of the line. The step stays at this level until the DUT is encountered after which it either stays the same for the matched load, jumps to the source voltage level for the open, or drops to zero for the short.

The next two waveforms separate the incident from the reflected portion. Here we see that the incident waveform is the step common to all three cases and that the reflected waveform is either a positive step in the case of the open, a negative step in the case of the short, or zero in the case of the matched load.

TDR users mentally remove the incident step from their thinking when they view a TDR waveform. In fact, it is sometimes common to simply shift the incident step edge slightly off the screen to the left to see essentially only the reflected waveform. This can be done when qualitative measurements are being performed. In the case of certain quantitative measurements, like impedance, it is helpful to *calibrate* the system. This is often done by applying a single short or open standard, calibrating the voltage drop or rise, and then assuming that the result of a short is an inverted version of the open or vice versa.

One thing worth noting here is that the DUT interface is shown in Figure 11.9 as denoting the time in the voltage waveform corresponding to the DUT, but this is not the waveform at the DUT itself. All TDR waveforms are sampled near the

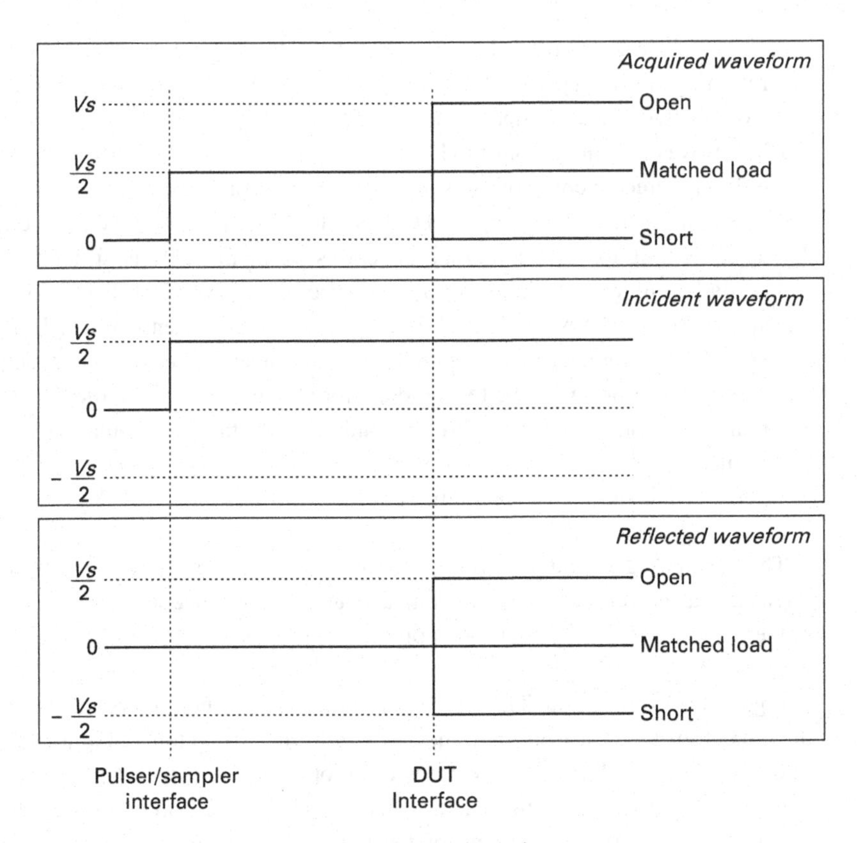

Fig. 11.9 TDR concept and incident and reflected step separation.

pulser. This means that the time in the waveform corresponding to the DUT is for a round-trip, meaning that the time corresponds to the time between the launch of the incident wave and the return of the reflected wave.

11.5 Basic method for TDR-based S-parameter measurement

Since the VNA is directly sampling frequency, the determination of frequency domain S-parameters is straightforward. The frequency content of the reflected waveform is divided by the frequency content of the incident waveform, which is one frequency for each measurement.

In TDR, all frequencies are launched at once and all frequencies are received in a single acquisition. By separating and converting the incident and reflected time domain waveforms to the frequency domain through the DFT, the equivalent S-parameter calculation is performed. In fact, once the DFT of the incident and reflected waveforms is computed, all other calculations, like calibration, proceed exactly like the VNA.

Traditionally in TDR-based S-parameter measurement, the incident portion of the waveform is not considered. As mentioned previously, if we examine Figure 11.9 we

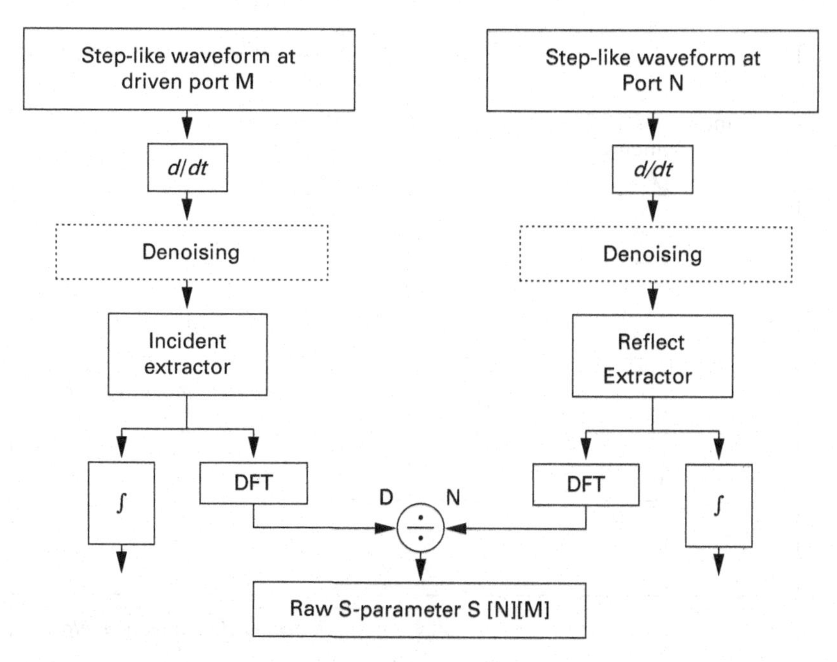

Fig. 11.10 TDR S-parameter calculation block diagram.

see that if the portion of the acquired waveform containing the incident edge is simply removed from the waveform (i.e. the time up to and just after the incident edge is removed) then the resulting waveform resembles a rescaled version of the reflected waveform. Traditionally, the step-like waveform with the incident portion gated off is used to compute the frequency content using methods provided by [12] or [13]. Using these methods, the incident frequency content is assumed to be unity (not a perfect step, but a perfect impulse) which is not really a problem as downstream calibration will take care of this. The disadvantage, however is that the calibration must take care of changing pulser conditions like frequency content and, most important, skew. Another lesser consideration is that using traditional methods, the error terms contain the confusing step frequency content that drops at 20 dB/decade[†]. In other words, traditional methods cause the error terms to look different to the error terms produced by the VNA.

Here we present an alternate method as shown in Figure 11.11. Here we see that the step-like waveform at both the driven and measured ports are differentiated (i.e. the first difference is calculated in discrete terms). They then undergo an optional denoising step, for example as provided in [14]. Then the incident portion is extracted from the driven port waveform and the reflected portion is extracted from the measurement port waveform. It should be apparent that by computing the frequency content of the now

[†] The way to understand this effect is to realize that the DFT of the impulse response of a system is essentially the frequency response of a system and tends to be mostly flat. The step response is the integral of the impulse response and therefore the response drops in frequency content as a system with a pole at zero frequency.

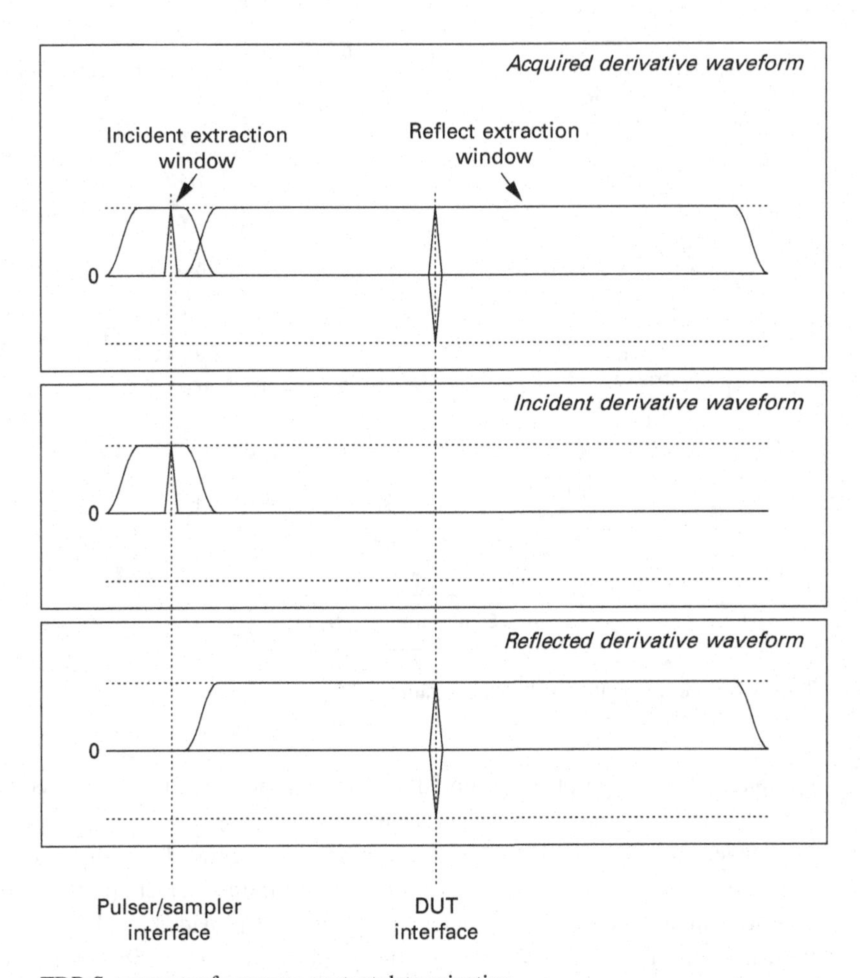

Fig. 11.11 TDR S-parameter frequency content determination.

separated portions, we have frequency domain versions of both. The preferred method is to utilize the DFT, FFT, or most preferably, the CZT [15], [16]. The CZT allows arbitrary end frequency and frequency spacing. The best way to treat the incident and reflected waveforms prior to conversion to the frequency domain is to zero out the incident and reflected portions of the original waveform. In other words, the incident waveform is created by zeroing out the reflected portion and the the reflected waveform is created by zeroing out the incident portion of the voltage waveform acquired. This operation can be understood as a windowing operation as shown in Figure 11.11. Because of the derivative and windowing, both incident and reflected waveforms have no edge discontinuities, but also very important is the fact that the two waveforms retain exactly the same length and sample timing. This alleviates the need for special handling of the incident and reflected waveforms.

As mentioned previously, the frequency-dependent incident waveform content is in the rising edge of step. This is seen more clearly by looking at the derivative waveforms as shown in Figure 11.11. Here it is clearly seen that the incident waveform occurs at an early time and that the entire incident waveform can be formed by simply extracting the beginning portion where the first impulse occurs. This impulse is so large and recognizable that it is not interesting to dwell on algorithms used for finding and extracting it. The waveform portion that occurs after the incident impulse is assumed to contain only reflected waveform. Note that this waveform portion may contain reflections due to interactions between returning waves and the pulser/sampler as discussed previously. While not described here, mathematical analysis shows that this is accounted for using ordinary VNA calibration algorithms. Note that in TDT, the entire waveform consists of the reflected portion.

The next and final step is to divide the frequency content of the incident portion into the reflected portions. This produces what is termed a raw S-parameter. It is an S-parameter because it is a frequency domain vector of ratios of reflected waves to incident waves. We say it is raw because it is not yet calibrated. We will say nothing here about calibration because at this step the results produced are similar to those produced by a VNA (see Chapter 8 for details on two-port VNA calibration).

To summarize:

- The raw S-parameters were computed using a true ratio of reflect to incident and the frequency domain content was computed from a derivative waveform.
- The method presented here is entirely insensitive to skew and requires only rudimentary efforts to place the incident edge.
- Because of the derivative action, the error terms produced in calibration look similar to those found in VNA calibrations.

In fact, the main difference between the S-parameters produced by TDR using these methods and by the VNA is the dynamic range (i.e. the amount of signal in the incident waveform relative to the amount of noise). By computing the derivative we have normalized the step frequency content which drops at 20 dB/decade, but have simultaneously reshaped the noise such that it increases with increasing frequency. This reshaping has not changed the ratio of the two and the fact remains that TDR generally has SNR, and therefore a dynamic range that drops at 20 dB/decade.

11.6 Summary of key distinctions between TDR and VNA

To summarize the key distinctions between the TDR and VNA:

(i) In TDR, the pulser produces all frequency content for the incident waveform in a single acquisition. In other words, while the VNA sweeps frequencies, TDR produces all frequency content in every pulse. As we will see, it is the broadband nature of the incident wave that leads to dynamic range degradation in TDR-based S-parameter measurement.

(ii) In TDR, the frequency content incident on the DUT is frequency dependent. More specifically, because the incident wave is present in the rising edge of the step-like waveform, the power drops approximately in a relationship that is inversely proportional to frequency.

(iii) TDR does not require directional couplers. This is because all the incident energy is present in the rising edge and with care, the incident and reflected waveform can be separated in time. The lack of directional couplers means that TDR does not suffer from dynamic range degradation at low frequency like the VNA.

(iv) TDR makes direct measurements in the time domain, whereas VNA measurements, while often employing samplers, are directly measuring in the frequency domain (because each acquisition is a sinusoid at a single frequency).

11.7 Dynamic range calculations

Since SNR or dynamic range is so important in TDR, it is useful to derive it and to highlight the features that improve and detract from the dynamic range.

Within time domain instruments, we acquire step waveforms, therefore we start with an acquired signal defined according to:

$$w[k] = s[k] + \varepsilon[k]. \tag{11.1}$$

In (11.1), $w[k]$ is a sample of the step waveform actually acquired, $s[k]$ is a sample of the step portion containing the signal of interest, and $\varepsilon[k]$ is a sample of the noise signal that we assume to be white, normally distributed, uncorrelated noise.

The signal content in the step is in the form of the frequency content of the derivative, so the derivation must consider this. Since during calculation we don't know the difference between the noise and the step, we must take the derivative of both. We will be approximating:

$$\frac{d}{dt}w(t) = \frac{d}{dt}[s(t) + \varepsilon(t)] = \frac{d}{dt}s(t) + \frac{d}{dt}\varepsilon(t) = x(t) + \frac{d}{dt}\varepsilon(t). \tag{11.2}$$

In (11.2), $x(t)$ represents the true desired input signal in the form of an impulsive wave front which is approximated as a discrete-time waveform with a sample $x[k]$, and $\frac{d}{dt}\varepsilon(t)$ is the time derivative of the noise signal which is approximated as a discrete-time waveform with a sample $\varepsilon'[k]$ which will be described in the following.

We are interested in these two signals in the frequency domain:

$$\mathbf{X} \approx \mathcal{F}\{x(t)\} = \mathrm{DFT}(\mathbf{x}),$$
$$\mathbf{E}' \approx \mathcal{F}\left\{\frac{d}{dt}\varepsilon(t)\right\} = \mathrm{DFT}(\varepsilon').$$

We calculate the dynamic range, for each frequency, as an SNR (ratio of signal strength and expected noise value):

$$SNR[n] = \frac{X[n]}{\overline{\overline{E'[n]}}}. \tag{11.3}$$

In order to calculate the SNR, we calculate the frequency content of each of these components separately and take the ratio. We start with the noise component. A noise signal ε which contains only uncorrelated, normally distributed, white noise, has a mean of 0 and a standard deviation of σ, which is the same as saying it has an rms value of σ. We have K points of this signal $\varepsilon[k]$, $k \in 0 \ldots K-1$.

If we calculate the discrete-Fourier-transform (DFT) of this noise signal, we obtain $N+1$ frequency points $N = K/2$, $n \in 0 \ldots N$:

$$E[n] = \frac{1}{K} \sum_k \varepsilon[k] e^{-j2\pi \frac{nk}{K}},$$

where the frequencies are defined as

$$f[n] = \frac{n}{N} \frac{F_s}{2} \tag{11.4}$$

and F_s is the sample rate.

By the definition of the rms value and by the equivalence of noise power in the time-domain and frequency-domain, we know that

$$\sqrt{\frac{1}{K} \sum_k \varepsilon[k]^2} = \sigma = \sqrt{\sum_{n=0}^{N_{bw}} \left(E[n] \frac{2}{\sqrt{2}} \right)^2}, \tag{11.5}$$

where N_{bw} is the last frequency bin containing noise due to any band limiting effects.

We define an average value for the noise E_{avg} that satisfies the following relationship:

$$\sqrt{\sum_n \left(E_{avg} \frac{2}{\sqrt{2}} \right)} = \sigma = \sqrt{N_{bw} \left(E_{avg} \frac{2}{\sqrt{2}} \right)^2}.$$

Therefore

$$E_{avg} = \frac{1}{\sqrt{\frac{f_{bw}}{F_s/2}}} \frac{\sigma}{\sqrt{K}},$$

where f_{bw} is the frequency limit for the noise calculated by substituting N_{bw} for n in (11.4).

We, however, are taking the derivative of the signal. The derivative in discrete terms is defined as

$$\frac{d}{dt} \varepsilon(t) \approx \varepsilon'[k] = \frac{\varepsilon[k] - \varepsilon[k-1]}{T_s},$$

where $T_s = 1/F_s$ is the sample period. Using the same equivalence in (11.5), we have

$$\sqrt{\frac{1}{K} \sum_k (\varepsilon'[k])^2} = \sigma' = \sqrt{\sum_{n=0}^{N_{bw}} \left(E'[n] \frac{2}{\sqrt{2}} \right)^2}. \tag{11.6}$$

Using the Z-transform equivalent of the derivative in the frequency domain, and an average value for the noise in it, it can be shown that

$$\sqrt{\frac{1}{K}\sum_{k}(\varepsilon'[k])^2} = \sigma' = \sqrt{\sum_{n}\left(\frac{\left|1-e^{-j2\pi\frac{f[n]}{F_s}}\right|}{Ts}E_{avg}\frac{2}{\sqrt{2}}\right)^2}.$$

Therefore, the average noise component at each frequency is given by

$$\overline{E'[n]} = \frac{\left|1-e^{-j2\pi\frac{f[n]}{F_s}}\right|}{T_s}E_{avg}\frac{2}{\sqrt{2}}.$$

We can make an approximation that gives us a further insight by expanding the numerator term in a series expansion

$$\left|1-e^{-j2\pi\frac{f}{F_s}}\right| = \frac{2\pi f}{F_s} + \mathcal{O}\left[\left(\frac{f}{F_s}\right)^3\right],$$

which allows us to approximate the noise component as

$$\frac{2\pi f[n]/F_s}{T_s}E_{avg}\frac{2}{\sqrt{2}} = \overline{E'[n]} = \frac{2\pi f[n]\sigma\sqrt{2}}{\sqrt{K}\sqrt{\frac{f_{bw}}{F_s/2}}}. \tag{11.7}$$

Note here that σ corresponds to the noise in the step waveform, not the noise in the derivative waveform, and the noise shaping for the derivative action is accounted for in (11.7).

Now that we have the noise component of dynamic range, we move to the signal component.

Without regard to the rise time or the frequency response of the step, which we will consider later, we define the signal such that, in the discrete domain, the integral of the signal forms a step

$$\mathbf{s}[k] = \mathbf{s}[k-1] + \mathbf{x}[k]T_s,$$

where $\mathbf{x}$ is an impulse such that $x[0] = A/T_s = A \cdot F_s$ and is zero elsewhere such that $\mathbf{s}$ forms a step that rises to amplitude A at time zero and stays there. $\mathbf{X} = \text{DFT}(\mathbf{x})$ and therefore the signal components at each frequency are defined as

$$X[n] = \frac{A}{T_s} = A \cdot F_s.$$

Again, to gain further insight, we define

$$K \cdot T_s = K/F_s = T_d, \tag{11.8}$$

where T_d is the acquisition duration (i.e. the amount of time in the acquired waveform). Therefore

$$X[n] = \frac{A}{T_d}.$$

Using (11.3), the ratio can therefore be expressed as

$$SNR[n] = \frac{X[n]}{E'[n]} = \frac{A\sqrt{K}\sqrt{f_{bw}}}{T_d 2\pi f[n]\sigma\sqrt{F_s}}.$$

Since these are voltage relationships, we can express the SNR in dB as

$$\text{SNR}(f) = 20\log\left(\frac{A\sqrt{K}\sqrt{f_{bw}}}{2T_d\pi f\sigma\sqrt{F_s}}\right) = 10\log\left(\frac{A^2 K f_{bw}}{4T_d^2\pi^2 f^2\sigma^2 F_s}\right) \tag{11.9}$$

and using (11.8), finally

$$\text{SNR}(f) = 10\log\left(\frac{A^2 f_{bw}}{T_d 4\pi^2 f^2\sigma^2}\right). \tag{11.10}$$

We would like to express the noise in dBm, so we have

$$\mathcal{N}_{dBm} = 20\log(\sigma) + 13.010 = 10\log\left(20\sigma^2\right)$$

and therefore

$$\sigma^2 = \frac{10^{\frac{\mathcal{N}_{dBm}}{10}}}{20}. \tag{11.11}$$

Substituting (11.11) in (11.10)

$$\text{SNR}(f) = 10\log\left(\frac{20A^2 f_{bw}}{4T_d\pi^2 f^2 10^{\frac{\mathcal{N}_{dBm}}{10}}}\right) = 10\log\left(\frac{20A^2 f_{bw}}{4T_d\pi^2 f^2}\right) - \mathcal{N}_{dBm}.$$

Then, to clean things up, we extract some constants

$$10\log\left(\frac{20}{8\pi^2}\right) \approx -6$$

and therefore

$$\text{SNR}(f) = 10\log\left(\frac{2A^2 f_{bw}}{T_d \cdot f^2}\right) - \mathcal{N}_{dBm} - 6.$$

Now let's consider some other factors. First, there is a frequency response of the pulse, and a frequency response of the sampler. These responses can be aggregated into a single response. Since, in decibels, it is simply the frequency response of the step calculated by taking the DFT of the derivative of the step (isolating only the sampled incident waveform) and calculating in dB, this value can simply be added to the dynamic range. Similarly, we account for cabling and fixturing which we aggregate into a single response, in decibels of $F(f)$. The signal must traverse the path through the cabling and fixturing twice:

$$\text{SNR}(f) = 10\log\left(\frac{2A^2 f_{bw}}{T_d f^2}\right) - \mathcal{N}_{dBm} + P(f) + 2F(f) - 6. \tag{11.12}$$

Next, we consider the effects of averaging. Averaging the waveform by an amount M achieves a 3 dB reduction in noise with every doubling of M. This leads to an improvement in the dynamic range by

$$20\log\left(\sqrt{M}\right) = 10\log(M). \tag{11.13}$$

The form of (11.13) allows it to be inserted directly into the numerator in (11.12)

$$\text{SNR}(f) = 10\log\left(\frac{2A^2 f_{bw} M}{T_{dur} f^2}\right) - \mathcal{N}_{dBm} + P(f) + 2F(f) - 6. \tag{11.14}$$

We really don't want to consider the dynamic range in terms of a number of averages and instead prefer to consider the amount of time we are willing to wait. The number of averages taken in a given amount of time is given by:

$$M = \frac{F_{sa} T_w}{T_d F_{se}}. \tag{11.15}$$

In (11.15), we now need to distinguish what is meant by sample rate. F_{se} becomes the equivalent time sample rate and replaces what we previously called F_s. F_{sa} is the actual rate that samples are acquired at in the acquisition system and T_w is the amount of time over which acquisitions are taken. Substituting (11.15) in (11.14), we obtain the complete dynamic range equation shown in (11.16).

$$\text{SNR}(f) = 10\log\left(\frac{2A^2 f_{bw} F_{sa} T_w}{T_d^2 f^2 F_{se}}\right) - \mathcal{N}_{dBm} + P(f) + 2F(f) - 6. \tag{11.16}$$

TDR-based S-parameters dynamic range equation

11.8 Dynamic range implications

The dynamic range equation (11.16) has several implications worth discussing. First the obvious ones. Regarding frequency, the dynamic range drops at 20 dB per decade (or 6 dB per octave). This can be considered as the effect of the drop-off in frequency components of a step. If the waveform utilized could be an impulse, this effect could be avoided. This effect is counteracted by the expression $P(f)$ which accounts for practical step responses.

Next is the obvious fact that the dynamic range is strongly dependent on the step size. It goes up by 6 dB for every doubling of the step amplitude, although the high frequency content is also accounted for in $P(f)$ (which is not concerned with the difference between pulser or sampler response). In other words, $P(f)$ is used to account for the product of the pulser energy content and the sampler response.

The dynamic range is directly proportional to the random noise and also losses in the cabling and fixturing, but this is also counteracted by a high sample rate. The dynamic range goes up by 3 dB for every doubling (or 10 dB for every ten times increase) in either the actual sampler sample rate or the time one waits for acquisitions to transpire.

The dynamic range is strongly affected by the length of the acquisition in time as indicated by the squared term T_d in the denominator. The reason why it is squared is two-fold. One effect is the amount of noise let into the acquisition. Remember that the actual signal – the incident wavefront – is contained in a very small time location, yet the noise is spread over the entire acquisition. As the acquisition length increases, the amount of noise increases with no increase in signal. If one knew where to look in the waveform, the effect of long acquisitions could be counteracted by limiting or gating of the waveform in the time domain. The second effect is the effect on averaging. Longer acquisitions take more time to acquire.

Now some more complicated considerations that are not necessarily obvious. First is the effect of the bandwidth limit f_{bw} on the noise. In many cases, noise in equivalent time sampler arrangements is essentially white. This is especially true if a major source of the noise comes from quantization effects in the ADC. This means that all the noise power is present up to the Nyquist rate $F_{se}/2$. In this case, $f_{bw} = F_{se}/2$ and these terms cancel so the dependence on noise bandwidth and equivalent time sample rate disappears from the equation and the dynamic range is completely independent of the equivalent time sample rate. This may seem counter-intuitive because increasing the sample rate causes more noise to fall outside the spectrum of interest due to even noise spreading, but this effect is fully counteracted by the increase in acquisition time and therefore the decrease in the number of acquisitions that can be averaged. In the case where the trace noise is specified with a bandwidth limit (as in most cases), the dynamic range is actually penalized by $10\log\left(f_{bw}/\left(F_{se}/2\right)\right)$, which seems unfair until you consider that unless the Nyquist rate is set exactly equal to this limit frequency, then acquisitions are needlessly oversampled (needless in theory, not necessarily in practice due to aliasing considerations). To make a proper comparison of band limited and non-band limited noise, one must compare using this adjustment.

From (11.16) therefore, after consideration, we see that there are a few basic ways to improve the dynamic range in TDR measurements. These are:

(i) Increase the amplitude of the step.
(ii) Increase the actual sample rate of the system.

(iii) Decrease the acquisition length.
(iv) Decrease the portion of the acquisition over which reflections are considered.
(v) Decrease the noise in the pulser/sampler hardware.
(vi) Increase the frequency response of the sampler and the frequency content of the pulser.
(vii) Decrease the length and losses in the cabling and fixturing.

All of these methods have been utilized to varying degrees in many TDR-based S-parameter measurement instruments with 11.8 and 11.8 involving improved hardware (as increasing the amplitude generally causes linearity problems). One particularly interesting technique that effectively accomplishes 11.8 in an algorithmic fashion is the use of wavelet de-noising techniques for lifting reflections from the noise [14].

11.9 Systematic errors and uncertainty due to measurement noise in a network analyzer

Non-idealities in the source, receiver, and various interconnections (like direction couplers, internal switches, and cables) introduce systematic errors in the measurements made by network analyzers. Such systematic errors are modeled in different ways and the model parameters are calculated by performing a calibration before making the DUT measurements. The model is referred to as the error-term model. The coefficients of the error-term model are collectively referred to as the error terms. Once the error terms have been determined, the uncorrected DUT measurements (referred to as raw measurements) are then combined with the error terms and the S-parameters of the DUT are calculated. Calibration algorithms for two-port and n-port VNAs have been described in Chapters 8 and 9, respectively. An algorithm to calculate the S-parameters of a multi-port DUT is described in [17]. The algorithm is general enough so that it works with any kind of model for the systematic errors. An important issue that should be considered is how the systematic error correction interacts with the noise in the measurement system. The dynamic range is not as high in the TDR-based network analyzer as it is in the frequency-based network analyzer. In such a system, both the calibration measurement as well as the raw DUT measurements are corrupted by noise. The error terms calculated by such noisy measurements are different from the actual systematic errors. In this section, we provide a method to determine the interaction of the error terms and noise in the raw DUT measurement when the final DUT S-parameters are calculated. We consider only the one-port DUT here. For more detailed information refer to [18, 19].

11.9.1 Error propagation for a one-port DUT

Figure 11.12 is a setup for measuring the S-parameters of a one-port DUT. Γ_{dut} represents the one-port S-parameters of the DUT. S_{ij} represents the S-parameters corresponding to the error terms of an error-term model of choice. Γ_{msd} represents the uncalibrated or raw measurement of the DUT. The S_{ij} are calculated by some calibration method. As an

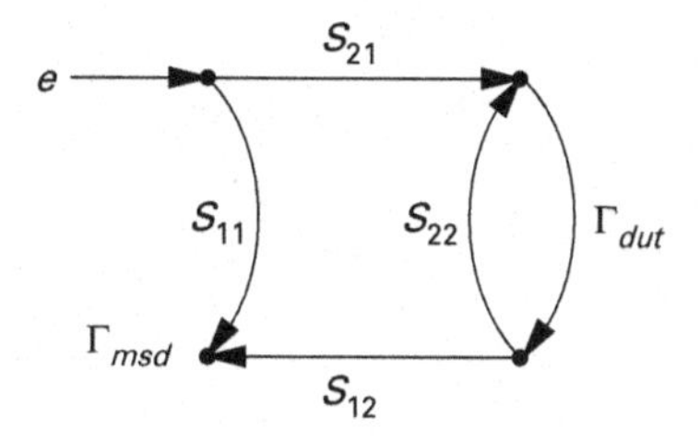

Fig. 11.12 One-Port DUT and error terms.

example, for the SOLT calibration technique, the S_{ij} are calculated by connecting the known short, open, and load calibration standards.

For the model in Figure 11.12, S_{11} corresponds to E_d – the directivity error term; S_{21} is chosen as one; S_{12} corresponds to E_r – the reflection error term; and S_{22} corresponds to E_s – the source match error term.

The expression for the raw measured DUT S-parameters can be derived from the signal flow diagram

$$\Gamma_{msd} = \frac{S_{11} - \Gamma_{dut}(S_{11}S_{22} - S_{21}S_{12})}{1 - \Gamma_{dut}S_{22}}. \tag{11.17}$$

Equation (11.17) can be modified to obtain the expression for DUT S-parameters from the raw DUT S-parameter measurement and the knowledge of error terms:

$$\Gamma_{dut} = \frac{\Gamma_{msd} - S_{11}}{\Gamma_{msd}S_{22} - (S_{11}S_{22} - S_{21}S_{12})}. \tag{11.18}$$

Equation (11.18) is the expression for the DUT S-parameters when there is no measurement noise and the error terms are known exactly. We would like to consider the effects of measurement noise on the DUT S-parameters. To simplify the analysis, we will consider the case when only the raw DUT measurement is noisy. Any noise in the uncalibrated DUT measurement will cause an uncertainty in the calculation of the S-parameters of the DUT. Suppose ε is the uncertainty in measuring Γ_{msd} and $\delta\Gamma$ is the uncertainty in calculating the S-parameters of the DUT, then,

$$\Gamma_{dut} + \delta\Gamma = \frac{\Gamma_{msd} + \varepsilon - S_{11}}{(\Gamma_{msd} + \varepsilon)S_{22} - (S_{11}S_{22} - S_{21}S_{12})}.$$

Substituting the expression for Γ_{msd} from (11.17), and after some algebraic manipulation, we have

$$\delta\Gamma = \varepsilon\frac{(1 - \Gamma_{dut}S_{22})^2}{\varepsilon S_{22}(1 - \Gamma_{dut}S_{22}) + S_{21}S_{12}}. \tag{11.19}$$

As expected, if there is no uncertainty in the raw measured DUT S-parameters, i.e. if $\varepsilon = 0$, then $\delta\Gamma = 0$, i.e. there is no uncertainty in the calculated DUT S-parameters (assuming that the known error-terms represented the true systematic errors). Also, if the error terms were such that there was no systematic error, i.e. $S_{11} = S_{22} = 0$ and

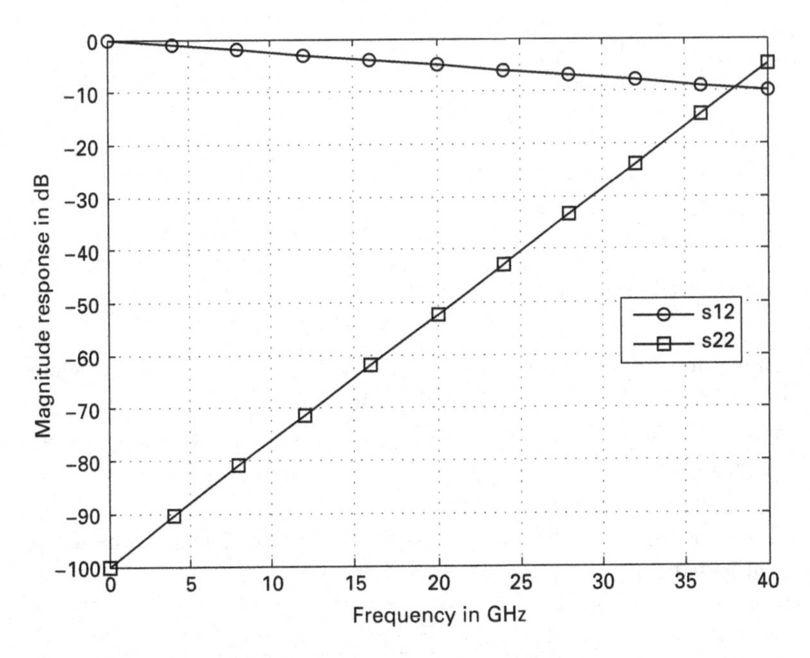

Fig. 11.13 S_{22} and S_{12} example.

$S_{21} = S_{12} = 1$, then $\Delta\Gamma = \varepsilon$, i.e. there is no uncertainty propagation and the uncertainty in measurement is translated as the uncertainty in the calculated DUT S-parameters.

When none of the above trivial cases is true, i.e. there is an uncertainty in measuring the raw DUT S-parameters, and the error terms are non-trivial, then (11.19) translates the uncertainty in DUT measurement to the uncertainty in Γ_{dut} calculation. As an example consider $\varepsilon = 0.01$, and further consider S_{22} and S_{12} as shown in Figure 11.13.

For $S_{21} = 1$ and Γ_{dut} corresponding to an ideal short (i.e. $\Gamma_{dut} = -1$), and if the uncertainty in raw DUT measurement is ε, the uncertainty in DUT S-parameters can now be calculated using (11.19). Figure 11.14 shows the effects of uncertainty propagation for a non-ideal case. Here the trace with circles is the actual Γ_{dut}, the trace with triangles is the DUT S-parameters with an uncertainty of $\pm\varepsilon$ (i.e. with ideal error-terms), while the trace with crosses is the DUT S-parameters with uncertainty for the non-trivial case.

There are multiple points to be noted for the case described above:

1. The uncertainty expression in (11.19) is a function of the uncertainty in the raw DUT measurements, i.e. one must know what the uncertainty is in order to determine the uncertainty in the DUT S-parameters. In general the actual uncertainty in the measurement is not known, but a probability distribution of the uncertainty due to noise is known. The problem then becomes estimating the distribution of the uncertainty in the DUT S-parameters.
2. Although it is not directly evident from (11.19), S_{22} plays an important role in the uncertainty propagation. As an example, instead of choosing an S_{22} as shown

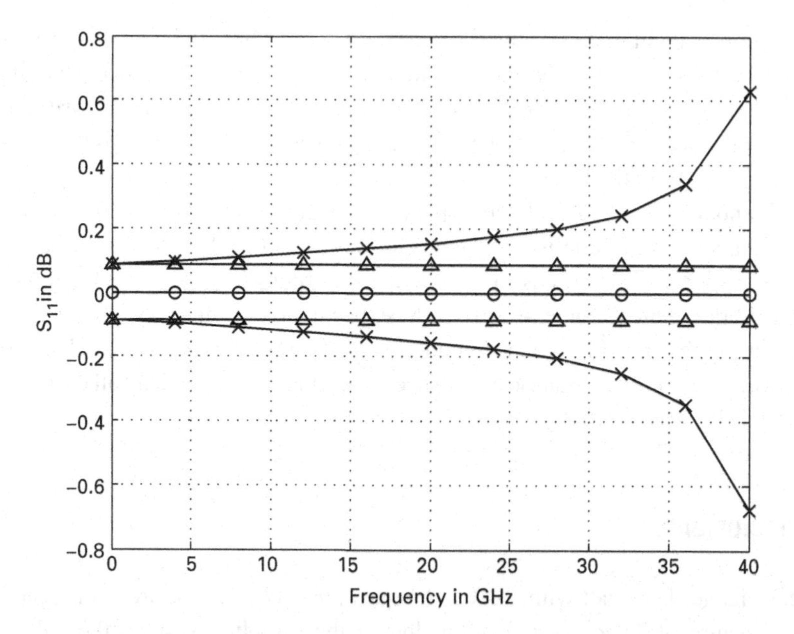

Fig. 11.14 $\Gamma_{dut} \pm \delta\Gamma$ for $\Gamma_{dut} = -1$, $\varepsilon = \pm 0.01$, $S_{21} = 1$, S_{12} and S_{22} as shown in Figure 11.13.

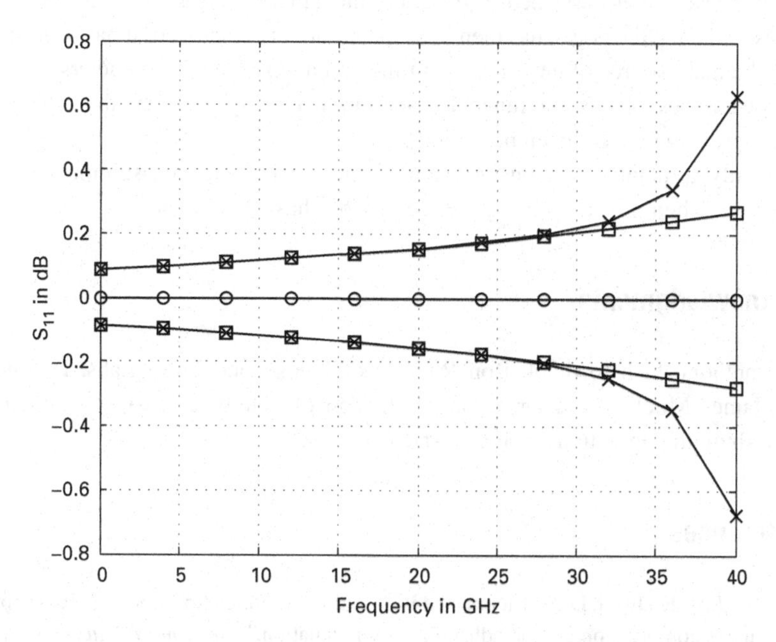

Fig. 11.15 $\Gamma_{dut} \pm \delta\Gamma$ for $\Gamma_{dut} = -1$, $\varepsilon = \pm 0.01$, $S_{21} = 1$, $S_{22} = 10^{-5}$, and S_{12} as shown in Figure 11.13.

in Figure 11.13, if we have an S_{22} that is 100 dB down throughout the frequency of interest (almost an ideal S_{22}), then the uncertainty in the DUT S-parameters is shown by the curve with squares in Figure 11.15. For comparison purposes, the older uncertainty is shown with crosses.

It is evident that in the high-frequency region, where the two S_{22} differ, the uncertainty is significantly higher for the non-ideal S_{22}. The reason for this increase in amplification is the sign of $\Gamma_{dut}S_{22}$. In the example provided above, the two are of opposite signs, making the numerator larger than one and thereby increasing the uncertainty in the DUT.

3. It should be noted that the expression in (11.18) is the uncertainty in DUT S-parameters only due to uncertainty in the raw DUT S-parameter measurement. A more general case needs to include the uncertainty propagation due to noise in the calibration measurements as well. As expected, the complexity of the math increases with different calibration techniques and the number of ports in the DUT. Complete software for the evaluation of the uncertainty, taking into account all the contributions, actually exists [18, 19].

11.10 Conclusions

This chapter has dealt with TDR techniques for network measurements. The hardware architecture of TDR instruments including the sampling system, the pulser, and the timebase have been described. The main element that effects the accuracy of time domain measurements, the noise, has then been quantitatively discussed. It has been shown how it is the main source of uncertainty in time domain derived S-parameters.

The sources of dynamic range degradation in TDR systems and the key design areas for improvement have been presented.

Finally, a quantitative consideration of how S-parameter measurements are affected by noise, when it is the primary source of error, has been given.

Acknowledgments

The authors wish to thank Ron Ramsey of Picosecond Pulse Labs, Dr. Steve Ems, Dr. James Mueller, and Dr. Leonard Hayden of Teledyne LeCroy for their input in describing the operation of pulsers and samplers.

References

[1] M. J. W. Rodwell, D. M. Bloom, and B. A. Auld, "Nonlinear transmission line for picosecond pulse compression and broadband phase modulation," *Electronics Letters*, vol. 23, p. 109, Jan. 1987.

[2] R. J. Baker, D. J. Hodder, B. P. Johnson, P. C. Subedi, and D. C. Williams, "Generation of kilovolt-subnanosecond pulses using a nonlinear transmission line," *Measurement Science and Technology*, vol. 4, pp. 893–895, 1993.

[3] J. R. Andrews, B. A. Bell, and E. E. Baldwin, "Reference flat pulse generator – Technical note," *National Bureau of Standards, Boulder, CO. National Engineering Lab*, Oct 1983. Report Number NBS-TN-1067.

[4] A. Agoston, J. B. Rettig, S. P. Kaveckis, J. E. Carlson, and A. E. Finkbeiner, "Dual channel time domain reflectometer," July 1988. U.S. Patent 4 755 742.

[5] A. Agoston and J. E. Carlson, "Fast transition flat pulse generator," July 1988. U.S. Patent 4 758 736.

[6] M. Kahrs, "50 years of RF and microwave sampling," *IEEE Trans. Microw. Theory Tech.*, vol. 51, pp. 1787–1805, June 2003.

[7] R. Miller, "Waveform translator for DC to 75 GHZ oscillography," June 2001. U.S. Patent 6 242 899.

[8] S. Ems, S. Kreymerman, and P. J. Pupalaikis, "Time domain reflectometry in a coherent interleaved sampling timebase," September 2010. U.S. Patent Application 12/888 550.

[9] R. L. Graham, D. E. Knuth, and O. Patashnik, *Concrete Mathematics: a foundation for computer science*. Addison-Wesley Professional, 1994.

[10] L. A. Hayden and V. K. Tripathi, "Characterization and modeling of multiple line interconnections from TDR measurements," *IEEE Trans. Microw. Theory Tech.*, vol. 42, pp. 1737–1743, September 1994.

[11] D. A. Smolyansky and S. D. Corey, "PCB interconnect characterization from TDR measurements," *Printed Circuit Design Magazine*, May 1999. TDA Systems App. note PCBD-0699-02.

[12] W. L. Gans and N. S. Nahman, "Continuous and discrete Fourier transform of step-like waveforms," *IEEE Trans. Instrum. Meas.*, vol. IM-31, pp. 97–101, June 1982.

[13] A. M. Nicolson, "Forming the fast Fourier transform of a step response in time domain metrology," *Electron. Lett.*, vol. 9, pp. 317–318, July 1973.

[14] P. Pupalaikis, "Wavelet denoising for TDR dynamic range improvement," in *DesignCon*, IEC, February 2011.

[15] P. Pupalaikis, "The relationship between discrete-frequency S-parameters and continuous-frequency responses," in *DesignCon*, IEC, February 2012.

[16] M. T. Jong, *Methods of Discrete Signal and System Analysis*. McGraw-Hill, 1982.

[17] P. Wittwer and P. J. Pupalaikis, "A general closed-form solution to multi-port scattering parameter calculations," in *72nd ARFTG Conference Digest*, p. 137, 2008.

[18] A. Ferrero, M. Garelli, B. Grossman, S. Choon, and V. Teppati, "Uncertainty in multiport S-parameters measurements," *Microwave Measurement Conference (ARFTG), 2011 77th ARFTG*, pp. 1–4, June 2011.

[19] METAS VNA Tools II [Online]. Available: http://www.metas.ch/metasweb/Fachbereiche/Elektrizitaet/HF/VNATools/VNATools.html.

Part IV

Nonlinear measurements

12 Vector network analysis for nonlinear systems

Yves Rolain, Gerd Vandersteen, and Maarten Schoukens

12.1 Introduction

The measurement of the nonlinear behavior of microwave systems and components has evolved a lot over the last years. Starting from instrument prototypes, vector network analyzers for nonlinear systems (NVNA) have now entered the product lines of all the major instrumentation vendors. The major challenge for the scientific community is to embed these devices in the mainstream design and characterization of nonlinear devices and circuits.

As the NVNA is still young, most currently active professionals did not experience NVNA technology during their education or their career. Therefore, it is extremely important to clearly define what can be expected from an NVNA. There is a need for an explanation of what an NVNA is and is not. Explaining the limitations of the NVNA technology is also extremely important, as this can avoid false expectations and deceptions.

This text has the ambition to take a small step in this direction. This is why much effort is spent in the first sections of this chapter in drawing the big picture around the NVNA. Our hope is that this might help practitioners to position the NVNA and to obtain some intuition about the actual measurements the NVNA makes.

The remainder of the text explains the ideas behind the different instruments that have NVNA capability. The setups are very different, but the measurements they make are very similar. The key idea is that to characterize a nonlinear device under test, one needs to measure the complete spectrum (amplitude and phase) of all the port quantities (waves or voltages and currents) that are present at all the ports of the device.

Remember that "A journey of a thousand miles begins with a single step." To avoid the reader becoming overwhelmed by new jargon and concepts, we will start from the S-parameter formalism and the linear time-invariant (LTI) system framework to outline the similarities and the differences with the nonlinear framework.

12.2 Is there a need for nonlinear analysis?

12.2.1 The plain-vanilla linear time-invariant world

S-parameters have been the driving force behind RF and microwave design and characterization of the last 40 years [1]. Their ability to describe a distributed circuit that is inherently complex and hard to understand in an intuitive way proves to be an efficient design tool that can also validate a design or a circuit.

Unfortunately, S-parameters also have their limitations. The basic assumption for their validity is that the circuit or system under test remains linear and time-invariant [1]. Put in layman's words, this means that the superposition principle holds: the response of a system to a sum of two inputs is the sum of the responses to the individual signals and the response scales proportional to the input(s).

Common sense tells us that this assumption is never valid in general. When the input power is increased without bounds, any practical system will break down and therefore is not LTI. Linearity always comes at a price, which is the acceptance of the small-signal operation paradigm. This type of operation assumes that the input signal is small enough to ensure that the response of the system stays close to linear.

Taking a step backwards to see the general picture leads to the striking conclusion that even our most basic tools are not always valid. They come with a set of assumptions that we have to meet to obtain reliable results. Even if this was probably very clear to practitioners in the early days of S-parameters, the wide dissemination and the general success of S-parameter-based design and characterization has diluted the feeling that these hypotheses do indeed matter.

12.2.2 Departure from LTI

The push of portable telecommunication towards power-efficient designs has continuously weakened the validity of the linearity assumption for practical designs with a long battery lifetime. The S-parameter framework first broke at the output of power amplifiers, where S_{22} was no longer power-independent at the higher power levels [2].

The engineer's way to overcome this problem is to extend the LTI framework to include the new situation. Keeping the input power constant was not sufficient to restore the reliability of the results. The predictive power could be restored at the cost of splitting S_{22} into two contributions. The first one is proportional to the incident wave at the output port. This is the normal S_{22} term. The addition of a $\widetilde{S}_{22}$ term that is proportional to the complex conjugate of the incident waves at port 2 solved the problem. This resulted in the so-called hot-S_{22}. This was the first breach in the LTI based S-parameter characterization framework. It was followed by many others.

12.2.3 Measuring a non-LTI system

This new situation leads to new challenges in the measurement world. Wave ratios provided by the S-parameters no longer suffice to characterize a system: it becomes mandatory that waves are measured separately and accurately. This results in a shift of measurement paradigm from a relative measurement (which is pretty easy) to an absolute measurement (which is pretty hard).

But more is needed to measure nonlinear systems. The LTI framework ensures that measurements can be performed one frequency at a time without jeopardizing the quality of the characterization. The response to any (periodic) signal can be obtained by the Fourier series decomposition of the input signal and the superposition of the responses to the individual sine waves in this Fourier series. The most complex experiment one

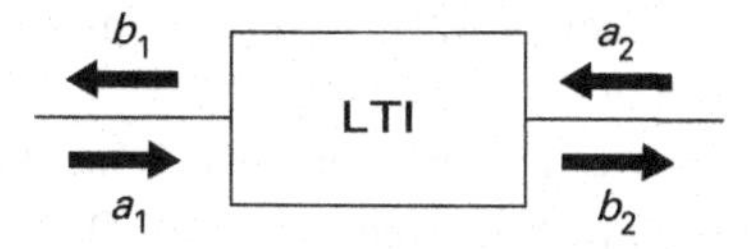

Fig. 12.1 Definition of the waves around a two-port device.

needs to set up for a complete characterization of a LTI system boils down to a sine wave test at one single frequency. There is no need for the measurement of information at different frequencies at the same time [3].

Once one departs from LTI systems, this is no longer true. Consider the most simple departure of linearity for a DUT: a polynomial static cubic nonlinearity. The equations governing the output waves of the ideally matched DUT are given below:

$$\begin{cases} b_2(t) = \alpha a_1(t) - \beta\alpha a_1^3(t) \\ b_1(t) = \gamma a_2(t). \end{cases} \tag{12.1}$$

The incident waves $a_i\,(t)$ and the reflected waves $b_i(t)$ at port $i = 1, 2$ are defined as in Figure 12.1. The constants α, β, and $\gamma \in \mathbb{R}$. A user then measures S_{21} with a standard VNA. The VNA performs a frequency sweep on the DUT. It is excited by a sine wave at the standard power level that is applied at the input port of the device to obtain the dynamic $S_{21}(\omega)$ response.

A novice instrumentation user is tempted to believe that this measurement does indeed represent the behavior of the device. It is easy to be fooled by a measurement that has a high signal-to-noise ratio and is very repeatable. If a second measurement is then taken at a different power level, it will result in a different behavior. Nonlinearity of the device can be a possible explanation, and a skilled instrumentation engineer will grasp a spectrum analyzer to view the complete spectrum.

VNA measurements hence are both extremely powerful but their outcome is very dependent on the validity of the LTI hypothesis. An engineer's solution to this problem is to measure the validity of the LTI hypothesis separately. The VNA is therefore extended to allow a power sweep at one frequency. Nonlinearity of the device will then result in a deviation from a constant gain versus the power. Designers need to be able to assess the magnitude of these perturbations, especially for high-performance designs.

12.2.4 Figures of merit to characterize the nonlinearity

To enable an easy comparison of the order of magnitude of the nonlinear disturbance for different systems, Figures of Merit (FOMs) have been introduced. The 1 dB compression point registers the (input or output) power level at which the actual gain is reduced by 1 dB compared to the linear gain. The FOM reduces a complete function of the power to a single number. Therefore it is clear that the comparison of this FOM over different systems cannot be completely fair. As a measurement addict, it is tempting to measure the complete gain versus power dependency instead. This is pretty easy to realize even with a simple and classical VNA, as the 1 dB compression point is obtained from (relative) gain measurements taken at the fundamental frequency only. Is the measurement instrument for the nonlinear behavior a simple extension of the classical VNA?

To show that this is not the case, let's move to a different FOM, the Third Order Intercept (TOI) point. Things are a bit more involved here [4]. The TOI is defined as the intersection of the extrapolation of the linear gain at the fundamental tone at f_0 and the extrapolation of the harmonic response taken at $3f_0$. As this result is obtained as the intersection of two extrapolated curves, it is again clear that it is not a really fair measure of the nonlinearity either. It does not take into account the saturation of the distortion at higher power levels. Again, the instrumentation engineer is tempted to measure the response curve at the fundamental and the third harmonic in a power sweep. This seems to be a good way to obtain more and better information about the nonlinearity.

This time the measurement is not a simple extension of the VNA measurement. The TOI requires the combination of measurements taken at two *different* frequencies: f_0 and $3f_0$. This requires either a spectrum analyzer measurement or a VNA that is capable of measuring at the fundamental and at a harmonic frequency simultaneously while exciting at the fundamental only.

In practice, the intermodulation product is also often used to measure the nonlinearity. The intermodulation product is measured using a 2-tone excitation signal (two sine waves whose frequency is really close), as this allows the measurement of the nonlinearity for narrow-band systems where the harmonic distortion lies out of the pass band of the device and is therefore attenuated. The measurement then is always performed with a spectrum analyzer.

12.3 The basic assumptions

The characterization of all nonlinear systems is both a much too ambitious and a pretty foolish goal. This can most easily be felt if this goal is translated to a totally different field of science, namely animal biology. It takes a lifetime to understand even a small part of the behavior of an elephant. Many biologists have spent their careers trying to understand this. A biologist will therefore certainly never try to describe the biology of the non-elephants. Even if this example looks stupid, taking one step back shines a different light on the problem. Because the LTI framework does not meet our demands, we are tempted to replace it by its complement, the class of nonlinear systems.

This is certainly not a very smart choice, as each of us can think of a system that is not linear and behaves in a totally crazy way: chaotic systems, uncertain systems, and systems that contain hysteresis are all nonlinear, but are also all very different from our well-known LTI class. They may be so different that they are almost impossible to use in a practical design.

To be successful, we will therefore extend the class of systems in a more directed way. We will consider systems that are close to the LTI behavior, but allow for saturation effects and large-signal operation [5].

Selecting a system class is not enough to enable a correct measurement of the behavior of the system. Even for a practical S-parameter measurement, there are conditions imposed on the excitation signal. Imposing small-signal operation is an often tacit assumption for a VNA. For a NVNA, we will investigate which assumptions are needed to obtain the measurements we are after.

12.3.1 Restricting the class of systems: PISPO systems

The class of the LTI systems is defined as the class of systems that obey the superposition principle. To expand this class gradually, we will remove the superposition principle, and replace it by a criterion that is sensible for the systems that one encounters in practical applications.

Consider for example a power amplifier in a telecommunication link. In small-signal operation, the amplifier output is a sine wave when fed by a (small signal) sine wave. When we increase the amplitude of the input, the shape of the wave starts to deviate gently. Increasing the input power increases the distortion in the wave, but it maintains its periodicity. We can formalize this period-maintaining behavior as follows.

DEFINITION 12.1. *A system that obeys the period-maintaining principle belongs to the PISPO class of systems.*

What is a PISPO system?

The PISPO class is used to extend the LTI class in the context of this chapter. Therefore, it is important that we clearly understand its properties. Intuitively, we see a link between the PISPO class and the Volterra systems [5]. Volterra systems also allow for a gradual and gentle departure from linearity. The key idea behind the Volterra model is the extension of the impulse response of a LTI system to a multilinear impulse response, the kernel function. For a second-order nonlinearity, the Volterra kernel $h_2(t_1, t_2)$ links the second-order output signal $y_2(t)$ to the input signal $u(t)$ as follows:

$$y_2(t) = \int_{-\infty}^{+\infty} \int_{-\infty}^{+\infty} h_2(\tau_1, \tau_2) u(t - \tau_1) u(t - \tau_2) d\tau_1 d\tau_2. \qquad (12.2)$$

A good introduction to Volterra systems can be found in [5]. The problem is that Volterra systems have a bad reputation because of their poor convergence properties for strongly nonlinear systems. Fortunately, PISPO systems do not suffer this problem. Using a least-squares fit, a Volterra system can model all PISPO systems with an approximation that has a perfectly regular behavior. The solution stems from the fact that the PISPO system approximates the hard nonlinearity in a mean-squared sense. Even if the system under test is a discontinuous static nonlinearity, the PISPO class provides a least-squares approximation for it.

Intuitively, one is tempted to believe that the PISPO class only contains systems that do not modify the frequency content of the excitation signals: namely amplifiers and attenuators. More often than not however, we also need to characterize systems that translate the frequency content, such as mixers or detectors.

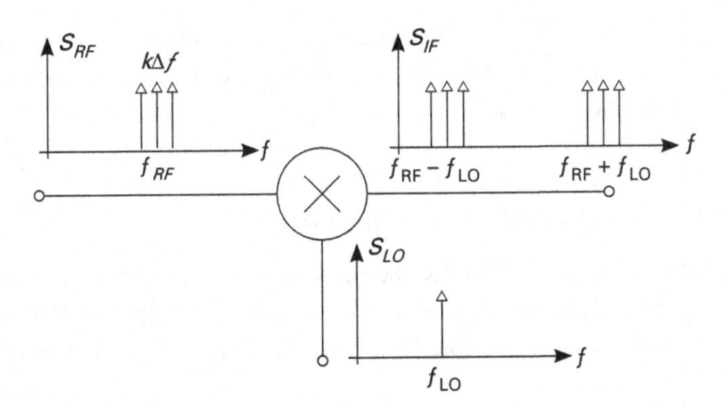

Fig. 12.2 A PISPO mixer as a three port.

A PISPO mixer

A frequency translating system is not a two-port DUT, but rather a three-port one. Consider the case of an ideal mixer [6] as in Figure 12.2. The RF port is excited by a multiple-tone periodic signal that has a carrier frequency f_{RF} and a modulation period T_{RF}. The LO port is excited by a sine wave having a frequency f_{LO}. The power spectra of the inputs are represented in Figure 12.2.

To determine whether or not this system belongs to the PISPO class, we need to be able to check whether a periodic excitation leads to a periodic output signal $s_{IF}(t)$ that has the same period. It turns out that this is not so trivial as it looks.

Mathematically speaking, the input signal can only be a periodic signal if there exists a joint period for the signal pair $s_{LO}(t)$, $s_{RF}(t)$. This requires that the two signals simultaneously repeat perfectly after the common period T_{in}:

$$\begin{cases} s_{LO}(t+nT_{in}) & = s_{LO}(t) \\ s_{RF}(t+nT_{in}) & = s_{RF}(t) \end{cases} \qquad n \in \mathbb{Z} \tag{12.3}$$

This condition can then be translated into a condition on the periods T_{RF} and T_{LO} of the two input signals:

$$T_{in} = kT_{RF} = lT_{LO} \qquad k, l \in \mathbb{N} \tag{12.4}$$

which means that the period of the input signals taken separately needs to be commensurate. In the frequency domain, this can be reformulated to the more commonly used requirement for commensurate frequencies:

$$\Delta f_{in} = \frac{\Delta_{RF}}{k} = \frac{\Delta_{LO}}{l}. \tag{12.5}$$

As a result, we can only determine whether the system belongs to the PISPO class if we can obtain one common frequency grid with spacing Δf_{in} for both signals

$$\begin{cases} f_{RF}(k) & = (l_{RF} + n_{RF}(k))\,\Delta f \\ f_{LO}(k) & = (l_{LO} + n_{LO}(k))\,\Delta f \end{cases} \tag{12.6}$$

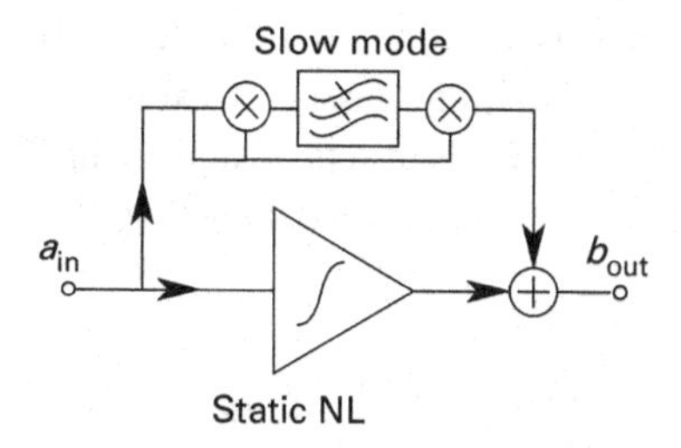

Fig. 12.3 Slow modes in a PISPO framework.

with $l_{RF}, l_{LO}, n_{RF}(k), n_{RF}(k) \in \mathbb{N}$, and $\Delta f \in \mathbb{R}$. This joint period is now used as the period of the input signal when we try to determine if the period of the input and the output waves do match.

PISPO looks beyond static nonlinearity

Many nonlinear models and measurement approaches somehow rely on the presence of a static nonlinearity [5]. This hypothesis also comes naturally into play for many applications. In practical circuitry, where active elements are connected to off-chip components, the linear dynamics introduced by the connections are often an order of magnitude slower than the on-chip dynamic effects. As a consequence, it is very tempting to assume that the nonlinearity is static or quasi-static. In practice, this proves to be a strong hypothesis, as very slow time constants (at a time scale of microseconds) appear around a signal that has a GHz frequency (time constants at the timescale of nanoseconds or less).

The origin of the slow time constants can be understood by the following simple example. Consider an amplifier that operates under compression. Besides its input and output ports, the device is also connected to the outside world via its DC bias port. This connection's impedance is not very relevant at RF frequencies, because the RF signal is carefully blocked by design at this port. However, the IF impedance is also known to have a significant contribution in the RF operation of the device [7], [8]. This impedance influences the slow modes of the amplifier, which are dynamic effects that appear around the RF frequency, but have time constants at the IF time scale. How is this possible?

To explain the slow modes we consider a very simple model, that uses the nonlinearity introduced before in (12.1). This is used to illustrate the behavior of the main path of the amplifier. To model the path that connects to the DC bias port, an additional parallel path is introduced as shown in Figure 12.3. Note that the resulting system is still a PISPO system.

12.3.2 Influence of the excitation signal

From here on, only PISPO systems are considered. Looking back to the LTI framework, it is clear that both the system and the excitation signal have to obey restrictions. Now that the system restrictions are made clear, it is time to take a look at the influence of the excitation signal on the system behavior.

It is evident that it is neither possible nor useful to build a different model for each different excitation signal. We will therefore delimit classes of excitation signals for which the system behaves in a similar way. This leads to sets of signals that are grouped based on their power spectrum or power spectral density and their PDF.

Does the signal choice matter?

To illustrate the change in behavior that results when changing the excitation signal, a series of excitation signals is fed to a PISPO system. In the first series of tests, the system is assumed to be the static nonlinear system, described in (12.1). In a thought experiment, we excite this system with a sine wave. The sine wave is said to have a frequency of 1 a.u. and an amplitude that is large enough to excite the nonlinearity up to high compression levels. Note that the frequency of the sine wave does not matter here as the system is assumed to be perfectly static. Plotting the input and output time signals in an X-Y plot yields the plot of Figure 12.4. This type of plot is sometimes called a Lissajous figure. This shows the nonlinearity of the DUT that is operated in very deep compression. Power levels even start to decrease with increasing input power! When the frequency of the excitation is changed, the response of the device remains unaltered.

In a second thought experiment, we cascade the static system of (12.1) with a bandpass filter. This filter mimics the dynamics of a real RF system. Here, we have chosen a fourth-order bandpass Butterworth response with a pass band ranging from 1 a.u. to 2 a.u. The amplitude response of the filter is given in Figure 12.5.

The response of the tandem connection to the sine wave excitation used in the first experiment is shown in Figure 12.6. There is a clear difference between these responses, due to the dynamics of the system. If the system were an LTI system, it would now be fully characterized.

We then perform a second series of experiments to show the dependence of the system response on the signal class. We will now excite the static and the dynamic system again with a different excitation signal, namely a narrow-band multisine signal. The signal consists of 512 equally spaced spectral lines of equal amplitude located in a frequency band ranging from 0.8 to 0.98 a.u.

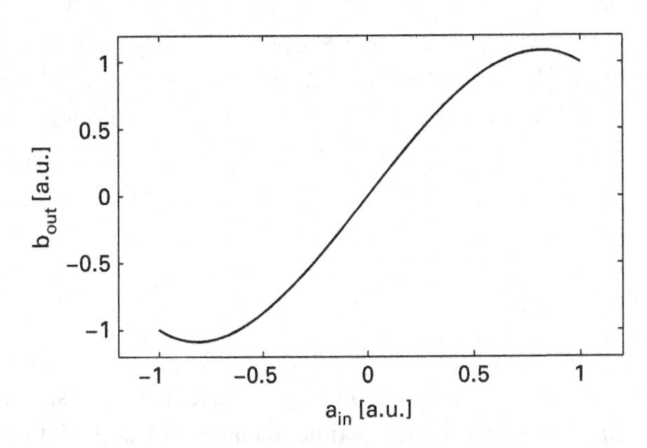

Fig. 12.4 Response of a static PISPO system to a sine wave (response is in arbitrary units [a.u.]).

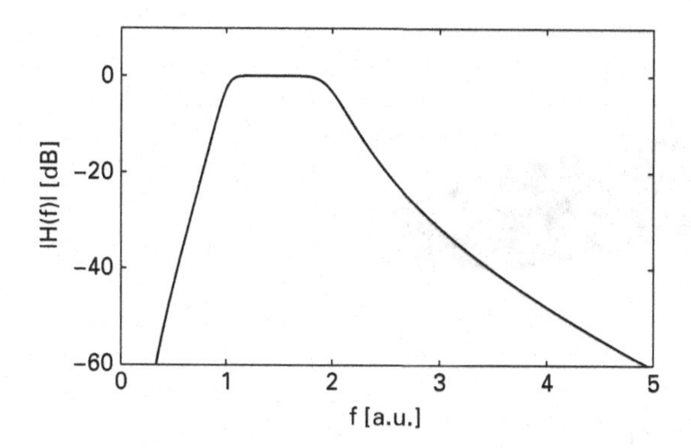

Fig. 12.5 Frequency response of the cascaded LTI system.

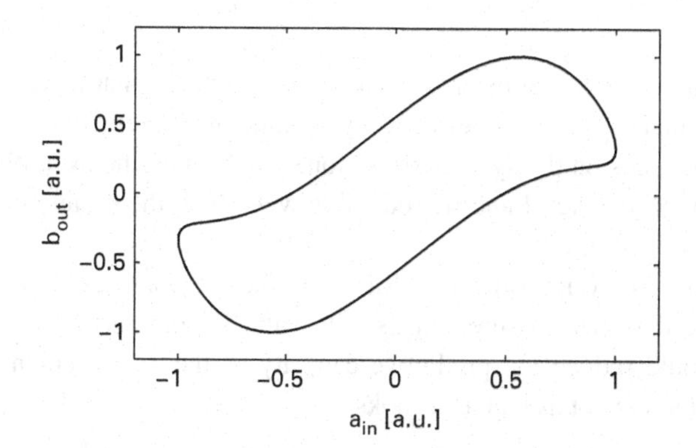

Fig. 12.6 Response of a dynamic PISPO system to a sine wave (response is in a.u.).

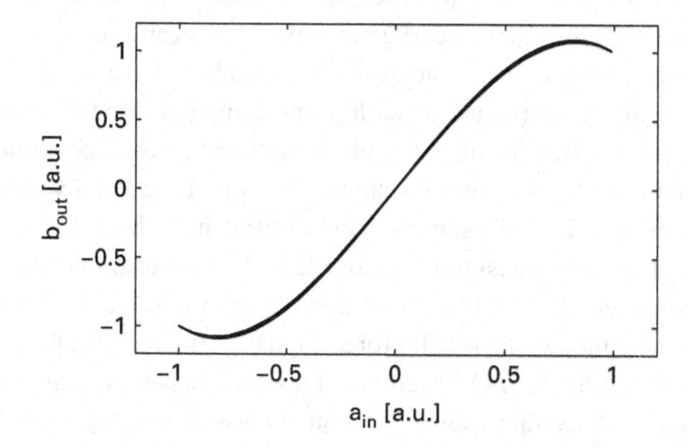

Fig. 12.7 Response of a static PISPO system to a multisine wave (response is in arbitrary units [a.u.]).

As shown in Figure 12.7, the response of the static system to a multisine or a sine wave signal is perfectly identical. There is no dependence on the properties of the signal. On the contrary, the response of the dynamic system, shown in Figure 12.8, no longer resembles the response obtained for the sine wave!

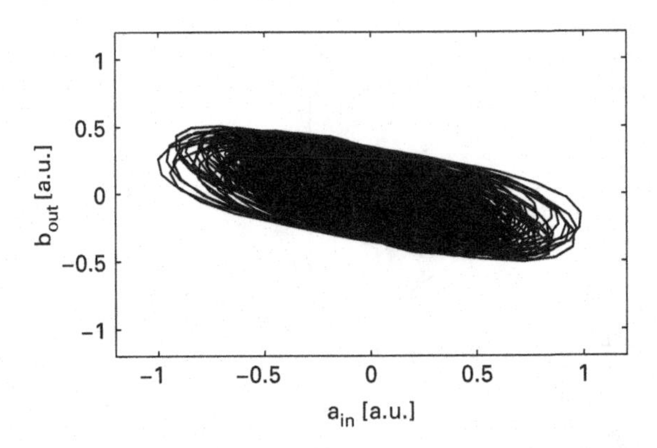

Fig. 12.8 Response of a dynamic PISPO system to a multisine wave (response is in arbitrary units [a.u.]).

What does that tell us about nonlinear measurements? It shows that there is a major difference between an LTI system and a PISPO system: the measurement of the system response no longer depends on the system class alone, but also on the excitation signal used. A measurement is only valid and reproducible when both the system class AND the excitation signal are specified.

This certainly looks like very bad news. Based on the previous results alone, it is tempting to conclude that the measurement is only valid for the particular excitation signal used. This would reduce the predictive capacity of the measurement to zero. Fortunately, the situation is not as bad as it looks.

Specifying the class of signals

A measurement is often the response to an engineer questioning the behavior of the system that operates or will operate in a certain context. This means that there is a lot of prior knowledge present about the possible excitation signals of the device.

Most applications will be constructed under limiting assumptions on the power spectrum or the power spectral density of the applied excitation. For telecommunication applications for example, a transmitter is designed to support a certain standard. This fixes the power spectral density of the signals to the spectral mask that is associated with that particular modulation. When designing a mixer LO driver for example, the spectrum of the signal is bounded with known bounds on the frequency and the amplitude.

We know that a general model that is valid for all signals is out of reach because of the lack of a general theory for the PISPO system. As it does not make sense to measure the behavior of the system outside this application bound class of signals, we will restrict the signals to have a fixed power spectral density and will sweep the power of the signal over the allowable power range.

What if the aim of the measurement itself is the identification of a model for the system? The challenge there is to capture the behavior that matters to the model and the use one wants to make of that model. Any model relies on basic hypotheses to be valid. The challenge in the selection of a signal class is to make the class narrow enough to meet these hypotheses and wide enough to remain applicable in practice.

The excitation signal for a single experiment needs to have a fixed power spectrum. For a sine wave excitation, this leaves us with a class containing a single signal: a sine of fixed amplitude and frequency.

For a modulated signal, there is a larger range left to choose the input signal from. For a fixed level of the total signal power, some type of modulation signals have a fixed power spectrum and a data-dependent phase spectrum. Others have a data-dependent power and phase spectra.

In the context of a measurement, this data dependency is conceptualized as a random variation of the phase and the amplitude of the excitation signal over a set of possible values. A single measurement is then performed on one realized signal in this class.

The power spectral density alone is not sufficient to define a signal class. This can intuitively be understood by the following thought experiment: consider two signals with the same power spectral density, but a different behavior in the time domain. The first signal is a swept sine. The second signal is a Gaussian noise source with a fixed power spectrum. When these waveforms with equal power excite a nonlinear system, the response of the system will be quite different. The level of the nonlinear contributions in the output signal can be up to an order of magnitude higher for the swept sine signal. To understand this behavior, we will look at the histogram of the time signal. This measured quantity represents a sampled version of the Probability Density Function (PDF) of the signal. The PDF describes the distribution of the different amplitude levels present in the signal (both signals are normalized to contain the same power). The histogram is shown in Figure 12.9 for signals that are 128 000 samples long.

The PDF of the signals has an almost inverse behavior. The Gaussian noise signal spends most of its time at low amplitude levels. Therefore, it excites the nonlinearity gently most of the time. From time to time, a peak value appears. The swept sine signal, on the other hand, spends most of its time at high amplitude levels. The nonlinearity is therefore strongly addressed during the major part of the excitation signal. This increases the level of the nonlinearity to much higher values than for the Gaussian noise signal.

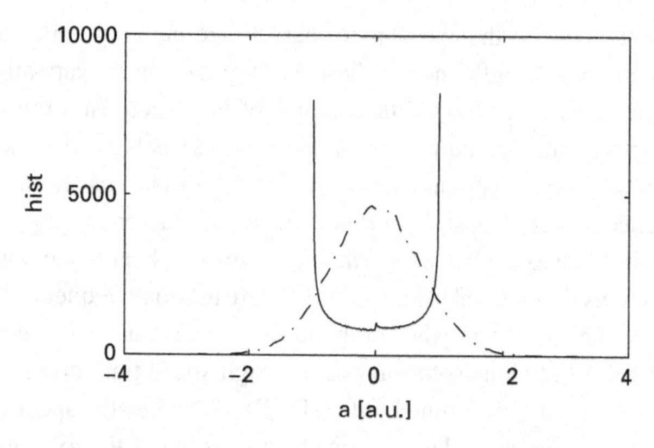

Fig. 12.9 Histogram of the swept sine (full line) and the Gaussian noise (dash-dot line) signal of equal power. The horizontal axis represents the amplitude in the time domain, the vertical axes the occurrence count.

This example clearly demonstrates that the PDF of the signal severely influences the nonlinear distortion too.

12.3.3 The definition of the nonlinear operating point

Repeatable measurements are the cornerstone of any characterization and/or modeling effort. To characterize a nonlinear system, the conditions that are required to enforce the repeatability of a characterization are much more involved than in the LTI case. First, one needs to restrict the class of systems that are to be measured. In this chapter, we restrict the class of systems that, when excited with a periodic signal, produce a periodic output that has the same period: the PISPO class. In addition to this, the class of input signals is to be reduced too. The signal class is reduced to signals with a fixed power spectrum and a fixed PDF.

It is the coupling of a fixed system class and a fixed signal class that enables one to obtain repeatable measurements for a nonlinear system. This shift in paradigm looks artificial at first glance. Note however the large similarity between this way of working and the setting of an operating point in an S-parameter transistor characterization. To stress this concept, this coupling of a fixed system and a signal class is called the *nonlinear operating point* throughout this work.

12.4 Principle of operation of an NVNA

This section is devoted to the general principles that govern the operation of an NVNA. First, we will look for an ideal instrument that is capable of nonlinear characterization. Next, we will shortly touch on the requirements imposed by the use of the discrete Fourier transform to obtain the spectral measurements. Finally, the challenges posed by the calibration of the NVNA are covered.

12.4.1 Introduction

Now that the class of the systems that we want to characterize and the possible test signals are defined, we can find out the influence of these assumptions on the capabilities of the instrumentation. The major issue lies in the absence of the superposition principle for the DUT: if the response to a certain class of excitation signals is to be known, the measurements have to be taken with an excitation that belongs to this class.

Engineering practice learns to start from known techniques to create something new. Can existing instruments be extended, adapted, or combined to handle the nonlinearity?

The nonlinearity mixes the spectral information of more than one frequency to generate the output response. Hence measuring the complete spectrum in one single measurement is the way to go. The most obvious solution lies in a high-speed time-domain measurement of the sampled wave data, combined with a DFT to calculate the spectrum [9, 10].

The behavior of the nonlinearity depends on the properties of the excitation signal. When an arbitrary signal can be generated, the excitation can be adapted accurately to the characterization needs. An instrument that combines a time domain data acquisition and an arbitrary waveform generator exists at IF frequencies. It is called a FFT analyzer or a DSA [9]. The challenge we face is to port the functionality of this instrument to the RF domain!

In the microwave frequency range, signals can be acquired in the time or the frequency domain. Classical frequency-domain instruments are a VNA or a spectrum analyzer. In time domain measurements, a real-time oscilloscope or a sampling oscilloscope [11], [12] is commonly used. Signals can be generated by a sine wave generator, a modulated generator, or an arbitrary waveform generator. Clearly, an oscilloscope and an arbitrary waveform generator or a modulated generator can mimic a DSA. We show below that the VNA can also deliver solutions, albeit at the cost of hardware extensions.

Microwave sources have for a long time been limited in their ability to generate modulated waveforms: see Chapter 3. In the last years however, increasing numbers of microwave sources have provided at least some modulation capability. Sources are continuously moving away from CW generation alone. They provide a (complex) modulation over ever-increasing bandwidths. The kind of signals generated by such a source is extremely suited to mimicking the behavior of real telecommunication signals while maintaining the capability to generate purely periodic waveforms. This is a most welcome feature for an easy and correct transformation of the waveforms to the frequency domain. Some acquisition systems require a trigger signal to operate properly. This is a time marker (often a block pulse) that repeats at the period of the modulated waveform. This trigger defines a fixed point in the period of the modulated signal. We see below that this feature is mandatory for some of the instrumentation setups to work properly.

In all these RF measurements, the error correction and assessment is a key issue. It is clear that not all measurements require precision at the level of a standards lab. But even if the demand for accuracy is modest, it still needs to be reached. This requires the existence of a calibration and a verification procedure of some kind. We will also have a quick look at this aspect of the nonlinear characterization.

The ideal microwave instrument for the nonlinear characterization measures complete wave spectra without distortion. It has a flexible signal generator, that can simultaneously impose the spectral content and the PDF of the excitation signal with a high spectral purity. All these features are also perfectly synchronized in frequency and in phase through the use of a common reference clock (labeled CLK in Figure 12.10). This ideal instrument needs now to be approximated with real setups. As a new device deserves a new name, this device will be called the NVNA from now on.

12.4.2 Basic requirements for nonlinear characterization

A PISPO system is a system with a mixed behavior. Its nonlinearity aspect produces waveform shapes that are best characterized in the time domain, while its dynamic behavior is best visualized in the frequency domain. The NVNA therefore needs to measure waves in a configuration that allows an easy transformation between both domains.

To transform measurements between the time and frequency domain, the DFT is used. Most often, we will choose to work with the FFT algorithm that is both numerically efficient and stable.

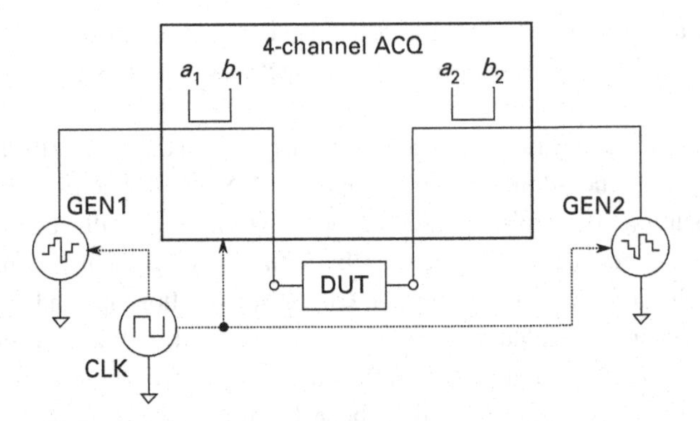

Fig. 12.10 The general NVNA setup. Note that CLK is a common reference clock used to avoid spectral leakage.

The power of the DFT can hardly be over-estimated, but being a real-world method it comes with a set of hypotheses that have to be met accurately to avoid problems. The DFT is prone to two types of errors in the spectral domain.

First, the signals to be transformed have to obey the Shannon-Nyquist theorem. Practically speaking, this theorem states that the discrete-time representation of a waveform is only unique (and valid!) if the signal is sampled fast enough. Sampling at a lower rate results in spectral folding or aliasing. Spectral components then appear at frequencies that are different to the original frequency. Aliasing is avoided by external filtering before the conversion of the signal to discrete-time. Sometimes, we will violate this hypothesis on purpose to obtain very large bandwidth measurements.

Requirement 1 The measured signals are band-limited to avoid unexpected aliasing of the spectra.

Secondly, the DFT is also prone to spectral smearing or spectral leakage. Leakage turns discrete spectral lines in a distribution of spectral power over a (larger) number of spectral lines. This ruins the spectral resolution, and results in errors in the spectrum that can be as large as 30%. Leakage avoidance is theoretically simple to explain but it is hard to implement. A leakage-free spectrum is obtained when the measurement time window is an integer multiple of the period of the excitation signal,

$$T_{acq} = mT_s = nT_{exc} \quad m, n \in \mathbb{R}, \tag{12.7}$$

with T_{exc} the period of the excitation signal and T_{acq} the time span of the measurement. This requires synchronization between the acquisition and the generator of the NVNA. This is seen if (12.7) is transformed to the frequency domain:

$$F_{acq} = \frac{F_s}{m} = \frac{F_{exc}}{n} \quad m, n \in \mathbb{R}. \tag{12.8}$$

The bad news here is that the DFT is extremely sensitive to the presence of errors in this equation, even if they are very small.

How can one impose this requirement with high accuracy in practice? This is only possible if the time reference (the reference clock) used by all the parts of the instrument is the same. The question is to know whether this condition is also sufficient. This can best be analyzed using a small example.

Consider an NVNA that excites a DUT with a sine wave at a frequency of 1 GHz. The wave signals are acquired at a sampling rate $F_s = 1/T_s = 5\,\mathrm{GHz}$. Generator and acquisition both run from the same 10 MHz reference clock (this is common for instrumentation). Assume that 500 samples are acquired, then

$$\begin{cases} T_{acq} = \dfrac{500}{5\,\mathrm{GHz}} = 0.1\,\mu\mathrm{s} \\ T_{exc} = \dfrac{1}{1\,\mathrm{GHz}} = 1\,\mathrm{ns}, \end{cases} \tag{12.9}$$

and the condition (12.7) is met. When we use a scope and a sine wave generator to perform the actual measurements, leakage is very likely to remain present. The scope and the generator multiply the reference frequency by a factor of 500 and 100, respectively. Such a high multiplication factor can be obtained by a phase-locked loop. The device frequencies will be accurately locked to the reference, but the uncertainty on the phase (also known as phase noise) will be increased by a factor that is roughly proportional to the frequency multiplication. As a consequence, a (slow) drift of the phase of the acquisition's F_s with respect to the generator's F_{exc} becomes very hard to avoid in practice. This results in the presence of leakage again, especially if the period of the signal becomes larger.

Requirement 2 Leakage-free measurements are obtained when the sampling frequency F_s of the acquisition and the frequency F_{acq} of any generated spectral line obey $F_{acq} = \frac{F_s}{m} = \frac{F_{exc}}{n}$ $m, n \in \mathbb{R}$. This requires that all the frequencies in a measurement are commensurate to a fundamental frequency. All these frequencies are also phase-coherent to the reference frequency of the instrument or the setup.

12.4.3 A calibration for nonlinear measurements

The measurement quality of a RF instrument is always determined by its calibration. The S-parameter calibration calibrates S-parameters, which means that it removes the systematic errors from the measured wave ratios. As a nonlinear characterization requires the knowledge of waves rather than wave ratios we expect that the S-parameter calibration needs an extension. In which sense is the calibration to be extended?

As a "nonlinear" measurement is no longer a ratio of waves taken at one single frequency, the S-parameter calibration of the VNA will no longer completely ensure correct measurements. Trouble can be expected because the S-parameter calibration assumes both the VNA and the DUT to be LTI systems. The linearity of the VNA enables a calibration to be performed one frequency at a time. The linearity of the DUT calls for the calibration of wave ratios only. As explained in Chapter 9, one entry in the compensation matrix can therefore be freely chosen.

For the NVNA, we will maximally re-use the already existing calibration. Thereto, we assume:

Assumption 1: The acquisition part of the NVNA is an LTI system

This is not a strong assumption. It can quite easily be met if we provide attenuation before the acquisition channels of the NVNA. Imposing the linearity is then a matter of keeping the power budget at the input of the channel in the zone where that channel combines a good SNR and a good linearity. This has to be imposed for each measurement.

Assumption 1 implies that the relation between measured and calibrated waves is LTI. The calibration can therefore still be performed one frequency at a time, and remains independent between frequencies. We will again use a practical example. Consider a classical 8-term calibration as in [13]. The wave correction at an angular frequency of ω is

$$\begin{bmatrix} a_1 \\ b_1 \\ a_2 \\ b_2 \end{bmatrix} = \begin{bmatrix} \mathbf{X}_1(\omega) & 0 \\ 0 & \mathbf{X}_2(\omega) \end{bmatrix} \begin{bmatrix} b_{m1} \\ a_{m1} \\ b_{m2} \\ a_{m2} \end{bmatrix}, \tag{12.10}$$

where the suffix m denotes a measured wave. A numerical suffix denotes a calibrated wave. All quantities are complex numbers ($\in \mathbb{C}$). The one-port correction matrices $\mathbf{X_1}(\omega)$ and $\mathbf{X_2}(\omega)$ are defined as

$$\mathbf{X_{1,2}} = \begin{bmatrix} l_{1,2} & -h_{1,2} \\ k_{1,2} & -m_{1,2} \end{bmatrix} = k_{1,2}\mathbf{Y_{1,2}} \tag{12.11}$$

at each frequency separately. The explicit dependence on the frequency is removed from this expression to reduce notational burden. These matrices will be used extensively in the following chapters. We can now rewrite (12.10) as follows:

$$\begin{bmatrix} a_1 \\ b_1 \\ a_2 \\ b_2 \end{bmatrix} = k_1 \begin{bmatrix} \mathbf{Y}_1 & 0 \\ 0 & \frac{k_2}{k_1}\mathbf{Y}_2(\omega) \end{bmatrix} \begin{bmatrix} b_{m1} \\ a_{m1} \\ b_{m2} \\ a_{m2} \end{bmatrix}. \tag{12.12}$$

The problem is that the outcome of any classical S-parameter calibration only determines the matrix without k_1. Hence, only seven of these eight error coefficients are known. Since S-parameters are defined as ratios between waves, $k_1(\omega)$ can be freely chosen in the S-parameter calibration. As long as a wave ratio is calculated, $k_1(\omega)$ does not matter: it appears in the numerator and the denominator of the wave ratio and gets factored out.

For the TOI as defined earlier, response measurements at ω_0 and $3\omega_0$ are to be combined. The values $k_1(\omega_0)$ and $k_1(3\omega_0)$ can no longer be factored out and hence their ratio appears in the result. The function $k_1(\omega_0)$ therefore needs to be "measured" by an additional calibration step: the so-called "absolute" calibration.

We will determine the complex function $k_1(\omega)$ in two successive steps: first the amplitude of the function is obtained; next the phase function is characterized. Once this complex function is known at all the test frequencies, the correction of the raw data boils down to a matrix multiplication as is shown in (12.10).

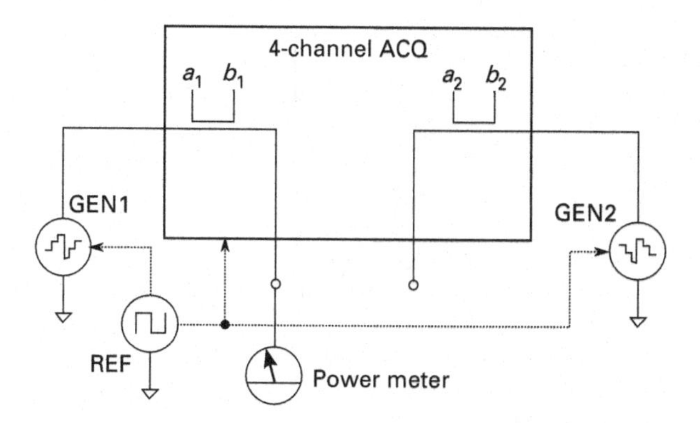

Fig. 12.11 Measurement for the power calibration.

Measuring $|k_1(\omega)|$

We start from the wave correction relation (12.10). Since the correction equations between port 1 and port 2 are decoupled (the off-diagonal blocks are zero), we can determine $|k_1(\omega)|$ by a one-port measurement as in Figure 12.11. The exact (but unknown) incident power at the DUT can be determined from a_1 and b_1. Thereto, consider the incident power to the DUT at port 1

$$P_{in} = |a_1|^2 - |b_1|^2 = |l_1 b_{m1} - h_1 a_{m1}|^2 - |k_1 b_{m1} - m_1 a_{m1}|^2. \tag{12.13}$$

After substitution of (12.11), we can rewrite this equation as follows:

$$P_{in} = |k_1|^2 \left(|{}^{l_1}/{}_{k_1} b_{m1} - {}^{h_1}/{}_{k_1} a_{m1}|^2 - |b_{m1} - {}^{m_1}/{}_{k_1} a_{m1}|^2 \right). \tag{12.14}$$

In this expression, everything can be calculated besides $|k_1(\omega)|$ and P_{in}. However, if it is possible to connect a power meter at the input port, $|k_1(\omega)|$ can be found since we then measure the power $P_m = P_{in}$,

$$|k_1|^2 = \frac{P_m}{\left(|{}^{l_1}/{}_{k_1} b_{m1} - {}^{h_1}/{}_{k_1} a_{m1}|^2 - |b_{m1} - {}^{m_1}/{}_{k_1} a_{m1}|^2 \right)}.$$

Everything is measured (at the same frequency) in this equation besides $|k_1(\omega)|^2$. The magnitude of the calibration function is therefore known. Note that the procedure outlined here is the same as the one that will be used for the real-time load pull system, as will be shown in Chapter 14.

Measuring $\angle k_1(\omega)$

Measuring the phase of $k_1(\omega)$ is the most complex part of the calibration, as here the phase difference between spectral components at different frequencies ω must be calibrated. Intuitively starting along the lines of a LTI reasoning, it might not be very clear why there is a need for this phase spectrum calibration at all.

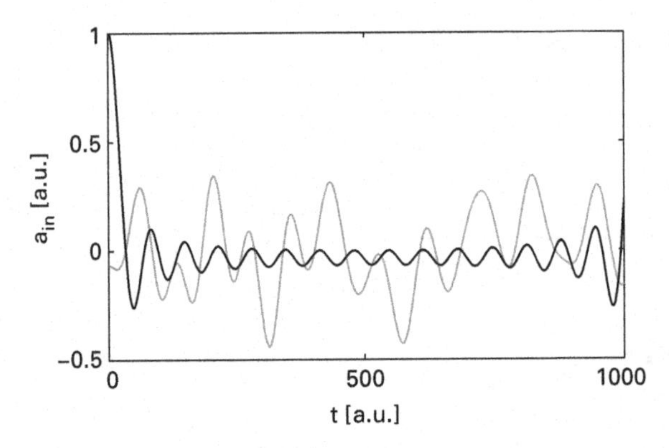

Fig. 12.12 Two multisine signals in a.u. with the same power spectrum and a different phase spectrum. The black line corresponds to zero phase, the gray one to a random phase spectrum.

To show the need for this alignment of spectral lines at different frequencies, we consider a thought experiment based on a simple multiple-tone (multisine) signal. The signal consists of 16 spectral lines, that all carry an equal power. For the first signal of Figure 12.12, the phase of all spectral components is set to zero (full line). This signal is pulse shaped and has an extremely high peak amplitude. The second signal has a phase spectrum that is randomly selected for each line. The phase is drawn from a uniform distribution ranging from $[0, 2\pi[$. The signal now looks very much as a random noise signal, even though the power spectrum is still the same. It is clear that the first signal will excite the nonlinearity in a totally different way than the second one.

What we see here, is that the phase spectrum of a multisine signal influences the shape of the signal in the time domain. Therefore, it is very important to measure the phase characteristic accurately when the properties of the nonlinearity are to be quantified.

Why is this information not measured by a normal VNA? A VNA measures one frequency at a time and calculates only wave ratios. It can only obtain the phase difference between sinusoidal waves that have the same frequency. Hence, there is no way to measure the phase spectrum: the phase spectrum measures the phase difference between spectral lines at different frequencies.

How are we going to calibrate the phase spectrum? The idea again is pretty simple to understand. We will create a signal that is repeatable over a long period of time and is very well known. This signal will then serve as a calibration element. It will be fed to one of the ports of the NVNA and will be measured. The known difference between the measured and the standard's phase will then be used to correct the measurement.

One of the problems that we face is that the phase standard signal has to contain a spectral line for each frequency that is involved in the measurement that we want to calibrate. Since this requires a wideband signal, some kind of a pulse-shaped signal will be used.

To make all this more practical, we will develop the idea for a sine wave excitation. Following the same lines, the method applies to modulated excitation signals using an appropriate calibration signal, as shown in Figure 12.13. When a sine wave excites

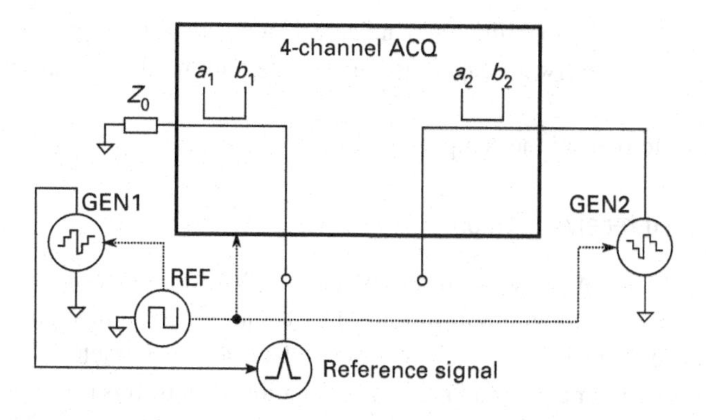

Fig. 12.13 Measuring the phase spectral standard.

a PISPO system, the output waves have the same period as the sine wave. A comb generator fed with a sine wave will create a large number of harmonics of the sine wave and therefore can act as a reference signal: it contains all the requested frequencies. This comb generator is often called "the golden diode" in the context of nonlinear characterization.

Again, we start from the error correction equation (12.10). The exact (but unknown) reflected DUT wave $b_1(l\omega_0))$ now contains N harmonic components simultaneously. The wave b_{m1} is then also measured at the same N harmonic components and we obtain a set of N complex equations:

$$b_1(l\omega_0) = k_1(l\omega_0)\,b_{m1}(l\omega_0) - m_1(l\omega_0)\,a_{m1}(l\omega_0) \quad \text{for } l = 1 \ldots N. \tag{12.15}$$

This can be rewritten to introduce the calibration coefficients as before:

$$b_1(l\omega_0) = k_1(l\omega_0)\left\{b_{m1}(l\omega_0) - \frac{m_1(l\omega_0)}{k_1(l\omega_0)}a_{m1}(l\omega_0)\right\} \quad \text{for } l = 1 \ldots N. \tag{12.16}$$

We require that the phase spectrum of the multiple tone wave is equal to the (exactly known) excitation of the standard signal in the reference plane of port 1 to calculate the phase $\angle k_1$:

$$\angle k_1(l\omega_0) = \angle b_1(l\omega_0) - \angle\left(b_{m1}(l\omega_0) - \frac{m_1(l\omega_0)}{k_1(l\omega_0)}a_{m1}(l\omega_0)\right) \quad \text{for } l = 1 \ldots N \tag{12.17}$$

This delivers the phase difference that is to be used to compensate for the dynamic errors in the spectrum of the measured signals.

12.5 Translation to instrumentation

We are now ready to analyze the behavior of the existing instruments and instrumentation setups. The perspective that we take here is to compare the setups on a purely technical basis. We will compare the capabilities and indicate their positive and negative properties. We will not include the software capability or the integration into existing simulation packages.

We have chosen to present the setups in chronological order.

12.5.1 Oscilloscope-based receiver setups

The instrumentation setup of the oscilloscope-based NVNA is given in Figure 12.14. The oscilloscope can either be a real-time [11] or a sampling oscilloscope [12]. Two generators are needed in this setup if we want to be able to characterize a two-port dynamic PISPO system for a particular class of excitation signals. It is no longer allowed to switch off the input source when a reverse excitation is to be applied, as this modifies the input signal and/or impedance and hence the PISPO approximation of the DUT can also change.

Note the presence of a common time reference (a reference clock). It is used to avoid frequency slipping between the timebase of the scope and the period of the generated signals.

The oscilloscope-based setup is a time-domain setup. As the waveforms are acquired in discrete time directly in the time domain, all the spectral lines in the signal are measured simultaneously and therefore they are automatically phase-aligned in the spectral domain. To obtain distortion-free waveforms, it is therefore sufficient to correct for timebase distortion and for the own (mainly linear) dynamics of the channel.

For a real-time oscilloscope, the samples are acquired at a rate that is equal to the sampling rate of the scope. As this requires extremely fast ADCs, the sampling oscilloscope has been proposed to reduce the conversion speed while maintaining the sampling rate for periodic signals [12].

For a sampling oscilloscope, the discrete time signal consists of samples that are acquired one at a time and each in a separate period of the signal. Sampling reduces the actual sample conversion speed to a value that is up to more than a hundred times slower than the actual sampling speed of the data record. The shape of the waveform is left untouched (for an ideal setup) as shown in Figure 12.15. The gray line represents the RF signal. In this example, the signal is sampled using a sample period that is a little

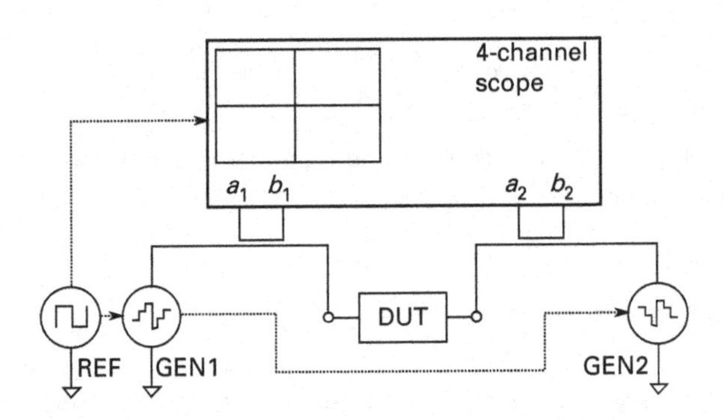

Fig. 12.14 The scope-based NVNA.

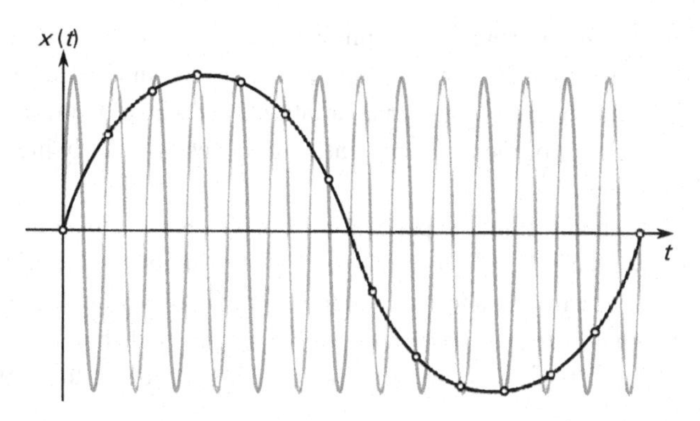

Fig. 12.15 Sampling of a waveform

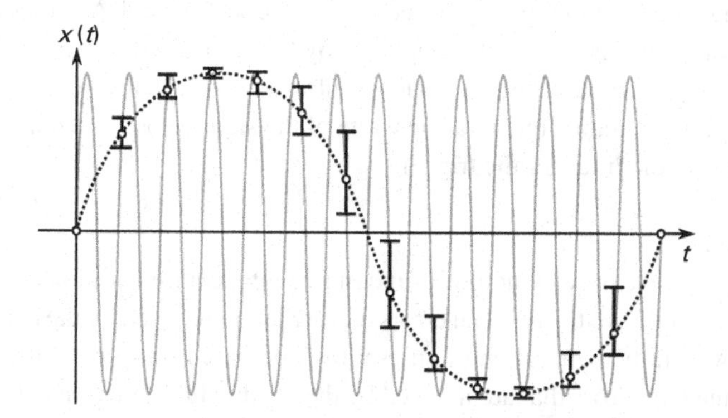

Fig. 12.16 Influence of a random variation of the trigger point position on the measurement.

larger than the period of the underlying signal. The measured samples are represented by the circles on the plot. The black line represents the (unknown) continuous time representation of the sampled signal. Visual comparison of the black and the gray lines shows the same signal, but represented on a stretched time axis. Sampling did indeed convert the signal to a lower frequency without distortion of the shape of the wave!

The AD conversion speed for sampling scopes becomes lower and therefore the conversion can be made more accurate. However, there is no "free lunch" in instrumentation! The key hypothesis needed to obtain a correct measurement is that the signal has to remain stationary during the acquisition of the complete signal (all samples in the trace). Time variations in the behavior of the system must therefore be slow. Theoretically we will assume a quasi-static behavior to reduce the errors to zero. Practically, the non-stationary variation of the DUT can only become visible on a timescale that is an order of magnitude slower than the acquisition time of a wave. For fast variations of the temperature for example, this can sometimes be a limitation.

To obtain good measurements, the stability of the position of the trigger point used by the scope to determine the periodicity of the signal becomes extremely important. To illustrate this, a thought experiment is again welcome. Starting from the ideally sampled signal of Figure 12.16, we introduce a small (with respect to the sampling period) random variation of the trigger point during the sampling process. This is what we call timing jitter; it introduces an uncertainty in the sampling instant results, and hence the measured signal varies too. The magnitude of the signal variation is indicated by the error bars on Figure 12.16. The sensitivity of the measurement to timing errors clearly increases with the derivative of the signal.

At first sight, this high and signal-dependent sensitivity seems to be a disadvantage. However, assume now that we want to minimize the measurement errors induced by the trigger uncertainty. The trigger acts as a window comparator whose timing precision increases with the signal derivative. The most stable trigger will therefore always be obtained when the trigger level is set to correspond to the portion of the signal with a maximal rate of change.

For modulated signals, the signal shape in the time domain is so complex that it is impossible to trigger the sampling scope directly from the wave signal. As a consequence, a frame synchronization signal is to be generated by the modulated source to define the trigger point. It is preferable to select a block-pulse-shaped signal (rather than a sine wave, for example) to obtain a maximal slew rate of the signal close to the trigger point and hence minimize the jitter for the trigger.

Calibration issues

In the following we discuss why nonlinear measurements require additional power and phase calibration to remove the residual errors that are left untouched by the S-parameter calibration. However, the calibration of an oscilloscope-based setup is a little bit more tricky: besides the LTI errors that are induced by the bandwidth limitations of the setup, the scope also requires a calibration to compensate for errors in the time grid of the acquired samples: the timebase correction [14].

The timebase correction is the most complex problem in the calibration of a scope. Often, the timebase suffers from nonlinear distortion (as a function of the sampling instants) or even from the presence of jumps. As a consequence, the data points in the acquired discrete time signal are no longer perfectly equidistant in time. This introduces a kind of phase modulation. Even if the waveforms to be measured are perfectly periodic at the expected frequencies, significant leakage errors are created in the DFT spectra. This timebase distortion comes in two types: a systematic distortion that is independent of the measurement realization and a stochastic component, the time jitter.

The fast and dirty approach to compensate for timebase errors is to use a nonrectangular time window to get rid of the leakage. This introduces a significant measurement error. The clean way to circumvent the problem is to calibrate the timebase of the scope.

Two classes of calibration method exist in the general literature. The first class performs prior characterization of the timebase errors, measuring a sine wave signal. It compensates for the systematic timebase error, leaving the jitter untouched [15], [14]. Maximum-likelihood methods that operate completely in the time domain allow for a distorted input waveform and are available in this context [16]. They do not impose requirements on the shape of the timebase distortion.

The second class of methods performs the correction during the measurement of the signals [17]. These require a clean sine source and two additional acquisition channels to be present in the setup. A sine wave and a cosine wave (obtained by a 90° hybrid) of appropriate frequency are then fed to the additional channels while the device waves are measured. A simultaneous acquisition of the sine wave, the cosine wave, and the device waves then allows for compensation of the systematic timebase distortion and the part of the jitter that is common to all the acquisition channels.

Conclusion

The oscilloscope-based NVNA is the intuitive solution to the measurement of a PISPO DUT. It captures the complete time signal directly in the time domain, thereby reducing the number of processing steps needed to obtain the data. Ideally speaking, it should be sufficient to take the DFT of the measured samples to obtain the raw data. In practice, the timebase errors introduce an additional complication in the processing. Their removal using a timebase calibration is certainly possible, but is rather complicated and pretty involved.

12.5.2 Sampler-based receiver setups

The main disadvantage of the oscilloscope-based devices lies in the presence of the timebase errors. In order to avoid these, it is tempting to replace the timebase by something new. The problem of the timebase is twofold: the stability of the trigger point (equivalent time) and the lack of coherence between the signal source and the timebase jeopardize the measurement quality.

We will replace the timebase by a sampling clock generator to impose an equally spaced sampling grid that is phase coherent with the generator waves by construction. Good measurement engineering practice teaches that measurement speed and accuracy seldom come together. The sampling or reduction of the AD conversion speed therefore is an attractive measurement option. Ideally, this reduces the leakage problems in the discrete time signal to zero and increases the measurement resolution [18–21].

To find out how to get a timebase replacement, it is important to understand the source of the potential problems. Sampling scopes rely on the detection of the position of the trigger point to generate their sampling grid. The position of the trigger is very noise sensitive because it relies on a point based decision. The trigger results from a comparison between two signals at a single instant in time. There is no possibility of averaging the noise in that type of circuitry. What we would like to obtain is that the position of the trigger point would be determined using the information present in a complete period of the signal to be acquired.

This calls for a PLL type of solution. The noise can then be reduced drastically by the narrow bandwidth of the PLL. The use of a phase locked loop to synchronize an acquisition to the generated signal was already used in the early days of network analysis. It is the idea behind the vector voltmeter. The phase accuracy that is obtained from the PLL in the sampling process results in time accuracy levels for equivalent time sampling that are better than for the real-time scope.

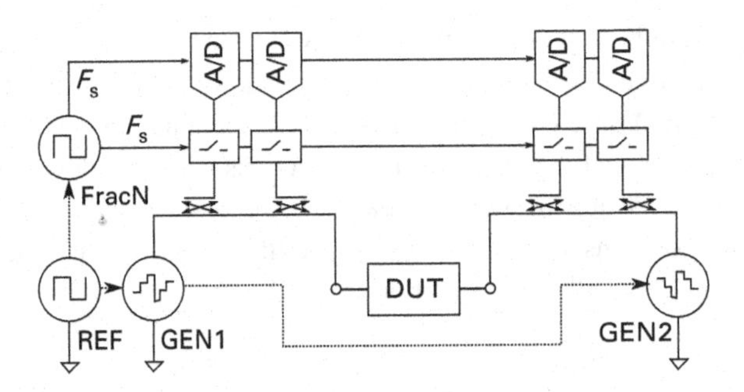

Fig. 12.17 The sampler-based setup.

What we aim for here is different. We also need to obtain a very flexible setting of the sampling clock to accurately define the frequency ratio between the acquisition sampling clock and the generator frequency grid. Hence, we would rather go for a fractional N synthesizer, that contains a PLL, to realize the high-resolution clock frequencies that are needed for the measurements.

The instrument that is obtained after the transformation of the timebase is shown in Figure 12.17. Note that the clock frequency that is generated by the FracN synthesizer is much lower than the bandwidth of the RF signal. The spectrum of the discrete time signal measured by the DFT after the ADC will be periodically repeating over F_S. The measured discrete time signal is real-valued, and this calls for an auto-conjugate spectrum. All the spectral lines that are present in the RF signal will therefore fold down in the Nyquist band (frequency between 0 and $F_s/2$). As long as the different spectral lines do not overlap, we will therefore be able to reconstruct the original signal.

Aliasing: a narrow-band solution to broadband measurements

To make this folding process more practical and to develop the intuition of the reader who is not so familiar with aliasing and playing with sampling frequencies, we will consider a series of example measurements that show the capabilities of the sampling converter measurement. We will increase the complexity of the experiments gradually to show the potential of the method.

In the first example, we show the very basics of aliased measurements using a sine wave excitation for a PISPO system. Although this case might seem trivial, it enables us to show the properties of the aliased measurements. We can start from the sampling scope waveforms obtained above and look for an equivalent measurement obtained by the sampling converter.

The second example extends the sine wave result to the measurement of the response of a PISPO system to a modulated excitation signal. We use a narrow-band modulated signal to show the differences and the similarities between the original RF signal and its IF converted replica.

Finally, we briefly show that the conversion of a wideband multiple tone signal is also possible. The additional complication encountered here stems from the large difference in shape between the time-domain signal before and after the conversion. We show that the spectral lines of the RF signal can still be recovered, albeit that their frequencies will be completely scrambled. Frequency engineering allows us to unscramble the measured spectrum and to restore the original waveform.

The frequency engineering behind aliasing

Before we look at an example, it is important to take a look at the generation of the clock frequencies that are used in the setup. Assume that the reference clock (REF in Figure 12.17) is a $F_{ref} = 10\,\text{MHz}$ clock. This is a common value for the reference frequency that is used by an RF instrument. The fractional-N synthesizer (labeled FracN in Figure 12.17) transforms this reference frequency into the sample frequency F_s

$$F_s = \frac{p}{q} F_{ref} \qquad p, q \in \mathbb{N}. \tag{12.18}$$

The ratio p/q is designed such that its maximum value is modest (for example 2) but its resolution is very high (for example better than 1 Hz). Both the samplers and the ADCs of the instrument run from the same sampling frequency F_s. While this is not mandatory, it allows us to obtain a much cleaner explanation.

In a measurement, the samplers and the ADCs acquire N_{acq} successive data samples. The time span over which data are acquired is therefore $N_{avq}/F_s = N_{acq}T_s$. After taking a DFT, this results in a folded spectrum that has a spectral resolution that is equal to $\Delta F_{acq} = F_s/N_{acq}$.

The easy case: down sampling the PISPO response for a CW excitation

Now, we consider a CW measurement of a PISPO system to understand the basic operation of the frequency aliasing method. The GEN1 source in Figure 12.17 generates a sine wave of frequency F_{sine}. After passing the DUT, the output wave contains a number of harmonic components of that excitation. Assume for simplicity that harmonics are only present at $2F_{sine}$ and $3F_{sine}$. The spectral mask of the input and output waves is shown in Figure 12.18. We select the sampling frequency F_s such that the aliased wave measured by the ADC is a time-dilated version of the RF wave. The time dilatation g results in a

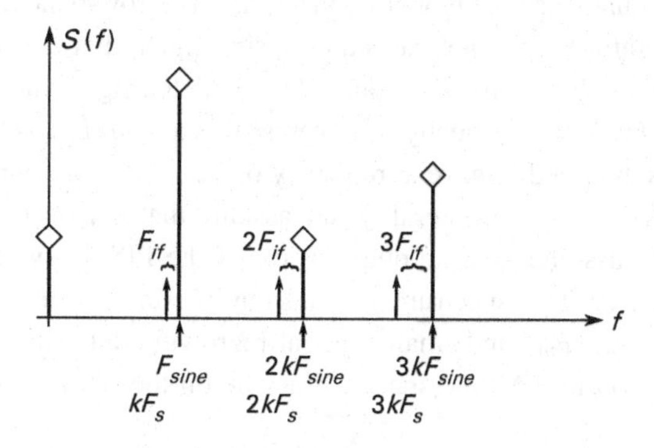

Fig. 12.18 Spectra of the DUT operating under CW excitation. The input sine wave frequency is F_{sine}. The conversion frequency and the IF frequency are shown for the different harmonics.

waveform that varies more slowly: $a_{sine}(t) = a_{IF}(gt)$, as was the case in Figure 12.15. The relation between the spectral lines in the RF wave (fundamental frequency F_{sine}) and the folded waveform (fundamental frequency F_{if} in Figure 12.18) can be expressed in the frequency domain. It can be reduced to:

$$\begin{cases} S(F_{IF}) & = S(F_{sine}) \\ S(2F_{IF}) & = S(2F_{sine}) \\ S(3F_{IF}) & = S(3F_{sine}) \end{cases}. \tag{12.19}$$

To obtain this, we select F_s and k such that $F_{sine} = kF_s + F_{IF}$. The fundamental IF frequency is chosen low enough to ensure that all the harmonics are converted to a frequency below the Nyquist frequency. Here, we choose $F_{IF} < \frac{F_s}{6}$. The spectral line at the fundamental frequency F_{sine} appears after folding in the discrete time spectrum at a frequency of F_{IF}. Since we convert the mth harmonic of F_{IF} with the same sampling frequency F_s, the harmonic response appears at a frequency of mF_{IF}, as requested. The measured response waveforms obtained by the sampling converter and the real-time oscilloscope are therefore identical up to a time-stretching.

To obtain measured waveforms that are free of spectral leakage, we must ensure that we always measure a complete period of the IF waveforms. This requires that the frequency of the folded (IF) lines lies on an integer multiple of the spectral resolution of the DFT:

$$F_{IF} = l\frac{F_s}{N} \qquad l \in \mathbb{N}. \tag{12.20}$$

Conclusion: The measurement of the response of a PISPO system to a sine wave excitation is as easy with a sampling converter based instrument as it is with a real-time oscilloscope. Both instruments will (ideally) measure the same waveform, without any distortion up to a linear scaling of the time axis. The time-domain shape of the RF and IF waveforms match perfectly. The spectra will also be equal up to the frequency compression factor.

Converting a narrow-band modulated signal

Now, let us complicate the situation a little more by considering a narrow-band modulated excitation signal. The generator GEN1 generates a multisine signal that consists of M spectral lines that are spaced ΔF_{ms} apart. The complete modulated signal has a modest bandwidth with respect to F_s. Mathematically, we express this as $M\Delta F_{ms} \ll F_s/2$. The first line of the multisine is assumed to have a frequency of F_{ms}. To avoid unnecessary notational clutter, and without loss of generality, we assume that $F_{ms} = k_{start}\Delta F_{ms}$ where again $k_{start} \in \mathbb{N}$. We assume that the output wave of the PISPO system under test is as shown in Figure 12.19. This spectrum contains out-of-band spectral distortion components around $2F_{ms}$ and $3F_{ms}$ and small spectral regrowth contributions (small white rectangles around F_{ms}) only. All the spectral lines lie on the same spectral grid that is spaced ΔF_{ms} apart.

The sampling process again compresses all the spectral lines that are present in the RF spectrum in an IF band ranging from $F = 0$ to $F = F_s/2$ using the relation $F = kF_s + F_{if}$.

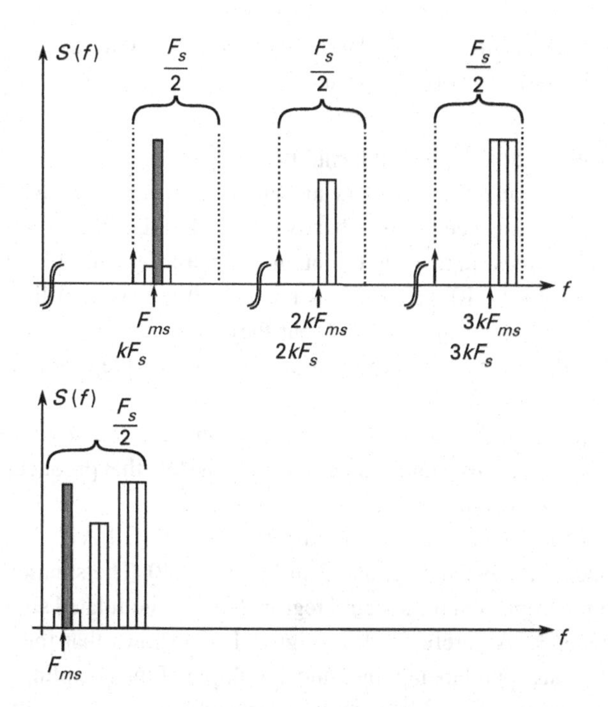

Fig. 12.19 Spectral mask of the modulated output signal. The modulated input waveform is drawn in dark gray.

The narrow bandwidth of the modulation spectrum assures that both the modulation and the spectral regrowth around the fundamental frequency F_{ms} are all aliased using the same value of k. Hence, the modulated lines around each carrier appear in the same order and with the same spacing in the IF and in the RF spectrum. The same is true for the modulation around the harmonics at $2F_{ms}$ and $3F_{ms}$. This folding process is illustrated in the frequency domain in Figure 12.19.

The IF spectrum obtained by the sampling converter is no longer exactly equal to the spectrum of the real-time scope up to a compression of the frequency axis. The frequency grid of the modulation ΔF_{ms} is the same in the IF and the RF spectra. However, the frequency spacing between the fundamental and its harmonics has been compressed. This means that the time waveforms of the IF and the RF signal are no longer the same; but, the waveforms are not totally different either. One can show that these waveforms share the same envelope. Again, we obtain a leakage-free measurement whenever we measure an integer number of periods of the IF wave.

Conclusion: The measurement of the response of a PISPO system to a narrow-band modulated excitation with distortion no longer yields exactly the same waveform for a sampling converter and a real-time oscilloscope. The instruments will (ideally) measure different signals that share the same envelope, without any distortion. The measured spectra will (ideally) be exactly equal around each carrier, but the spacing in between the carriers will be compressed. One can exactly reconstruct the RF signal based on the measurements of the sampling converter if the spacing between the carriers is restored to its original value.

Converting a wideband multiple tone signal with harmonics

In the third example, we consider the case of a wideband modulation [22]. The bandwidth of the modulation signal now exceeds the IF bandwidth of the ADC. In this case, things are more involved. It becomes impossible to obtain a spectrum for the IF wave that is identical or even similar to the RF spectrum as before. All the spectral lines that are present in the RF spectrum are still measured, but they no longer appear in the same order in the IF and the RF spectra. We can illustrate this on a simple example, again to avoid notational clutter.

Consider a multisine signal that consists of three lines that are spaced more than the sampling frequency F_s apart. This signal excites a PISPO system that produces a second and a third harmonic response only.

For the sake of illustration, we assume that the lines are spaced $\Delta F_{ms} = F_s + \delta_{ms}$ apart. The excited lines are labeled 1, 2, and 3 in Figure 12.20. The smaller diamond labeled lines in the figure represent the spectral regrowth of the modulated signal around the fundamental frequency. A careful look at Figure 12.20 shows that the frequency difference between the diamond labeled lines and a multiple of the sampling frequency F_s differs from line to line. After folding, all the diamond labeled lines will therefore appear at a different IF frequency.

The spectral lines that lie around $2F_s$ are shown in Figure 12.21. The lines that lie around the fundamental are superimposed on this plot artificially to show the interdependence between the IF frequencies. The overlay is constructed such that if a fundamental line is located at $f_{fund}(k) = kF_s + \delta_k$, this line is overlaid with $f_{2F_s}(k) = 2kF_s + \gamma_k$. Again, a closer look at Figure 12.21 shows that the spectral lines that are included in the second harmonic response will fold to other frequencies than the lines around the fundamental frequency.

The result of the complete frequency folding is illustrated in Figure 12.22. Note that all the diamond labeled lines appear twice in the measured spectrum: once at a frequency f and once at a frequency $F_s - f$. For the excited lines this was made visible by the labels 1, 2, and 3 and mirrored 1, 2, and 3. The second contribution is always the complex conjugate

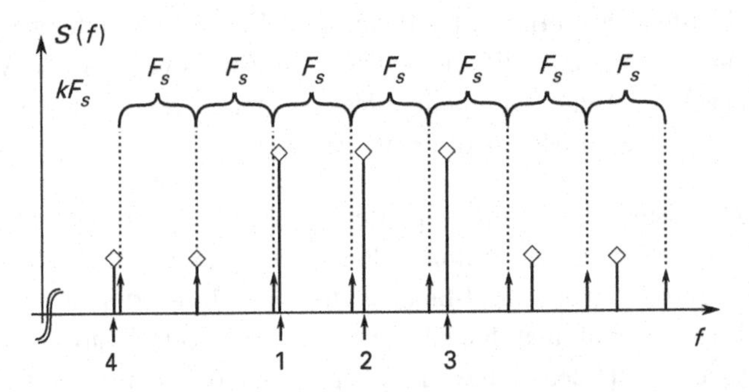

Fig. 12.20 The response measured around the fundamental frequency F_{ms}. Excited lines are labeled $1, 2, 3$.

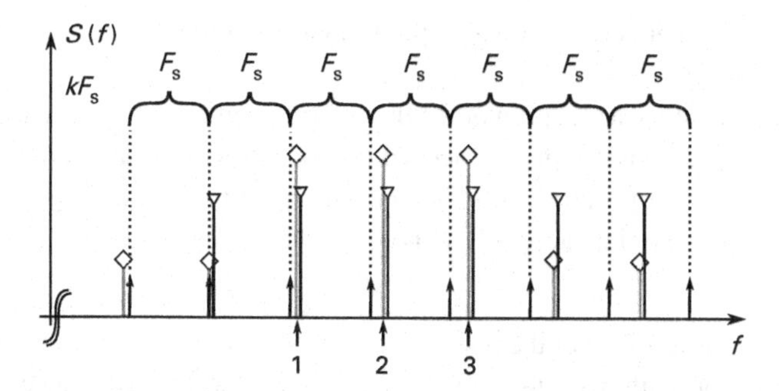

Fig. 12.21 The response measured around twice the fundamental frequency $2F_{ms}$. The fundamental lines are superimposed on this spectrum, such that kF_s overlaps $2kF_s$. This shows the difference in frequencies between the fundamental and second harmonic response.

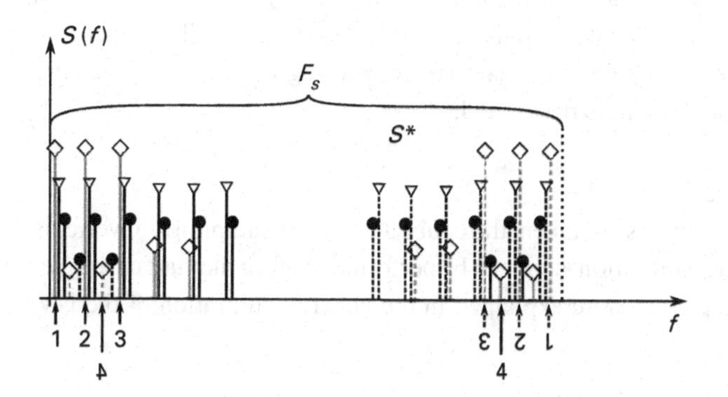

Fig. 12.22 The IF spectrum of the wideband multisine. Lines labeled with a diamond lie close to the fundamental F_{ms}, triangle labeled lines lie close to $2F_{ms}$ and dot labeled lines lie around $3F_{ms}$. Dashed lines represent the complex conjugate of corresponding full lines.

of the first one. This additional mirroring symmetry around the Nyquist frequency $F_s/2$ is imposed by the real-valued character of the time signals.

There is one additional important detail that is visible in Figure 12.22. Some spectral lines, such as line 4, initially fold to a frequency that is larger than the Nyquist frequency $F_s/2$. Of course, these lines also appear mirrored and complex-conjugated in the lower half of the IF band (line with mirrored 4). If the spectrum is to be measured without errors, this complex conjugate has to be accounted for in the reconstructed measured spectrum.

Hence the measurement still contains all the spectral lines contained in the original signal. A proper choice of the sampling frequency can still impose that the spectral lines that have different RF frequencies end up at different IF frequencies too. Of course, this does not come free. The price that one has to pay is that the spectral resolution that is required in the IF domain is increased. Longer measurement records are therefore to be acquired and processed. In the example considered here, we have to provide room for three more lines in between the lines of the fundamental tones, which increases the resolution by a factor of 4.

Finally, we can now show a blueprint of the IF spectrum that is measured. It is clear that a significant amount of housekeeping is needed to reconstruct the waveforms properly. The IF lines have to be repositioned to their original position in the original RF grid (taking complex conjugates into account) if one is to obtain a distortion-free time domain waveform.

Conclusion: The measurement of the response of a PISPO system to a broadband modulated excitation with distortion does not directly yield the same spectrum for a sampling converter and a real-time oscilloscope. The spectra measured by the two instruments still contain the same spectral lines if a proper sampling frequency is selected to avoid overlap of the RF lines in the IF domain. The time-domain waveforms obtained by a direct transformation of the IF and RF spectra to the time domain will be totally different however. The original RF time waveform can still be (perfectly) reconstructed if each spectral line is carefully replaced (watch out for the complex conjugates) to the frequency it originated from before the conversion to the time domain is performed.

Calibration issues

Once the measurements are taken, the calibration of the sampling converter is pretty easy to realize. All the calibration steps can be performed before the start of the measurements, and only the steps that were explained in the general calibration of the NVNA have to be taken.

Conclusion

The sampling converter has the advantage that it is freed from the timebase problems and the consecutive spectral leakage problems that can be present in the oscilloscope-based receiver. However, this comes at a price. For sine wave measurements and modulated measurements of low bandwidth (with respect to the sampling frequency) the sampling frequency can be selected such that the sampled waveforms are time-stretched copies of the RF waveforms. When a wideband modulation is applied, the measurement remains possible, but comes at the cost of increased housekeeping to determine the origin of the measured spectral lines.

12.5.3 VNA-based setups

Up to now, the setups have all been measuring directly in the time domain or the equivalent time domain. Put in a nutshell, this means that all the spectral lines that are present in the signal are acquired simultaneously. The main advantage of this way of working lies in the fact that the different lines are synchronized by construction of the measurement.

This situation changes completely whenever a linear VNA is used to acquire the data [23, 24]. A VNA can be conceptualized as a frequency-domain acquisition device. A modern VNA measures the four waves of a two-port device simultaneously, but does so for one frequency at a time. The measurements that are taken at different frequencies are not synchronized. Put in a different way, one can state that there is no common time reference for the measurements that are taken at the different frequencies. Formally speaking, this means that the measurements contain an additional phase indeterminacy,

$$\begin{bmatrix} A_1(\omega_k) \\ B_1(\omega_k) \\ A_2(\omega_k) \\ B_2(\omega_k) \end{bmatrix} = e^{j\varphi(\omega_k)} H_e(\omega_k, \omega_1) \begin{bmatrix} A_1(\omega_1) \\ B_1(\omega_1) \\ A_2(\omega_1) \\ B_2(\omega_1) \end{bmatrix}. \tag{12.21}$$

In this expression, $\varphi(\omega_k)$ is an unknown phase-shift that varies in a random way from measurement to measurement and hence from frequency to frequency. The complex matrix $H_e(\omega_k, \omega_1) \in \mathbb{C}^{4\times 4}$ describes the exact relationship between the waves. This means that this is the relationship that would be obtained if the measurement were performed by an ideal real-time scope. The suffix one in ω_1 labels one test frequency that serves as a reference. This reference can be freely chosen. It is clear that the phase-shift φ_k is unimportant for S-parameter measurements. Whenever a ratio of waves is taken at one single frequency, this term disappears.

Whenever the measured quantity contains spectral rays at different frequencies, this phase must be known. As an example, consider again the amplifier as above. To determine the third-order intermodulation contribution at the third harmonic one calculates the influence of the third-order term:

$$e^{j\varphi(3\omega_0)} B_2(3\omega_0) = e^{j\varphi(\omega_0)} H A_1^2(\omega_0) A_1^*(\omega_0), \tag{12.22}$$

with H the complex gain associated with the Volterra kernel. The phase of the intermodulation product is scrambled by a different unknown phase term in the right and the left hand sides of the equation. This results in a phase indeterminacy of the intermodulation components.

How can we get around this? If we could have a trigger signal that is common to all the single frequency experiments, the problem would be solved. To see this, assume that the trigger signal consists of a perfect Dirac impulse that indicates the start of the period of the generated wave. Assume also that the internal sources of the VNA are perfectly synchronized to that Dirac impulse. A VNA measurement is then only started when the signal arrives at exactly the same point in the modulated waveform. As a result, $\varphi(\omega_k) = 0$ for each test frequency and the measured spectra coincide perfectly with the exact ones.

Now we can translate this idea into a real device: the synchronizer. This device generates a periodic impulse train out of a periodic signal. As far as the system is concerned, there are a number of alternatives that could be tempting when designing a synchronizer. The first and most obvious choice is to use a comb generator. This idea was first successfully implemented in [23]. When a step recovery diode [25] is fed by a sine wave, it will ideally generate a large impulse at a fixed position along the sine wave. When tuned and packaged properly, this generates tens to hundreds of harmonics over a bandwidth that is wide enough to cover current instrumentation needs. A second possibility is to use a nonlinear transmission line. Proper design allows the steepening of one of the edges

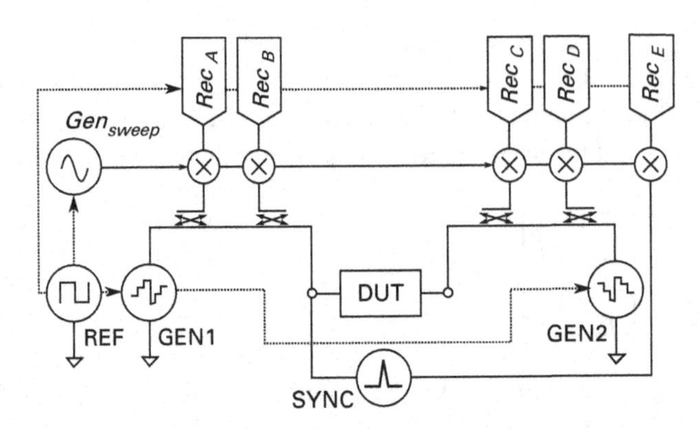

Fig. 12.23 The VNA-based setup.

(positive or negative going edge) of the applied signal to extremely high slew rates. The advantage of this approach is that these nonlinear transmission lines can easily be matched. This enables us to use input waveforms that vary over a bandwidth that is much larger than the input bandwidth of the comb generator. A third possible alternative is to use an extremely wideband pulse-shaping amplifier that conceptually acts as an ideal comparator for the applied input signal. From a user perspective, the actual implementation does not matter, as the only demand is that a known spectral line is present at each frequency of interest.

The synchronizer needs to be fed by a reference signal that is somehow related to the period of the input signal generated by the source. If the NVNA uses the built-in sources of the VNA, this is quite easy to realize. One of the sine wave sources is then split and fed in parallel to the synchronizer and the DUT, as shown in Figure 12.23. When a modulated signal is used as an excitation, the synchronization is to happen on the period of the modulated signal.

Calibration issues

The phase calibration of the VNA-based setup follows the general case as is explained in Section 12.4.3. It requires a second synchronizer besides the "measurement synchronizer" that is shown to be needed for the measurement itself. The second signal is needed because the measurement synchronizer is connected to an additional receiver path. This path is not calibrated during the calibration of the VNA. The measurement of the additional reference signal enables the compensation for all the phase errors of that additional signal path.

Conclusion

The VNA-based setup has the advantage that it can rely on a very large installed base of network analyzers. As the VNA is the most popular RF measurement device of the last half century, it also can count on an RF community that is well skilled and feels comfortable using it. However, the VNA-based setup comes at a cost: it requires the presence of an additional receiver and a synchronizer device. This device generates a reference signal,

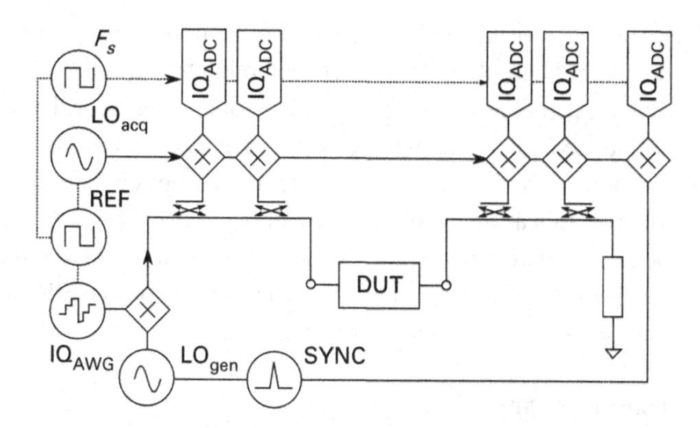

Fig. 12.24 The IQ modulation-demodulation based setup.

hence it is an active device and it needs regular re-calibration to keep the accuracy of the measurement. The disadvantage is that a malfunctioning of the synchronizer is hard to notice in normal operation and therefore it remains a potential source of measurement errors.

12.5.4 IQ-modulator based setups

The last setup in the list mimics the architecture of a receiver/transmitter pair to realize a measurement device [26,27]. The structure of the setup is shown in Figure 12.24. The main idea behind this setup is to measure modulated signals with a moderate modulation bandwidth. The generator consists of an IQ-modulated source. The IQ-modulated signal can be expressed as a complex time signal as

$$a_{1c}(t) = m(t)\, e^{j\omega_c t}, \tag{12.23}$$

where the modulation envelope of the signal is called $m(t)$, while the carrier of the IQ signal is labeled $c_{gen}(t)$.The signal that is fed to the DUT is then

$$a_1(t) = \Re(a_{1c}(t)) + \Im(a_{1c}(t)). \tag{12.24}$$

This signal passes through the DUT. Its output $b_2(t)$ is then measured and is demodulated by the signal $c_{acq}(t)$. Ideally, the carrier signals at the generation and the acquisition are equal $c_{acq}(t) = c_{gen}(t)$. For some generators, the presence of a static phase error cannot be avoided in practice and $c_{acq}(t) = c_{gen}(t)\, e^{j\varphi_{acq}}$. This is assumed in the remainder of this section.

To make things more practical, we again use an example to make the argumentation more concrete. We use the same example as for the VNA setup. We measure the third-harmonic response of the simple PISPO system defined before, and obtain:

$$e^{j\varphi_{acq}(3\omega_{gen})} B_2(3\omega_0) =^{j\varphi_{acq}(\omega_{gen})} H\, A_1^2(\omega_0)\, A_1^*(\omega_0), \tag{12.25}$$

for any spectral line $B_2(3\omega_0)$ located in the third-harmonic frequency band and $A_1(\omega_0)$ located in the fundamental band. Any time a new demodulation frequency is selected, the phase difference φ_{acq} can change randomly. This is due to the hybrid nature of the IQ instrument: time and frequency domain approaches are indeed combined in a single setup. When the IQ-modulation frequency of the generator or the acquisition is changed, a random phase error can appear in a way that is identical to the VNA setup, unless the generator and the acquisition are made phase-coherent by construction. All the spectral rays in the demodulated envelope are automatically synchronized because the acquisition behaves as a real-time scope for the complete envelope signal.

Phase calibration considerations

This kind of measurement setup simplifies the phase calibration of modulated measurements quite a bit. The phase calibration of a modulated wideband signal is hampered by signal-to-noise ratio issues: each line that can potentially be measured has to be energized on a frequency grid that has a line density that is set by the frequency resolution of the modulated signal. A huge number of spectral lines that will never be used have to be present in the reference signal anyway to maintain general applicability. In the case of the IQ modulator, this is no longer the case. The calibration of the envelope measurements requires a timebase correction and a calibration of the acquisition in the IF domain. The calibration between the harmonics calls for a comb generator that only needs to carry energy at the harmonics of the fundamental IQ-modulation frequency. Of course, like the VNA, the IQ modulator requires an additional measurement channel to measure this reference signal simultaneously with the other acquired waves.

Conclusion

The IQ-modulator setup attempts to combine the advantages of the frequency and the time domain approaches. It avoids the timebase problems and the triggering problems of the RF oscilloscopes, but the ability to measure the complex envelope in the time domain makes the measurements quite easy to set up. The analysis bandwidth of the envelope is narrow enough to allow direct conversion to a discrete time signal in the envelope domain and thus this device avoids the complications of the sampling receivers with spectral folding. It also avoids the frequency by frequency sweep inherent in the frequency-domain methods.

To measure the response between the fundamental and harmonic bands, the IQ-modulator uses a frequency-domain approach of sweeping the frequency and therefore also needs a synchronizer if the phase relation of the carrier and its harmonics is to be measured. This increases the complexity of the setup and requires an additional receiver, exactly as with the VNA setup.

12.6 Conclusion, problems, and future perspectives

We arrive now at the end of this bird's eye view of the measurement of the nonlinear PISPO systems operating at microwave frequencies. Of course, there is much more to tell and there are many more issues and problems. They are discussed extensively in the literature, in a stream of new and exciting papers. Our goal here is neither to be complete nor encyclopedic. Triggering the curiosity of the reader and providing a glimpse into the world of nonlinear characterization using the NVNA seems already very ambitious.

Where to go from here? It is our belief that we are now at a point where the basic measurement capability that allows us to characterize the most commonly used systems is indeed present.

What are the next stepping stones and where will they lead to? Of course, prediction of the future is hard and a matter of opinion. Our belief is that the NVNA will become really useful in the next decade, as it slowly finds its way into mainstream RF design. An NVNA-like device will allow us to port and extend the available nonlinear design theory into practice. It will allow us to close the design loop from a nonlinear point of view. However, there are a number of very challenging issues that remain to be solved. There is a lot more work needed to demonstrate the usefulness of the nonlinear information in a design world that is still mainly dominated by S-parameters.

References

[1] S. F. Adam, "*Microwave Theory and Applications.*" Englewood Cliffs, NJ: Prentice-Hall, 1969.

[2] D. Rytting, "Network analyzers from small signal to large signal measurements," *PROC. 67th ARFTG Conference, 2006* , pp. 11–49, 16-16, June 2006.

[3] Y. Rolain, W. Van Moer, G. Vandersteen and J. Schoukens, "Why are nonlinear microwave systems measurements so involved?" *IEEE Trans. Instrum. Meas.*, vol. 53, no. 3, pp. 726–729, June 2004.

[4] P. Wambacq, "*Symbolic analysis of large and weakly nonlinear analog integrated circuits,*" Ph.D. dissertation, K.U. Leuven, Belgium, Jan. 1996.

[5] M. Schetzen, "*The Volterra and Wiener Theories of Nonlinear Systems.*" New York: Wiley, 1980.

[6] S. A. Maas, "*Nonlinear Microwave Circuits*". New York: Wiley, 1988.

[7] W. Van Moer and Y. Rolain, "Measuring the sensitivity of microwave components to bias variations," *IEEE Trans. Instrum. Meas.*, vol. 53, no. 3, pp. 787–791, June 2004.

[8] A. Alghanim, J. Benedikt and P. J. Tasker, "Investigation of electrical base-band memory effects in high-power 20W LDMOS transistors using IF passive load pull," *3rd International Conference on Information and Communication Technologies*, vols 1–5, Syria, 2008.

[9] E. Oran Brigham, "*The Fast Fourier Transform.*" Englewood Cliffs, NJ: Prentice-Hall, 1974.

[10] J. Schoukens and R. Pintelon, "*Identification of Linear Systems.*" London, U.K.: Pergamon Press, 1991.

[11] M. Sipila, K. Lehtinen and V. Porra, "High-frequency periodic time domain waveform measurement system," *IEEE Trans. Microw. Theory Tech.*, vol. 36, no. 10, pp. 1397–1405, Oct. 1988.

[12] G. Kompa and F. van Raay, "Error-corrected large-signal wave-form measurement system combining network analyser and sampling oscilloscope capabilities," *IEEE Trans. Microw. Theory Tech.*, vol. 38, no. 4, pp. 358–365, Apr. 1990.

[13] A. Ferrero and U. Pisani, "Two-port network analyzer calibration using an unknown 'thru'", *IEEE Microwave Guided Waves Lett.*, vol. MGWL-2, pp. 505–507, Dec. 1992.

[14] J. Verspecht, "Broadband sampling oscilloscope characterization with the nose-to-nose calibration procedure: A theoretical analysis," in *Proc. 1994 IEEE Instrumentation Measurement Technology Conf.*, Hamamatsu, Japan, May 1994, pp. 526–529.

[15] J. Verspecht and K. Rush, "Individual characterization of broadband sampling oscilloscopes with the nose-to-nose calibration procedure," *IEEE Trans. Microw. Theory Tech.*, vol. 43, no. 2, pp. 354–374, Apr. 1994

[16] G. Vandersteen and R. Pintelon, "Maximum likelihood estimator for jitter noise models for HF sampling scopes," *IEEE Transactions on Instrumentation and Measurement*, vol. 49, no. 6, pp. 1282–1284, Dec. 2000.

[17] P. D. Hale, C. M. Wang, D. F. Williams, K. A. Remley and J.D. Wepman, "Compensation of random and systematic timing errors in sampling oscilloscopes," *IEEE Transactions on Instrum. Meas.*, vol. 55, no. 6, pp. 2146–2154, Dec. 2006.

[18] W. Van Moer, Y. Rolain and A. Geens, "Measurement based nonlinear modeling of spectral regrowth," *IEEE Trans. Instrum. Meas.*, vol. 50, no. 6, pp. 1711–1716, Dec. 2001.

[19] W. Van Moer, Y. Rolain and J. Schoukens, "An automatic harmonic selection scheme for measurements and calibration with the nonlinear vectorial network analyser," *IEEE Trans. Instrum. Meas.*, vol. 51, no. 2, pp. 337–341, Apr. 2002.

[20] W. Van Moer and Y. Rolain, "Proving the usefulness of a 3-port nonlinear vectorial network analyser through mixer measurements," *IEEE Trans. Instrum. Meas.*, vol. 52, no. 6, pp. 1834–1837, Dec. 2003.

[21] K. M. Youngseo, P. Roblin, Myoung Sukkeun, J. Strahler, F. De Groote, and J. P. Teyssier, "Multi-harmonic broadband measurements using a large signal network analyzer," *PROC. 75th ARFTG*, pp. 1–6, 28-28, May 2010.

[22] M. El Yaagoubi, G. Neveux, D. Barataud, T. Reveyrand, J.-M. Nebus, F. Verbeyst, F. Gizard, and J. Puech, "Time domain calibrated measurements of wideband multisines using a large-signal network analyzer," *IEEE Trans. Microw. Theory Tech.*, vol. 56, no. 5, pp. 1180–1192, May 2008.

[23] U. Lott, "Measurement of magnitude and phase of harmonics generated in nonlinear microwave two-ports," *IEEE Trans. Microw. Theory Tech.*, vol. 37, no. 10, pp. 1506–1511, Oct. 1989.

[24] P. Blockley, D. Gunyan and J. B. Scott, "Mixer-based, vector-corrected, vector signal/network analyzer offering 300 kHz-20 GHz bandwidth and traceable phase response," IEEE *MTT-S International Microwave Symposium Digest*, pp. 4 pp., 12-17 June 2005.

[25] "Microwave Components 0.001-40 GHz," Herotek, Datasheets [Online]. Available: www.herotek.com.

[26] M. Thorsell and K. Andersson, "Fast multiharmonic active load–pull system with waveform measurement capabilities," *IEEE Trans. Microw. Theory Tech.*, vol. 60, no. 1, pp. 149–157, Jan. 2012.

[27] H. Jie, K. G. Gard, N. B. Carvalho and M. B. Steer, "Dynamic time-frequency waveforms for VSA characterization of PA long-term memory effects," *ARFTG Conference, 2007 69th*, pp. 1–5, 8-8, June 2007.

[28] T. Van den Broeck and J. Verspecht, "Calibrated vectorial nonlinear network analyzer," in *Proc. IEEE MTT-S*, San-Diego, CA, 1994, pp. 1069–1072.

13 Load- and source-pull techniques

Valeria Teppati, Andrea Ferrero, and Gian Luigi Madonna

13.1 Introduction

Chapter 12 has explained how S-parameter measurements in small-signal conditions are not adequate to characterize an active device for a relevant number of applications. In large-signal conditions the performance of the active device under test (e.g. output power, gain, and efficiency) does not only depend on the chosen bias point and the excitation frequency, but also on the input power level and the loading conditions at the input and output ports.

The "Rieke diagrams" were already used in the early 1940s to show how the performances of microwave tubes and oscillators vary as a function of load impedance [1]. The oscillating frequency and output power used to be plotted on the Smith chart as contours at constant level. The modern term of "load-pull" comes from these early applications, where a change of the load impedance would have "pulled" the output oscillator frequency.

More precisely, the term "source and load-pull systems" refers nowadays to the set of instrumentation (at microwaves and at lower frequencies) needed to address two main issues:

- setting and monitoring the DUT loading conditions,
- measuring the DUT performance of interest.

Chapter 12 has already described in detail the most modern large-signal measurement techniques to address the second point. However, it will be clear in this chapter that the two problems are strictly correlated, as the solutions chosen for the former have a considerable impact on the latter. In particular, considerations of cost and test duration, complexity of the measurement setup, as well as the availability of equipment and trained personnel might pose strong constraints on the load-pull techniques that can be used. Eventually, the choice of the final solution needs to balance these aspects with the required load-pull functionalities and accuracy.

Complete and automated source and load-pull systems appeared in the early 1970s [2,3]. They are today able to quickly characterize microwave devices under large-signal excitation and they are used for a wide range of applications:

- matching network design, to achieve optimal performance from a device operating in nonlinear conditions (typically, for amplifier or oscillator design);

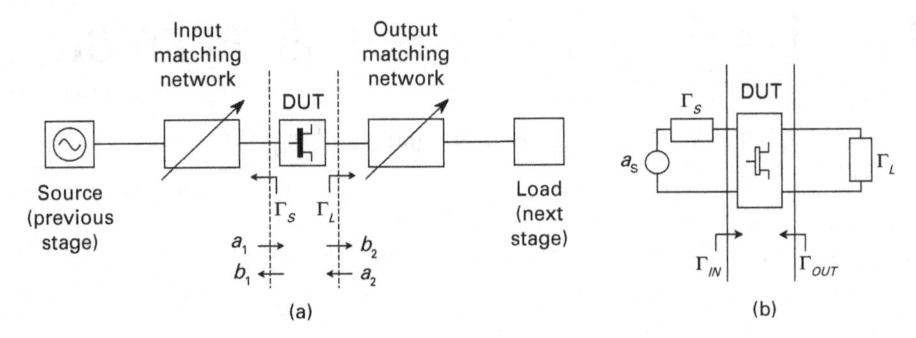

Fig. 13.1 Definition of the quantities of interest. ©2001 IEEE. Reprinted, with permission, from [7].

- large-signal device modeling, where load-pull data is used directly in a behavioral model, or indirectly extract and tune a device model;
- technology process development, where load-pull measurements are used to determine the effects of varying process parameters on device large-signal performances;
- device reliability study and identification of failure conditions (e.g. ruggedness testing of modules, where the device is measured with a specified output VSWR at all phases);
- quick device evaluation for production, where specific load and source impedances are set for a single pass/fail test.

This list is not exhaustive, as source and load-pull can be in general applied to the evaluation of any nonlinear device. Further examples can be found in literature for mixer design [4], oscillator measurements [5], and diode characterization [6].

To explain the basic principle, let us consider the situation in Figure 13.1(a). It refers to the typical example of an amplifier design, where a microwave transistor is connected to the previous and following stages by two linear and *tunable* matching networks.

The basic design target is to find the proper load reflection coefficients Γ_L and Γ_S to meet the required specifications. If the transistor operated in linear conditions, the solution could be found with the only knowledge of its S-parameters. For instance, the maximum output power would be achieved by designing the output matching network so that $\Gamma_L = \Gamma_{OUT}^*$. If the transistor were considered unilateral (i.e. $S_{12} = 0$), then the condition would simply be $\Gamma_L = S_{22}^*$.

In large-signal conditions, the transistor nonlinearities play a fundamental role and the optimum loading conditions may be significantly different from the linear case [8], [9]. Load-pull systems allow finding the proper load values experimentally.

In the simplest implementation, the active device is driven by the microwave source at a single frequency and its performances are measured while physically changing Γ_L (for *load-pull*) or Γ_S (for *source-pull*). The monitored quantities are typically:

- input and output power (P_{IN} and P_{OUT});
- operating and transducer gain (G_{OP} and G_T), along with the corresponding compression;
- PAE or drain efficiency;
- noise figure and noise parameters.

Moreover, by driving the device with two tones or properly modulated signals, intermodulation or ACPR measurements can be performed to investigate the linearity of the amplifier.

In the simplest case, it is sufficient to control the source or load impedance on a relatively small bandwidth around the excitation frequency. This is true for a relatively large number of applications. However, if the active device is driven into strong nonlinear conditions (e.g. as in the case of high-PAE amplifiers [10, 11]), the spectral content of the output signal can be relevant at harmonic frequencies, too; the corresponding load conditions can significantly affect the device performance [12, 13], and *harmonic source and load-pull* systems are therefore used to experimentally investigate these effects.

A first classification of the source/load-pull systems refers to the techniques used to control the reflection coefficients Γ_S and Γ_L. *Passive* load-pull systems use mechanically tunable elements (the so-called "tuners"), while *active* systems synthesize the desired loads electronically. Reflection coefficients with magnitude up to unity can be reached at the DUT ports by the active techniques, thus overcoming the limitations of the passive tuners due to fixture and probe losses (for details on fixturing and probing issues see Chapter 2). Passive tuners are described in Section 13.2, while active load-pull systems are the subject of Sections 13.3 and 13.4. Fundamental and harmonic load tuning are discussed, along with the advantages that combining passive and active tuners can have for some specific applications.

A second classification of the load-pull techniques refers to the DUT measurement principle. *Non-real-time* techniques are typically simpler; they are used only with passive systems and they are based on the tuner pre-characterization. *Real-time* systems can exploit all types of loads – active, passive, or a combination of them – and their measurement accuracy does not rely on the mechanical repeatability of the tuner position. The peculiarities of real-time and non-real-time systems are described in detail in Section 13.5.

Section 13.6 focuses on the most recent advances in the load-pull technology, which combine harmonic load-pull, mixed-mode signals, and time domain waveform measurement. These techniques require test setups that are usually not available off-the-shelf; nevertheless, they are becoming increasingly important in the R&D labs for accurate nonlinear device modeling and to assist in the design of advanced microwave nonlinear circuits.

One final remark: in this chapter we generically refer to *load-pull* systems. *Source-pull* measurements use, in principle, the same techniques to synthesize the source reflection coefficient; however, the presence of a generator term in the source equivalent circuit poses additional challenges in the accurate measurement of the source reflection coefficient Γ_S. This is the topic of Section 13.7.

13.2 Setting the load conditions: passive techniques

Passive techniques are based on mechanical, tunable devices (usually referred to as *tuners*) to generate the required impedance at the ports of the DUT.

13.2.1 Basics

The most common tuner type is the "slide screw" tuner. As shown in Figure 13.2, it is based on a slab line, consisting of two parallel ground planes with a center main line, plus a reflective element (a conductive "probe"). When the tuner is used as the output matching network of Figure 13.1(a), the load reflection coefficient Γ_L is controlled by setting the position of the tuning probe along the longitudinal and vertical axes of the slab line. When the probe is completely retracted, the line impedance – typically 50 Ω – is not perturbed. When the probe moves closer to the main line, an impedance step is introduced, corresponding to the probe length. In particular, the presence of the probe reduces the impedance of the corresponding slab-line portion.

The mismatch introduced by the probe peaks at the frequency where the probe length corresponds to $\lambda/4$, where λ is the wavelength in the slab-line. In order to increase the frequency range, commercially available tuners [14, 15] typically provide two or three probes.

In a very first approximation, changing the probe position along the vertical direction causes a change of the reflection coefficient magnitude, while the movement along the longitudinal axis changes the reflection coefficient phase. The reflection coefficient control is no longer straightforward if two or more probes are simultaneously used, due to their combined effect on the slab line.

The movement can be manually set by micrometer positioners, or it can be automatically controlled by precise stepper motors. Manually controlled tuners are usually simpler and cheaper. Automated tuners, however, allow reduced measurement time and greater accuracy thanks to the precise stepped motors, making them nowadays the preferred solution [16].

Ideally the probes should be non-contacting, touching neither the ground planes nor the center conductor. This enables the motors to move the probes quickly and precisely, with no perceptible wear or drift over time, thus providing longer tuner life and excellent

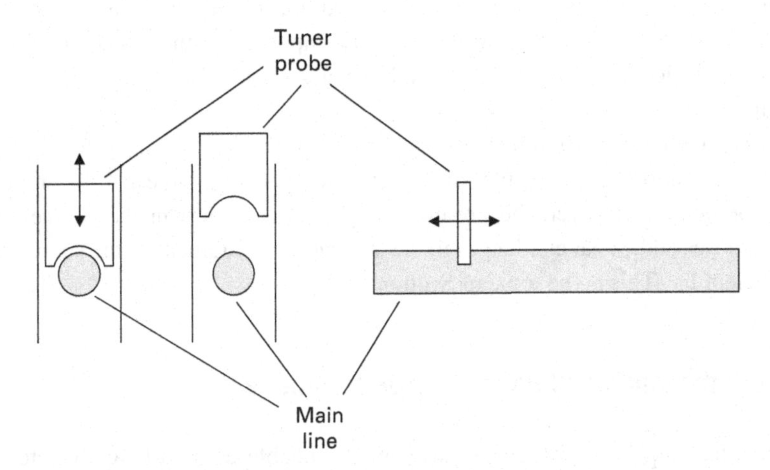

Fig. 13.2 A slide screw tuner, with a conductive probe moving in two directions within a slab line.

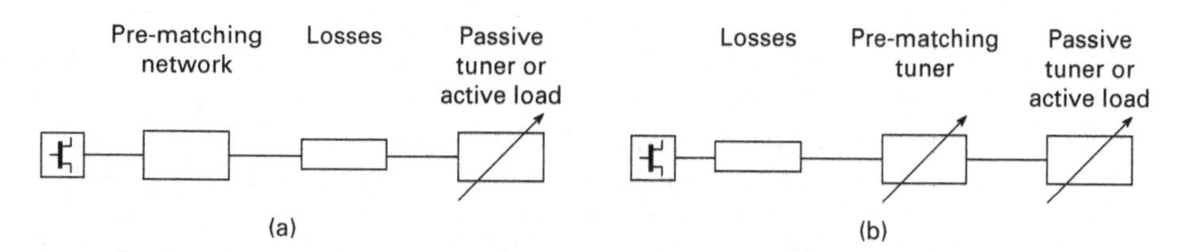

Fig. 13.3 Pre-matching at device level (a) or at measurement system level (b).

repeatability. Some mechanical tuners realize a sliding contact between the probe and the ground planes. Contacting probes are easier to design, but at the cost of shorter tuner life, slower operation, and worse repeatability.

Due to the tuner's intrinsically passive nature, the synthesized reflection coefficients are limited in magnitude by the unavoidable losses of the test setup (due to cables, on-wafer probes, etc.). Highly reflective loads cannot be realized at the DUT reference planes, especially in the on-wafer environment. For instance, consider that an insertion loss of 1 dB between the tuner and the probe tip transforms an ideal $|\Gamma| = 1$ at the tuner output into a $|\Gamma| = 0.8$ at the DUT port. Even if tuner losses are completely removed, the actual DUT load can be unsuitable for highly mismatched devices, especially at higher frequencies (where losses are larger), thus precluding the investigation of interesting regions of the Smith chart.

In order to overcome the problem, some fully passive solutions are available [17], based on pre-matching networks at the device level – as shown in Figure 13.3(a) – or on pre-matching tuners at the measurement system level – as shown in Figure 13.3(b). The most recent versions of the mechanical passive tuners, integrate programmable pre-matching capabilities [18]. As shown in Figure 13.3, in both pre-matching configurations it is possible to use an active load instead of a tuner. This is treated in Sections 13.3 and 13.4.

13.2.2 Harmonic load-pull with passive tuners

As mentioned in Section 13.1, harmonic load-pull involves the control of the load reflection coefficient at a finite (usually small) number of harmonically related frequencies. Figure 13.4 shows three different methods to use passive tuners for harmonic load-pull, based on [19, 20], namely:

- cascaded tuners,
- triplexers with normal tuners,
- stub resonators.

The first solution refers to Figure 13.4(a): it simply cascades different tuners to increase the number of degrees of freedom and allow fundamental and harmonic load tuning. It increases the bench complexity and losses, but at the same time exhibits a greater flexibility for frequency control and does not need any special hardware. Obviously, no independent harmonic control is possible since the movement of one of the tuners affects the impedance at both fundamental and harmonic frequencies.

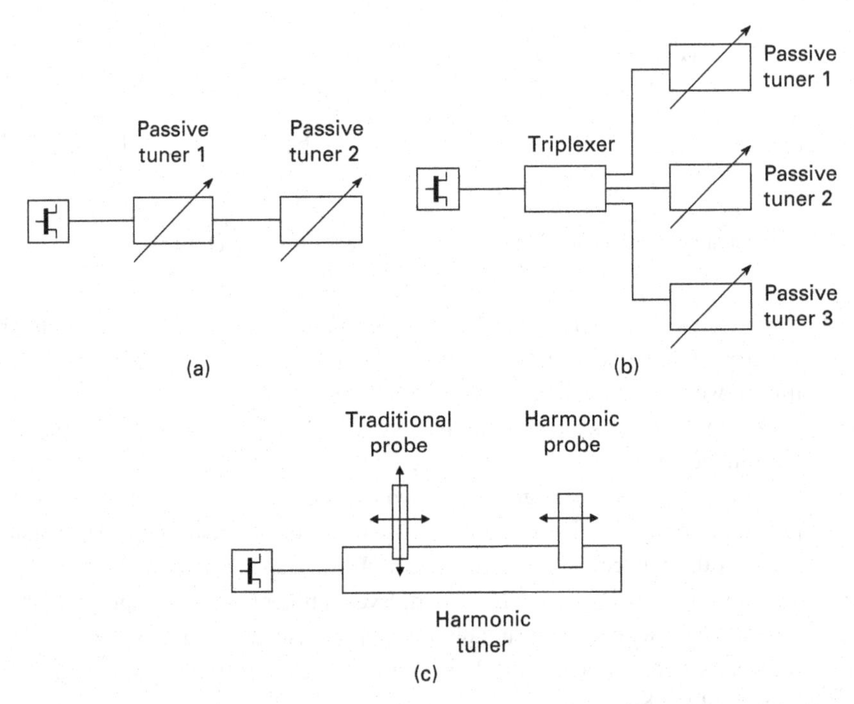

Fig. 13.4 Different types of harmonic tuning with passive tuners. Cascaded tuners (a), triplexer solution (b), and stub resonator technique (c). ©2007 IEEE. Reprinted, with permission, from [21].

The second solution is shown in Figure 13.4(b). It uses different traditional tuners for each harmonic, with a filter triplexer to separate the fundamental and harmonic signals, so that they may be tuned independently [19]. In this way it is easier to change bands and the entire Smith chart may be covered with independent controls, but the insertion loss of the triplexer considerably limits the attainable reflection coefficient magnitude.

Finally, harmonic tuners with a resonant probe are shown in Figure 13.4(c). They are a compact, elegant solution that in principle allows high reflection coefficients along with independent harmonic control. They are, however, relatively narrow-band and they require cumbersome procedures to change the operating band (i.e. disassembling the tuner, changing the slug, repeat the pre-characterization with a VNA). This might result in reduced repeatability and reflection coefficient control.

13.3 Setting the load conditions: active, open-loop techniques

The passive tuners described in Section 13.2 provide a simple, mostly effective and economic way to control the load conditions at the lower microwave frequencies and for connectorized DUTs. However, in the presence of larger losses (i.e. at higher frequencies and for on-wafer applications) passive load-pull systems do not allow to reach highly reflective loading conditions. The problem is especially evident for harmonic tuning,

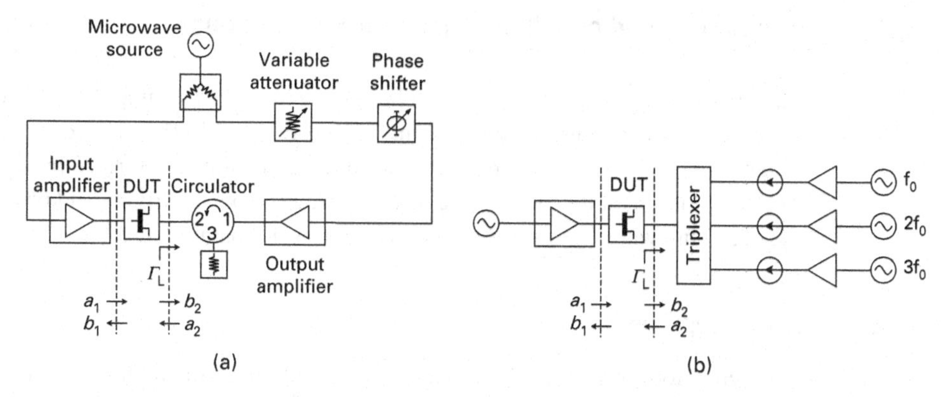

Fig. 13.5 Active, open loop load-pull, with single source (a) [28], and with multiple sources (b) [29]. ©2001 IEEE. Reprinted, with permission, from [7].

since the optimum harmonic termination is typically on the edge of the Smith chart [22]. Active-load systems were originally introduced at the end of the 1970s as a solution to this problem; these days they are commercially available in different forms [23–27]. They probably represent the most reliable scheme for microwave and millimeter wave load-pull test-sets.

A simple way to electronically synthesize a load reflection coefficient is to inject a coherent signal into the DUT output. The concept can be explained by referring to Figure 13.5. If the DUT delivers an out-going traveling wave b_2, controlling the in-going wave a_2 simply translates into setting the load reflection coefficient Γ_L to the value a_2/b_2. The a_2 signal can be taken from the same input source (as shown in Figure 13.5(a)) or from other external signal sources, coherent with the excitation signal (see Figure 13.5(b)).

This technique was originally introduced by Takayama in 1976 [28] and it has been widely used in industrial and research environments [29, 30]. It can be easily extended to harmonic tuning, with the help of frequency multipliers [30], or with the use of additional sources tuned to the desired harmonic frequencies [29], as shown in Figure 13.5(b).

In the case of Figure 13.5(a), the synthesized load is controlled by the variable phase-shifter and attenuator, and it is constant as long as the DUT outgoing wave b_2 does not change with respect to a_2. For instance, the attenuator setting must be continuously adjusted during an input power sweep, to compensate for the output power change.

In general, computer-controlled measurements are mandatory to achieve a constant load. Iterative algorithms continuously monitor the actual load and properly control the settings of the attenuator and phase-shifters in the configuration of Figure 13.5(a) or the settings of the microwave sources in the configuration of Figure 13.5(b). This increases the measurement time, as well as the possibility of failures that are potentially destructive for the DUT (e.g. as in the case of load reflection coefficient magnitudes higher than unity).

13.4 Setting the load conditions: active-loop techniques

The *active loop* is a closed-loop technique for microwave impedance synthesis in load-pull measurements. Originally introduced in 1982 [32], it improves the control of the load impedance compared to the open-loop techniques described in Section 13.3. In contrast to them, it generates the in-going wave a_2 by amplifying and phase-shifting the out-going wave b_2, instead of generating it by a separate source.

13.4.1 Active loop: basics

The active-loop principle is shown in Figure 13.6(a), which shows its implementation at the output of a generic DUT. The out-going wave b_2 produced by the DUT itself is coupled through a directional coupler, attenuated, phase-shifted, amplified, and re-injected back into the DUT as a_2. The "losses" block at the DUT output represents all the attenuation between the DUT reference plane and the loop directional coupler input, e.g. due to the fixture, probes, and cable losses. Under the assumption that all the components are ideally matched to the same reference impedance, the synthesized reflection coefficient Γ_L can be approximated by:

$$\Gamma_L = l^2 \cdot k \cdot \alpha \cdot G \cdot e^{-j\phi}, \tag{13.1}$$

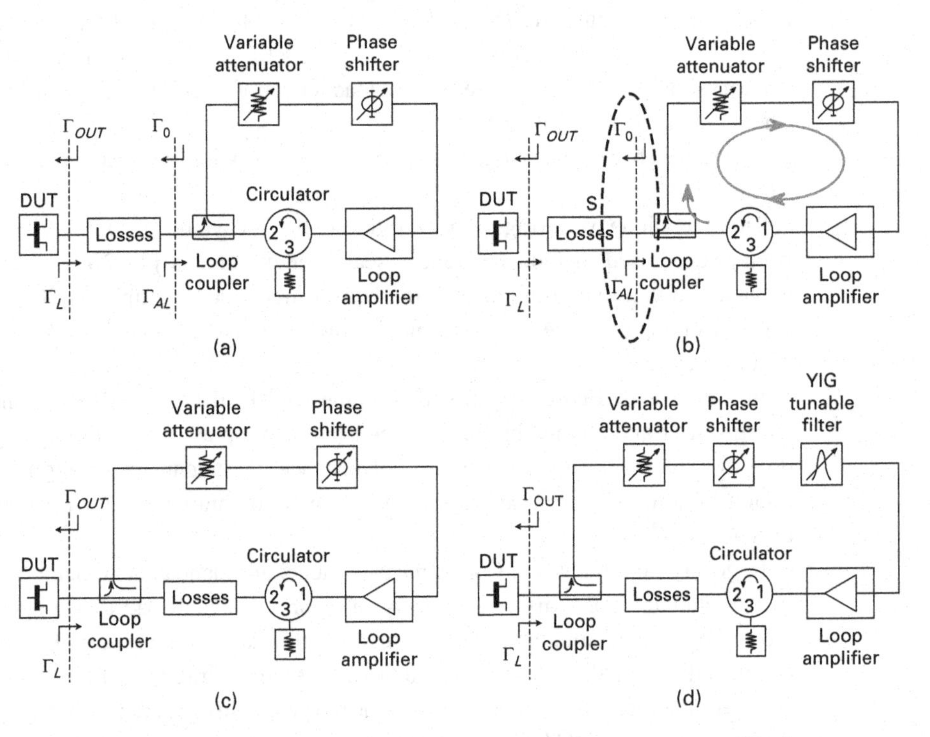

Fig. 13.6 Active-loop principle schematic (a), the two possible oscillation types (b), and further improvements to avoid oscillations (c), (d).

where l represents the losses between the DUT and the active loop (effectively as a gain lower than one), k is the directional coupler coupling factor, α represents the variable attenuation setting, G is the amplifier gain (all expressed in real, linear units), while ϕ is the value of the loop phase-shift. The variable attenuator is used to change the Γ_L magnitude, while the phase-shifter is used to change its phase. These two controls act separately on the magnitude and on the phase, respectively, through α and ϕ of (13.1).

Additionally, it is clear from (13.1) that Γ_L does not depend in first approximation on the output power of the DUT, i.e. no adjustments of the variable attenuator are needed during automated input power sweeps. Along with the independent control of the reflection coefficient magnitude and phase, this property makes the active-loop technique advantageous over the open-loop methods in terms of simplicity, safety, and ease of use.

However, active loops have two main drawbacks: namely, their potential instability and the relatively large phase-shift inside the loop bandwidth. These are addressed in the next sections.

13.4.2 Stability analysis of the active loop

The active loop of Figure 13.6(a) could be unstable for two main reasons [33].

The first one is related to the finite isolation of the loop directional coupler. This creates a feedback path from the output of the loop amplifier back into the attenuator and phase-shifter, indicated as a gray line in Figure 13.6(b). Oscillations can arise if the loop gain exceeds unity – which is possible because of the gain G. Calling I the directional coupler isolation, this "internal" loop gain is equal to $I \cdot \alpha \cdot G$.

We will now find a condition to avoid this type of instability. The maximum required gain G is obtained when the synthesized $|\Gamma_L|$ is 1. In that case $G = \frac{1}{\alpha l^2 k}$. As $\alpha \leq 1$ always, then:

$$G \geq \frac{1}{l^2 k}. \tag{13.2}$$

To avoid oscillations, the loop gain must be lower than 1, i.e. $I \cdot \alpha \cdot G < 1$. This is always verified if $G < \frac{1}{I}$. From (13.2), we have

$$\frac{1}{l^2 k} \leq G < \frac{1}{I}. \tag{13.3}$$

This means that to avoid oscillations it must hold that

$$\frac{1}{l^2} < \frac{k}{I} = D, \tag{13.4}$$

where D is the directional coupler directivity.

In conclusion, the larger the losses between the DUT and the active loop, the higher the coupler isolation must be.

A second – and often more critical – reason for the loop instability is related to the reflection coefficient Γ_O that the active loop sees at its input. For this reason, this issue is sometimes referred to as the "external" loop stability. Referring again to Figure 13.6(b) (dashed line), instability can arise when

$$|\Gamma_O| \cdot |\Gamma_{AL}| \geq 1. \tag{13.5}$$

By calling S the scattering matrix of the "losses" block, the quantity Γ_O is:

$$\Gamma_O = S_{22} + \frac{S_{12}S_{21}\Gamma_{OUT}}{1 - S_{11}\Gamma_{OUT}}. \tag{13.6}$$

The condition for the stability is then:

$$\left| S_{22} + \frac{S_{12}S_{21}\Gamma_{OUT}}{1 - S_{11}\Gamma_{OUT}} \right| \cdot |\Gamma_{AL}| < 1, \tag{13.7}$$

which is always verified if:

$$|S_{22}||\Gamma_{AL}| + \left| \frac{S_{12}S_{21}\Gamma_{OUT}}{1 - S_{11}\Gamma_{OUT}} \right| \cdot |\Gamma_{AL}| < 1. \tag{13.8}$$

Unfortunately, the quantity $|\Gamma_{AL}|$ can be greater than one, because it must compensate for the losses between the DUT and the active loop. In particular, if the "losses" block is not well matched – i.e. $|S_{22}|$ is considerably different to zero – and if the required $|\Gamma_{OUT}|$ is high (close to unity), this condition could be false and oscillations could be triggered.

A possible remedy consists of moving the directional coupler as near as possible to the DUT output. This solution was patented in 1999 [34] and it is shown in Figure 13.6(c). Ideally, the stability condition is now

$$|\Gamma_{OUT}| \cdot |\Gamma_L| < 1. \tag{13.9}$$

In principle this is always true, as $|\Gamma_{OUT}| < 1$ and the optimum $|\Gamma_L|$ of interest cannot be larger than unity. Moreover, the reflection coefficient at the DUT reference plane becomes:

$$\Gamma_L = l \cdot k \cdot \alpha \cdot G \cdot e^{-j\phi}, \tag{13.10}$$

while the "internal" loop gain is now $I \cdot \alpha \cdot G \cdot l$.

Moving the losses inside the loop therefore has multiple advantages:

- it improves the "external" loop stability, according to (13.9);
- it improves the "internal" loop stability, as its loop gain is multiplied by a factor l, with $l < 1$;
- finally, the loop amplifier gain G needed to obtain the same Γ_L is lower, as can be seen by comparing (13.1) and (13.10).

13.4.3 Practical active-loop implementations

Various practical implementations of the active-loop-based load-pull systems are described in the literature [29, 30], [35–37] and the first systems have begun to be commercially available in recent years [23–27].

In Section 13.4.2, the stability of the active loop was analyzed at its operating frequency – i.e. the frequency at which the loop is intended to synthesize the desired reflection coefficient. However, the loop is stable only if the same criteria are fulfilled at all the frequencies inside the pass-band of the various loop components (attenuator, phase-shifter, and amplifier) – particularly in the lower range, where the loop gain G is usually higher.

The common solution consists of attenuating the loop gain outside the frequencies of interest as much as possible, by adding a narrow band-pass filter centered at the loop operating frequency. In particular, Yittrium Iron Garnet (YIG) filters combine very narrow bandwidth (around 0.5% of the central frequency) with excellent tuning capabilities. As shown in Figure 13.6(d), the YIG filter is usually located after the attenuator and phase-shifter to minimize its input power.

Besides improving the loop stability, the YIG filter in the active loop enables a number of interesting features.

First of all, it naturally makes the loop frequency-selective. Inside the YIG filter pass-band the synthesized load is controlled in magnitude and phase by the attenuator and phase-shifter settings (as described in Section 13.4.1). Outside the YIG pass-band, the synthesized load is a match, as long as the circulator is well matched and the loop coupler has a very low coupling factor (between −25 and −30 dB, so that the YIG filter out-of-band high input SWR has no effect). This is a great advantage with respect to basic passive tuners, that have uncontrolled and unpredictable behavior in frequency. As an important side effect, an active loop can be designed to cover several octaves (by a proper choice of attenuators, phase-shifters, and amplifiers) and its operating frequency can be run-time reconfigured by controlling the YIG coil DC current.

Moreover, if the YIG coil current control is fine enough, it is possible to use it to slightly tune the phase of the YIG filter S-parameter S_{21}. This phase-shift control is faster and more repeatable than any mechanical phase-shifting. Figure 13.7(a–b) shows the typical response of a YIG filter, magnitude and phase, while varying the main coil current. In Figure 13.7(c–d), the responses are normalized to one of them, taken as a reference. At 3 GHz central frequency, for example, a total phase variation of 160–170° is possible, while the S_{21} magnitude is still in the 3 dB bandwidth. Modern active load-pull test-sets exploit this electronic phase-shifting capability to considerably speed up the measurements.

Given the active-loop frequency selectivity, the realization of *harmonic active loads* is straightforward. To achieve fundamental and harmonic tuning at the same time, or multiple harmonic tuning, more loops can be combined using wideband components, as in the example of Figure 13.8. In this implementation, all the passive components and the loop amplifier must have at least a bandwidth of one octave. The loads at fundamental and harmonic frequencies are controlled in a completely independent way. As we have seen

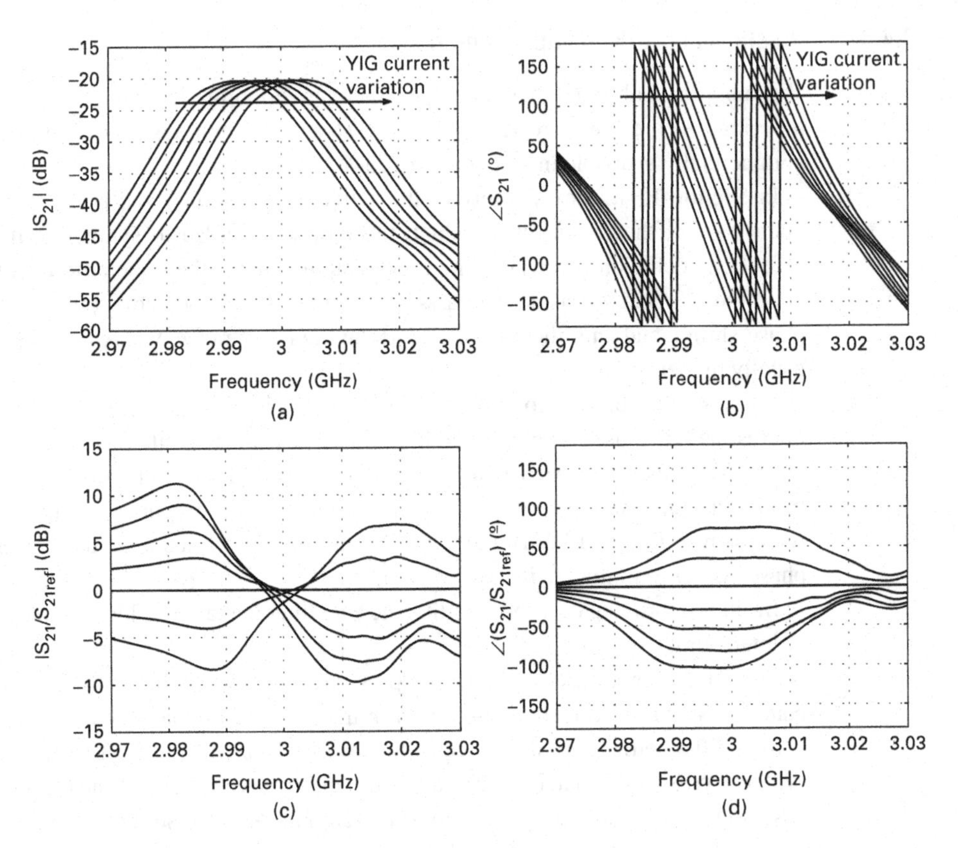

Fig. 13.7 Typical YIG filter response versus main coil current, magnitude (a), and phase (b). In (c) and (d), the same YIG responses are normalized with respect to the central one.

in Section 13.2.2, the realization of harmonic loads with passive tuners is possible, but more complicated, due to the intrinsic wideband characteristics of these passive devices.

Similarly, more loops can be combined through hybrid couplers to obtain *differential/common-mode active loads* [38–41], as shown in Figure 13.9. If the microwave hybrids are ideal, the two loops independently tune the differential and common modes. Harmonic differential- and common-mode loads are obtained simply by changing the YIG filter frequency. Differential/common-mode loads are the basis of *mixed-mode load-pull* systems, which are described in Section 13.5.3. Mixed-mode load-pull techniques are of increasing importance, as differential active devices (e.g. transceivers and amplifiers) are being extensively used in many applications for the reduction of the effects of external disturbances.

13.4.4 Wideband load-pull

Besides CW signal stimuli, modern applications also require testing devices under two-tone excitations or wideband modulated signals, such as for multicarrier W-CDMA applications.

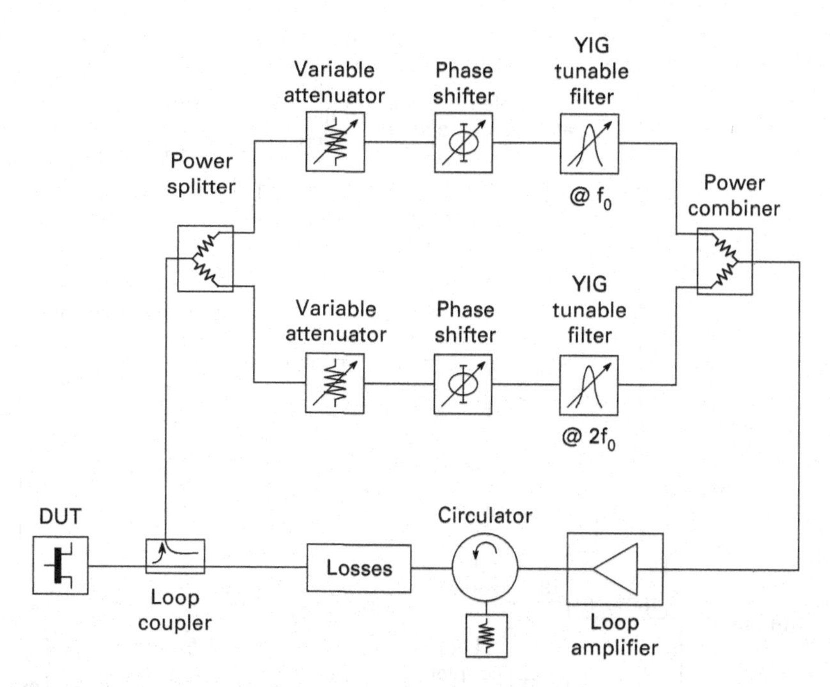

Fig. 13.8 Fundamental and harmonic tuning through active loop. In this implementation all the passive components and the loop amplifier need to have a bandwidth of one octave. ©2007 IEEE. Reprinted, with permission, from [21].

Unfortunately, the electrical delay introduced by a passive tuner and the connecting cables makes the phase of the reflection coefficients presented to the DUT reference planes not constant in frequency. Active techniques also suffer from the same problem, especially if a YIG filter is present – as in the active loop. Under wideband excitation, this means that the reflection coefficient varies inside the signal bandwith. At harmonic frequencies, the in-band phase change of Γ_L is even larger. This is usually not acceptable, as it does not correctly represent the loading conditions of the device when used in a real circuit. In this case the variations of Γ_L and Γ_S with frequency are much smaller, due to reduced dimensions of the circuit components (often approximated as lumped elements).

Active techniques can be modified to overcome this problem, and Chapter 14 is dedicated to the detailed description of such wideband systems.

13.4.5 Combining passive tuners and active techniques

The concept of combining passive tuners and active loops has been well known since the late 1970s [19]. There are several ways to mix the two techniques and a vast literature on the topic; see for example [19, 33, 35–37].

The idea is that, even when using a passive element (a tuner or a taper as in [36], or a sliding short circuit as in [35]), it is always possible to inject an additional signal to "boost" the effective reflection coefficient seen by the DUT. This additional signal can

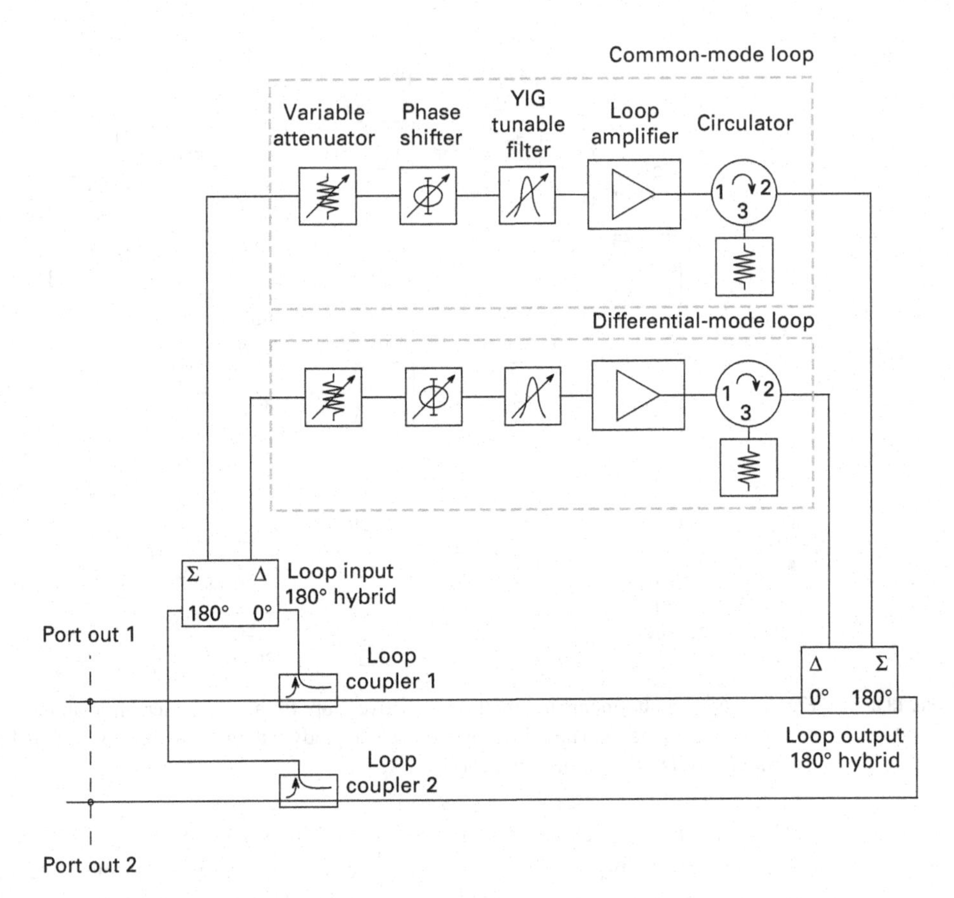

Fig. 13.9 Differential active loop implementation. ©2006 IEEE. Reprinted, with permission, from [38].

come from the same microwave source used for the device excitation (as in the case of the open-loop techniques of Section 13.3) or from the device output itself (thus creating a loop, similar to that described in Section 13.4). It can be coupled through directional couplers, combiners, or circulators, and can be amplified or not.

Some examples are shown in Figure 13.10. The simple configuration of (a) is exploited in [36], where the passive element is a taper. The hybrid-load configuration with a directional coupler of (b) is described in [35], where the passive element is a sliding short. The hybrid load with feedback loop of (c) appears in [37] (without amplifier) as well as in [19]. Finally, the hybrid configuration with active loop of (d) is described in detail in [33].

In all cases, the advantages of the hybrid loads are the following:

- higher reflection coefficients than using a passive element alone;
- less power needed to reach the same reflection coefficient, as compared to the active load alone [33, 36].

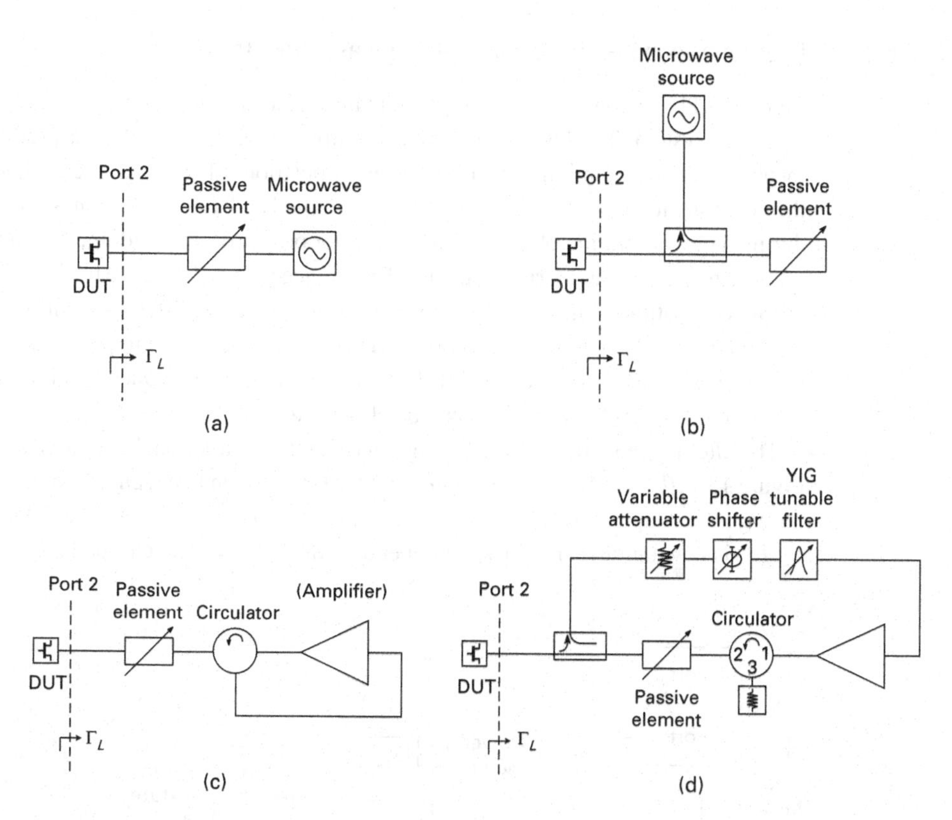

Fig. 13.10 Simplest passive/active hybrid load [36] (a). Hybrid load configuration with a directional coupler [35] (b). Hybrid load with feedback loop [37, 19] (c). Hybrid configuration with active loop [33] (d).

13.5 Measuring the DUT single-frequency characteristics

Sections 13.2 to 13.4 dealt with the problem of presenting a desired load reflection coefficient Γ_L at the DUT ports, at single and multiple frequencies, as well as in broadband cases. We now focus on the measurement of the DUT quantities, including the synthesized loading conditions. Many of the concepts already described in Chapter 12 for generic nonlinear measurements are resumed here and analyzed from the perspective of the practical load-pull test-sets.

Regarding the DUT measurement technique, there are two main types of load-pull systems:

- *non-real-time* systems, which rely on the off-line pre-characterization of the tunable loads;
- *real-time* systems, which measure the DUT loading conditions on-line, by using properly calibrated instrumentation (usually, but not exclusively, a vector network analyzer).

13.5.1 Real-time vs. non-real-time load-pull measurements

Figure 13.11(a) shows a typical implementation of a basic *non-real-time* load-pull measurement system. The DUT input and the output power levels are measured by power meters while sweeping the tuners in different positions. The S-parameter matrix of all the components between the power meters and the DUT reference planes are measured during a pre-calibration phase. In this way, after proper de-embedding, the power meter measurements can be referred to the DUT reference planes. Since the tuners are included in this calibration path, they must be measured for all the different positions that will be used during the actual measurements. This is a time-consuming procedure, but since the tuner is placed very near the DUT, the effect of losses is reduced and the reflection coefficient at the reference plane can be close to unity.

The alternative is to swap the tuning device and the directional couplers, as shown in Figure 13.11(b), to obtain a *real-time* system. The relation between the waves measured through the couplers (a_{m1}, b_{m1}, a_{m2}, b_{m2}) and the waves at the reference plane (a_1, b_1, a_2, b_2) is now unique, and does not depend on the load or source impedances. If a VNA

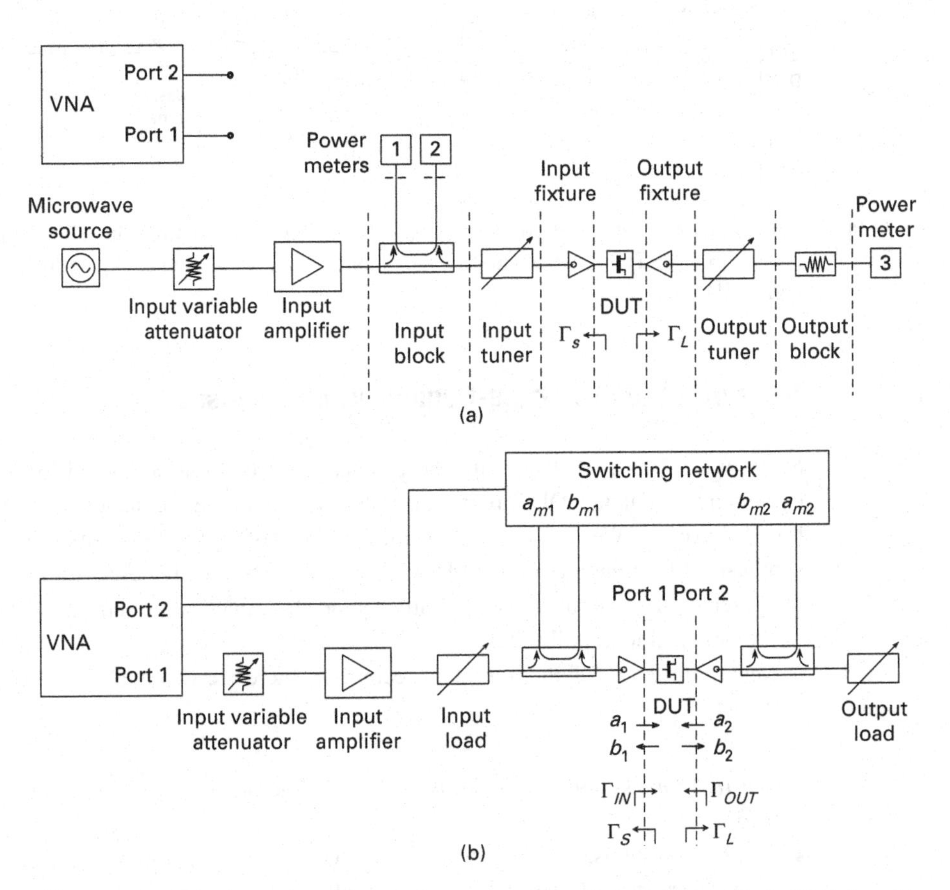

Fig. 13.11 Simplified scheme of a non-real-time pre-calibrated load-pull system (a), and of a VNA based, real-time load-pull system (b). ©2007 IEEE. Reprinted, with permission, from [21].

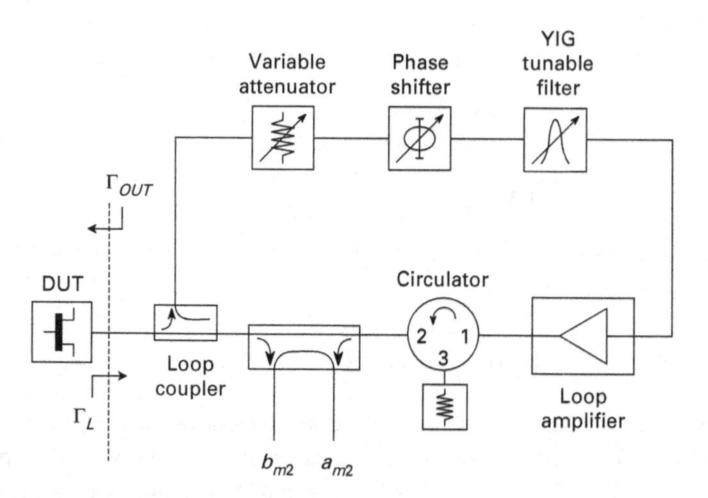

Fig. 13.12 Active loop in a real-time load-pull system.

is connected to the couplers, it becomes possible to apply a calibration procedure to perform accurate measurements at the DUT reference planes.

Originally, the real-time configuration could be used only with active loads, since the coupler losses set a severe limitation on the tuner reflection coefficients. In particular, the best configuration to improve the loop stability is the one shown in Figure 13.12, where the measurement couplers are placed inside the loop (see also Section 13.4.2).

More recently, the introduction of very low loss, ultra-wideband directional couplers [42,43] with an insertion loss in the range 0.1–0.2 dB up to 15 GHz, has made it possible to use the real-time configuration with passive tuners.

An analysis of the uncertainty contributions and a comparison of the uncertainty budget for real-time and non-real-time techniques up to 40 GHz has been presented in [44]. Experiments and simulations have shown how real-time techniques are far more accurate than non-real-time, especially as the frequency increases.

The calibration procedure typically used for real-time systems is an extension of the traditional ones for S-parameter measurements. Since it deserves some explanation, it will be described in detail in the next section.

13.5.2 Calibration of real-time systems

The error-correction theory for load-pull network analyzer-based measurements has been described mainly in three fundamental papers by Tucker *et al.* [45], Hecht [46], and Ferrero *et al.* [47]. Even though it can be seen as a subset of the calibration techniques for the nonlinear vector network analyzers described in Chapter 12, it is here summarized to understand its implications to the setup of a real-time load-pull system.

The error coefficients of a traditional, two-port VNA calibration are the eight elements of the matrices $\boldsymbol{X_1}$ and $\boldsymbol{X_2}$ defined as [48]:

$$\begin{pmatrix} a_1 \\ b_1 \end{pmatrix} = \boldsymbol{X_1} \begin{pmatrix} b_{m1} \\ a_{m1} \end{pmatrix} \tag{13.11}$$

and

$$\begin{pmatrix} a_2 \\ b_2 \end{pmatrix} = \mathbf{X_2} \begin{pmatrix} b_{m2} \\ a_{m2} \end{pmatrix}, \tag{13.12}$$

where

$$\mathbf{X_{1,2}} = \begin{pmatrix} l_{1,2} & -h_{1,2} \\ k_{1,2} & -m_{1,2} \end{pmatrix} = k_{1,2}\mathbf{Y_{1,2}}. \tag{13.13}$$

If all the eight error coefficients are known, it is possible to find the actual power at the DUT input reference plane as

$$P_{in} = |a_1|^2 - |b_1|^2 = |l_1 b_{\mathrm{m}1} - h_1 a_{\mathrm{m}1}|^2 - |k_1 b_{\mathrm{m}1} - m_1 a_{\mathrm{m}1}|^2.$$

Only seven out of these eight coefficients are the outcome of any classical S-parameter calibration, since the S-parameters are defined as ratios between waves. In particular, the coefficients have to be normalized with respect to k_1 and only the quantities l_1/k_1, m_1/k_1, h_1/k_1 and k_2/k_1, l_2/k_1, m_2/k_1, h_2/k_1, are known. The input power P_{in} can be computed only if the last coefficient k_1 is known, using the following formula:

$$P_{in} = |k_1|^2 \left(|l_1/k_1 \cdot b_{\mathrm{m}1} - h_1/k_1 \cdot a_{\mathrm{m}1}|^2 - |b_{\mathrm{m}1} - m_1/k_1 \cdot a_{\mathrm{m}1}|^2 \right) = |k_1|^2 P_{in}^n.$$

If it is possible to connect a power meter at the input port, the magnitude of the coefficient k_1 can be found during the calibration. If P_m is the reading of the power meter, it holds that:

$$P_{in} = P_m = |k_1|^2 P_{in}^n, \tag{13.14}$$

and therefore the error coefficient k_1 is computed as

$$|k_1|^2 = \frac{P_m}{P_{in}^n}. \tag{13.15}$$

This is a straightforward solution for connectorized DUTs, where a power meter is likely to be available with the same type of connector as the DUT. However, the technique needs some adaptation for on-wafer measurements, as the direct connection of an accurate power sensor at the DUT reference planes is not possible.

The technique described in [47] does not use any on-wafer power sensor; it is shown in Figure 13.13. It is based on the assumption that the output reflectometer has typically an on-wafer port (port 2) and a connectorized one (port 3, e.g. a *coaxial* port). Thus, the reflectometer measurements $a_{\mathrm{m}2}$ and $b_{\mathrm{m}2}$ have a fixed relationship with port 2 waves (a_2, b_2), and a different – but still easy to find – relationship with the port 3 waves (a_3, b_3).

The calibration is performed in three conceptual steps.

1. First, a traditional, two-port, on-wafer calibration (e.g. TRL) is performed at the DUT ports, with the system configured as in Figure 13.13(a).
2. Second, a generic two-port DUT (for instance, a thru) is connected to the on-wafer ports and a one-port SOL calibration is performed at the coaxial reference plane port 3, as shown in Figure 13.13(b).

Calibration phase I: S-parameter calibration

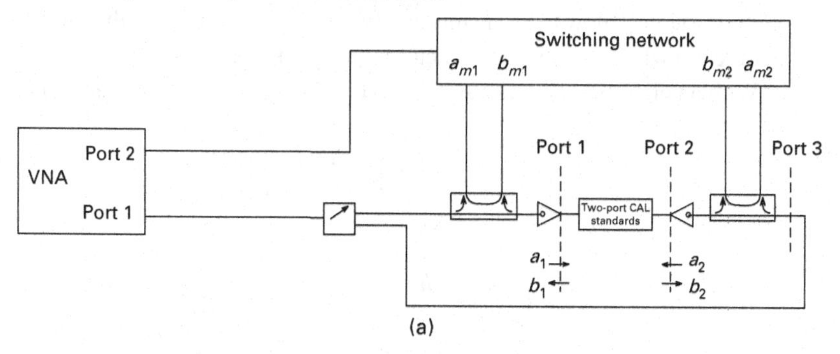

(a)

Calibration phase II: coaxial port 3 calibration

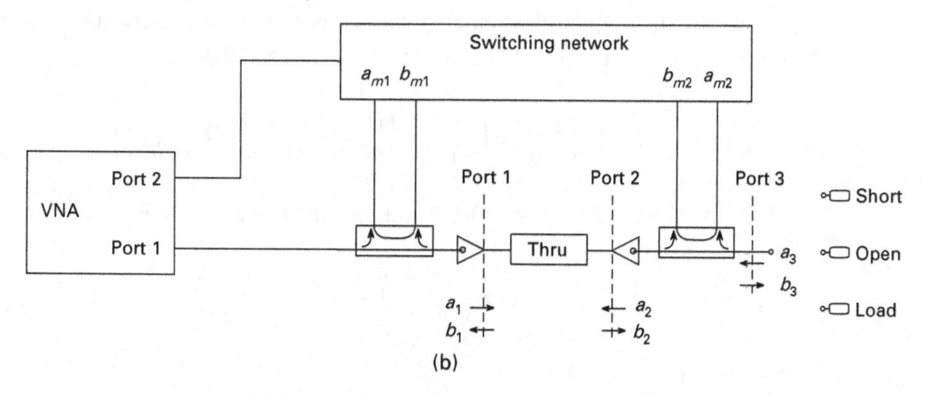

(b)

Calibration phase III: power meter measurement at port 3

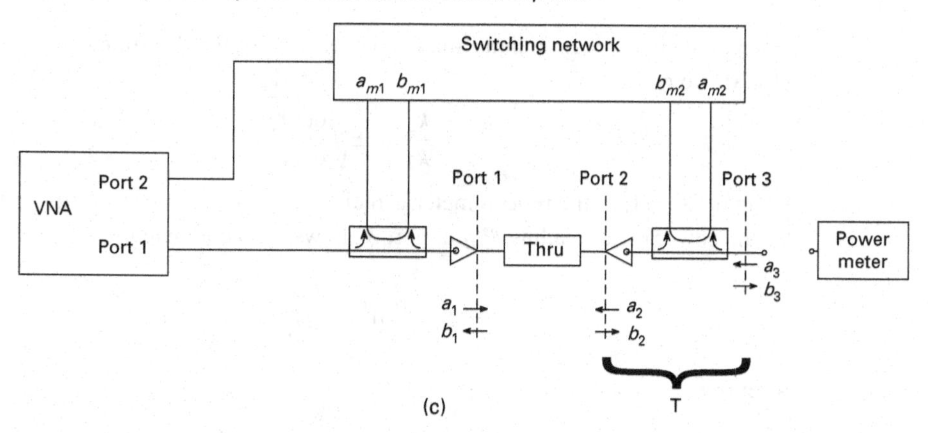

(c)

Fig. 13.13 Steps for the real-time load-pull system calibration.

3. Finally, the power level at the port 3 reference plane is measured by a coaxial reference power meter, as in Figure 13.13(c).

In other words, this procedure replaces the on-wafer power reading at port 1 with a coaxial power measurement at port 3, at the price of an additional one-port calibration at the coaxial port. The device connected between ports 1 and 2 during the coaxial port calibration should allow the transmission of some power to excite the output reflectometer, but the knowledge of its S-parameter is in principle not needed.

With an error box notation similar to (13.11) and (13.12) we can write

$$\begin{pmatrix} a_3 \\ b_3 \end{pmatrix} = \boldsymbol{X_3} \begin{pmatrix} b_{m2} \\ a_{m2} \end{pmatrix}, \tag{13.16}$$

where

$$\boldsymbol{X_3} = \begin{pmatrix} l_3 & -h_3 \\ k_3 & -m_3 \end{pmatrix} = k_3 \begin{pmatrix} \frac{l_3}{k_3} & -\frac{h_3}{k_3} \\ 1 & -\frac{m_3}{k_3} \end{pmatrix} = k_3 \boldsymbol{Y_3}. \tag{13.17}$$

Matrix $\boldsymbol{Y_3}$ can be obtained from the one-port calibration procedure in Figure 13.13(b).

However, the output reflectometer is a two-port, reciprocal network. Its transmission matrix $\boldsymbol{T}$, shown in Figure 13.13(c), imposes the condition

$$\begin{pmatrix} a_2 \\ b_2 \end{pmatrix} = \begin{pmatrix} T_{11} & T_{12} \\ T_{21} & T_{22} \end{pmatrix} \begin{pmatrix} a_3 \\ b_3 \end{pmatrix} = \boldsymbol{T} \begin{pmatrix} a_3 \\ b_3 \end{pmatrix}. \tag{13.18}$$

By substituting (13.16) in (13.18) and then in (13.12) we obtain

$$\boldsymbol{X_2} \begin{pmatrix} b_{m2} \\ a_{m2} \end{pmatrix} = \boldsymbol{T X_3} \begin{pmatrix} b_{m2} \\ a_{m2} \end{pmatrix} \tag{13.19}$$

and therefore

$$\frac{k_3}{k_2} (\boldsymbol{Y_2})^{-1} \boldsymbol{T Y_3} = \boldsymbol{I}, \tag{13.20}$$

where $\boldsymbol{I}$ is the (2×2) identity matrix. By extracting the determinant of both sides of (13.20) we get

$$\frac{k_2}{k_3} = \pm \sqrt{\frac{\det \boldsymbol{Y_3}}{\det \boldsymbol{Y_2}}}, \tag{13.21}$$

being $\det \boldsymbol{T} = 1$, as the reflectometer is reciprocal.

The power meter reading P_{mt} at port 3 allows us to compute $|k_3|$:

$$|k_3|^2 = \frac{P_{mt}}{P_{mt}^n}, \tag{13.22}$$

where

$$P_{mt}^n = |l_3/k_3 \cdot b_{m2} - h_3/k_3 \cdot a_{m2}|^2 - |b_{m2} - m_3/k_3 \cdot a_{m2}|^2 \tag{13.23}$$

is computed from the error coefficients in $\boldsymbol{Y_3}$ and the network analyzer readings a_{m2}, b_{m2} when the power meter is connected.

In conclusion we obtain:

$$|k_2| = \sqrt{\left|\frac{\det \boldsymbol{Y_3}}{\det \boldsymbol{Y_2}}\right| \frac{P_{mt}}{P_{mt}^n}}. \tag{13.24}$$

This eventually allows us to compute k_1 and to solve the power calibration problem, because the ratio k_2/k_1 is known from the two-port S-parameter calibration.

13.5.3 Mixed-mode, harmonic load-pull systems

The first experiments on harmonic load-pull measurements were performed in the early 1990s [30, 49, 13]. In [49] the system was based on six-port-reflectometer measurements and the influence of the harmonic termination at $2f_0$ on the output power was for the first time demonstrated for a MESFET device. Similarly, the influence of harmonic terminations on the transistor PAE was demonstrated in [13] by using a VNA-based receiver. In the following years the technique was further refined and many innovative results obtained.

This section describes one of the most recent achievements of active-loop-based load-pull techniques [38] and demonstrates the potentials of active harmonic load-pull for differential device characterization (in particular, a Bluetooth transceiver). The complete measurement system is shown in Figure 13.14. The active differential load is the one already shown in Figure 13.9, with the possibility of tuning the loads at fundamental or harmonic frequencies by simply changing the YIG tuning frequency. Again,

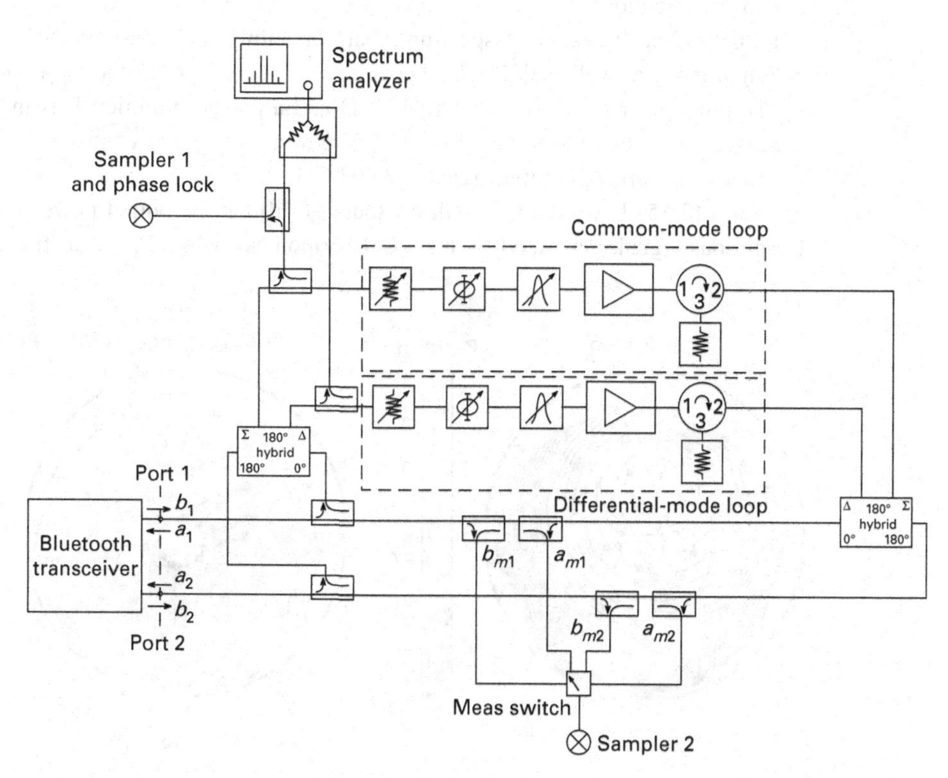

Fig. 13.14 Differential/common-mode harmonic load-pull system. ©2006 IEEE. Reprinted, with permission, from [38].

these features can be achieved only by active-loop-based systems, as passive tuners cannot independently tune the differential- and common-mode loads at fundamental and harmonics.

The measurement system consists of two reflectometers, a VNA, and a spectrum analyzer. The measured performances are:

$$\begin{aligned} \Gamma_D &\equiv b_D/a_D \\ \Gamma_C &\equiv b_C/a_C \\ P_D &\equiv |a_D|^2 - |b_D|^2 = |a_D|^2(1 - |\Gamma_D|^2) \\ P_C &\equiv |a_C|^2 - |b_C|^2 = |a_C|^2(1 - |\Gamma_C|^2), \end{aligned} \tag{13.25}$$

where the differential- and common-mode quantities are defined as:

$$\begin{pmatrix} a_D \\ a_C \end{pmatrix} \equiv \frac{1}{\sqrt{2}} \begin{pmatrix} 1 & -1 \\ 1 & 1 \end{pmatrix} \begin{pmatrix} a_1 \\ a_2 \end{pmatrix} \tag{13.26}$$

$$\begin{pmatrix} b_D \\ b_C \end{pmatrix} \equiv \frac{1}{\sqrt{2}} \begin{pmatrix} 1 & -1 \\ 1 & 1 \end{pmatrix} \begin{pmatrix} b_1 \\ b_2 \end{pmatrix}. \tag{13.27}$$

The system can be calibrated with any classical load-pull calibration at ports 1 and 2, as shown in Section 13.5.2.

Furthermore, the use of a spectrum analyzer allows the measurement of spurious common-mode as well as differential power ratios at all the frequencies of interest.

The transceiver includes an internal VCO and a power amplifier. It is mounted on a connectorized evaluation board (including power supply and control signals) and it generates a differential output signal at 2.402 GHz.

Figure 13.15 shows the load-pull contours of differential output power and pulling (a spurious signal, generated by the VCO harmonics), when Γ_C at all frequencies is

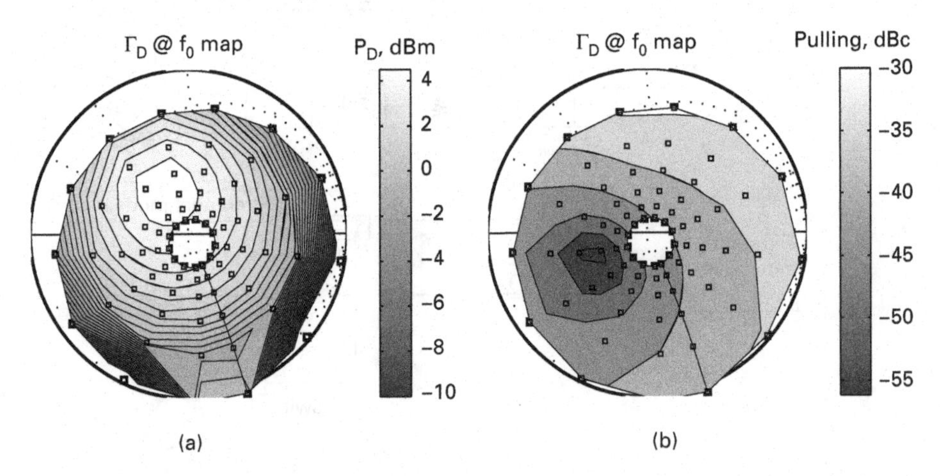

Fig. 13.15 Γ_D @ f_0 load-pull contour map of differential output power (a) and pulling (b). ©2006 IEEE. Reprinted, with permission, from [38].

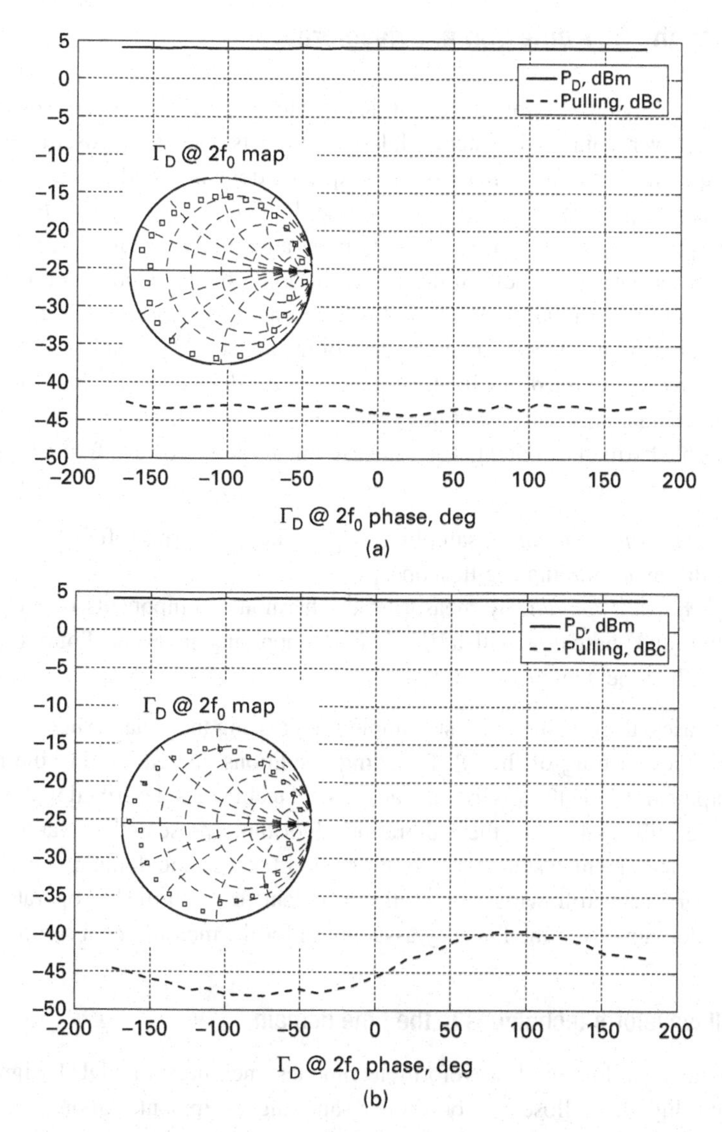

Fig. 13.16 Effect of differential-mode (a) and common-mode (b) harmonic load on pulling and output differential power (Γ_D @ $f_0 = 0.3 \angle 115°$). ©2006 IEEE. Reprinted, with permission, from [38].

kept constant to a value close to zero. These are typical load-pull maps, that allow the identification of optimum Γ_D @ f_0 loads for output differential power or pulling.

The influence of the harmonic differential termination on output power and pulling is shown in Figure 13.16. While sweeping the harmonic loads, Γ_D @ f_0 is kept constant at the optimum value for output differential power, i.e. $0.3 \angle 115°$. This was the first experimental verification of the effect of Γ_C @ $2f_0$ (around 10 dB variation) and of the very limited influence of Γ_D @ $2f_0$ on the pulling signal. This demonstrates the innovation potential of harmonic mixed-mode load-pull measurements.

13.6 Measuring the DUT time domain waveforms

As seen in the previous section, the quantities of interest for load-pull are typically input and output power, gain, PAE, intermodulation products and ACPR, oscillator pulling, etc. – i.e. quantities mainly defined in the frequency domain. In addition, the measurement of the complete voltage and current time domain waveforms at the DUT input and output ports is being given increasingly more attention. As examples, time domain waveform information can help in the design of high-efficiency power amplifiers [50, 51], in building and validating more sophisticated nonlinear models [52, 53], and – in general – in reaching a deeper understanding of the active DUT behavior [54].

As extensively discussed in Chapter 12, a nonlinear device excited at a fundamental frequency f_0 in large-signal conditions generates distorted waveforms, i.e. signals with non-negligible harmonic content. Their measurements at microwave frequencies can be carried out in two ways:

- directly in the *time domain*, by sampling the periodic waveforms (often "sub-Nyquist") as in traditional sampling oscilloscopes;
- in the *frequency domain*, by measuring the harmonic components of the signal in magnitude and phase (e.g. with a VNA), and by applying an inverse Fourier transform to obtain the time domain waveforms.

In both cases, the measurement still implies addressing the usual issues for load-pull characterization: setting of the DUT loading conditions and calibrating the measurement equipment. While the passive and active load techniques described in the previous sections are still applicable, the calibration algorithms presented so far need to be extended to implement systematic error correction for the time domain waveforms. As will be clear in the following, it is still convenient to perform this operation in the frequency domain, given the linearity assumption for the measurement setup.

13.6.1 Load-pull waveform techniques in the time domain

A first family of load-pull waveform techniques measures the DUT signals with high-speed digital oscilloscopes or similar sampling equipment. Among the possible instrumentation, it is worth mentioning the Microwave Transition Analyzer [55] (now discontinued, but still used in research laboratories) and the Large Signal Network Analyzer [56, 57], which can be considered as the evolution of the MTA: very similar in principle, but much faster, with an increased dynamic range and more channels. Still, any sampling oscilloscope with at least two channels can be generally employed in a time domain waveform load-pull setup.

In the first implementations, the sampling oscilloscope was used in a very simple way, as shown in Figure 13.17(a). It was possible to monitor the output voltage on the oscilloscope screen while varying the load, typically with passive techniques. No vector error correction was possible (as only two out of four DUT waves were measured) and the load could be measured with a VNA during a pre-calibration phase. Of course, this solution was used to provide a very quick, but only qualitative evaluation tool.

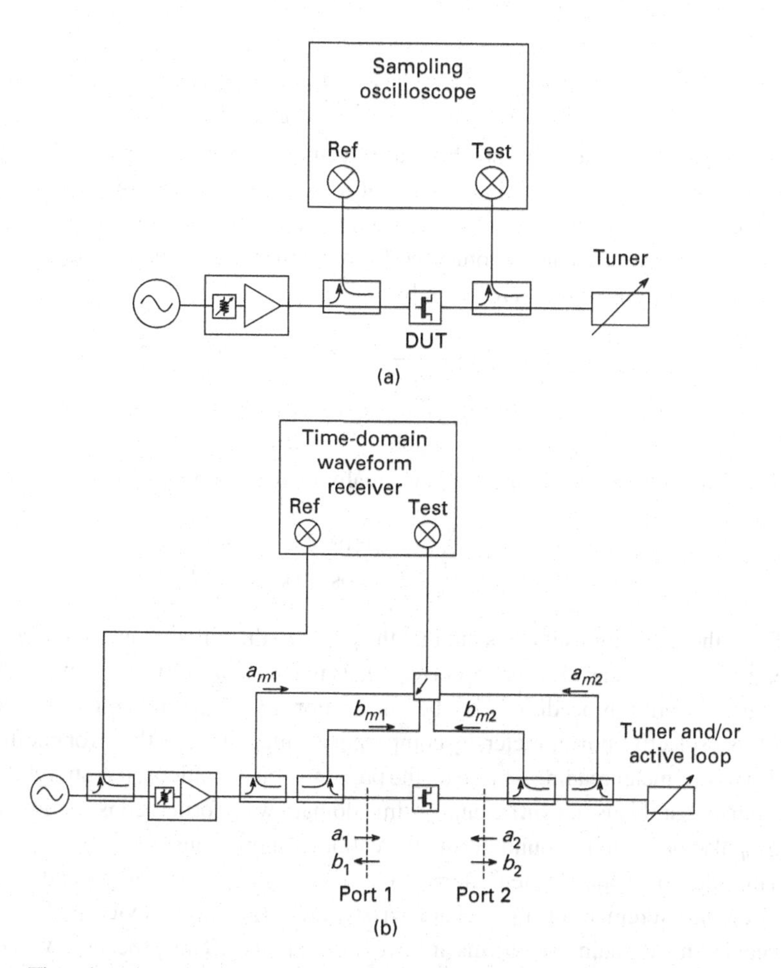

Fig. 13.17 Time domain waveform load-pull in its simplest implementation (a) and in the vector-corrected configuration (b). ©2008 IEEE. Reprinted, with permission, from [58].

A more complete solution is shown in Figure 13.17(b). The time domain receiver "test" channel is used to measure all the four incident and reflected waves by a microwave multiplexing switch. Obviously, if the time domain receiver had four test channels, the four waves could be measured simultaneously, the switch would be no longer needed, and the measurement would be much faster.

As anticipated, it is still convenient to apply the systematic error-correction procedure in the frequency domain. In particular, the sampled waveforms are first transformed in the frequency domain via FFT. Since the signals are periodic, they can be represented by a discrete set of phasors. We will refer to them as $a_{m1,n}$, $a_{m2,n}$, $b_{m1,n}$, and $b_{m2,n}$, where $n = 1, 2, ...$ represents the order of the different harmonic components.

The error model is still the same one as described in Section 13.5.2, here reported for completeness:

$$\begin{pmatrix} a_{1,n} \\ b_{1,n} \end{pmatrix} = \boldsymbol{X_{1,n}} \begin{pmatrix} b_{m1,n} \\ a_{m1,n} \end{pmatrix} \tag{13.28}$$

and

$$\begin{pmatrix} a_{2,n} \\ b_{2,n} \end{pmatrix} = \boldsymbol{X_{2,n}} \begin{pmatrix} b_{m2,n} \\ a_{m2,n} \end{pmatrix}, \tag{13.29}$$

where $a_{1,n}, a_{2,n}, b_{1,n}$, and $b_{2,n}$ are the phasors at the DUT reference planes. The frequency dependence of the two error coefficient matrices is made explicit by the n subscript.

If the error coefficients are known for all the frequencies of interest, the phasors at the DUT reference planes can be computed from the measured signals. The corresponding voltage and current phasors are given by

$$\begin{aligned} V_{i,n} &= \frac{|Z_{ref}|}{\sqrt{\Re\{Z_{ref}\}}}(a_{i,n} + b_{i,n}) \\ I_{i,n} &= \frac{|Z_{ref}|}{Z_{ref}\sqrt{\Re\{Z_{ref}\}}}(a_{i,n} - b_{i,n}) \end{aligned} \tag{13.30}$$

and the time domain waveforms are eventually reconstructed as

$$\begin{aligned} v_i(t) &= \textstyle\sum_{n=1}^{N} |V_{i,n}| \cos(2\pi k f_{0t} + \angle V_{i,n}) \\ i_i(t) &= \textstyle\sum_{n=1}^{N} |I_{i,n}| \cos(2\pi k f_{0t} + \angle I_{i,n}) \end{aligned} \tag{13.31}$$

From these equations it turns out that the phase of the different harmonic components (i.e. $\angle V_{i,n}$, $\angle I_{i,n}$) has to be error-corrected, as much as the corresponding magnitudes.

The calibration procedure described in Section 13.5.2 already allows the use of the VNA as a selective power meter, by computing the magnitude of the error coefficient $k_{1,n}$ with a power meter reading. Based on the previous considerations, it is straightforward to recognize that this is not sufficient for time domain waveform reconstruction; the phase of $k_{1,n}$ also needs to be found during the calibration procedure.

This additional step – often referred to as *phase calibration* – is generally performed with the measurement of a pre-characterized DUT (a *golden* device), which produces traceable time domain waveforms at its reference planes. This procedure was originally introduced in 1989 [59], and the principle is still under improvement[60]. NIST traceability is obtained by measuring the golden device with a sampling oscilloscope, previously calibrated using the "nose-to-nose" [61] technique.

13.6.2 Load-pull waveform techniques in the frequency domain

A second family of load-pull waveform techniques uses a network analyzer to vectorially measure (i.e. in magnitude and phase) the harmonic components of the signals in the frequency domain, instead of relying on time domain sampling. Given the high frequency-selectivity of a VNA with respect to a broadband oscilloscope, this theoretically results in a much higher dynamic range and measurement accuracy.

However, a VNA cannot be directly used to measure load-pull time domain waveforms, mainly because the coherency of the phase measurement is not maintained during the frequency sweep. Some modifications have to be introduced in the measurement setup. We will try to explain this concept with the help of Figure 13.18.

It is well known that the basic, conceptual block of a VNA is the vector voltmeter. This is basically a super-heterodyne receiver with two channels (namely, *test* and *reference*),

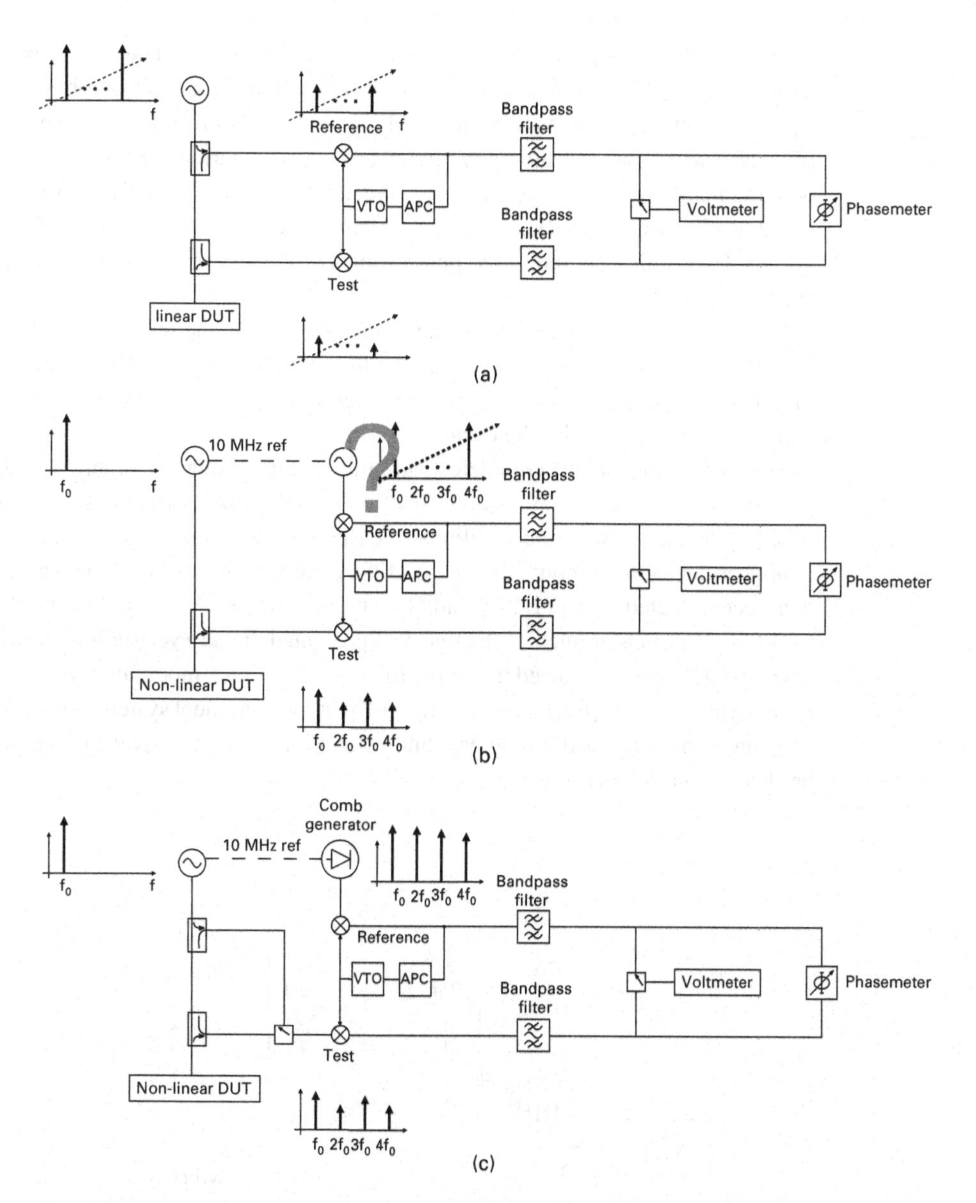

Fig. 13.18 Simplified scheme of a vector voltmeter (a). In (b) the reference signal sweeps through the various harmonics of the distorted signal under test, and it is not possible to have a stable reading of the phase of the harmonics. In (c), the reference signal is taken from a comb generator having fixed harmonic phases; now the measurement of the distorted signal is possible.

which is tuned to measure the test and reference phasor magnitude at a certain frequency, along with the phase of the test phasor with respect to the reference one.

For reflection coefficient measurements, the reference channel typically measures a signal coupled from the excitation signal, as depicted in Figure 13.18(a). If the measurement and excitation frequency changes, the phase of this reference signal changes randomly, which generally is not an issue. The same setup could be used to measure the harmonics of a distorted test signal, generated by a nonlinear DUT one-by-one. In principle, this could be done as in Figure 13.18(b), by simply changing the reference signal frequency with a second, auxiliary microwave source, sharing the same reference clock with the main one. In practice however, the auxiliary source usually does not maintain the same phase reference (i.e. the same "origin of the time axis") while sweeping its frequency. Eventually, the measured phase values of the test signal harmonic components are not meaningful.

In order to correctly measure the harmonic phases, a stable reference signal, containing *all* the harmonics of interest, must be fed into the reference channel, as shown in Figure 13.18(c). This "comb" signal can be generated by any kind of nonlinear device and must be coherent with the excitation source.

The calibration of the complete load-pull system, shown in Figure 13.19, can be performed in the same way as described in Section 13.6.2 for sampling-oscilloscope-based systems, i.e. with the additional measurement of a pre-characterized, traceable nonlinear device to compute the phase of the error coefficient $k_{1,n}$. The waveforms are then reconstructed using (13.30) and (13.31). In 2005, a VNA-based load-pull system, exploiting this measurement technique, was presented. It had a very large bandwidth (300 kHz–20 GHz) which allowed the reconstruction of complex modulated waveforms, with 80 dB dynamic range [62]. More recently, a 4-port measurement system, with differential- and common-mode load tuning capabilities and time domain waveform measurements has been proposed [41].

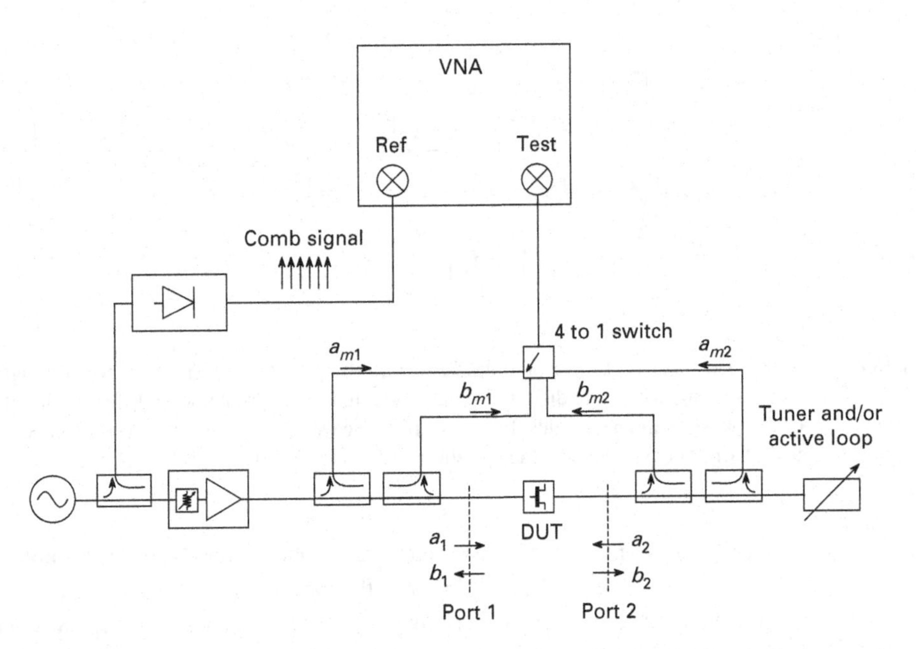

Fig. 13.19 Time domain waveform load-pull with frequency domain receiver and phase lock on a comb generator. ©2008 IEEE. Reprinted, with permission, from [58].

13.6.3 Other calibration approaches

The load-pull waveform techniques described in the previous sections rely on the use of a *golden* nonlinear device (e.g. a golden diode) to calibrate the phase of the harmonic components. This can be avoided if an extra sampling oscilloscope/MTA/LSNA channel is available during calibration.

In fact, the golden device typically used in phase calibration is nothing but a *transfer standard*: it transfers the accuracy of the oscilloscope that has characterized it to the load-pull system. If, however, an additional calibrated time domain receiver is available (i.e. a calibrated channel of an oscilloscope, MTA or LSNA), it can be directly used for the phase calibration. This method is more robust with respect to the transfer standard and it removes the effects of the connection repeatability. The calibration accuracy and traceability are then directly related to those of the auxiliary time domain receiver, which must be calibrated at its reference plane (e.g. by a "nose-to-nose" calibration [61] or with the NIST pulse standard).

Similar to the setup described in Section 13.5.2, this method makes use of an auxiliary port (port 3) during calibration. The S-parameter and power calibration steps are the same as described in Figure 13.13(a–c). The only difference is the receiver type, which is no longer a VNA but it can be any kind of time domain receiver. The step for phase calibration is shown in Figure 13.20. The source generates a CW signal, whose frequency is stepped through the values of interest (i.e. the intended fundamental and harmonic frequencies). At the generic k-th frequency, the measurements of the raw waves $a_{m1,n}$, $b_{m1,n}$, $a_{m2,n}$, and $b_{m2,n}$ are acquired simultaneously with the voltage $V_{3,k}$ at port 3 by the auxiliary time domain receiver channel.

Since a SOL calibration is already performed at port 3 according to Figure 13.13(b), the reflection coefficient $a_{3,n}/b_{3,n}$ of the calibrated auxiliary channel can be computed, and $b_{3,n}$ is obtained by (13.30) (with $i = 3$). Moreover, the elements of the scattering matrix $\mathbf{S_n}$ of the passive network between ports 2 and 3 in Figure 13.20 can be easily computed.

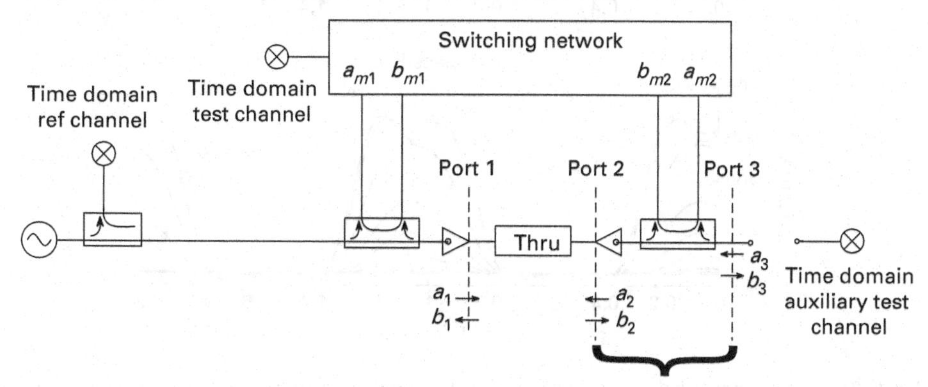

Fig. 13.20 Phase calibration using a calibrated auxiliary time domain receiver.

By comparing the $b_{2,n}$ wave at the DUT reference plane 2 obtained with the two receivers:

$$b_{2,n} = b_{3,n} \frac{S_{12,n}}{S_{22,n} \frac{a_{2,n}}{b_{2,n}} - \Delta_n} = k_{2,n} b_{m2,n} - m_{2,n} a_{m2,n}, \tag{13.32}$$

where Δ_m is $S_{11,n}S_{22,n} - S_{12,n}S_{21,n}$, it is possible to find the phases $\angle k_{2,n}$.

The calibration problem is eventually solved by computing $\angle k_{1,n}$, since the complex ratio $k_{2,n}/k_{1,n}$ is known from the S-parameter calibration at ports 1 and 2.

13.6.4 Measurement examples

Figure 13.21 shows the time domain drain currents of a microwave FET in common source configuration during a power sweep, for two different classes of operation – namely, class A and B [58, 63]. As an example, this type of plot allows the observation of overshooting of the instantaneous current values that could degrade the transistor efficiency, and to reshape them by properly choosing the fundamental and harmonic loading conditions [54, 64].

Similarly, the plots in Figure 13.22 show the trajectory of the instantaneous point in the I_D-V_{DS} plane for different power levels (solid lines). The transistor DC characteristics are superimposed for reference (dotted lines). These plots allow the observation of potential loading conditions that could bring the transistor to operate in unsafe areas of the I-V plane for a certain class of operation.

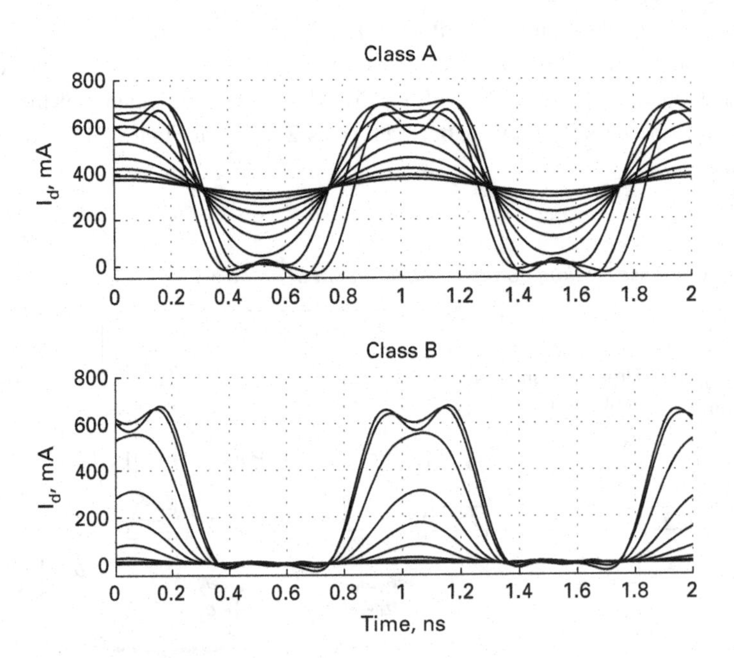

Fig. 13.21 Time domain drain current waveforms for a FET, biased in class A and class B. ©2008 IEEE. Reprinted, with permission, from [58].

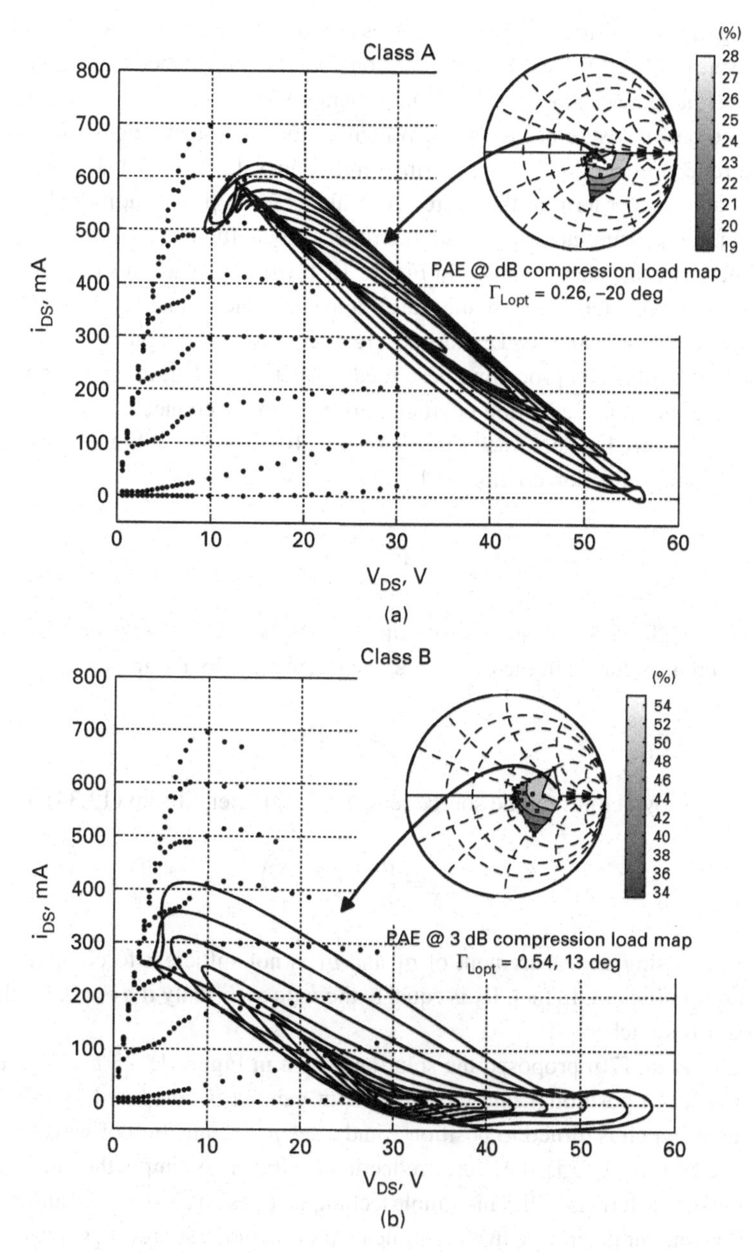

Fig. 13.22 Dynamic load lines for increasing power, on the best load for PAE, in class A (a) and B (b). ©2008 IEEE. Reprinted, with permission, from [58].

13.7 Real-time source-pull techniques

Source-pull systems have important applications in low noise amplifier design, where the lowest noise figure and, in general, the optimal transistor noise parameters are found by applying different source impedance values [65, 68] (for this topic, see also Chapter 10). Moreover, it is well known that the harmonic source impedance can heavily influence the power-added efficiency in power amplifiers [69].

Source-pull measurements use, in principle, the same techniques to synthesize the source reflection coefficient as described in Sections 13.2 to 13.4. However, the presence of a generator term in the source equivalent circuit poses additional challenges in the accurate measurement of the source reflection coefficient Γ_S. We will refer to the simplified test-set scheme shown in Figure 13.11(b) to explain this concept. The VNA and two reflectometers measure the waves at the reference planes of the DUT, while two tuners set the source and load conditions, respectively, at the input and output ports.

After the calibration procedure described in Section 13.5.2, the reflectometer is able to measure the input reflection coefficient of the circuit connected to its test port. For example, the port 1 reflectometer in Figure 13.11(b) allows calibrated measurement of the DUT input reflection coefficient Γ_{IN} as

$$\Gamma_{IN} = \frac{b_1}{a_1}. \tag{13.33}$$

Equation (13.33) defines the relationship – set by the DUT – between the waves at the input reference plane. On the other side, the microwave source imposes

$$a_1 = a_S + \Gamma_S b_1 \tag{13.34}$$

where Γ_S is, by definition, the source reflection coefficient. From (13.34), it results

$$\Gamma_S = \frac{a_1}{b_1}\left(1 - \frac{a_S}{a_1}\right). \tag{13.35}$$

Therefore, a single measurement of a_1 and b_1 is not sufficient to compute the source reflection coefficient. In fact, Γ_S is equal to the ratio a_1/b_1 only if $a_S = 0$, i.e. the internal generator is switched off.

Hughes *et al.* [70] proposed the solution shown in Figure 13.23(a). First, the source switch is set to position 1 and the DUT input gamma is computed by (13.33). Then, the source switch is turned to position 2 and a second acquisition of waves a_1 and b_1 is performed. From (13.35), the source reflection coefficient is simply the ratio $\Gamma_S = a_1/b_1$, since the source term is null. This simple technique relies on two basic assumptions. First, the DUT is not unilateral, so that a significant portion of the source signal from port 2 can reach the input reflectometer. Moreover, the reflection coefficient of the source switch Γ_{SW} does not change while turning the switch from position 1 to 2.

An entirely different approach is described in [71] and shown in Figure 13.23(b). Here the signal from the microwave source is summed with the wave reflected by the tuning element and injected into the DUT. The reflectometer is used in an unconventional configuration (referred to as *reverse*) and it directly monitors the tuner coefficient Γ_T. After a proper calibration procedure, Γ_S is directly available, but – this time – it is the DUT reflection coefficient Γ_{IN} that cannot be determined.

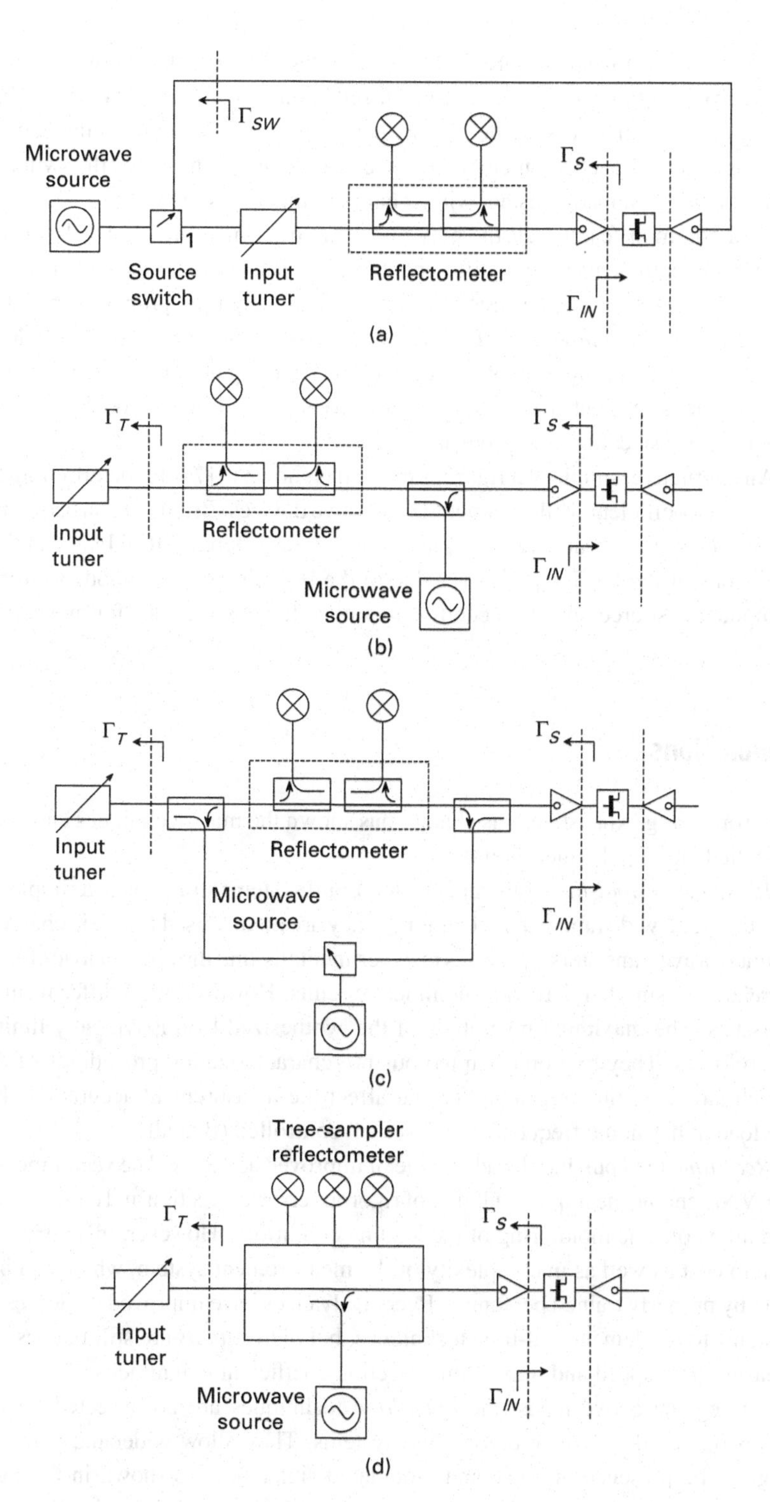

Fig. 13.23 Existing solutions for source reflection coefficient measurement. ©2001 IEEE. Reprinted, with permission, from [7].

The method shown in Figure 13.23(c) solves the latter problem in two steps [72]. First, the microwave signal is injected before the reflectometer and the DUT input characteristic is computed. Then, it is switched immediately after, and the source reflection coefficient is measured by the reflectometer in the reverse configuration. Again, the switch reflection coefficient is assumed constant while changing the switch position.

As a common feature, all the previous techniques measure the DUT and the source reflection coefficients by two different steps. For fast and automatic characterization of active devices, that can be time consuming. The technique proposed in [73] is based on the concept of a *three-sampler reflectometer* (see Figure 13.23(d)), which allows the simultaneous determination of source and DUT input gamma. This technique is indeed fast and accurate, but it is based on an unconventional error model and it requires a special-purpose calibration procedure.

An alternative, simple, yet rigorous method is shown in [74] for determining the source reflection coefficient while characterizing active devices. Briefly, it consists in measuring the waves at the input reference plane under two different DUT bias conditions. The variations of the DUT input waves due to the bias change give enough information to compute the source reflection coefficient with sufficient accuracy for most applications.

13.8 Conclusions

Far from being exhaustive, this chapter has shown the main techniques and issues of the so-called "load-pull" measurements.

Passive, *non-real-time*, fundamental load-pull systems still provide a simple and robust solution for a wide range of applications. They are mainly used for basic characterization of microwave transistors, as well as to experimentally find the optimal loads for the design of relatively simple microwave nonlinear circuits. However, they suffer from a number of issues. The maximum magnitude of the synthesized load is typically limited by the setup losses. They rely on often tedious pre-characterization procedures of the tuners, which have a limited repeatability and affect the measurement accuracy [44]. Finally, the load at harmonic frequencies is not well controlled (if at all).

Real-time load pull has the advantage of improved accuracy. The vector measurements of a VNA enable the implementation of rigorous error-correction and calibration methods and allow on-line monitoring of the loading conditions. However, this introduces a first step in cost as well as in complexity of the measurement system, which can be handled only by properly trained personnel. Recent advances have improved the losses in passive systems to implement real-time techniques, but *active loads* are still the best option for measurement speed and maximum reflection coefficient magnitudes.

Among the active loads, the *open-loop* techniques are not affected by oscillation issues that could occur in *active-loop* systems. They allow wideband characterization (e.g. in the presence of wideband modulated signals), as is shown in Chapter 14, but they require a more complicated control to synthesize and maintain the desired load. However, load setting is much easier and more robust with active-loop techniques, which also allow straightforward control of common vs. differential-mode load impedances in mixed-mode measurements.

If the DUT operates in strong nonlinear conditions, such as for high-efficiency amplifiers, *harmonic load-pull* becomes important to keep the loading conditions at harmonic frequencies under control. Similarly, if the device is used in wideband conditions (i.e. with bandwidths of some MHz) it is important that the load remains constant within the input signal bandwidth to resemble the real operating conditions of the device. This topic is discussed in detail in Chapter 14. In both cases, however, the system measurement setup is complicated and no off-the-shelf solutions are available on the market.

When the complete signal waveforms at the DUT input and output ports are needed (e.g. for nonlinear model extraction), load-pull *time domain waveform* techniques are applied. *Time domain sampling-based* techniques allow fast waveform acquisition but suffer from limited dynamic range related to the broadband oscilloscope input stages. *VNA-based* techniques measure the different harmonic components in the frequency domain separately, which generally slows down the measurement. However, the high dynamic range instrinsic to the VNA narrow-band measurement principle makes this technique suitable for millimeter-wave applications. As a final consideration, load-pull waveform techniques provide the most comprehensive tool for thorough large-signal device characterization, but always at the price of a considerable test setup complexity, which makes them mainly suitable for advanced R&D labs.

References

[1] D. Hamilton, J. Knipp, and J. Kuper, *Klystrons and Microwave Triodes*. New York: McGraw-Hill, 1948.

[2] G. Heiter, "Characterization of nonlinearities in microwave devices and systems," *IEEE Trans. Microw. Theory Tech.*, MTT-21, pp. 797–805, Dec. 1973.

[3] J. Cusack, S. Perlow, and B. Perlman, "Automatic load contour mapping for microwave power transistors," *IEEE Trans. Microw. Theory Tech.*, MTT-22, pp. 1146–1152, Dec. 1974.

[4] D. Le and F. Ghannouchi, "Multitone characterization and design of FET resistive mixers based on combined active source-pull/load-pull techniques," *IEEE Trans. Microw. Theory Tech.*, MTT-46, pp. 1201–1208, Sept. 1998.

[5] F. Ghannouchi and R. Bosisio, "Source-pull/load-pull oscillator measurements at microwave/mm wave frequencies," *IEEE Trans. Microw. Theory Tech.*, MTT-41, pp. 32–35, Feb. 1992.

[6] E. Davis, "Rieke diagrams for avalanche diodes show performance as a function of load," *Proc. IEEE*, 55, pp. 1521–1522, Aug. 1967.

[7] A. Ferrero, V. Teppati, G. Madonna, and U. Pisani, "Overview of modern load-pull and other nonlinear measurement systems," in *58th ARFTG Conference Digest*, San Diego, USA, Nov. 2001.

[8] S. C. Cripps, "A theory for the prediction of GaAs FET load-pull power contours," in *IEEE MTT-S Intl. Microwave Symp. Dig.*, Boston, MA, May 1983, pp. 221–223.

[9] S. C. Cripps, "Old-fashioned remedies for GaAs FET power amplifier designers," *IEEE MTT-S Newsletters*, pp. 13–17, Summer 1991.

[10] F. Sechi, "High efficiency microwave FET power amplifier," *Microwave J.*, pp. 59–63, Nov. 1981.

[11] I. Bahl, E. Griffin, A. Geissberger, C. Andricos, and T. Brukiewa, "Class-B power MMIC amplifiers with 70 percent power-added efficiency," *IEEE Trans. Microw. Theory Tech.*, MTT-37, pp. 1315–1320, Sept. 1989.

[12] D. Snider, "A theoretical analysis and experimental confirmation of the optimally loaded and overdriven RF power amplifier," *IEEE Trans. Electron Devices*, ED-14, pp. 851–857, Dec. 1967.

[13] A. Ferrero and U. Pisani, "Large signal 2nd harmonic on wafer MESFET characterization," in *36th ARFTG Conf. Dig.*, Monterrey, CA, Dec. 1990, pp. 101–106.

[14] Maury Microwave Corporation [Online]. Available: http://www.maurymw.com.

[15] Focus Microwaves Inc. [Online]. Available: http://www.focus-microwaves.com.

[16] F. Sechi, R. Paglione, B. Perlman, and J. Brown, "A computer controlled microwave tuner for automated load pull," *RCA Review*, 44, pp. 566–572, Dec. 1983.

[17] J. F. Sevic, "A sub 1 Ω load-pull quarter wave prematching network based on a two-tier TRL calibration," *IEEE Trans. Antennas Propag.*, 47, no. 2, pp. 389–391, Feb. 1999.

[18] C. Tsironis, "Prematched programmable tuners for very high VSWR testing," in *IEEE μAPS Microwave Application & Product Seminars*, Anaheim, CA, June 1999.

[19] R. Stancliff and D. Poulin, "Harmonic load-pull," in *IEEE International Microwave Symposium Digest*, Orlando, FL, Apr. 1979, pp. 185–187.

[20] Focus Microwaves, "An affordable harmonic load pull setup," *Microwave J.*, pp. 180–182, Oct. 1998.

[21] V. Camarchia, V. Teppati, S. Corbellini, and M. Pirola, "Microwave measurements part II: nonlinear measurements," *IEEE Instrum. Meas. Mag.*, 10, no. 3, pp. 34–39, June 2007.

[22] P. Colantonio, F. Giannini, and E. Limiti, "Nonlinear approaches to the design of microwave power amplifiers," *International Journal of RF and Microwave Computer-Aided Engineering*, 14, pp. 493–506, Nov. 2004.

[23] Verspecht-Teyssier-DeGroote s.a.s. [Online]. Available: http://www.vtd-rf.com.

[24] Antwerta Microwave [Online]. Available: http://www.anteverta-mw.com.

[25] NMDG NV [Online]. Available: http://www.nmdg.be.

[26] Mesuro Ltd. [Online]. Available: http://www.mesuro.com.

[27] High Frequency Engineering [Online]. Available: http://www.hfemicro.com.

[28] Y. Takayama, "A new load-pull characterization method for microwave power transistor," in *IEEE MTT-S Intl. Microwave Symp. Dig.*, Cherry Hill, NJ, June 1976, pp. 218–220.

[29] J. Benedikt, R. Gaddi, P. J. Tasker, and M. Goss, "High-power time domain measurement system with active harmonic load-pull for high-efficiency base-station amplifier design," *IEEE Trans. Microw. Theory Tech.*, 48, no. 12, pp. 2617–2624, Dec. 2000.

[30] F. Ghannouchi, F. Larose, and R. Bosisio, "A new multiharmonic loading method for large-signal microwave and millimeter-wave transistor characterization," *IEEE Trans. Microw. Theory Tech.*, 39, pp. 986–992, June 1991.

[31] J. Nebus and J. V. P. Bouysee, J. M. Coupat, "An active load-pull setup for the large signal characterization of highly mismatched microwave power transistors," in *Instrumentation and Measurement Technology Conference*, 18–20 May 1993, pp. 2–5.

[32] G. Bava, U. Pisani, and V. Pozzolo, "Active load technique for load-pull characterization at microwave frequencies," *Electronic Lett.*, 18, pp. 178–179, Feb. 1982.

[33] V. Teppati, A. Ferrero, and U. Pisani, "Recent advances in real-time load-pull systems," *IEEE Trans. Instr. Meas.*, 57, no. 11, pp. 2640–2646, Nov. 2008.

[34] A. Ferrero, "Active load or source impedance synthesis apparatus for measurement test set of microwave components and systems," *U.S. Patent 6 509 743*, 12 June 2000.

[35] J.-M. Nebus, J.-M. Coupat, and J.-P. Villotte, "A novel, accurate load-pull setup allowing the characterization of highly mismatched power transistors," *IEEE Trans. Microw. Theory Tech.*, MTT-42, no. 2, pp. 327–332, Feb. 1994.

[36] Z. Aboush, J.Lees, J. Benedikt, and P. Tasker, "Active harmonic load-pull system for characterizing highly mismatched high power transistors," in *IEEE MTT-S Symp. Dig.*, Long Beach, CA, 12–17 June 2005, pp. 1311–1314.

[37] F. Ghannouchi, M. Hashmi, S. Bensmida, and M. Helaoui, "Loop-enhanced passive source and load-pull technique for high reflection factor synthesis," *IEEE Trans. Microw. Theory Tech.*, MTT-58, no. 11, pp. 2952–2959, Nov. 2010.

[38] A. Ferrero and V. Teppati, "A novel active differential/common-mode load for true mixed-mode load-pull systems," in *International Microwave Symposium*, San Francisco, CA, USA, June 2006, pp. 1456–1459.

[39] M. P. van der Heijden, D. Hartskeerl, I. Volokhine, V. Teppati, and A. Ferrero, "Large-signal characterization of an 870 MHz inverse class-F crosscoupled push-pull PA using active mixed-mode load-pull," in *Radio Frequency Integrated Circuits (RFIC) Symposium, 2006 IEEE*, San Francisco CA, USA, June 2006, pp. 389–392.

[40] V. Teppati, M. Garelli, V. Camarchia, A. Ferrero, and U. Pisani, "Advanced load-pull techniques – from single-ended to multiport/differential measurement systems," in *European Microwave Week Workshop WSW6*, Munchen, Oct. 2007, pp. 105–120.

[41] V. Teppati, A. Ferrero, M. Garelli, and S. Bonino, "A comprehensive mixed-mode time domain load- and source-pull measurement system," *IEEE Trans. Instrum. Meas.*, 59, no. 3, pp. 616–622, Mar. 2010.

[42] V. Teppati, M. Goano, A. Ferrero, V. Niculae, A. Olivieri, and G. Ghione, "Broad-band coaxial directional couplers for high-power applications," *IEEE Trans. Microw. Theory Tech.*, MTT-51, no. 3, pp. 994–997, Mar. 2003.

[43] V. Teppati and A. Ferrero, "A new class of non-uniform, broadband, non-symmetrical rectangular coaxial-to-microstrip directional couplers for high power applications," *IEEE Microwave and Wireless Components Letters*, 13, no. 4, pp. 152–154, Apr. 2003.

[44] V. Teppati and C. R. Bolognesi, "Evaluation and reduction of calibration residual uncertainty in load-pull measurements at millimeter-wave frequencies," *IEEE Trans. Instrum. Meas.*, 61, no. 3, pp. 817–822, Mar. 2012.

[45] R. Tucker and P. Bradley, "Computer-aided error correction of large-signal load pull measurements," *IEEE Trans. Microw. Theory Tech.*, MTT-32, pp. 296–300, Mar. 1984.

[46] I. Hecht, "Improved error-correction technique for large-signal load-pull measurements," *IEEE Trans. Microw. Theory Tech.*, MTT-35, pp. 1060–1062, Nov. 1987.

[47] A. Ferrero and U. Pisani, "An improved calibration technique for on-wafer large-signal transistor characterization," *IEEE Trans. Instrum. Meas.*, IM-47, pp. 360–364, Apr. 1993.

[48] A. Ferrero, U. Pisani, and K. Kerwin, "A new implementation of a multiport automatic network analyzer," *IEEE Trans. Microw. Theory Tech.*, MTT-40, pp. 2078–2085, Nov. 1992.

[49] F. Larose, F. Ghannouchi, and R. Bosisio, "A new multi-harmonic load-pull method for nonlinear device characterization and modeling," in *IEEE MTT-S Intl. Microwave Symp. Dig.*, 1, May 1990, pp. 443–446.

[50] D. Barataud, C. Arnaud, B. Thibaud, M. Campovecchio, J. Nebus, and J. Villotte, "Measurements of time domain voltage/current waveforms at RF and microwave frequencies based on the use of a vector network analyzer for characterization of nonlinear devices — application

to high-efficiency power amplifiers and frequency-multipliers optimization," *IEEE Trans. Instrum. Meas.*, IM-47, pp. 1259–1264, Oct. 1998.

[51] D. Barataud, M. Campovecchio, and J.-M. Nebus, "Optimum design of very high-efficiency microwave power amplifiers based on time domain harmonic load-pull measurements," *IEEE Trans. Microw. Theory Tech.*, 49, no. 6, pp. 1107–1112, June 2001.

[52] R. Gaddi, J. Pla, J. Benedikt, and P. Tasker, "LDMOS electro-thermal model validation from large-signal time domain measurements," *IEEE MTT-S Symp. Dig.*, 1, pp. 399–402, 2001.

[53] O. Jardel, F. D. Groote, T. Reveyrand, J.-C. Jacquet, C. Charbonniaud, J.-P. Teyssier, D. Floriot, and R. Quere, "An electrothermal model for AlGaN/GaN power HEMTs including trapping effects to improve large-signal simulation results on high VSWR," *IEEE Trans. Microw. Theory Tech.*, MTT-55, no. 12, pp. 2660–2669, Dec. 2007.

[54] D. Williams, J. Leckey, and P. Tasker, "A study of the effect of envelope impedance on intermodulation asymmetry using a two-tone time domain measurement system," *IEEE MTT-S Symp. Dig.*, 3, pp. 1841–1844, June 2002.

[55] Hewlett-Packard Co., *The microwave transition analyzer: a versatile measurement set for bench and test*. Rohnert Park, CA: HP Product Note 70820-1, 1991.

[56] J. Verspecht, F. Verbeyst, and M. V. Bossche, "Large-signal network analysis: Going beyond S-parameters," in *URSI 2002*, Maastricht, Netherlands, Aug. 2002.

[57] J. Verspecht, "Large-signal network analysis," *Microwave Magazine*, 6, no. 4, pp. 82–92, Dec. 2005.

[58] V. Teppati, A. Ferrero, V. Camarchia, A. Neri, and M. Pirola, "Microwave measurements part III: Advanced nonlinear measurements," *IEEE Instrum. Meas. Mag.*, 11, no. 6, pp. 17–22, Dec. 2008.

[59] U. Lott, "Measurement of magnitude and phase of harmonic generated in nonlinear microwave two-ports," *IEEE Trans. Microw. Theory Tech.*, MTT-37, no. 10, pp. 1506–1510, Oct. 1989.

[60] L. Gommé and Y. Rolain, "Design and characterisation of an RF pulse train generator for large-signal analysis," *Measurement Science and Technology*, accepted for publication, 2009.

[61] J. Verspecht, "Broadband sampling oscilloscope characterization with the "nose-to-nose" calibration procedure: a theoretical and practical analysis," *IEEE Trans. Instrum. Meas.*, 44, no. 6, pp. 991–997, Dec. 1995.

[62] P. Blockley, D. Gunyan, and J. Scott, "Mixer-based, vector-corrected, vector signal/network analyzer offering 300kHz–20GHz bandwidth and traceable phase response," in *IEEE International Microwave Symposium Digest*, Long Beach, CA, June 2005.

[63] V. Teppati, V. Camarchia, A. Ferrero, M. Pirola, and S. D. Guerrieri, "A comprehensive GaN HEMT characterization for power amplifier design," in *IEEE Topical Workshop on Power Amplifiers for Wireless Communications*, San Diego, USA, 16–17 Jan. 2006, pp. 1–2.

[64] P. Colantonio, F. Giannini, E. Limiti, and V. Teppati, "An approach to harmonic load- and source-pull measurements for high-efficiency pa design," *IEEE Trans. Microw. Theory Tech.*, MTT-52, no. 1, pp. 191–198, Jan. 2004.

[65] V. Adamian and A. Uhlir, "A novel procedure for receiver noise characterization," *IEEE Trans. Instr. Meas.*, IM-22, pp. 181–182, June 1973.

[66] G. Caruso and M. Sannino, "Determination of microwave two-port noise parameters through computer-aided frequency-conversion techniques," *IEEE Trans. Microw. Theory Tech.*, MTT-27, pp. 779–783, Sept. 1979.

[67] A. Davidson, B. Leake, and E. Strid, "Accuracy improvements in microwave noise parameter measurements," *IEEE Trans. Microw. Theory Tech.*, MTT-37, pp. 1973–1978, Dec. 1979.

[68] D. Le and F. Ghannouchi, "Noise measurements of microwave transistor using an uncalibrated mechanical stub tuner and a built-in reverse six-port reflectometer," *IEEE Trans. Instrum. Meas.*, IM-44, pp. 847–852, Aug. 1995.

[69] P. Colantonio, F. Giannini, E. Limiti, and V. Teppati, "An approach to harmonic load- and source-pull measurements for high-efficiency PA design," *IEEE Trans. Microw. Theory Tech.*, MTT-52, no. 1, pp. 191–198, Jan. 2004.

[70] B. Hughes and P. Tasker, "Improvements to on-wafer noise parameter measurements," in *36th ARFTG Conf. Dig.*, Monterrey, CA, Nov. 1990, pp. 16–25.

[71] D. Le and F. Ghannouchi, "Source-pull measurements using reverse six-port reflectomenters with application to MESFET mixer design," *IEEE Trans. Microw. Theory Tech.*, MTT-42, pp. 1589–1595, Sept. 1994.

[72] G. Berghoff, E. Bergeault, B. Huyart, and L. Jallet, "Automated characterization of HF power transistor by source-pull and multiharmonic load-pull measurements based on six-port techniques," *IEEE Trans. Microw. Theory Tech.*, MTT-46, pp. 2068–2073, Dec. 1998.

[73] G. Madonna, M. Pirola, A. Ferrero, and U. Pisani, "Testing microwave devices under different source impedances: a novel technique for on-line measurement of source and device reflection coefficents," in *IMTC/99 Conf. Proc.*, Venezia, Italy, May 1999, pp. 130–133.

[74] G. Madonna and A. Ferrero, "Simple technique for source reflection coefficient measurement while characterizing active devices," in *53nd ARFTG Conf. Dig.*, Anaheim, CA, June 1999, pp. 104–106.

14 Broadband large signal measurements for linearity optimization

Marco Spirito and Mauro Marchetti

14.1 Introduction

The recent introduction of high-performance modulation schemes (e.g. (W)-CDMA and OFDM) provides the capability of realizing high-data rate communication links (i.e. up to 100 Mbps from 20 MHz spectrum)[1]. The broadband nature of those signals together with the large difference between the peak and the average power across the modulation bandwidth requires a large number of spectral components to accurately represent the signal statistics. The modulated signal should be amplified by the transmitting chain without loss of information (i.e. low EVM) and with little out-of-band-distortion to avoid interference with adjacent transmitting channels. The quality of the communication link can be translated into specification parameters of the active element of the transmission chain, in the case of the PA, through the device IM_3 and ACPR level. In general, it is very difficult to link the technology parameters of an active device directly to its linearity performance, since the linearity achieved for a given PA is the result of its interaction with the surrounding circuitry. For this reason, most attempts to improve the linearity of PAs are currently made at the circuit level.

In order to properly compare different technologies (e.g. SiGe and III-V) or device technology generations, one must provide the optimum loading conditions, at fundamental, harmonic, and baseband frequencies, to the active device during the evaluation phase, ideally under the same driving signal of the final application. This measurement task is intrinsically complex since the broadband nature of the signal of interest conflicts with the narrow-band nature of the currently employed high dynamic range receivers (i.e. narrow IF bandwidth super-heterodyne receivers). Moreover, the capability of conventional load-pull setups (active and passive) allows the accurate control of the reflection coefficient only in a narrow frequency bandwidth. In this chapter we review the active load-pull techniques and test-benches that have been developed in the last years [2–4] to facilitate large signal measurements with digitally modulated (standard compliant) broadband signals. These test-benches allow optimization of the device large-signal performance, e.g. saturated output power, PAE as well as its linearity performance, such as IM_3 and ACPR, providing the capability to link technology generations with linearity performance.top

The chapter is organized as follows: first, the problem statement of electrical delay in load-pull systems is introduced. Second, the architecture of two broadband load-pull test benches (i.e. closed-loop and mixed-signal open-loop) and their various building blocks are analyzed. Then the power and linearity requirements for the amplifiers used in the active loads are analyzed. In conclusion, some experimental results presenting a linearity-optimized SiGe PA and a high power LDMOS are given.

14.2 Electrical delay in load-pull systems

When testing the large-signal performance of devices or circuits with modulated signals, it is important to control the reflection coefficient offered to the DUT, not only within the modulation bandwidth (i.e. at the fundamental frequency), but also within the frequency bands where the nonlinear device generates power, e.g. adjacent channels, harmonics

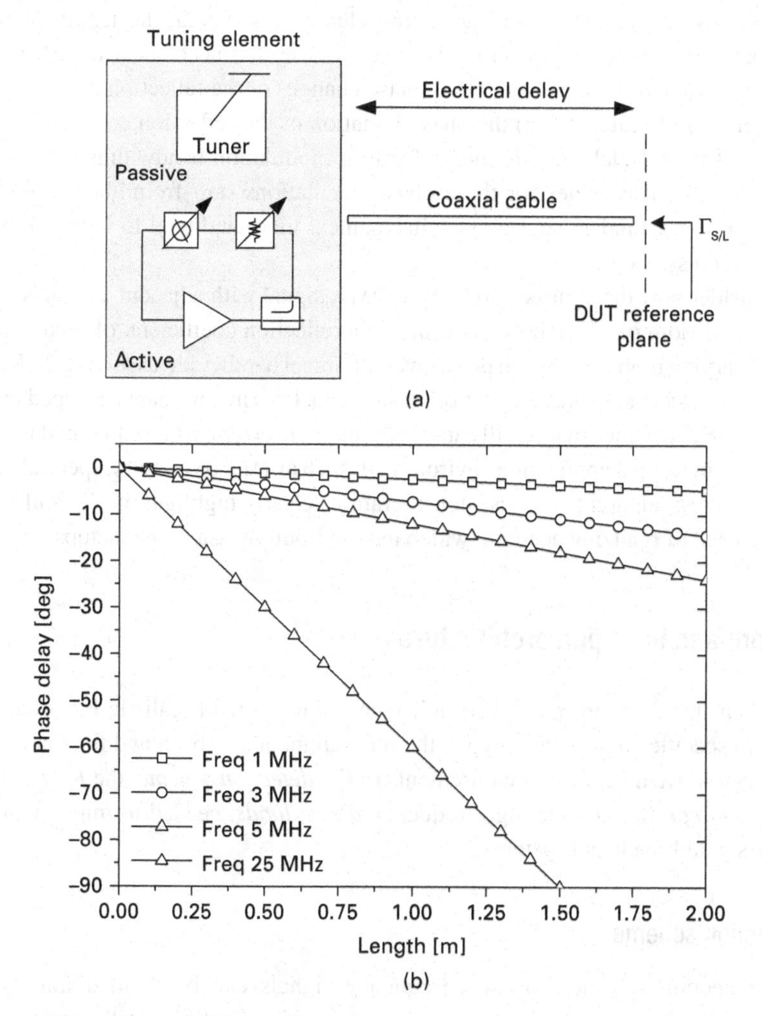

Fig. 14.1 (a) Phase delay caused by the electrical lengths of the cables present between the tuning element and the DUT (b) phase rotation of the reflection coefficient at the DUT reference plane as a function of cable length and signal bandwidth.

and baseband frequencies [3, 5]. The reason for this stringent requirement comes from the fact that load-pull can be an accurate predictive tool of the device performance in the application environment (i.e. IC or board level), only when the impedance provided by the measurement setup closely resembles those offered to the DUT in the final operating conditions.

In the final circuit implementations, matching networks are placed in close proximity to the DUT with distances in the order of a few hundred microns for ICs to a few millimeters for PCB assemblies. These distances are negligible, in terms of electrical lengths, with respect to the modulation bandwidths of the signals employed today, e.g. up to hundreds of MHz. However, in all conventional load-pull setups the actual tuning element (i.e. tuner or active loop) is always located at some distance from the DUT (Figure 14.1(a)), which is much larger than for any practical matching network. This distance, as well as any physical length within the tuning element itself (e.g. the length of the active feedback loop, or the position of the probe inside the mechanical tuner), yields very large electrical delays, causing rapid phase changes of the reflection coefficients versus frequency. In Figure 14.1(b) the phase variation of the reflection coefficient versus the length of the coaxial cable is shown for four modulation bandwidths (i.e. 1, 3, 5, and 25 MHz). Typical values for these phase fluctuations start from about a few degrees per MHz for a fundamental passive mechanical tuner and tend to be much higher for active-loop systems.

Consider now that a three carrier W-CDMA signal with adjacent channels provides a total bandwidth of 25 MHz. This results in a reflection coefficient, offered by a passive tuner, varying more than 50° in phase over the signal bandwidth (assuming 2°/MHz phase variation). Such large phase variation, which translates in non-realistic impedance conditions provided to the device, will cause measurement errors, when compared to the device response in its final application environment, such as IM_3 asymmetry, spectral re-growth, and PAE degradation [4]. These considerations clearly highlight the difficulties as well as the need of realizing accurate wideband load-pull measurement setups.

14.3 Broadband load-pull architectures

Test-benches that can make load-pull measurements with realistic broadband signals require specific choices, mainly for the measurement approach and the load control. In this section we analyze the requirements of the *detection scheme*, the *RF front-end*, the *baseband control*, and the high-frequency *active loads*, needed to implement accurate large-signal broadband systems.

14.3.1 Detection scheme

The detection schemes for high-frequency signals can be divided into two major categories: direct acquisition (i.e. real-time or sub-sampling oscilloscope based) and down-converted acquisition (i.e. heterodyne mixer or sampler-based architectures). While real-time oscilloscopes are starting to be available with high sampling rates and high-frequency front-end capabilities [6], i.e. above 26 GHz, the 8-bit vertical resolution in combination with the higher noise level (due to the large bandwidth where the noise is received) make such receivers unsuitable for device linearity characterization. Sub-sampling oscilloscopes, at the other end of the scale, provide large vertical resolution for the low-frequency ADC (i.e. better than 14 bits) with high input frequency capabilities (i.e. higher than 60 GHz) [7]. Nevertheless, the low sampling speed of the sampler (i.e. below 1 MHz) combined with the higher noise bandwidth, requires a long measuring time and large averaging to reach a dynamic range in the order of 70 dB. Moreover, when considering realistic modulated signals (i.e. more than 20 000 frequency bins in a 4 MHz span), the limiting factor is the memory depth of the instrument. For these reasons, large-signal load-pull test-benches employing this detection scheme have usually been confined to multitone excitations with a limited number of tones [8]. The rest of this section is devoted to a more in-depth analysis of heterodyne architectures that allow the detection of standard compliant modulated signals.

These large-signal test-benches rely on a tuned-receiver architecture to provide high signal sensitivity and large dynamic ranges. In such detection schemes, the high-frequency signals are translated to lower intermediate frequencies. Narrow-band filtering of the down-converted signal is employed to reduce noise and increase the detection dynamic range. When employing broadband signals, narrow-band filtering is avoided and the full ADC bandwidth or an external spectrum analyzer is used to acquire the modulated signal. The signal down-conversion to an intermediate frequency can be achieved with either high-frequency mixers or sub-sampling-based systems.

When employing high-frequency mixers, the LO is swept over the signal harmonics down-converting bandwidths up to few GHz for commercially available mixers, with an input third-order intercept point above 20 dBm [9], [10]. Such components provide a high system linearity so that the intermodulation distortion products of linear amplifiers can be properly detected. Note that the harmonic distortion products should be at least 18 dB below the harmonic distortion level of the DUT to guarantee low linearity measurement uncertainties [11]. Recalling that the relation between the IM_3 products expressed in dBc (ΔIM_3) and the OIP_3 is given by [12]:

$$OIP_3 = P_{out} + \frac{\Delta IM_3}{2}, \tag{14.1}$$

the mixer OIP_3 needs to be at least 9 dB higher than the device under test OIP_3.

Moreover, when used in combination with wideband ADCs [13] these detection schemes allow the down-conversion and sampling of a large portion of the spectrum around the carrier frequency and the harmonics (i.e. three carrier W-CDMA signals with adjacent and alternate channels providing a total bandwidth of 35 MHz).

Sub-sampling based systems employ a sampler down-conversion, driven by a precise low-frequency signal (typically 10–25 MHz) phase locked to the 10 MHz crystal that provides the frequency reference to the signal driving the DUT. The sampling pulses are created by a step recovery diode [14] or a nonlinear transmission line [15] and allow the down-conversion of the entire system RF bandwidth to the ADC acquisition band. When considering modulated signals, the frequency of the LO signal driving the samplers must be chosen properly to avoid overlapping of different tones on the same baseband frequency. The minimum frequency windows, and hence LO frequency, to properly down-convert a modulated signal with N_{SSB} single-sideband tones and considering N_H harmonics of interest, is given by [16].

$$f_{BW} = f_{LO_{min}} = (2N_H + 1)(2N_{SSB} + 1)\, f_{RES}, \tag{14.2}$$

where f_{RES} is the required resolution frequency, which is the inverse of the required measurement time. Consider for example the measurement of a modulated signal with the following test conditions: a standard compliant W-CDMA signal including the upper and lower adjacent channels, resulting in a 15 MHz wide signal (i.e. 5000 tones per channel assuming a 1 KHz spacing between tones) and a measurement time of 0.5 msec (i.e. translating into a 500 Hz frequency resolution).

When we substitute these values in (14.2), we obtain:

$$\begin{aligned} N_H &= 3 \\ N_{SSB} &= 3 \times 2500 \\ f_{RES} &= 500Hz \\ f_{BW} &= f_{LO_{min}} \approx 52.5MHz. \end{aligned} \tag{14.3}$$

Note the 0.5 msec window chosen here is given as an example representing a reasonable measurement time to allow for sufficient averaging to reach the required measurement dynamic range. These results indicate that when choosing realistic measurement times and memory depths (i.e. f_{RES} not too small) typical sampler architectures (i.e. f_{LO} 10–25 MHz) are not indicated for measurements of standard compliant modulated signals. For this reason, in the rest of the chapter we will only consider mixer-based architectures.

14.3.2 RF front-end

When measuring devices are matched for optimum linearity, it is worth noting that there will be a dramatic difference between power levels in the fundamental and harmonic frequency bands. This is due to the high OIP_2 and OIP_3 of such devices.

Using a single mixer to down-convert the entire frequency band poses severe challenges to an accurate detection for the following reasons:

- using an attenuator to optimize the power level at the mixer RF port will increase the system noise floor (see Figure 14.2(a)), preventing the measurement of the "low-power" harmonic components;
- when no attenuation is used, the nonlinearities of the mixer itself, which is over-driven at the fundamental tone, prevent the correct measurement of these "low-power" harmonic components (see Figure 14.2(b)).

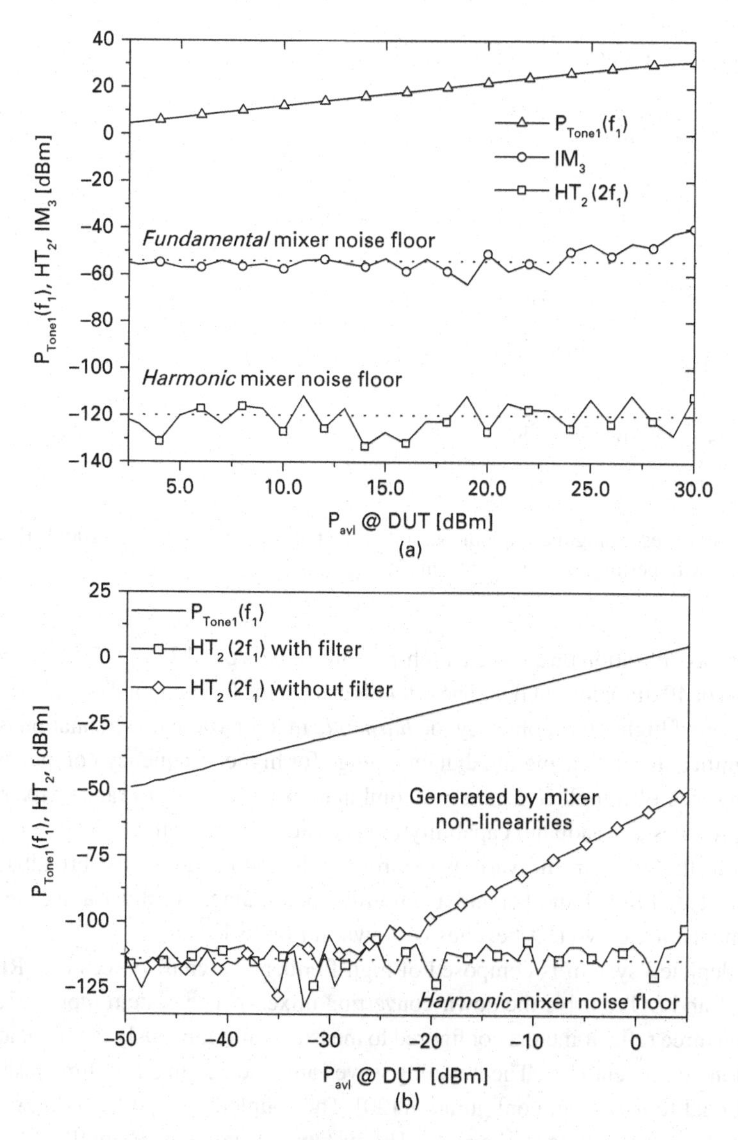

Fig. 14.2 Two-tone measurement on calibration Thru ($f_c = 2.14$ GHz, $\Delta f = 0.2$ MHz), showing the corrected measured power at the fundamental; (a) amplitude of fundamental, IM_3 and HT_2 components measured by the HP 8510 (high power levels), (b) 2nd-harmonic, with and without high-pass filter (low power levels). © [2004] IEEE. Reprinted, with permission, from [17].

To overcome this restriction, a multi-branch mixer configuration was first introduced in [18]. Using this approach each of the waves (i.e. a and b) coupled out by the input and output reflectometers can be routed to two mixers; see Figure 14.3. A *fundamental* mixer, used to measure the fundamental frequency band, employing an attenuator to maximize the system dynamic range for the high power (i.e. fundamental) tones. All the higher harmonics are measured by the *harmonic* mixer, using a high-pass filter in the signal path. Figure 14.2 (b) [17] presents the calibrated data for a two-tone power

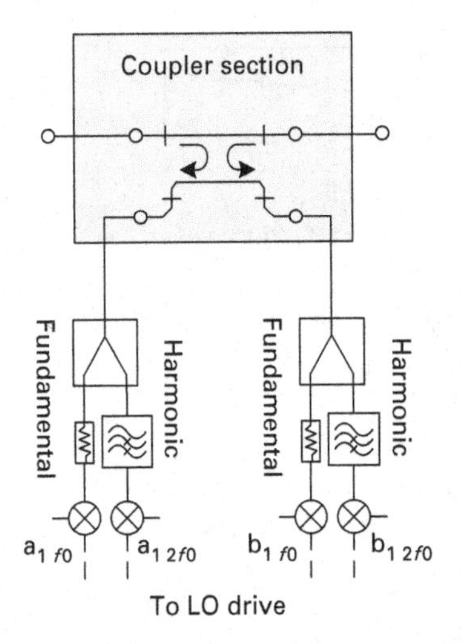

Fig. 14.3 Block scheme describing the implementation of the multi-branch mixer. © [2006] IEEE. Reprinted, with permission, adapted from [3].

sweep on a calibration thru using a high-pass filter (pass band 3.0–26.5 GHz) providing a rejection of 40 dB below 3 GHz. The filter blocks out the fundamental signal, avoiding the generation of higher harmonics by the *harmonic* mixer (where no attenuation is present) and significantly increasing the dynamic range for higher-frequency components.

The choice of a high IP_3 mixer in combination with a multi-branch mixer configuration provides a broadband capability to accurately measure the reflection coefficients offered to the DUT in the various control bands. Two large signal broadband setups were presented in [3] and [4] and a simplified block diagram, illustrating the common components in the two test-benches is shown in Figure 14.4.

The depicted system is composed of high-frequency signal sources (i.e. RF to drive the DUT and LO to drive the down-converting mixers). The system input RF section is based on three reflectometers configured to measure simultaneously the input and source reflection coefficient [19]. The traveling waves are detected in a real-time fashion using a traditional four-coupler configuration [20]. The coupled $a-$ and $b-waves$ are down-converted to a lower IF to be digitized. The system presented in [3] employs the HP 8510 receiver unit to process both the high-frequency and low-frequency (baseband) signal components. All the signals have to be down-converted (RF) or up-converted (baseband) to the first IF frequency of the instrument (i.e. 20 MHz). This is done through RF and baseband (BB) mixers. Note in Figure 14.4 the baseband mixers are not explicitly shown. The LO synthesizer provides the required down-converting or up-converting signal. Mechanical switches are used to route these IF signals to the HP 8510 mainframe. The system described in [4] employs wideband AD converters (100 MS/s sampling frequency) to digitize the IF signals. This architecture enables the direct measurement of the device reflection coefficients over a wide bandwidth of 40 MHz in a single data acquisition. With this hardware, wider bandwidths, up to 120 MHz, and the frequency content in the harmonic bands can be measured by stepping the frequency of the LO that

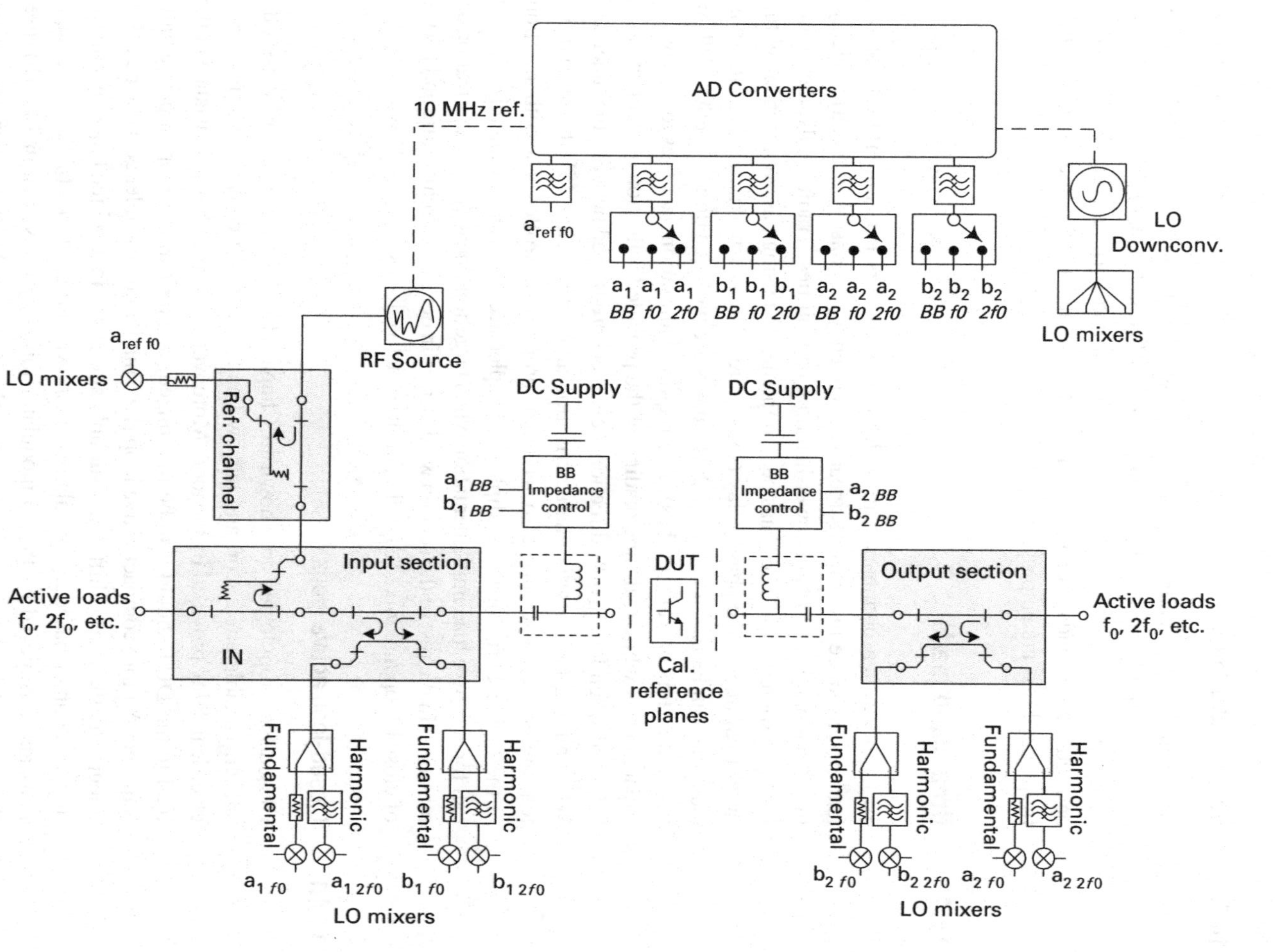

Fig. 14.4 Load-pull architecture for linearity optimization under broadband stimuli. © [2006] IEEE. Reprinted, with permission, adapted from [3].

drives the down-converting mixers. The large bandwidth of the AD receiver allows the baseband components to be measured directly without using up-converting mixers. In this configuration, IF electronic switches route the different signals to the receiver.

14.3.3 System calibration

Broadband system calibrations follow those of traditional large-signal setups and are, usually, a combination of the techniques described in [21] and [22], performed at all the frequency tones of interest at the fundamental and harmonic bands. Interpolation can be used when standard compliant modulated signals (composed by a large number of frequency bins) are employed.

14.4 Broadband loads

To facilitate measurements with wideband excitations it is important to optimize the load performance in terms of linearity, frequency response (i.e. electrical delay within the signal bandwidth), and power-handling capabilities. Traditionally, the tuning of the load and source impedances has been implemented using passive mechanical tuners. In order to minimize the phase delay (see Figure 14.1) and maximize the reflection coefficient that can be provided to the DUT, tuners are mounted directly at the DUT interface (i.e. connectors or wafer probes). This requires calibration data-files to retrieve the tuner S-parameters while real-time reading of the provided reflection coefficient can only be achieved when low-loss bi-directional couplers are placed between the tuner and DUT. When aiming for linearity characterization and optimization [5], the second harmonic load and source impedance also need to be properly controlled. The path insertion losses at the higher harmonics (i.e. diplexer, coupler sections, and eventually wafer-probes) call for an active load implementation when targeting linearity characterization and optimization. In the rest of the section we describe in detail the architecture and performance of closed-loop and mixed-signal open-loop active loads.

14.4.1 Closed-loop active loads

Closed-loop topologies are shown in Chapter 13. The use of an off-the-shelf phase-shifter, variable attenuator, and narrow-band filter often comes at the price of a large electrical delay provided by the loop. Moreover, since the physical distance of the active load to the DUT should also be minimized, compact active-loop implementations are required. A more compact active loop (i.e. smaller electrical length) can be achieved by integrating the phase-shifter and variable attenuator in a unique block by employing 90 degrees coupler-based IQ modulators, as shown in Figure 14.5 (a) [23]. The impedances that can be achieved by the IQ modulator topology are shown in 14.5 (b), presenting the provided reflection coefficients over the IQ plane. At RF frequencies (i.e. below 10 GHz) coupler-based IQ modulators can be realized using low-cost PCB technologies and PIN diodes. In Figure 14.6 [3], a board-level implementation of such an IQ modulator, operating at 2.14 GHz, is shown. The in-phase and quadrature resistances are tuned via

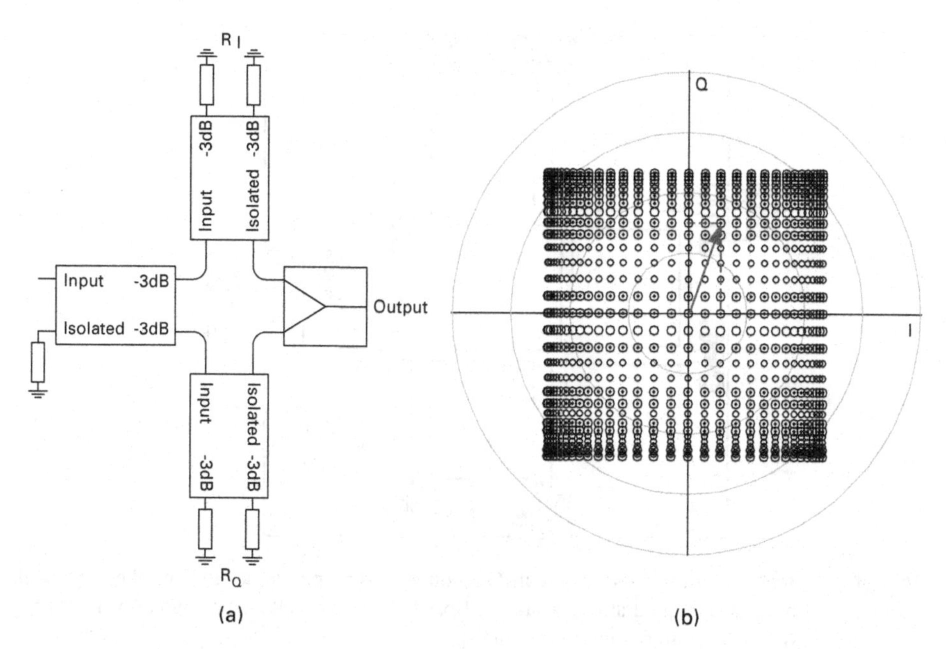

Fig. 14.5 (a) Simplified block diagram of an IQ modulator based on quadrature hybrids, (b) simulation of the IQ modulator constellation showing the I and Q signal required for a given phase delay and amplitude attenuation (arrow).

current-controlled PIN diodes. The operation of the modulator can be summarized as follows. The first 90 degrees coupler splits the signal into two orthogonal components (i.e. I and Q). The two signals are now amplitude-adjusted by changing the variable resistances (i.e. the PIN diodes) at the output ports of the two quadrature power splitters. As an example, when the terminating resistances are set to 50 Ohm no power is reflected and the I or Q signal provided at the input of the power combiner is zero; when the terminating resistances are low (high), the signals are strongly reflected and provided to the power combiner with phase inversion (no phase inversion). The final combination between the I and Q components achieves a 360-degree phase-shift and variable attenuation. A fixed attenuator at the input of the IQ board can be used to optimize the maximum power at the input of the modulator in order not to overdrive the diodes and thus avoid linearity degradation. During calibration, the IQ modulator can then be stepped in different positions in order to characterize the loop response and allow the user to request a specific reflection coefficient at the DUT reference plane.

Note that since the 90 degree couplers are implemented using transmission lines, higher-frequency implementations will be smaller (i.e. 61 × 65 mm at 4.28 GHz versus 96 × 101 mm at 2.14 GHz), which partially compensates the increased electrical delay provided by the fixed dimensions interconnects to the DUT.

As mentioned before, to minimize the overall $\angle\Gamma$ variation with frequency, the electrical length of the various sections of the active load-pull system must be kept as short as possible. In this respect, the small dimensions of the coupler-based IQ modulators allow

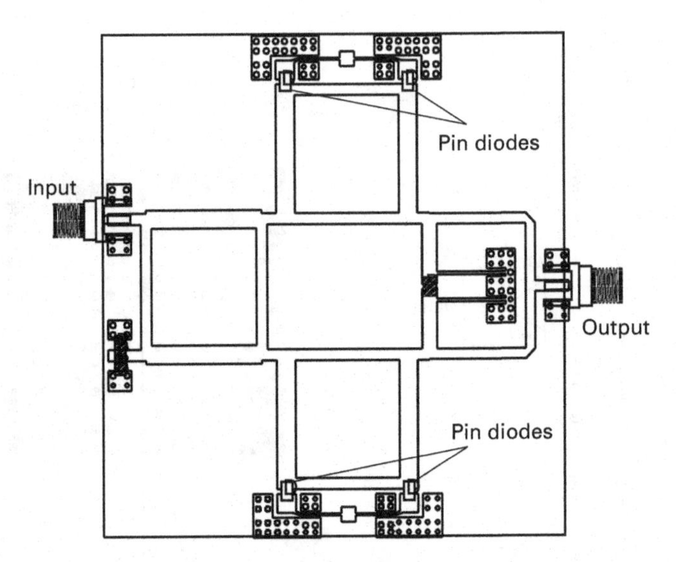

Fig. 14.6 Board-level implementation of IQ modulator using pin diodes to control electronically the in-phase and quadrature resistance. Board dimension is 96 x 101 mm (WxH). © [2006] IEEE. Reprinted, with permission, from [3].

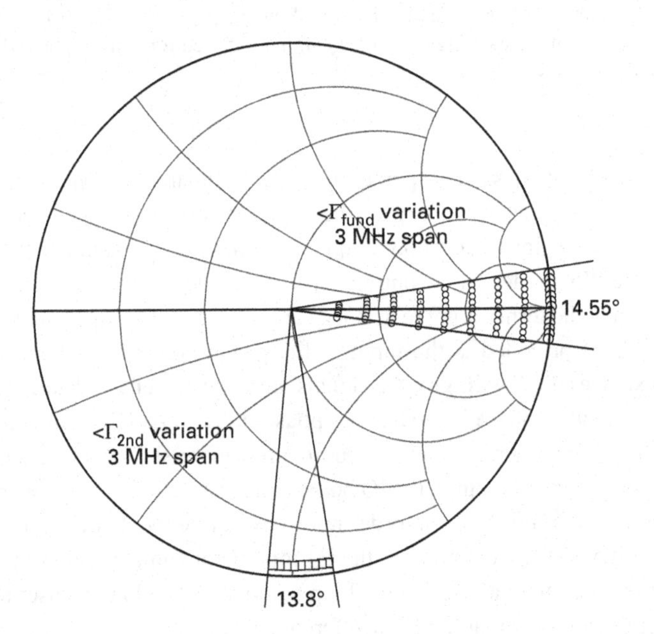

Fig. 14.7 $\angle\Gamma$ variation with frequency at the DUT reference plane for active loop topology at the output (frequency span 3 MHz). © [2006] IEEE. Reprinted, with permission, from [3].

for very compact active loops. The integration of a compact coupler-based IQ modulator placed in close proximity to the DUT [3] achieves the $\angle\Gamma$ variation with frequency shown in Figure 14.7. The impacts of the individual contributions to the $\angle\Gamma$ variation with frequency of the various system components are summarized in Table 14.1.

Table 14.1 Active Loop $\angle\Gamma$ variation

	Active Loop [°/MHz]
	input/output
probe + cable	0.9 / 0.9
coupler section	1.2 / 1.0
diplexer	0.4 / 0.4
active loop f_0	2.55 / 2.55
active loop $2f_0$	2.3 / 2.3
Total f_0	5.05 / 4.85
Total $2f_0$	4.8 / 4.6

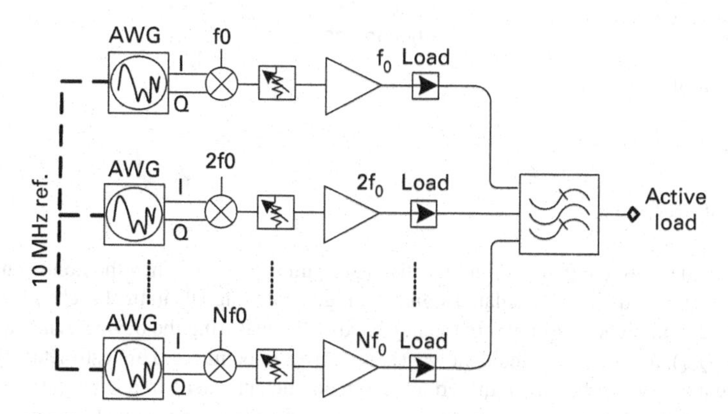

Fig. 14.8 Simplified block diagram of the wideband active loads with phase-coherent frequency up-conversion. © [2008] IEEE. Reprinted, with permission, from [4].

14.4.2 Mixed-signal active loads

In Chapter 13, wideband open-loop techniques were briefly introduced, in this section we discuss in details an open-loop concept based on signals generated by baseband AWGs and up-converted, to the desired RF frequency, using IQ mixers [4]. Although conceptually simple, this method requires high speed and high dynamic range DACs to generate the a-waves with high dynamic range (i.e. 70 dB spurious free dynamic range) and with a wide bandwidth to cover the needs of modern communication signals (i.e. larger than 100 MHz to cover a three carrier W-CDMA signal with adjacent and alternate channels at the third harmonic). Moreover, the signal injection at the various ports needs to be phase-coherent at both the RF frequencies as well as at the baseband.

The mixed-signal architecture described in [4, 24] integrates all the AWGs in a PCI extensions for instrumentation (PXI) express platform (enabling 5.5 GByte/s transfer capacity to the PXI backplane [13]), sharing the same timebase, making them fully synchronized. The source signal and all injection signals needed to synthesize the user-defined reflection coefficients at the DUT reference planes, are originating from fully synchronized (400 MS/s) arbitrary waveform generators, shown in Figure 14.8.

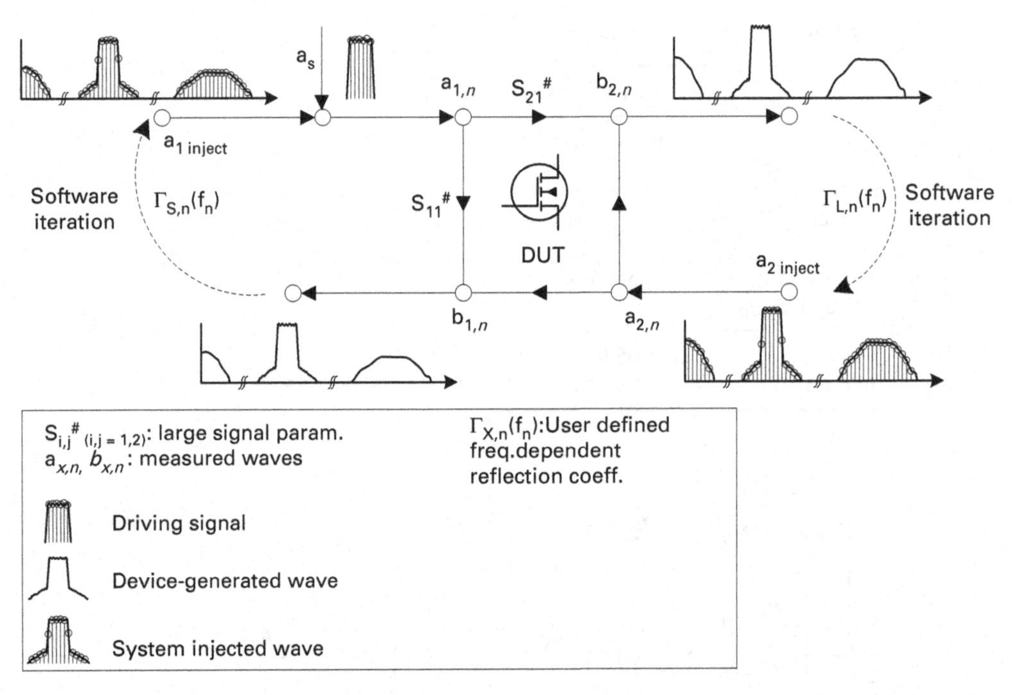

Fig. 14.9 Principle of the mixed-signal open-loop active load-pull approach. When the nonlinear DUT is excited with a user-defined modulated signal a_s, it generates signals in the baseband, fundamental, and higher harmonic frequency bands. By measuring the device-generated waves ($b_{1,n}$ and $b_{2,n}$), as well as the incident waves, the waves to be injected are estimated by successive iterations. When the required reflection coefficient versus frequency (at every controlled band) is achieved, the iteration has converged and the large signal parameters (e.g. PAE, P_{out}, IM_3, IM_5) are measured. © [2008] IEEE. Reprinted, with permission, from [4].

In order that the source and all the injected signals (fundamental and harmonics) are phase coherent, or in other words, are perfectly locked in phase and exhibit no phase drift among each other, IQ up-conversion is preferred over digital IF techniques [25]. By employing IQ up-conversion, a single synthesizer can be used to drive the local oscillator port of the IQ mixers (Figure 14.8). Frequency multipliers are used to obtain the LO signals driving the IQ mixers at the higher harmonic bands (i.e. a × N multiplier is used for generating the Nth harmonic). This approach guarantees that the active loads and the driving signal are fully phase coherent.

The principle of operation of the mixed-signal broadband architecture is shown in Figure 14.9. As in the classical open-loop approach, in the mixed-signal case only the content of the driving waveform (a_s) is known prior to the acquisition. All other injection signals ($a_{1inject,n}$ and $a_{2inject,n}$), containing all the frequency components of the signal of interest, need to be created from scratch, and any desired reflection coefficient behavior vs. frequency can be generated. This final result is obtained by iteratively adjusting the amplitude and phase of the injected waveforms independently at each frequency band of interest. To obtain a specific reflection coefficient, an injection signal, based on the linear relation shown in (14.4), is required at all the frequency components of interest.

$$a_{x,n}(f_n) = b_{x,n}(f_n) \cdot \Gamma_{x,n}(f_n), \tag{14.4}$$

where $a_{x,n}$ and $b_{x,n}$ are the incident and reflected waves at port x and harmonic index n, while $\Gamma_{x,n}$ represents the user defined reflection coefficient versus frequency at port x and harmonic index n. By monitoring the deviation of the measured reflection coefficient from the desired one for each frequency bin, the injected wave can be optimized and is found by subsequent iterations. The error checking (i.e. the distance between the obtained value and the required reflection coefficient value) and the optimization are done in the frequency domain, while the actual injection signals are loaded and acquired in the time domain. The open-loop approach guarantees that no sustained oscillations can occur. In practice, when the user-defined reflection coefficients (input or output) force the device to operate in an instable region, the system simply fails to converge. In all other situations the optimization algorithm converges as normal. Finally, computer controlled attenuators and high power amplifiers are placed in the signal path, after the IQ up-converters, in order to level the power of the injection signals. This approach makes full use of the maximum dynamic range of the AWGs at all times, which is an essential step to meet (in generation) the spectral requirements of modern wideband communication signals.

The IQ up-conversion approach requires, compared to other known signal generation techniques, relatively limited length of the data records (i.e. limited to the DAC up-converted bands) yielding a significant speed advantage, when standard models of complex modulated signals are employed in the measurements.

Figure 14.10 shows the functionality of the setup presented in [4], where a test signal composed of 161 sinusoidal tones in the bandwidth between 2060 MHz and 2220 MHz is fed to a calibration thru, while the output active load is set to provide an open condition over the whole 160 MHz bandwidth. The measured reflection coefficient at the output reference plane of the thru is plotted as a function of frequency. Figure 14.10 provides clear evidence that, using an open-loop mixed-signal technique, the desired reflection coefficient ($\Gamma_L = 1$) can be set, without any phase delay or amplitude unbalance, over a wide modulation bandwidth (i.e. 160 MHz).

Wideband Signal Generation

When working with complex modulated signals a good place to start is the modulation test standard [26, 27]. According to the standard a test signal is created, which consists of a finite sequence of IQ data samples specified in the time domain. In conventional lab instrumentation such as vector-signal generators, this sequence for a given standard (e.g. W-CDMA) is typically embedded in the instrument. In testing operation these signals are uploaded in the internal AWGs and up-converted with IQ mixers yielding the modulated RF signal. In practice, these test records are sequentially repeated yielding a large but finite number of discrete spectral components in the frequency domain. More precisely, the number of samples, in combination with the sampling speed at which the signal is generated, result in an effective frequency bin size (Δf_{AWG}), or frequency resolution of

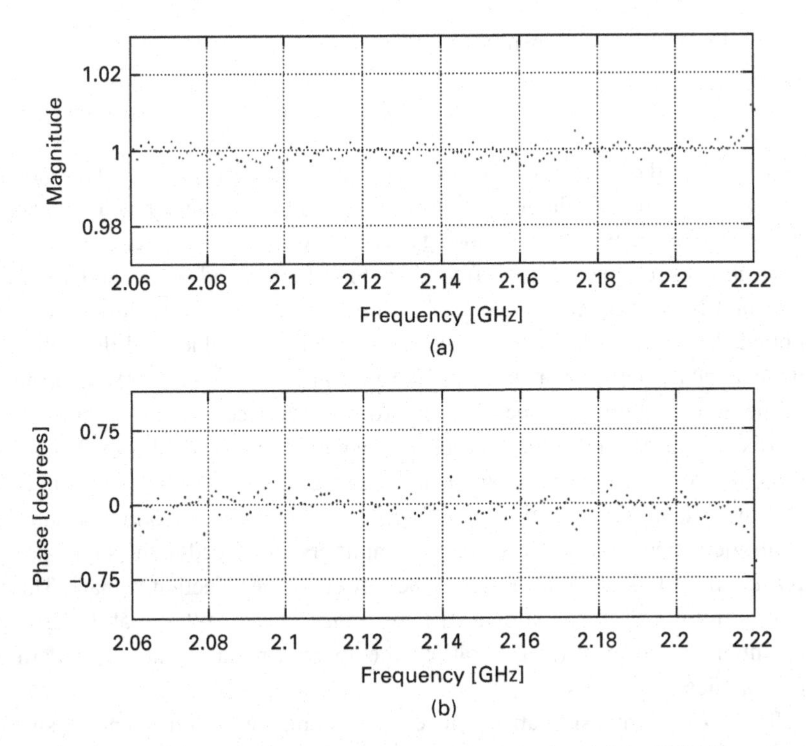

Fig. 14.10 Measured reflection coefficient at the output reference plane of the DUT for a signal composed of 161 sinusoidal tones in a 160 MHz bandwidth. © [2008] IEEE. Reprinted, with permission, from [4].

the generated signal,

$$\Delta f_{AWG} = \frac{fs_{AWG}}{N_{AWG}} = \frac{1}{T_{MOD}}, \tag{14.5}$$

where Δf_{AWG} represents the frequency bin size of the generated signal, fs_{AWG} and N_{AWG} are, respectively, the sampling frequency and the number of samples used by the AWGs to construct the waveform, and T_{MOD} is the time period of the source signal that is needed to meet the requirements of the modulation standard according to the given test model. To provide the reader with an example, a W-CDMA signal has a channel bandwidth of 5 MHz, a chip rate of 3.84 Mcps, 2560 chips/slot, and 15 slots/frame. When considering one frame the complex waveform is 10 ms long or in other words it will have a frequency resolution of 100 Hz if we then consider a single slot, the frequency resolution becomes 1.5 kHz. This frequency representation allows us to analyze modulated communication signals such as "classical" multi-tone signals, but now with a very large number of frequency tones (e.g. more than 23 000 frequency tones when considering a bandwidth of 35 MHz).

Figure 14.11 shows the frequency-binned spectral content of the I and Q signals to be delivered to the IQ modulator (Figure 14.11(a)). This block generates the RF source signal that drives the DUT with a given modulation (e.g. W-CDMA). Due to the ever present nonlinearities of the active device under test, the DUT-generated waves ($b_{1,fund}$

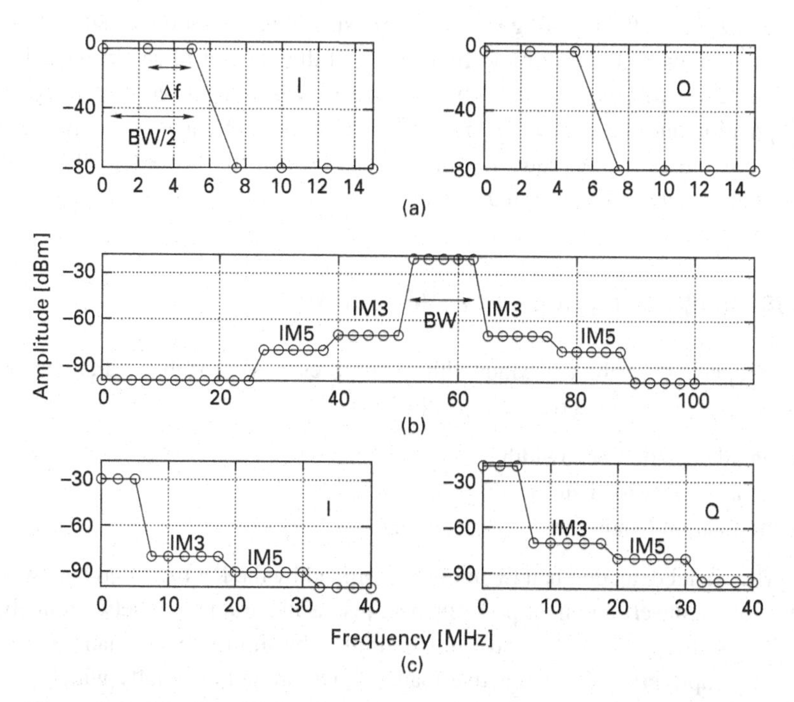

Fig. 14.11 Illustration of the generated and acquired signals in the proposed load-pull system. (a) Frequency-binned spectral content of the I and Q waveforms for generating the drive signal of the DUT. (b) Down-converted low IF representation of the spectrum in the fundamental band at the output of the DUT. (c) Spectral content of the I and Q waveforms for generating the active load injection signal to achieve the user-defined reflection coefficient over the fundamental band. © [2008] IEEE. Reprinted, with permission, from [4].

and $b_{2,fund}$) contain intermodulation sidebands besides the desired fundamental signal. Moreover, the spectral content generated by nonlinearities is also present in the baseband and harmonic frequency bands. When considering fundamental operation, the down-converted RF signal with intermodulation sidebands is given in Figure 14.11(b). In order to realize the desired reflection coefficients over the total bandwidth where spectral content is present, the I and Q injection signals must now include the third- and fifth-order intermodulation distortion (IM_3 and IM_5) sidebands (Figure 14.11(c)). Failing to provide the proper signal at the IM_3 and IM_5 frequency bands creates an unrealistic 50 Ω termination for those DUT-generated signals, invalidating the linearity performance evaluation. The I and Q baseband injection signal at all the controlled harmonics must also include the third and fifth harmonic to provide a realistic reflection coefficient in these frequency bands.

Note that, when combining high-speed AD converters [4] with mixed-signal active loads, the time span and hence the frequency bin size used for the data acquisition must be compatible with the generated test signal, as described by the following equation,

$$\Delta f_{AD} = \frac{f s_{AD}}{N_{AD}} = \frac{\Delta f_{AWG}}{k} = \frac{1}{k \cdot T_{MOD}}, \tag{14.6}$$

where, Δf_{AD} is the resulting frequency bin size of the acquired signals; fs_{AD} and N_{AD} are, respectively, the sampling frequency and the number of samples used by the AD converters; and k is an integer. For a correct measurement, the frequency bins of the acquisition and the generation should match; thus the frequency resolution of the AD converter should be set equal ($k = 1$), or an integer factor better (smaller frequency bin size) than that of the generated signals.

14.5 System operating frequency and bandwidth

To properly describe the bandwidth of the presented broadband architecture we will consider the following different bandwidths:

- signal detection bandwidth,
- signal generation bandwidth,
- modulation bandwidth.

The signal detection bandwidth is determined by the RF front-end bandwidth, i.e. by the down-converting mixer performance, just as in a traditional network analyzer.

The signal generation bandwidth is limited by the frequency handling capabilities of the amplifiers inside the active loads, which are commercially widely available up to 40 GHz, and are less widely available up to 110 GHz. Moreover, depending on the active-load topology (i.e. closed-loop or mixed-signal open-loop), discussed in the previous section, the operating frequency of the analog components within the active load defines the active-load frequency range, i.e. 90 degree coupler-based IQ modulators (intrinsically narrow-band) for the closed-loop topology and IQ mixer (available in wideband configurations) for the mixed-signal open-loop topology.

Finally, the maximum modulation bandwidth is determined, for the closed-loop configuration, by the system configuration (as shown in Table 14.1) and the user-specified maximum $\angle\Gamma$ variation with frequency that can be tolerated at the DUT reference plane. For the mixed-signal open-loop configuration, the maximum analog frequency of the arbitrary waveform generators is the only limit for the bandwidth over which the reflection coefficients can be controlled, since the $\angle\Gamma$ variation with frequency can be canceled by the mixed-signal open-loop approach.

14.6 Injection power and load amplifer linearity

After reviewing the load-pull architectures and the relative active-load topologies that allow us to perform large-signal characterization employing broadband signals, in this section we analyze the power and linearity requirements of the active-load amplifier to properly synthesize a specific Γ_L at the DUT reference plane.

To better analyze the problem, the active load can be described with its Thevenin equivalent as depicted in Figure 14.12, where E_{DUT} and Z_{DUT} and E_{SYS} and Z_{SYS} are the equivalent voltage source and output impedance of the DUT and of the measurement

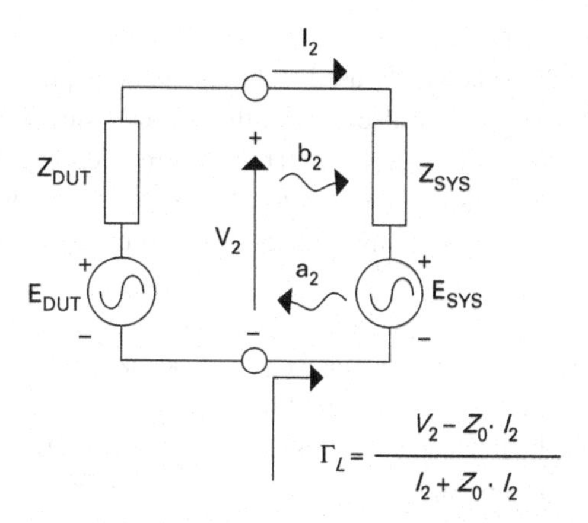

Fig. 14.12 Thevenin equivalent schematic of an active load-pull configuration. The load impedance offered to the DUT at the reference plane is varied by adjusting the equivalent voltage source E_{SYS} in amplitude and phase. The related power needed to synthesize specific impedances depends strongly on the equivalent system impedance (Z_{SYS}).

system, respectively. The equivalent voltage sources can be expressed in terms of the transmitted and incident waves to the DUT as,

$$E_{DUT} = \frac{b_{DUT} \cdot (Z_{DUT} + Z_0)}{\sqrt{Z_0}} \qquad b_{DUT} = \sqrt{2 \cdot P_{b_2} \cdot (1 - |\Gamma_{DUT}|^2)} \tag{14.7}$$

$$E_{SYS} = \frac{b_{SYS} \cdot (Z_{SYS} + Z_0)}{\sqrt{Z_0}} \qquad b_{SYS} = \sqrt{2 \cdot P_{a_2} \cdot (1 - |\Gamma_{SYS}|^2)}. \tag{14.8}$$

With reference to the schematic of Figure 14.12, the required injected power needed to achieve a certain Γ_L, or in other words a certain impedance $Z_L = V_2/I_2$, can be calculated as,

$$P_{a_2} = P_{b_2} \cdot \frac{(1 - |\Gamma_{DUT}|^2)}{(1 - |\Gamma_{SYS}|^2)} \cdot \frac{|Z_{DUT} + Z_0|^2}{|Z_{SYS} + Z_0|^2} \cdot \frac{|Z_L - Z_{SYS}|^2}{|Z_{DUT} + Z_L|^2}. \tag{14.9}$$

As is clear from (14.9), the required injected power not only depends on the output power of the DUT and the desired Γ_L, but also on the output impedance of the device. When considering high power devices, with output impedances in the order of a few Ω, the required injection power to cover the desired Smith chart area can be extremely high in a 50 Ω system (e.g. 2 to 10 times higher than the maximum output power of the DUT).

Applying (14.9) to the case of a high power amplifier with an output impedance of 2 Ω and an available output power of 200 W results in a required injection power larger than 2 kW to synthesize a load impedance of 1 Ω in a 50 Ω system. Clearly this represents a strong limitation of active loads. This is usually overcome by employing pre-matching circuitry, converting the 50 Ω impedance of the system to a value that is much closer to the output impedance of the DUT. This widely used technique (also applied in passive load-pull) not only reduces the losses, but also lowers the power requirement of the load injection amplifier [29, 30]. When using a pre-matching fixture reducing the system impedance to 10 Ω, the required injection power for the same load condition (i.e. 1 Ω) reduces from 2 kW to 360 W, while with a pre-match to 5 Ω the required injection power is only 142.2 W.

When considering multi-tone or modulated signals, the situation becomes more complicated, as the linearity of the injection amplifier also needs to be taken into account [3]. To study the linearity constraints on the injection amplifier, consider a two-tone test signal, for which the power injected by the load amplifier at the third-order intermodulation frequency products of the two-tone test signal is given by,

$$\begin{aligned} P_{a_2,IM3(dBm)} &= 3 \cdot P_{a_2,fund(dBm)} - 2 \cdot IP_{3,a_2} = \\ &= 30 \cdot \log[P_{b_2,fund(mW)} \cdot \frac{(1-|\Gamma_{DUT}|^2)}{(1-|\Gamma_{SYS}|^2)} \cdot \\ &\cdot \frac{|Z_{DUT}+Z_0|^2}{|Z_{SYS}+Z_0|^2} \cdot \frac{|Z_L - Z_{SYS}|^2}{|Z_{DUT}+Z_L|^2}] - 2 \cdot IP_{3,a_2} \end{aligned} \tag{14.10}$$

where $P_{b_2,fund}$ is the available power coming out of the DUT at the fundamental tones, and $P_{a_2,fund}$ and IP_{3,a_2} are the power injected by the load amplifier at the fundamental tones and its output third-order intercept point, respectively. A harmonic balance simulation with an Agilent ADS is performed using the simple schematic illustrated in Figure 14.13. In this schematic an amplifier component based on a polynomial model is used to simulate the DUT and the injection amplifier linearity. The same DUT as for the single tone considerations is used, with the same output impedance of 2 Ω, while its OIP_3 is set in this simulation to 63 dBm. For this device the output power is set equal to 50 W per tone, in order to achieve the same peak voltage as in the single tone case. These conditions yield an actual IM_3 of the DUT of −30.35 dBc. The results of the simulation are shown in Figure 14.14, where the apparent IM_3 of the DUT is plotted as a function of the decreasing OIP_3 of the injection amplifier, for different pre-matching conditions of the system impedance. The dotted line is the actual IM_3 level as would be achieved with a passive circuit. The dot-dash line represents the IM_3 level due to the P_{a_2,IM_3} as approximated by (14.10). From Figure 14.14, we can observe that the correct

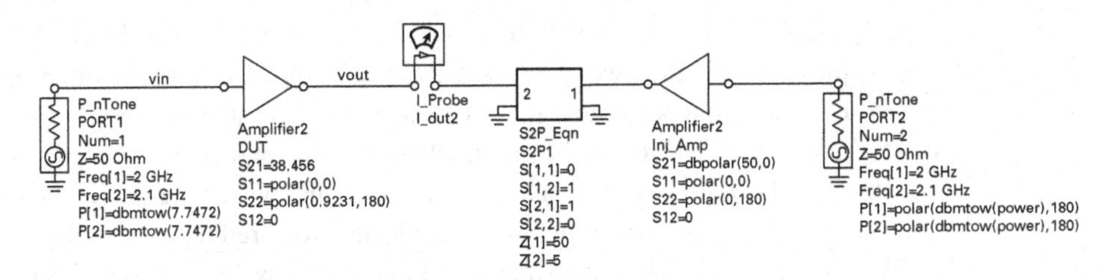

Fig. 14.13 ADS schematic for the evaluation of the required injection amplifier linearity by active load-pull.

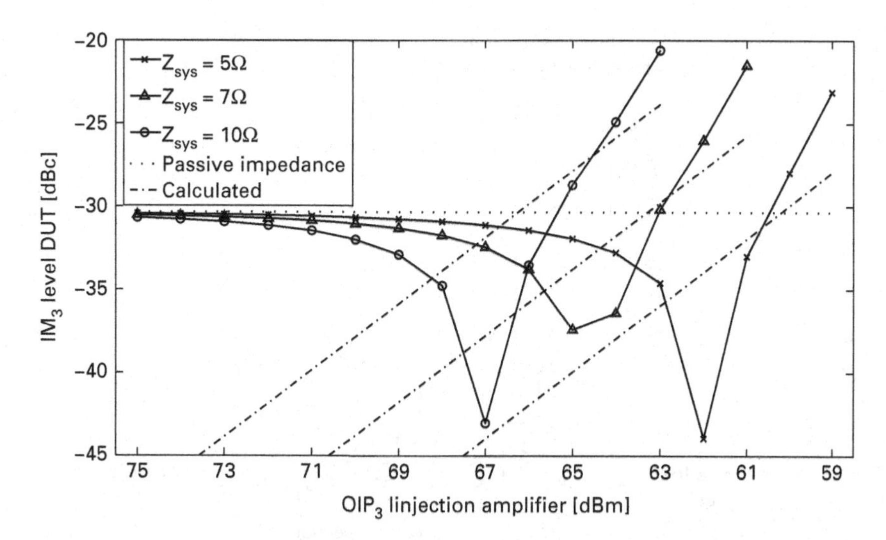

Fig. 14.14 Harmonic balance simulated IM_3 level of the DUT vs. decreasing OIP_3 of the injection amplifier for different impedance pre-match values. The dotted line is the actual IM_3 level as would be achieved with passive matching techniques. The dot-dash line represents the IM_3 level only due to the P_{a_2,IM_3} as approximated by (14.10). A polynomial model was used for the amplifier linearity. © [2010] IEEE. Reprinted, with permission, from [28].

IM_3 level is only achieved when the injection amplifier OIP_3 is sufficiently high. When the injection amplifier is less linear, it introduces significant IM_3 products that cause an error in the measurements, such as IM_3 increase or cancellation effects. Consequently, to have reliable linearity measurements in a load-pull setup, even when pre-matching is used, the injection amplifier (and thus its peak power) needs to be at least 10 times higher than the one of the DUT.

At this point it is important to note that while the peak power requirements apply for both closed-loop and mixed-signal active loads, the linearity requirements strictly apply only for closed-loop active loads. The reason for this is to be found in the iterative convergence approach employed by the mixed-signal approach. Since the iteration on the required reflection coefficient involves the signal channel as well as its adjacent channels (intermodulation products), this procedure compensates, in the injection signal, for most of the nonlinearities of the active-loop amplifier. For this reason it is difficult to indicate the linearity requirement for the mixed-signal approach in a closed formula since it also depends on the convergence algorithm used. In general it can be stated that such a technique allows the use of the active-load amplifier, even under modulated signal excitation, much closer to its compression point.

14.7 Baseband impedance control

The importance of controlling the baseband impedance, when aiming for optimum device linearity, was proven theoretically in [31] and [5] as well as experimentally in [32]

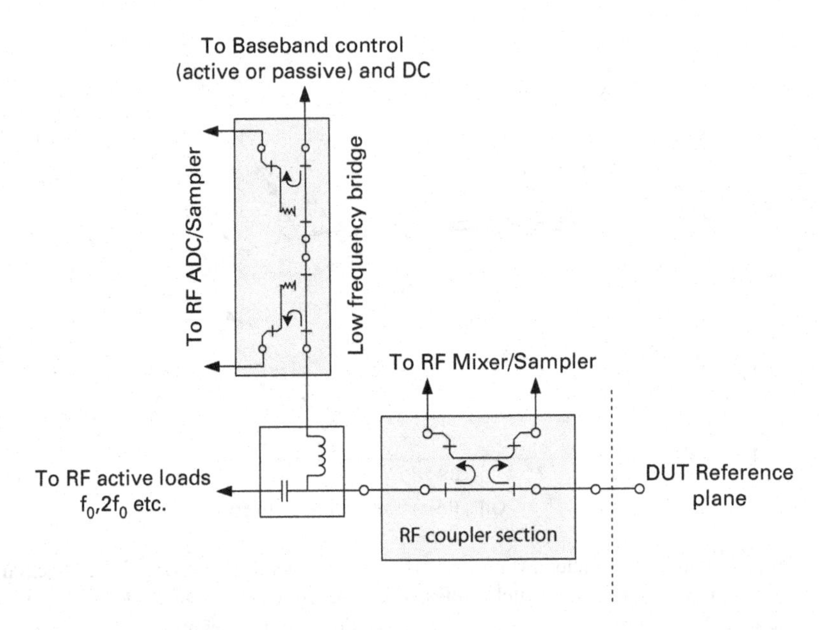

Fig. 14.15 Low-frequency detection bridges placed on the bias line to couple the low-frequency traveling waves to the broadband receiver, i.e. sampler or AD.

and [33]. When attempting to properly control the baseband impedance two difficulties arise when employing conventional high-frequency system architectures, namely:

1. how to obtain calibrated measurements of the baseband reflection coefficients,
2. how to obtain impedance control over the wideband (baseband) frequency.

The first point in the above list is due to the limited bandwidth of the reflectometers, often implemented as coupled line couplers, which are usually employed in the test sets of VNA and load-pull architectures (see Figure 14.4). These components provide a minimum operating frequency (linked to the dimension of the component) often in the order of few hundred MHz. To circumvent this limitation in [34] and [35] low-frequency (i.e. resistive bridges) couplers were employed in the bias line. Inserting the low-frequency detection bridges on the bias path, as shown in Figure 14.15, avoids high-frequency signals being routed through these components which would provide very high losses at RF frequencies. The low-frequency traveling waves at the output of the detection bridges can be directly sampled by the receiver, as was shown in [34], using a microwave sampling oscilloscope architecture or by the broadband AD employed in a heterodyne mixer based architecture as was presented in [35]. Using low-frequency bridges also allows us to employ the conventional 12 error-terms calibration techniques for the calibrated measurement of the baseband reflection coefficients. Moreover, the use of dedicated low-frequency detection bridges allows us to optimize the accuracy of the controlled baseband reflection coefficients, due to the high performance of the low-frequency bridges (i.e. high directivity). Finally, employing similar calibration techniques the ones used at RF allows us to share the same calibration standards for both the RF as well as the baseband calibration (e.g. short, open, load, and thru), reducing the overall calibration time.

In order to enable baseband impedance control in [34] an additional AWG generating a signal coherent (i.e. sharing the same clock) with the RF signal driving the device, was employed. As this is an open-loop topology on the baseband path, low-frequency amplifiers need to be employed. When standard compliant modulated signals are employed (i.e. W-CDMA) DC coupled amplifiers should be employed to provide a controlled impedance through the entire baseband frequency range. In [17] a simplified baseband impedance control was presented that employed a simple resistive switch bank. While this method allows only a resistive impedance control, when implemented in a small form factor (i.e. using SMD components and PCB dedicated layout) this approach allows a simple control of the baseband resistance over a broad frequency range (i.e. up to 5 MHz), which can be employed for device linearity improvement [3].

14.8 Broadband large signal measurement examples

In this section some measurement examples demonstrating the large-signal characterization capabilities of the broadband architectures presented in the previous sections given.

14.8.1 IMD asymmetries measurements

As was reported in [5], the out-of-band source termination can cause asymmetries between the upper and lower IMD products, shown in Figure 14.16, where the maximum

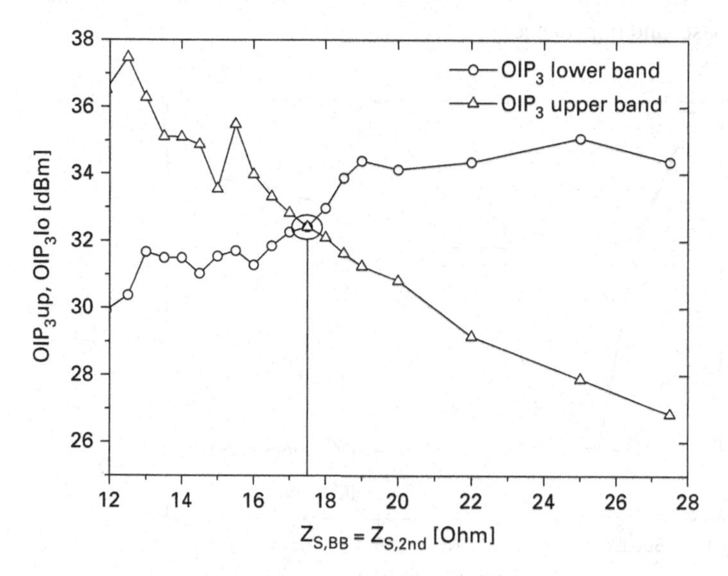

Fig. 14.16 Measured maximum OIP_3 levels for upper and lower IM_3 components versus resistive $Z_{S,BB} = Z_{S,2nd}$ using a swept I_{cq} bias conditions ($f_0 = 2.14$ GHz, $\Delta f = 0.5$ MHz). © [2006] IEEE. Reprinted, with permission, from [3].

measured upper and lower band OIP_3 (achieved at each I_c sweep) are plotted as a function of $Z_{S,BB} = Z_{S,2nd}$ [33]. This means that the IMD at $2f_1 - f_2$ has a different magnitude to that at $2f_2 - f_1$. Since the linearity performance is limited by the highest IMD level, this asymmetry leads to a degradation in linearity performance compared to the optimum symmetrical case. This asymmetry between the IM_3 signals versus tone spacing is often referred to as the *memory effect*. These memory effects can be divided into two classes: thermal memory effects (up to a few MHz) and electrical memory effects (caused mostly by the source and load termination, including the biasing network impedances). The mechanism generating these memory effects can be quite complex, and standard available models are quite often not able to predict them. This highlights the importance of properly characterizing these effects through measurements, which can provide the information required for their cancellation. A clear example of the large variation of IM_3 upper and lower tone power versus the tone spacing of the input signal is given in Figure 14.17. Here a bipolar device matched for optimum linearity was stimulated with a two-tone signal with an increasing tone spacing.

The importance of properly characterizing these intermodulation distortion mechanisms becomes clear when studying (digital) pre-distortion techniques to linearize the PA behavior. The information required from the measurements is the power versus frequency of the IMD components and their phases [36,37]. This has been obtained through different approaches, both in the frequency domain [36] using sinusoidal inputs as well as in the time domain [38] using a waveform with a complex time dependence. In most cases, however, this information was not coupled with full control of the source and load termination offered at the device under test. The broadband system architecture described in the previous sections provides the capability to characterize the intermodulation distortion as a function of tone spacing, while controlling the high-frequency termination and the baseband impedance.

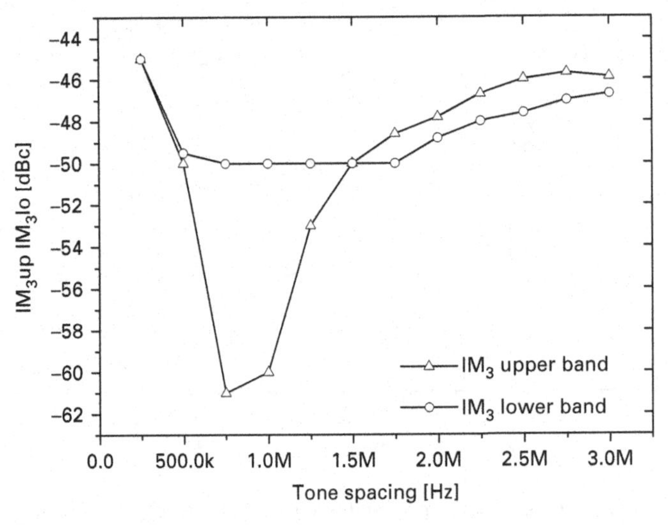

Fig. 14.17 Measured upper and lower band IM_3 components versus tone spacing for fixed bias and load conditions.

14.8.2 Phase delay cancellation

When employing mixed-signal loads it is possible to demonstrate, directly on the DUT performance degradation, the effects of phase delays when employing broadband signals. In [4], an NXP Gen 6 LDMOS device with a gate width of 1.8 mm is measured using wideband modulated signals in load-pull test conditions. The drain current and voltage are set to 13 mA and 28 V, respectively. First, the optimum fundamental load and source matching conditions are found using conventional single tone load-pull measurements, namely: $\Gamma_{L,f1} = |0.6| \angle 45°$ and $\Gamma_{S,f1} = |0.5| \angle 90°$. The input and output baseband impedances are set to enforce a short condition; and the input and output 2nd harmonics are set to an open circuit condition ($\Gamma_{L,f2} = \Gamma_{S,f2} = |0.95|$) to optimize the efficiency [39]. The electrical delay-free operation, provided by mixed-signal loads is compared with the results (under phase delay) which would be achieved by employing the closed-loop loads presented in Section 14.4.1. For the driving signal, a two-channel W-CDMA signal (centered at 2.135 GHz and 2.145 GHz) is chosen, and the input and output reflection coefficients are set to the above defined optimal conditions providing a phase delay given by the following two cases:

1. without electrical delay,
2. with an electrical delay of 4.85°/MHz for the fundamental source and load and 4.6°/MHz for the 2nd harmonic source and load.

Figure 14.18 illustrates the source and load matching conditions provided to the active device under test for the two different cases. Note that the filled markers represent the source and loading conditions for the two-carrier W-CDMA signal without any electrical delay, yielding points that are completely overlapping versus frequency in the Smith chart. As shown in Figure 14.18, for the case with electrical delay the fundamental load trajectory has been shifted such that the optimum matching condition is now centered at 2.135 GHz. This was required to avoid the unstable region of the active device.

The comparison is to the “best known case” of a closed-loop load, since in practical closed-loops there are amplitude variations within the control frequency band that are not accounted for. Moreover, oscillation conditions in closed-loop systems for very large bandwidths are difficult to avoid, due to the use of wideband loop filters. Passive loads with harmonic tuning will have a comparable or even worse phase variation of the reflection coefficients versus frequency, depending on the distance of the tuner from the DUT reference plane.

The measurement results are summarized in Table 14.2. There is significant performance degradation for the active device when measured with an electrical delay present in the reflection coefficients. This is also evident from Figures 14.19(a) and (b) which show the power spectral density at the device output reference plane for the fundamental and 2nd harmonic frequency bands. Note that a 5 dB output power drop and close to an 8% degradation of the PAE can be observed, when compared to the situation with no electrical delay.

Table 14.2 Measurement results comparison in the two cases with and without electrical delay

MEASUREMENT RESULTS		
	Without electrical delay	With electrical delay
PAE	24.2 %	16.3 %
POUT Ch. 1	20.3 dBm	20.5 dBm
POUT Ch. 2	20.6 dBm	15.4 dBm
ACLR1 Ch. 1	−43.9 dBc	−43.0 dBc
ACLR2 Ch. 1	−42.2 dBc	−41.6 dBc
ACLR1 Ch. 2	−42.1 dBc	−41.8 dBc
ACLR2 Ch. 2	−39.6 dBc	−39.2 dBc

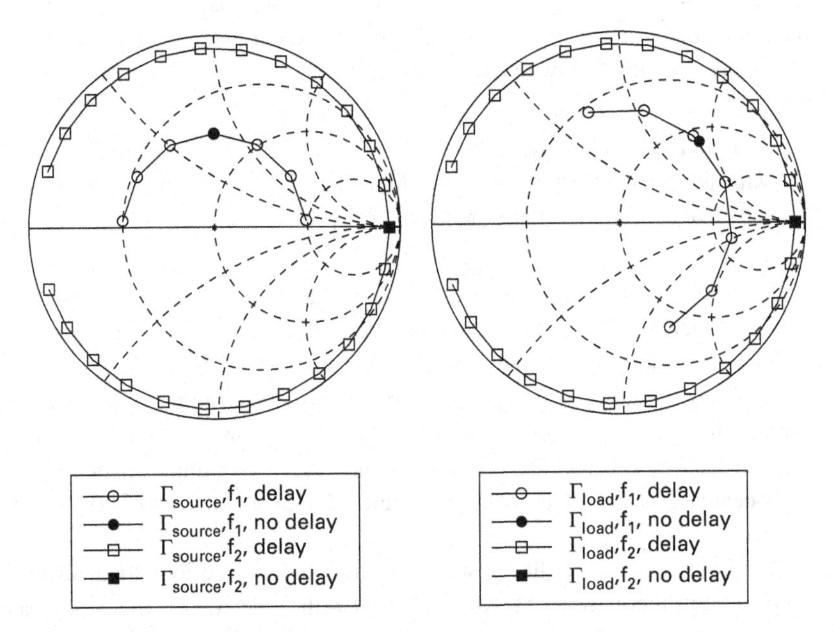

Fig. 14.18 Source and load reflection coefficients at the device reference plane in the fundamental (2.1225 GHz – 2.1575 GHz) and harmonic (4.245 GHz – 4.315 GHz) frequency range, with electrical delay (open symbols) and without electrical delay (filled symbols). © [2008] IEEE. Reprinted, with permission, from [4].

14.8.3 High power measurements with modulated signals

As previously explained in Section 14.6, in an active load-pull system, to synthesize a specific Γ_L at the DUT reference plane, it is necessary to inject a certain amount of power into the DUT. When working at high power levels with wideband signals an extra complication arises because the linearity of the injection amplifier needs to be taken into account.

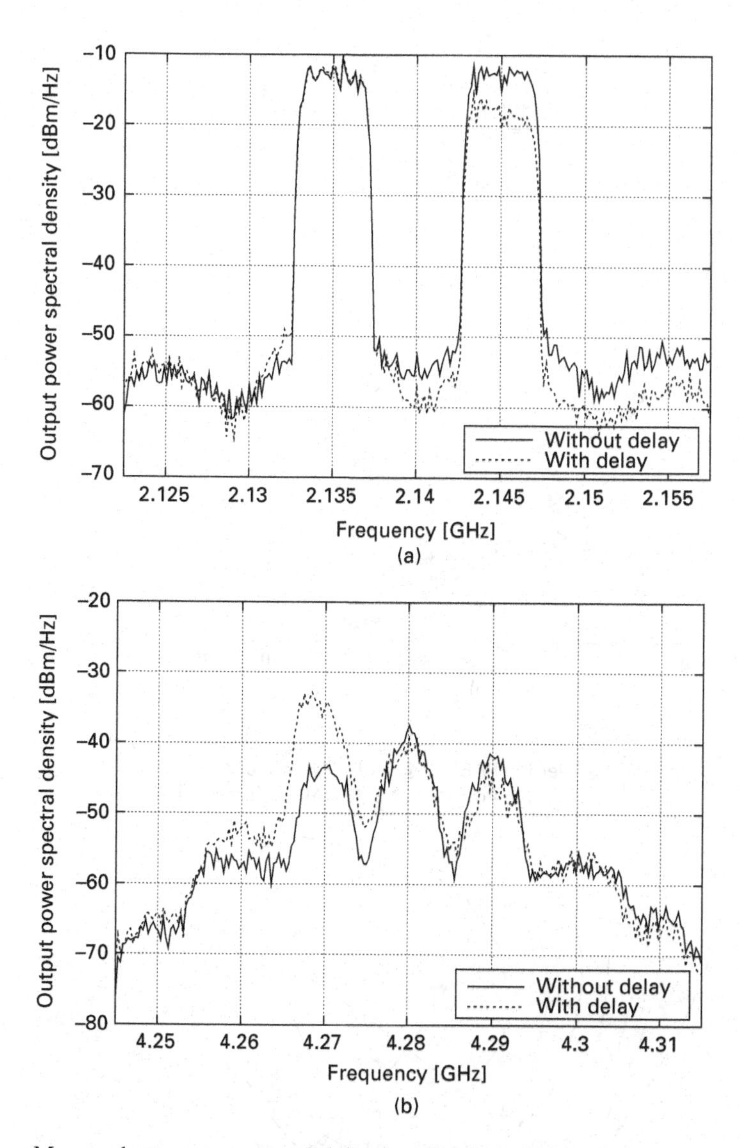

Fig. 14.19 Measured output power spectral density (dBm/Hz) vs. frequency (GHz) of a NXP GEN 6 LDMOS device (gate width 1.8 mm) in the proposed load-pull setup (a) at the fundamental frequency band using a 3 kHz resolution bandwidth. (b) at the 2nd harmonic frequency band using a 6 kHz resolution bandwidth. The measurement is shown for the two cases with (dashed line) and without electrical delay (drawn line). The reflection coefficients offered to the device under test are given in Fig. 14.18. © [2008] IEEE. Reprinted, with permission, from [4].

As was shown in Figure 14.14, to have reliable linearity measurements in an active load-pull setup, even when pre-matching is used, the injection amplifier linearity (and thus its peak power) needs to be at least 10 times higher than that of the DUT, which becomes extremely expensive and impractical when working with devices with peak envelope powers as high as 200 W, e.g. designed for base-station applications.

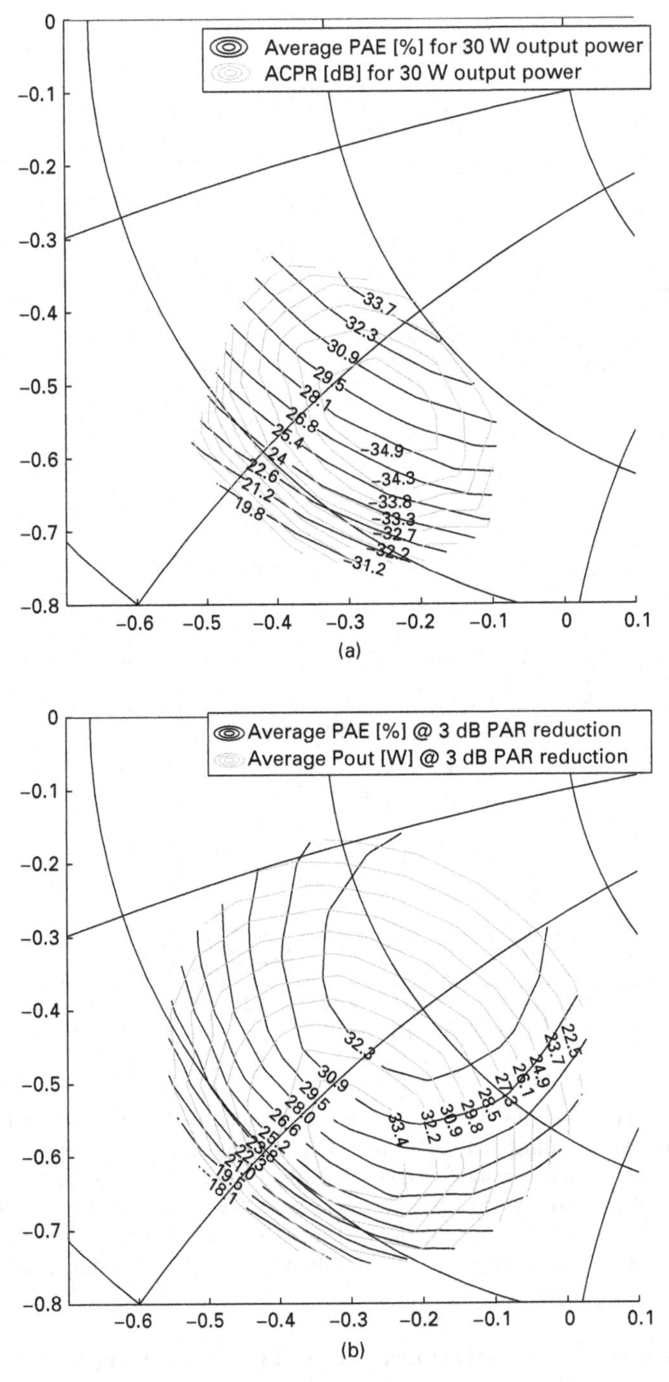

Fig. 14.20 (a) Load-pull contours, on a 5 ohm normalized Smith-chart, of average PAE and ACPR for an average output power of 30 W. (b) Load-pull contours, on a 5 Ω normalized Smith-chart, of average PAE and average output power at 3 dB of peak-to-average ratio reduction. The related peak to average power (PEP) is as high as 150 W. © [2010] IEEE. Reprinted, with permission, from [28].

In this section the capability of a mixed-signal architecture to work with complex modulated signals at power levels that are typically in use for base station applications (e.g. peak envelope power $\sim$ 200 W), is reviewed [28].

As previously mentioned, when using mixed-signal loads, an iterative process is performed to optimize the reflection coefficient of each individual frequency component of the wideband signal (e.g. 23 362 tones with 1.5 kHz spacing for a W-CDMA considering a total bandwidth of 35 MHz). Due to these iterations, the injection amplifier is basically pre-distorted for its own nonlinearities, allowing the use of an injection amplifier with a much lower linearity compared with what is typically required in closed-loop architectures (i.e. 10 times higher than that the of the DUT). Figure 14.20 shows the ACPR, average PAE, and output power for a single-channel W-CDMA signal at 2.14 GHz with a peak-to-average ratio of 9.5 dB, employing an injection amplifier with 200 W and an associated 60 dBm output IP_3. In these measurements, the nonlinearity of the injection amplifier does not affect the measurement results because the Γ_L is controlled to the user-defined value in and out of the band.

References

[1] Motorola, *Long Term Evolution (LTE): A Technical Overview*. Motorola technical white paper, 2010.

[2] H. Arthaber, M. Mayer, and G. Magerl, "A broadband active harmonic load-pull setup with a modulated generator as active load," in *Microwave Conference, 2004. 34th European*, 2, Oct. 2004, pp. 685–688.

[3] M. Spirito, M. Pelk, F. van Rijs, S. Theeuwen, D. Hartskeerl, and L. de Vreede, "Active harmonic load-pull for on-wafer out-of-band device linearity optimization," *IEEE Trans. Microw. Theory Tech.*, 54, no. 12, pp. 4225–4236, Dec. 2006.

[4] M. Marchetti, M. Pelk, K. Buisman, W. Neo, M. Spirito, and L. de Vreede, "Active harmonic load-pull with realistic wideband communications signals," *IEEE Trans. Microw. Theory and Tech.*, 56, no. 12, pp. 2979–2988, Dec. 2008.

[5] M. van der Heijden, H. de Graaff, and L. de Vreede, "A novel frequency-independent third-order intermodulation distortion cancellation technique for BJT amplifiers," *IEEE J. Solid-State Circuits*, 37, no. 9, pp. 1176–1183, Sept. 2002.

[6] (2012) Agilent technologies website. [Online]. Available: http://www.home.agilent.com/agilent/product.jspx?nid=-33821.0.00&cc=US&lc=eng

[7] (2012) Tektronix website. [Online]. Available: http://www.tek.com/oscilloscope/dsa8300-sampling-oscilloscope

[8] M. Akmal, V. Carrubba, J. Lees, S. Bensmida, J. Benedikt, K. Morris, M. Beach, J. McGeehan, and P. Tasker, "Linearity enhancement of GaN HEMTs under complex modulated excitation by optimizing the baseband impedance environment," in *Microwave Symposium Digest (MTT), 2011 IEEE MTT-S International*, June 2011, pp. 1–4.

[9] (2012) Miteq inc. website. [Online]. Available: http://www.miteq.com/

[10] (2012) Marki microwave website. [Online]. Available: http://www.markimicrowave.com/

[11] Agilent Technologies, *Optimizing Dynamic Range for Distortion Measurements*. Santa Rosa, CA: Agilent 5980-3079EN, 2000.

[12] D. M. Leenaerts, J. van der Tang, and C. S. Vaucher, *Circuit Design for RF Transceivers*. Boston, Dordrecht, London: Kluwer Academic Publishers, 2001.

[13] (2012) National instruments website. [Online]. Available: http://www.ni.com/data-acquisition/

[14] S. Vandenplas, J. Verspecht, F. Verbeyst, F. Vandamme, and M. Bossche, "Calibration issues for the large signal network analyzer (LSNA)," in *ARFTG Conference Digest, Fall 2002*, Washington, DC, Dec. 2002, pp. 99–106.

[15] K. Noujeim, J. Martens, and T. Roberts, "A frequency-scalable NLTL-based signal-source extension," in *41st European Microwave Conference (EuMC), 2011*, Manchester, UK, Oct. 2011, pp. 476–479.

[16] P. Roblin, *Nonlinear RF Circuits and Nonlinear Vector Network Analyzers: Interactive Measurement and Design Technique*. Cambridge UK: Cambridge University Press, 2011.

[17] M. Spirito, L. C. N. de Vreede, M. de Kok, M. Pelk, D. Hartskeerl, H. F. F. Jos, J. E. Mueller, and J. Burghartz, "A novel active harmonic load-pull setup for on-wafer device linearity characterization," in *Microwave Symposium Digest 2004*, Fort Worth, TX, June 2004, pp. 1217–1220.

[18] D. D. Poulin, J. R. Mahon, and J. P. Lanteri, "A high power on-wafer pulsed active load pull system," *IEEE Trans. Microw. Theory Tech.*, 40, pp. 2412–2417, Dec. 1992.

[19] G. Madonna, A. Ferrero, M. Pirola, and U. Pisani, "Testing microwave devices under different source impedance values - a novel technique for on-line measurement of source and device reflection coefficients," *IEEE Trans. Instrum. Meas.*, 49, pp. 285–289, Apr. 2000.

[20] G. P. Bava, U. Pisani, and V. Pozzolo, "Active load technique for load-pull characterisation at microwave frequencies," *Electronics Letters*, 18, pp. 178–180, Feb. 1982.

[21] D. Rytting, "Network analyzer error models and calibration methods," *White Paper, September*, 1998.

[22] A. Ferrero and U. Pisani, "An improved calibration technique for on-wafer large-signal transistor characterization," *IEEE Trans. Instrum. Meas.*, 42, no. 2, pp. 360–364, Apr. 1993.

[23] L. Devlin and B. Minnis, "A versatile vector modulator design for MMIC," in *Microwave Symposium Digest, 1990., IEEE MTT-S International*, May 1990, pp. 519–521, vol. 1.

[24] M. Squillante, M. Marchetti, M. Spirito, and L. de Vreede, "A mixed-signal approach for high-speed fully controlled multidimensional load-pull parameters sweep," in *Microwave Measurement Conference, 2009 73rd ARFTG*, Jun. 2009, pp. 1–5.

[25] W. C. E. Neo, J. Qureshi, M. J. Pelk, J. R. Gajadharsing, and L. C. N. de Vreede, "A mixed-signal approach towards linear and efficient N-Way doherty amplifiers," *IEEE Trans. Microw. Theory Tech.*, 55, no. 5, pp. 866–879, May 2007.

[26] "UTRA (BS) FDD; Radio Transmission and Reception," Technical Specification Group Radio Access Networks – 3rd Generation Partnership Project, 3GPP TS 25.104 V4.1.0 (2001-06), 2001.

[27] "Base station conformance testing (FDD)," Technical Specification Group Radio Access Networks – 3rd Generation Partnership Project, 3GPP TS 25.141 V4.1.0 (2001-06), 2001.

[28] M. Marchetti, R. Heeres, M. Squillante, M. Pelk, M. Spirito, and L. de Vreede, "A mixed-signal load-pull system for base-station applications," in *Radio Frequency Integrated Circuits Symposium (RFIC), 2010 IEEE*, May 2010, pp. 491–494.

[29] Z. Aboush, C. Jones, G. Knight, A. Sheikh, H. Lee, J. Lees, J. Benedikt, and P. Tasker, "High power active harmonic load-pull system for characterization of high power 100-watt transistors," in *2005 European Microwave Conference*, 1, Oct. 2005, p. 4.

[30] V. Teppati, A. Ferrero, and U. Pisani, "Recent advances in real-time load-pull systems," *IEEE Trans. Instrum. Meas.*, 57, no. 11, pp. 2640–2646, Nov. 2008.

[31] V. Aparin and C. Persico, "Effect of out-of-band terminations on intermodulation distortion in common-emitter circuits," in *Microwave Symposium Digest, 1999 IEEE MTT-S International*, 3, 1999, pp. 977–980, vol. 3.

[32] D. Williams, J. Leckey, and P. Tasker, "Envelope domain analysis of measured time domain voltage and current waveforms provide for improved understanding of factors effecting linearity," in *Microwave Symposium Digest, 2003 IEEE MTT-S International*, 2, June 2003, pp. 1411–1414, vol. 2.

[33] M. Spirito, M. van der Heijden, M. Pelk, L. de Vreede, P. Zampardi, L. Larson, and J. Burghartz, "Experimental procedure to optimize out-of-band terminations for highly linear and power efficient bipolar class-AB RF amplifiers," in *Proceedings of the Bipolar/BiCMOS Circuits and Technology Meeting, 2005*, Oct. 2005, pp. 112–115.

[34] A. Alghanim, J. Lees, T. Williams, J. Benedikt, and P. Tasker, "Using active IF load-pull to investigate electrical base-band induced memory effects in high-power ldmos transistors," in *APMC 2007, Asia-Pacific Microwave Conference, 2007*, Dec. 2007, pp. 1–4.

[35] M. Mirra, M. Marchetti, F. Tessitore, M. Spirito, L. de Vreede, and L. Betts, "A multi-step phase calibration procedure for closely spaced multi-tone signals," in *2010 75th ARFTG Microwave Measurements Conference (ARFTG)*, May 2010, pp. 1–5.

[36] J. Vuolevi, T. Rahkonen, and J. Manninen, "Measurement technique for characterizing memory effects in RF power amplifiers," in *Radio and Wireless Conference, 2000. RAWCON 2000. 2000 IEEE*, 2000, pp. 195–198.

[37] J. Vuolevi, J. Manninen, and T. Rahkonen, "Cancelling the memory effects in RF power amplifiers," in *The 2001 IEEE International Symposium on Circuits and Systems, 2001. ISCAS 2001.* 1, May 2001, pp. 57–60, vol. 1.

[38] P. Draxler, I. Langmore, T. Hung, and P. Asbeck, "Time domain characterization of power amplifiers with memory effects," in *Microwave Symposium Digest, 2003 IEEE MTT-S International*, 2, June 2003, pp. 803–806, vol. 2.

[39] D. Hartskeerl, I. Volokhine, and M. Spirito, "On the optimum 2nd harmonic source and load impedances for the efficiency-linearity trade-off of RF LDMOS power amplifiers," in *Radio Frequency integrated Circuits (RFIC) Symposium, 2005. Digest of Papers. 2005 IEEE*, June 2005, pp. 447–450.

15 Pulse and RF measurement

Anthony Parker

15.1 Introduction

Circuits exhibit a variety of operational traits that are far from the behavior presented in introductory circuit design textbooks. Transistor characteristics curves vary significantly depending on how they are measured and on the history of electrical conditions. The characteristics are not always repeatable, which raises the dilemma of the choice of which characteristic to base a design upon.

The central idea behind pulse measurements is that the high-frequency characteristics of a device are a function of a quiescent operating condition. Pulse techniques attempt to determine these in an invariable operating condition. If short enough pulses are used, a pulse measurement at a specific condition of operation gives the characteristics that a high-frequency signal would encounter. This is a simple idea, but there is a practical limit to how short the pulses can be, so it is then necessary to draw upon radio-frequency techniques to probe past higher-frequency anomalies in the characteristics.

Dynamic processes and interactions in active elements produce seemingly complicated electrical characteristics that are best explored with pulse and RF techniques. These processes can be traced to mechanisms of self-heating by power dissipation, bias-dependent change in trapped charges, and to impact ionization and breakdown.

This chapter covers a set of topics that provide a foundation for understanding the pulse measurement technique augmented with RF measurements. Pulse characterization techniques dovetail with RF and nonlinear techniques to explore transistor dynamics for small-signal and nonlinear applications.

15.2 Dynamic characteristics

Many devices exhibit characteristics that change with time, frequency, and with operating conditions such as temperature and terminal bias. All characteristics are affected, including terminal current, linearity, and charge state. The pulse measurement technique is one of the more powerful and insightful methods for characterizing these dynamics in transistors and circuits [1].

The characteristics vary considerably with the time taken to measure each point in the curves. This is well illustrated in Figure 15.1, which shows characteristics that are typical for GaAs pHEMT transistors. They are reproducible and are consistent from wafer to wafer in a mature fabrication process. The dilemma that this dynamic behavior presents is the question of what are the characteristics seen by signals in a circuit.

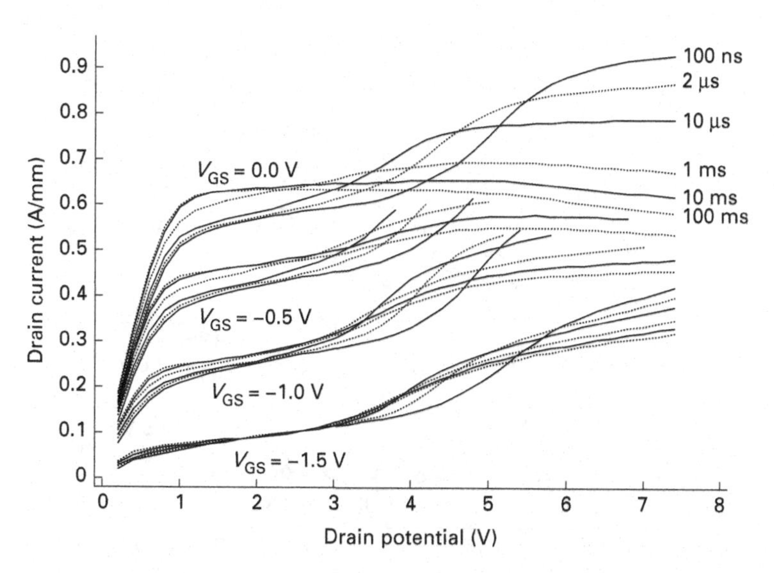

Fig. 15.1 Pulse characteristics of a 150μm GaAs/AlGaAs pHEMT with 2.0V pinch-off. The pulses emanate from a bias at 4.0 V on the drain and −2.0 V on the gate. Six pulse widths were used, with a 100ms pulse repetition period.

Pulse characteristics also change with the bias condition used between pulse measurements, as shown in Figure 15.2. This raises the question of what is the bias and how does it relate to the quiescent condition relative to the rate of change in characteristics. A large signal may see a fixed set of characteristics set by the average bias, or various characteristics corresponding to conditions set by the signal during its excursions along the load line. This is a question that relates to the characteristic frequencies of the processes involved.

In practice the electrical characteristics of field-effect transistors change with the timing of signals, bias condition, and frequency. From the perspective of high-frequency signals, the characteristics are time-variant, or a function of memory of previous signals.

Pulse measurements attempt to establish a history at a fixed bias and then measure the characteristics quickly enough, so the bias is not perturbed significantly. Each bias point has a corresponding set of static characteristics.

Changes in static characteristics with bias come from changes in physical processes associated with the operation of the transistor. Temperature and charge state at trapping sites are important factors, or state variables, that change with bias. The drain current is a time-invariant function of terminal potentials and state variables such as junction temperature and the potential of trapped charge.

Each state variable has its own response to the terminal conditions. From a broadband perspective, these responses are slowly varying signals that also control the characteristics of the transistor. From a narrow-band perspective, the responses are state variables dependent on past signals that control the static characteristics.

In many devices the variations caused by changes in state variables can be quite dramatic and affect basic performance parameters such as the intrinsic gain of a transistor, as shown in Figure 15.3.

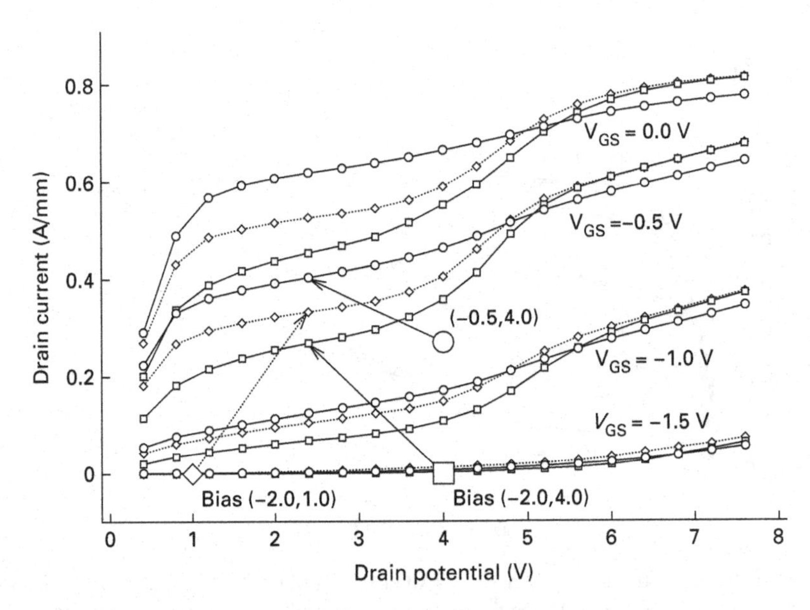

Fig. 15.2 Pulse characteristics of the 150 μm GaAs/AlGaAs pHEMT shown in Fig. 15.1 from three bias points. The 100 ns pulse characteristics vary considerably with the long-term bias condition established between pulses. The arrows show the transition from each bias point to ($V_{GS} = -0.5$, $V_{DS} = 2.4$).

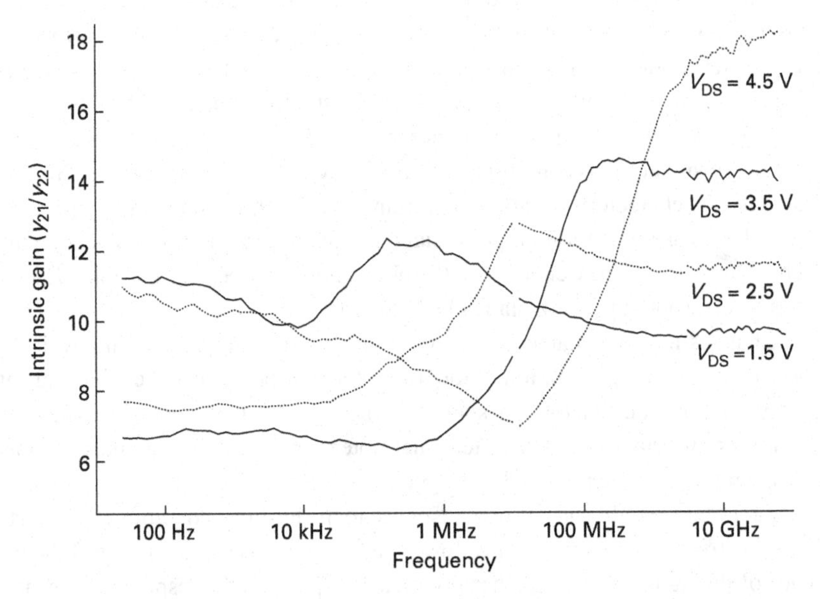

Fig. 15.3 Intrinsic gain of the 150 μm GaAs/AlGaAs pHEMT shown in Fig. 15.1. There is variation across several decades of frequency that depends significantly on drain bias. There is a break in the data at 10 MHz below which pulse measurements were used and above which RF measurements were used.

15.3 Large-signal isodynamic measurements

The principle behind a typical use of pulse measurements is maintaining a fixed bias relative to any *dynamic processes* that change the device's electrical characteristics. A nominal bias is held for a long period between very short pulses during which measurements are made. Ideally the bias period would be longer than the response times of the transistor's or circuit's dynamic processes and the pulses would be shorter than these response times. For a FET, drain current measurements during a set of pulses, each to a different potential, provide a pulsed drain-current characteristic. In the ideal case, there would be no response recorded from the transistor's dynamic processes, so the characteristics are considered to be *isodynamic*.

Isodynamic characteristics are those for constant state variables. Each set of state variable values has a corresponding isodynamic characteristic and variation of transistor characteristics with operating condition are a result of changes in the state variables. That is, the dynamic behavior of the transistor is described by variation of the state variables. The processes that link state variables to terminal or operating conditions are the dynamic processes of the transistor. True DC characteristics are those for which the state variables have reached steady state at each point.

An isodynamic pulse measurement is illustrated in Figure 15.4, which also shows a true DC measurement for comparison. The latter requires measurement after a long time at each point to ensure that all dynamic processes have reached a steady state. Traditional step-and-sweep measurements can be too fast for this and exhibit manifestations of transistor dynamics, which are also shown in Figure 15.4. The timing of true DC

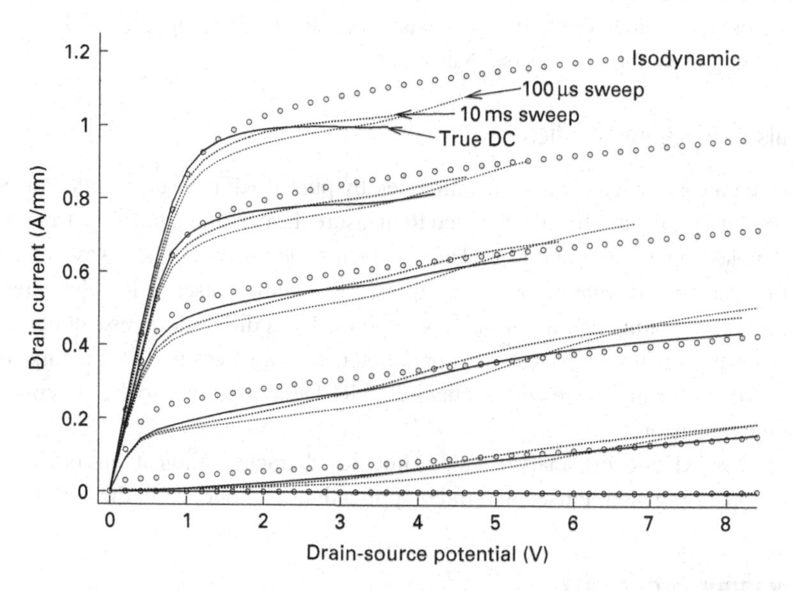

Fig. 15.4 Comparison of step-and-sweep measurements with estimates of isodynamic and true DC characteristics. The isodynamic characteristics are from a bias at $V_{DS} = 5$, $I_{DS} = 0.36$ A/mm where they overlap the DC characteristic. The gate-source potential from –2.0 to 0.5 V in 0.5V steps is the parameter.

measurements can be in the range 10-100 ms for Gallium Arsenide devices to many hundreds of seconds in wide band-gap devices, such as Gallium Nitride.

Reliance on DC data alone is problematic and leads to inconsistency between bias and small-signal characteristics. For example, the bent curves in step-and-sweep measurements suggest an apparent change in drain conductance that is an artifact of the measurement sequence and timing that is not observed during radio-frequency operation.

The DC and isodynamic characteristics are coincident at their common bias point but differ considerably at other points. In most radio-frequency applications the operating frequency of the transistor is higher than that of the response of its dynamic processes, so the isodynamic characteristics provide a better view of device operation. That is, signal excursions and small-signal parameters at radio frequency should be determined from the isodynamic characteristics rather than the DC characteristics.

15.3.1 Measurement outside safe-operating areas

The safe-operating area, defined by limits beyond which the device may be damaged, varies with signal condition. The safe area is bounded by: a maximum voltage; a maximum temperature and hence maximum power-time product; a maximum current; and a maximum current-time product. Exceeding these limits can cause breakdown or permanent physical alteration of the device.

Time constants associated with damage limitations allow pulse measurements to extend beyond the safe-operating area of DC measurements. Temperature rise can be slow enough to allow very high power levels to be achieved for short periods. After a short time, the device must be returned to a low power condition to cool down. The larger safe-operating area for shorter pulses is invaluable for investigating devices that may be used in pulsed applications, such as radar.

15.3.2 Pulsed-RF characteristics

Pulsed measurement can be accompanied by pulsed-RF measurements. Pulsed-RF network analyzers have been developed to measure the performance of the transistor during its pulsed operating condition. These systems measure the microwave characteristics of devices in isodynamic bias conditions. The result is a set of RF characteristics in a transient operating condition that corresponded to a different quiescent condition. This can be quite different from static S-parameters that are measured over a range of DC bias conditions because there will be changes in RF behavior linked to the dynamic processes in the transistor.

Pulsed-RF network analyzers also allow RF characterization at bias points outside the safe-operating area. Pulsed and radar applications can operate in these regions.

15.4 Dynamic processes

Characterizing and modeling the dynamics of transistor systems requires an understanding of the processes in the device that cause the observed variation. This understanding is essential for the design of pulse and linearity measurements, the interpretation of the resulting data, and for the development of circuit models that predict the observed behavior.

The two dominant mechanisms are temperature dependence and charge trapping. A description of these is presented in this section to provide the understanding required to interpret measurements and model transistor dynamics.

15.4.1 Temperature and self-heating

Temperature has a significant impact on the nature of all aspects of transistor operation. The dynamic processes, their response rates, and the current transport process all vary with temperature. These processes influence a transistor's current and hence power dissipation, which then affects temperature via a thermal feedback. This feedback is known as *self-heating* and it is an inescapable aspect of all dynamic processes in any transistor circuit.

Over a reasonable temperature range, say 250 K to 400 K, a linear approximation might be assumed. Across this temperature range, the drain current of a FET would follow

$$i_D = i_{DO}(1 - \lambda\,\Delta T), \tag{15.1}$$

where λ [K^{-1}] is a thermal coefficient at temperature T_N [K], and i_{DO} [A] is the current at temperature T_N [1,2]. The temperature rise ΔT [K] is the difference between T_N and the channel temperature. A temperature rise is produced when there is power dissipated in the channel, $p_D(t)$ [W], which is the product of the channel current, i_D, and potential, v_{DS} [V]. The temperature at any time, t [s], is

$$\Delta T(t) = \int p_D(\tau)\,h_T(t-\tau)\,d\tau, \tag{15.2}$$

which is a convolution of the power dissipation and the impulse response, $h(t)$, of the thermal path from the channel to ambient [3].

Thermal response

Power dissipation and temperature are related by the specific heat capacity, c [J/K·kg], and thermal conductivity, k [W/m·K], of the structure. Consider heat flow into a material, $\frac{dQ}{dt}$ [$\mathrm{W/m^2}$], through a small cross section to a heat sink. The product of area density and heat capacity, $\rho_A c$ [$\mathrm{J/K{\cdot}m^3}$], is a measure of the energy that can be stored in a region of material for a given temperature, which is analogous to charge in a capacitor for a given voltage. Thermal conductivity relates temperature difference or gradient through the region of material, ∇T [K/m], to the heat flow into it. The thermodynamic rate equation relates these quantities as

$$\frac{dQ}{dt} = \rho_A c \frac{d}{dt} T + k \nabla T. \tag{15.3}$$

The time constant of this thermal response is the ratio of the mass-heat capacity product and thermal conductivity, $\tau_T = \rho_A c / k$.

A small region in the vicinity of the channel of a transistor has dimensions in the order of fractions of microns with correspondingly small area density and high thermal conductivity. The rate of temperature rise of such a region is in the order of nanoseconds. The whole transistor and its surroundings form a distributed thermal path that draws heat from the channel. The rate of heat flow for the larger structure is slower because the net thermal conductivity is lower and mass is larger.

Transient measurements show heating to be a sub-first-order phenomenon. That is a gradual response over several decades of time, which contrasts a first-order response that occurs over about one decade of time. A detailed solution involves fractional calculus, which can confirm that the response of a regular distributed thermal path is of the order of one half. In the frequency domain, the response of a transistor's thermal path closely conforms to:

$$H_T(\omega) = \frac{R_T}{(1 - j\,\omega/\omega_T)^{n_T}} \frac{1}{(1 - j\,\omega/\omega_0)^{1-n_T}}, \tag{15.4}$$

where n_T is the order of the response, which will be near to or less than 0.5, and R_T [K/W] is the thermal resistance of the thermal path, which is the inverse of its thermal conductivity. The characteristic frequency of the response is ω_T, which is the inverse of the time constant of the whole path from the channel to ambient. This path includes the total mass and thermal conductivity of the device, so the characteristic frequency is in the relatively low Hz to kHz range. The channel region where heat is generated has a finite, albeit small, size that has a characteristic frequency, ω_o in (15.4), in the order of 10 to 100 GHz. Above this frequency, a first-order response is appropriate because the heat source is distributed throughout this region. The frequency response, $H_T(\omega)$, and impulse response $h_T(t)$, form a Fourier transform pair.

Since all aspects of the dynamics of transistors depend on temperature, the characterization and simulation of temperature variation with time is important. At any instant in time, there is an instantaneous channel temperature, which is a function of the history of power dissipation. In the steady state, the temperature rise in (15.1) is the product of the power dissipation and the thermal resistance, R_T (that is, $H_T(0)$). The reduction in drain current is then given by $i_D = i_{DO}(1 - \lambda\, R_T\, i_D\, v_{DS})$.

The dynamic response for time-varying current and voltage is a function of the time-varying temperature rise, which is a convolution of the time-varying power dissipation with the thermal impulse response, (15.2). Although the thermal response can then be inferred from a transient measurement of drain current, there will be some ambiguity because the initial thermal response in the channel region is too fast for 100 ns pulse equipment. The response can be more readily analyzed and measured in the frequency domain in terms of the characteristic frequencies and order, n_T, of the response [3, 4]. This approach to the extraction of heating parameters is discussed later in Section 15.7.3.

15.4.2 Charge trapping

Within the structure of transistors there are regions, or sites, that trap charge in mid-band energy states [5]. Mid-band states are always present at surfaces and interfaces and can be included in bulk regions to control or pin certain process parameters. Rather than moving between valence and conduction band, charges can move to the mid-band state. Once in the mid-band state, further movement is delayed, so the charge is trapped temporarily. Deep-level states are those close to the middle of the semiconductor band gap and these trap charges for longer time periods.

The classic manifestation of charge trapping is gate lag. This is an additional rise in drain current that occurs a few milliseconds after stepping the gate potential to turn-on. The size of the current increase and the delay vary with the initial bias and the destination of the gate step. Variations over several orders of magnitude in response time are observed, but the lag is a first-order response, that is a response over one or two decades of time. The lag can be extremely long, with tens of minutes being typical in devices fabricated with wide band-gate materials such as Gallium Nitride. Trapping in passivation layers can be responsible for long-term alteration of transistor characteristics. In high electric field conditions, such as at high drain-source potentials, the lag can be faster than the 100 ns resolution of pulse measurements. This is dealt with in more detail in Section 15.5.

Because the trapping process imposes an inherent delay, its influence on drain current is a function of the past bias conditions. This is a memory effect with a bias and frequency dependence related to the occupation and charging rates of the trap centers [6–8].

Trap rates

The extent and period of trapping is well described by capture and emission processes in terms of carrier concentrations and energy bands in the semiconductor [9]. Drawing an electric circuit analogy of a trap center provides a description that is readily understood by engineers working with FET circuits and that can be implemented in a circuit simulator [10, 11].

Charge in a trap center is analogous to charge in a capacitor, C_T [F]. The ionization potential of the trap, v_T [V], is analogous to the potential across the capacitor. The ionization potential is always restricted between zero, for neutral charge, and the potential, V_O [V], of the fully depleted trap. The latter is positive or negative depending on the ionization polarity of the trap.

The capacitor representing the trap center is charged by a nonlinear controlled current source given by

$$i_T = \omega_e(T) C_T \left[V_O - v_T - v_T \exp\left(\frac{q v_I}{kT}\right)\right], \tag{15.5}$$

where q [C] is the electron charge and v_I [V] is a control voltage that accounts for the change in Fermi level due to electric fields [4], and

$$\omega_e(T) = A_T T^2 \exp\left(-\frac{E_T}{kT}\right), \tag{15.6}$$

where $A_T T^2$ [s^{-1}] is an Arrhenius factor, k [eV/K] is the Boltzmann constant, and E_T [eV] is the trap's activation energy.

In the steady state with zero net current, the characteristic frequency for the trapping process is

$$\omega_E(v_I, i_J, T) = A_T T^2 \exp\left(-\frac{E_T}{kT}\right)\left[1 + \exp\left(\frac{q v_I}{kT}\right)\right]. \tag{15.7}$$

Control of the trap by transistor terminal currents is by v_I through appropriate functions.

Three characteristics of trapping can be noted. The first is that the trapping response has a single time-constant for fixed v_I. The second is that the time constant is set by the value of the target v_I and is independent of the initial bias. The third is that temperature significantly affects the trap response, so trapping effects include evidence of a simultaneous heating response.

15.4.3 Impact ionization

Observable impact ionization in a FET requires a sufficiently strong electric field, which is governed by the drain-source potential. Avalanche breakdown can result when a cascade of ionizations occurs where the additional carriers go on to generate even more.

Experiments show a reasonably logarithmic relationship between the impact ionization rate and the inverse of electric field strength [5]. This experimental evidence is the basis for the following expression for impact ionization rate, R_I, with slope and intercept corresponding to two fitting parameters A and B [V/m]:

$$R_I(F) = A\,e^{-B/F}, \tag{15.8}$$

where F is the electric field strength [V/m]. Increased temperature reduces the impact ionization rate relative to the electric field [12, 13].

In terms of the electron current from the source, the total current is thus:

$$i_D = \frac{i_{DS}}{1 - R_I}. \tag{15.9}$$

The hole current returning toward the source is $i_D R_I = i_{DS} \frac{R_I}{1-R_I}$ and some fraction of this will tunnel to the gate or surface. Tunneling is more probable when the gate bias is negative, which is attractive to the holes, but reduces exponentially with increasing gate potential [14]. The measured gate current from impact ionization, as shown in Figure 15.5, increases as the drain current increases, but varies with the gate potential because the tunneling probability varies.

The hole tunneling is easily measured with low-frequency semiconductor parameter analyzers. Pulse system, which has less dynamic range, only detects high levels associated with breakdown. However, the kink in the drain current characteristics is quite distinct in pulse measurements and has a response time consistent with a trapping process.

Measurement of the kink effect

A more substantial increase in drain current is produced by the positive potential of accumulated holes that have reached the surface, usually at the source end of the channel. Those that occupy trap states cause a response according to trap occupation and charging rates and the polarity of the trap. These rates are observed to increase dramatically with drain potential, such that the effect is much faster than even the shortest pulse measurements [15].

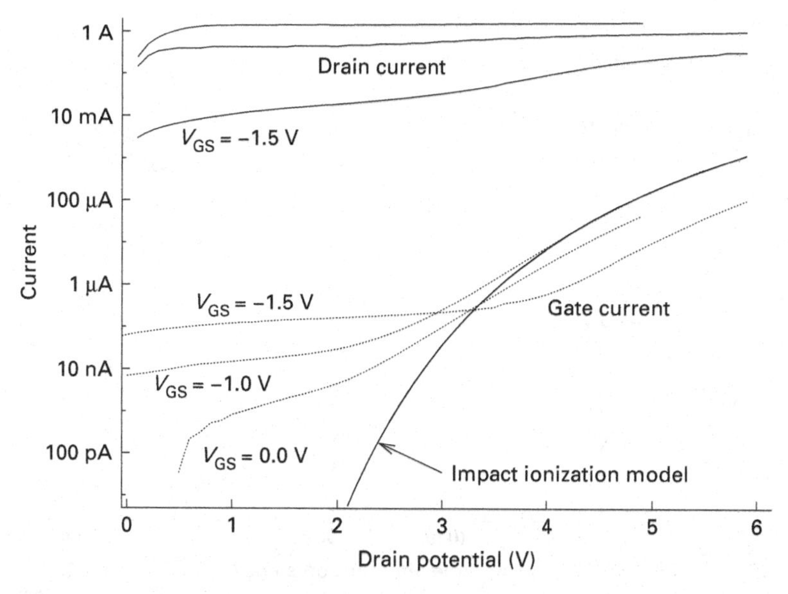

Fig. 15.5 Gate and drain DC characteristics of a typical pHEMT. Hole current generated by impact ionization is clearly evident in the gate current. The dashed line is the product of drain current and (15.8).

Thus, the increased drain current comes from positive feedback via a field effect. The transconductance of the FET amplifies the effect of a relatively low trapping potential, so the observed drain current increase is significantly greater than the contribution from ionization alone [6].

At higher power bias points the traps respond at increasingly faster rates. The effect is slow at modest drain-source potentials, so pulse measurements are able to observe a rise in drain current as the traps ionize. The position of the kink in the drain current moves to higher drain potentials as the width of the measurement pulses reduces, as shown in Figure 15.1. This is because the kink is centered at the drain potential where the trapping rate is comparable to that of the pulse length. That is, the traps have time to ionize on the high side of the kink, but not on the low side.

15.5 Transient measurements

The typical use of pulse characterization is to avoid dynamic processes in the transistor rather than to analyze them. This overlooks considerable information in a complete step response over a longer time period. The picture is completed by repeated measurements from different biases, which can be illustrated in an investigation of gate lag.

15.5.1 Measurement of gate lag

Gate lag is a delayed additional rise in drain current that occurs a short time after a FET is turned on, as discussed in Section 15.4.2. A measurement of gate lag is shown in

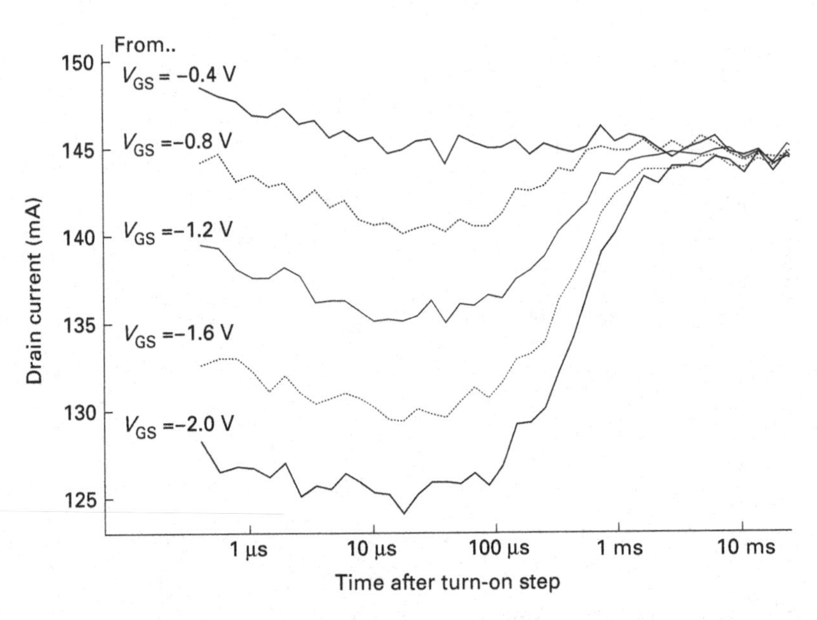

Fig. 15.6 Time domain response after stepping to zero gate-source potential, $v_{GS} = 0$ V, from various initial gate biases, V_{GS}, as annotated. The drain potential is fixed at $V_{DS} = 1.5$ V.

Figure 15.6. Here the device is switched to zero gate-source potential without changing the drain potential. After a few milliseconds the drain current settles to the same value irrespective of the initial gate potential, which is a reasonable expectation.

When the FET is switched on to zero gate-source potential, the current rises from the much lower value that it had prior to zero time. The current rise causes heating with a response over several decades in time, as described in Section 15.4.1. This produces the reduction in current over several decades of time that is common to all the turn-on transients.

The current rise at about 300 μs can be explained by a hole trap in the substrate. Before the transient the negative gate bias prior to the transient injects electrons into the substrate, which ionizes the trap. The extent of ionization increases with more negative gate bias, so the height of the gate-lag current rise is proportional to the initial gate bias. When the gate potential is stepped to zero, the trap potential increases as holes are captured through a first-order process with time constant set by the FET's terminal potentials.

The dependence on initial drain bias is shown in Figure 15.7. Increasing the drain bias offsets the influence of negative gate potential on the electric fields in the substrate.

The time constant of the gate lag is set by the destination point of the transient. The variation in timing with destination drain-source potential is shown in Figure 15.8.

A complication in the gate-lag characteristic is the temperature dependence of the capture process. This increases as the temperature does, so the responses interact. Consequently, the transitions in Figure 15.8 vary from the one-decade rise of a single time constant process to a faster response when heating is coincident.

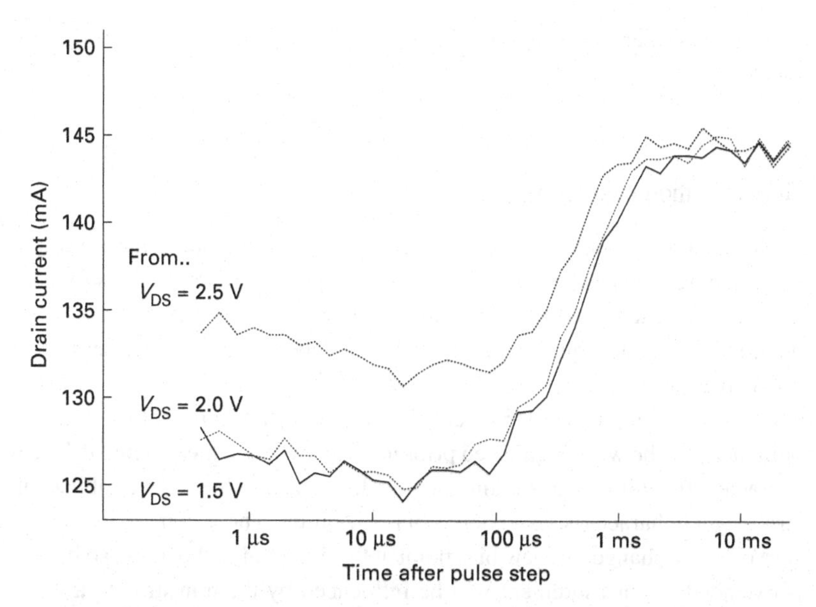

Fig. 15.7 Time domain response after stepping to $v_{DS} = 1.5$ V, $v_{GS} = 0$ V (the same destination as Fig. 15.6) from $V_{GS} = -2.0$ V and various drain potential biases, V_{DS}, as annotated. The curve for $V_{DS} = 1.5$ V is the same as the bottom curve in Fig. 15.6.

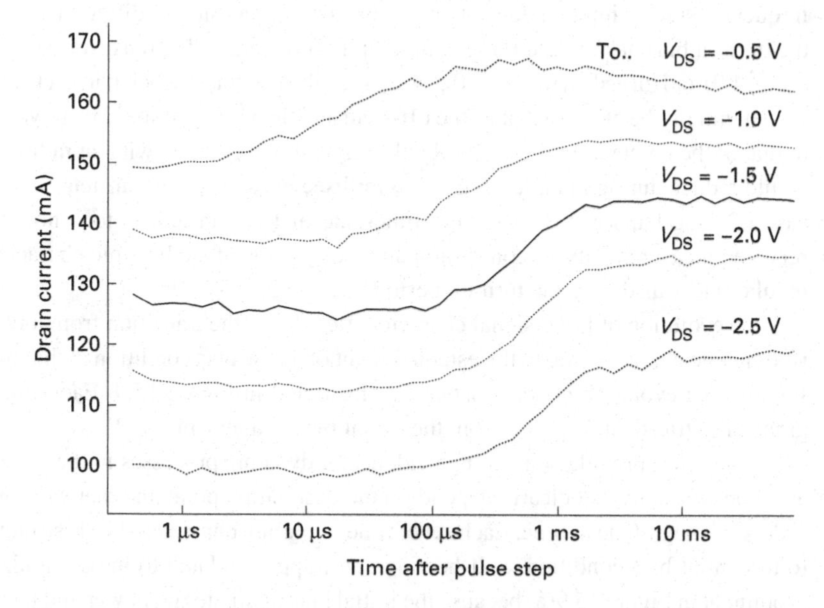

Fig. 15.8 Time domain response after stepping to $v_{GS} = 0$ V and various v_{DS} from $V_{GS} = -2.0$ V, $V_{DS} = 1.5$ V. The center curve is the same as the bottom curves in Figs 15.7 and 15.6.

To study the trap ionization process, the pulse measurement can be repeated with varying times at the initial bias [16]. That is, the gate lag is measured as a function of how long the trap is allowed to ionize. This is an inverse pulsing technique, where the bias time between pulses is varied and the pulses are set to a long period to measure the transient.

15.5.2 Time evolution characteristics

Each bias point about which pulse measurements are made establishes a temperature and trapping state. There is a unique set of isodynamic characteristics for each state, and therefore for each bias point. The three sets of pulse measurements corresponding to three bias points in Figure 15.2 illustrate the dependence of isodynamic characteristics on the bias state.

Two aspects of pulse and transient characteristics are worth noting. One is that the bias point needs to be well established prior to an isodynamic measurement. The isodynamic characteristic will vary depending on how long the bias has been established, that is, the large-signal characteristics undergo bias evolution. The second is that any pulse is the start of a step change to a new bias point at the potential of the pulse, so if the pulse width is extended the measurement will be influenced by the transition to a new bias point. Transient responses over a set of long pulses produce data that shows a time evolution of large-signal characteristics.

The concept of bias evolution is that an isodynamic characteristic, seen by a radio-frequency signal, immediately after turning on a transistor is different to that when the turn-on bias has settled. For example, in Figure 15.2 if the transistor was off at bias (–2.0, 4.0) and turned on to bias (–0.5, 4.0), then the isodynamic behavior of the transistor would initially be close to that of the off-state and then over a finite time it would evolve to that of the on-state. In this example, the large-signal current swing at radio frequency would reduce during the microsecond to millisecond switch-on transient. Except in the case of wide band-gap devices, the timescale of the transient is too short to permit repeated measurements by conventional pulse systems of isodynamic characteristics at regular intervals during the turn-on period.

Time evolution of large-signal characteristics shows the transition from a point on an isodynamic characteristic to the establishment of a new bias condition at that point. Two sets of time-evolution characteristics are illustrated in Figure 15.9. Each line in these graphs is a transient response from the initial bias to a new bias [17].

The surfaces provide convincing evidence of dynamic processes in the device. There are time constants that clearly depend on the destination point and that vary over many orders-of-magnitude in time. Each set of time-evolution characteristics is strongly linked to the initial bias condition. Self-heating and trapping related to impact ionization are prominent in Figure 15.9(a) because the initial bias dissipates no power and the trap state is at the pinch-off extreme. In Figure 15.9(b), the initial bias condition already establishes a degree of heating and a moderate trap state.

A notable feature of the data in Figure 15.9 is the dependence of time constants on the destination drain potential. There is a straight-line relationship between log-time and potential. This is explored from an alternative small-signal vantage in Section 15.8. Detailed information about specific dynamics is present in a select set of transient

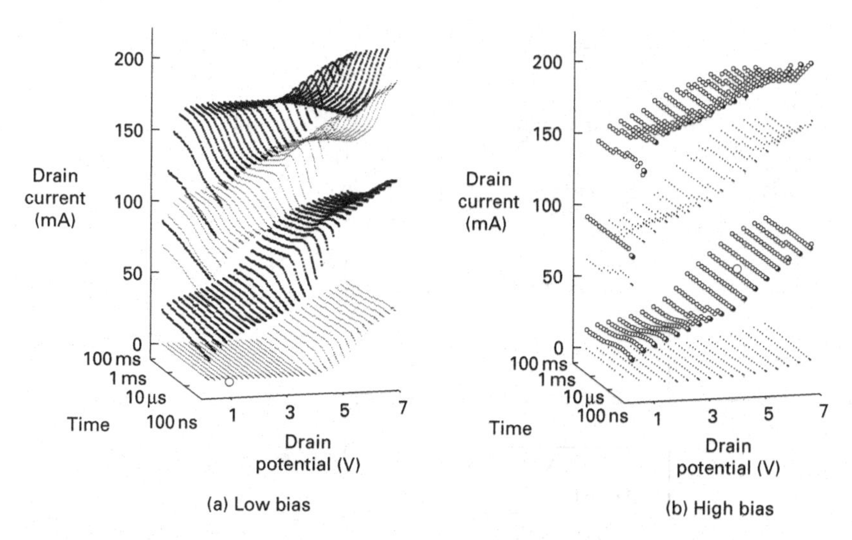

Fig. 15.9 Time-evolution large-signal characteristics of two pHEMTs. Each plot shows drain current versus time after a step from a fixed bias point (large o at 100 ns). There are four surfaces corresponding to gate-source potential from −1.5 to 0.0 V in 0.5V steps as the parameter.

measurements in the time-evolution data and from the dependence of these on a range of initial bias conditions.

15.6 Pulsed measurement equipment

There are many pulse measurement systems reported in the literature offering a variety of options. They can be assembled from individual instruments that provide bias and pulses, synchronize timing, and measure current and voltage. In advanced systems, pulsed radio sources and network parameter measurements are incorporated, particularly for measuring high power devices intended for pulsed-radar applications. Pulse techniques also facilitate nondestructive investigation of transistor breakdown regions [18]. A degree of sophistication is achieved with arbitrary pulse patterns and arbitrary control of initial bias and pulse timing [19, 20].

Commercial systems are available that are capable of sub-microsecond pulses. Examples are Accent Opto's Dynamic I(V) Analyser [21], Auriga Measurement System's Pulse IV/RF System, Focus Microwave's modular pulse system, and systems by Agilent Technologies, Keithly Instruments, and Amcad Engineering. These systems fill the instrumentation gap between semiconductor parameter analyzers and nonlinear vector network analyzers. Typical pulse systems offer a large-signal capability and speeds sufficient to give near isodynamic characteristics at low drain biases.

15.6.1 System architecture

The functional diagram of a pulsed measurement system, shown in Figure 15.10, includes both pulsed-I/V and pulsed-RF subsystems. Pulse and bias sources, voltage and current

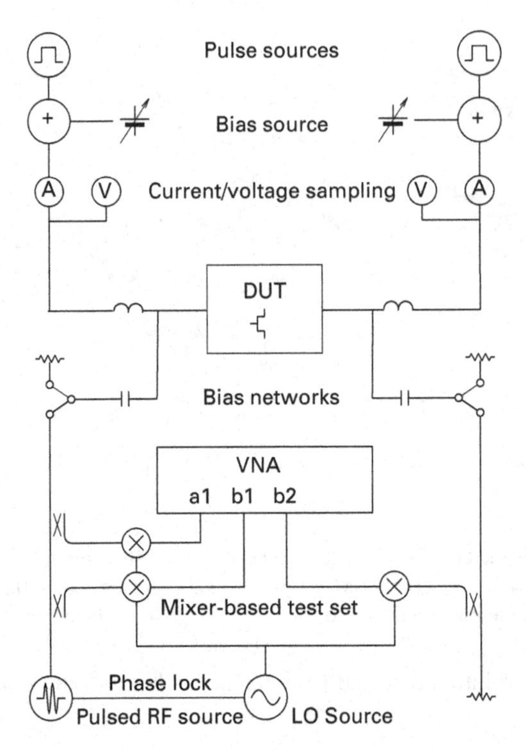

Fig. 15.10 Simplified diagram of a generic pulsed measurement system. Alternative connections provide load terminations when there is no pulsed-RF test set.

sampling blocks, and associated timing generators form the pulsed-I/V subsystem. A pulsed-RF source and mixer-based vector network analyzer form the pulsed-RF subsystem. The test device is connected directly to the pulsed-I/V subsystem, or to bias networks that connect the pulsed-RF subsystem or RF terminations.

Pulsed-I/V system

Steady-state semiconductor parameter analyzers provide a source-monitor unit for each terminal of the test device. The unit sources one of voltage or current while monitoring the other. In a pulsed measurement system, a pulsed voltage is added to a bias voltage and applied to the device. A precise measurement grid is rarely obtainable by pulse systems because of the transient response limitations of pulse equipment. Thus, actual terminal conditions, both voltage and current must be recorded. It is essential to recognize that the pulse data do not lie on a regular grid of values, so a naive plot of characteristics curves can be misleading because each line will not correspond to a constant control potential.

The position of the voltage and current sensors between the pulse source and the test device is affected by transmission line effects associated with the cabling between the sensing points. These affect the transient response and performance of the pulse system. An additional complication is introduced when the test device must be terminated for RF stability. A bias network is required but this introduces its own transient response to the measured pulses. The initial 100 ns transient in many pulsed measurements is dominated by the bias network.

Current can be measured by various methods, which trade between convenience and pulse performance. Hall-effect/induction probes placed near the test device can sense terminal current. These probes have excellent common-mode immunity but tend to drift and add their own transient response to the data. A stable measurement of current is possible with a series sense resistor. This requires a differential input with very good common-mode rejection at high frequencies.

Pulsed-RF system

Pulsed-RF test sets employ vector network analyzers with a wideband intermediate frequency (IF) receiver and an external sample trigger [20, 22]. The systems need two RF sources and a mixer-based S-parameter test set. One source provides a continuous phase reference for the mixers and samplers, while the other provides a pulse-modulated RF output.

The pulsed bias must be delivered through bias networks. During a pulsed-I/V measurement, the RF source is disabled and the RF test set provides terminations for the test device. Pulsed-RF measurements are made one pulse point at a time. With the pulsed bias applied, the RF source is gated for a specified period during the pulse and the network analyzer is triggered to sample the RF signals. The same pulse point is repeated to work through a required frequency list and averaging setting.

Bias networks

The bias network that connects the test device needs to provide stable high-frequency termination while passing pulse stimuli. In addition, there need to be current and voltage measurement ports. The trade-off between these requirements necessarily limits the maximum rise time of the pulse. If faster than 100 ns pulses are required, then the pulse source must be connected directly to the test device [23].

An enhanced bias network that allows reasonable length cables to the samplers is shown in Figure 15.11. The DC-blocking capacitor is reduced, so that it does not draw current for a significant portion of the pulsed bias. At the same time it provides adequate passage at RF frequencies. The isolating inductor must be small enough to pass the pulsed bias while providing adequate RF isolation. In the figure, the DC-blocking capacitor and isolating inductor values are an order of magnitude smaller than are those in conventional bias networks. The network provides a good RF path for frequencies above 500 MHz and does not significantly disturb pulses longer than 100 ns. Modifying the network to provide a RF path at lower frequencies will disturb longer pulses.

The pulsed bias is fed to the bias network in Figure 15.11 through a cable that introduces transmission line transients. A snubber is added to control these. The values shown are suitable for suppressing the 10 ns transients associated with a 1 m cable.

Voltage sampling in Figure 15.11 is through a frequency-compensated network that provides isolation between the RF path and the cable connected to the voltage sampling digitiser. Without this isolation, the capacitance of the cable would load the pulsed-bias waveform, significantly increasing its rise time. The voltage sample point should be as

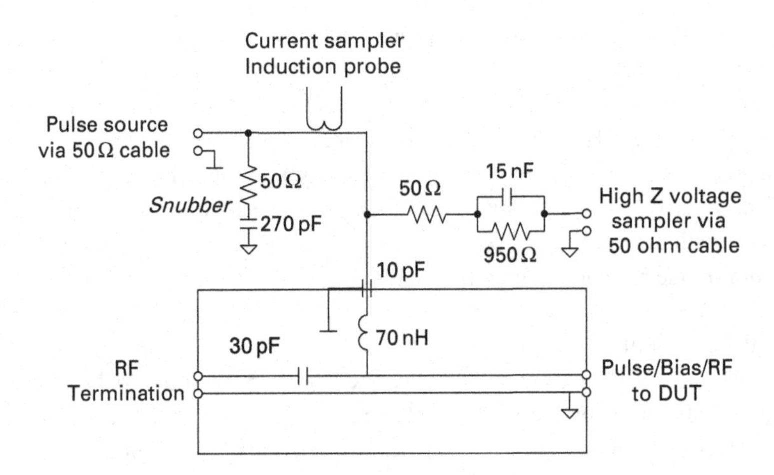

Fig. 15.11 An enhanced bias network that allows voltage and current measurement instruments to be connected via reasonably long cables. The bias network is designed to cut-off at 500 MHz, to allow pulses through the bias port to the DUT.

close as possible to the DUT to reduce the effect of reflected pulses. The network in this example sets a practical limit of about 15 cm on the length of the cable connecting the transistor under test to the bias network.

Induction current probes introduce their own time constants to the measurement that is visible in the time domain transient record. Current measurement with series sense resistors ameliorates this, but adds to the output impedance of the pulse source. Usually a capacitance of a few picofarads is associated with the sense or bias network that restricts the choice of resistance value for a specified rise time.

Series-resistor sensing requires a floating differential amplifier operating over the range of pulse potentials. The common-mode gain of the amplifier is higher for short time intervals, so some of the step change in potential is recorded as a current transient. Placing a sense resistor in the ground return is an alternative, but the transmission-line effects of the connection between the pulser and DUT need to be considered.

15.6.2 Timing

The most critical aspect of pulse measurement is sample timing. In many cases the sample will be gathered at some point in a time-dependent dispersion process, so it is important to consider the timing relative to the time-constants of these processes. In general, full information can only be gathered by a time domain pulse-profile measurement. Equally, the time spent establishing the quiescent condition before the pulse must be long enough that there are no residual effects from a previous pulse. At least two orders of magnitude less than the time constants of the dynamic processes is recommended.

Measurement equipment capable of pulsing to points on the I/V-plane in a random sequence provides a powerful display for verifying isodynamic timing. If the time at quiescence between pulses is insufficient, then the pulse measurement will be dependent upon the particular history of previous pulse points. Step-and-sweep sequencing generates a monotonic change in history, so dynamic effects are not obvious because adjacent points have similar pulse histories. However, if a random sequence is employed, the adjacent points will have different pulse histories and the corresponding effect of dynamic processes will be visibly different.

Interpolation and iteration

Often, measurements are desired at a particular pulse point or on a regular grid of points. For a target pulse voltage, the actual voltage at the transistor will usually be different due to interactions with the pulse amplifier output impedance and amplifier time constants, as well as cabling and bias network transients. These could be compensated for in advance with known current, but this current is being measured. This is why pulsed voltages need to be measured at the same time as the device currents.

If measurements are desired at specific voltage values, then one of two approaches can be used. Firstly, over successive pulses, the target voltage values can be adjusted to iterate to the desired value. This necessarily involves a measurement control overhead and can require considerable time for many points. The second approach is to establish a look-up table to calibrate the pulse setting to give the target voltage. For true step-response data, this correction needs to be adjusted throughout the pulse period to compensate for the transient response of the pulse source [24].

15.7 Broadband RF linearity measurements

The conventional wisdom was to consider dynamic processes to be slow enough to be irrelevant during the high-frequency operation of transistor circuits. Other than an anomalous shift of bias conditions, the slow processes do not affect performance at high enough operating frequencies, which need only be a few tens of MHz. Kinks, hysteresis, or memory seen in low-frequency and slow-pulse drain characteristics are not evident in the bias dependence of high-frequency small-signal parameters. Thus, a set of fast pulse measurements that provide drain characteristics at various bias points can be indicative of the high-frequency operation. Little information on the nature of slow dynamics is required to predict high-frequency small-signal parameters.

This is not true for low-signal, or weak nonlinearity scenarios. In particular, when there are two tones in the signal, any nonlinearity generates distortion products including intermodulation products near the signal frequencies. Intermodulation exhibits a significant variation with bias and this bias dependence changes with the difference between the frequencies used in a two-tone intermodulation measurement. Contemporary wisdom attributes this to slow processes in the transistor and low-frequencies impedances in the circuit [7, 25, 26]. If the difference-frequency between two tones resonates with the characteristic frequencies of the dynamic processes in a transistor, then these processes will be excited and affect the high-frequency characteristics. For certain bias conditions, the variations of intermodulation level and asymmetry between the upper and lower products can be several orders of magnitude. Thus generation of distortion and intermodulation is a crucial aspect of the slow dynamic processes in a transistor.

The requirement to control linearity at microwave frequencies is probably the most compelling reason to characterize the full dynamics of heating and trapping processes in transistor models. At the very least, a priori understanding of these processes is essential to the correct interpretation of pulse measurements. Even so, the contribution to linearity often spans frequencies beyond the reach of pulse equipment. It is necessary to draw on characterization techniques based on intermodulation and high-frequency small-signal measurements to bridge the gap.

15.7.1 Weakly nonlinear intermodulation

Generation of intermodulation can be analyzed with a weakly nonlinear model, which is often a small-signal model with nonlinear elements described by a Taylor series. An expression sufficient to illustrate intermodulation generation by slow dynamic processes in the transistor is:

$$i_d(v_g, v_d) \approx g_m\, v_g + g_m'\, v_g^2 + g_m''\, v_g^3 + g_{md}\, v_g v_d, \tag{15.10}$$

where the signal potentials are $v_g = v_G - V_D$ and $v_d = v_D - V_D$. The transconductance and its derivatives are given by $g_m = \frac{d}{dv_G} i_D$, $g_m' = \frac{1}{2}\frac{d^2}{dv_G{}^2} i_D$, and $g_m'' = \frac{1}{6}\frac{d^3}{dv_G{}^3} i_D$. A cross-conductance $g_{md} = \frac{1}{2}\frac{d}{dv_D}\frac{d}{dv_G} i_D$ is also included because it is key to understanding the interaction with slow dynamics. All the conductances are evaluated at $v_G = V_G$ and $v_D = V_D$.

(15.10) is a very simple description that neglects the drain conductance and any terms higher than third order. The bias drain current I_D is neglected, so that (15.10) gives the signal current, i_d, in terms of the signal potentials, $v_g = v_G - V_D$ and $v_d = v_D - V_D$.

A load impedance, Z_L, driven by the drain current, as shown in Figure 15.12, develops a drain signal given by

$$v_d \approx -g_m\, v_g\, Z_L - g_m'\, v_g^2\, Z_L - g_m''\, v_g^3\, Z_L - g_{md}\, v_g v_d\, Z_L, \tag{15.11}$$

where the value of Z_L varies with the frequency of each current component.

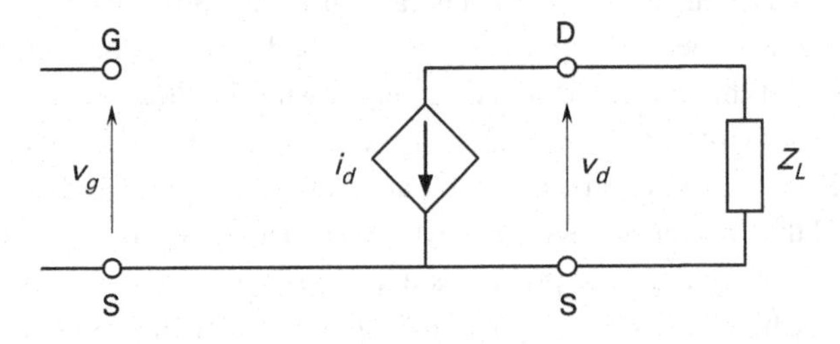

Fig. 15.12 Simple weakly nonlinear unilateral model of a FET driving a load impedance.

Intrinsic nonlinearity

An intrinsic level of intermodulation is produced by (15.10) when the gate is driven by a two-tone signal. For the case of tones at frequencies ω_1 and ω_2 the third-order intermodulation product will be at frequencies of $\omega_1 - \Delta\omega$ and $\omega_2 + \Delta\omega$. Drain currents at these frequencies will be generated by the third-order term, $g''_m\, v_g^3$, and by the second-order term, $g_{md}\, v_g\, v_d$, in (15.10) [27,28].

For the case of two tones close to ω such that $\Delta\omega = \omega_1 - \omega_2 \ll \omega$, the frequencies of the intermodulation currents will also be close to ω, so the drain potential for the intermodulation products will be proportional to the load impedance at ω, which is $Z_L(\omega)$.

The third-order intermodulation current associated with the cross-conductance term, $g_{md}\, v_g v_d$, comes from second-order components in v_d at $\pm\Delta\omega$ as well as at $\pm 2\omega$. The latter is usually not significant in slow-rate dynamics and the former comes from drain currents at frequencies $\pm\Delta\omega$ generated by the second-order term, $g'_m\, v_g^2\, Z_L$, in (15.11), so is proportional to the load impedance at the difference frequency, $Z_L(\pm\Delta\omega)$.

The amplitude of the drain potential's third-order intermodulation product, $V_d^{<\mathrm{IMD}>}$, in terms of the gate signal amplitude, V_g, is the sum of contributions from the transconductance and cross-conductance terms:

$$V_d^{<\mathrm{IMD}>} \approx -\frac{1}{4} {V_g}^3 Z_L(\omega) \left(g''_m - g_{md}\, g'_m\, Z_L(\pm\Delta\omega) \right). \tag{15.12}$$

This is an intrinsic level of intermodulation because slow dynamic processes have not yet been considered. It is proportional to $(g''_m + g_{md}\, g'_m\, Z_L(\pm\Delta\omega))$ where $Z_L(-\Delta\omega)$ and $Z_L(+\Delta\omega)$ affect the upper and lower intermodulation products at $\omega_1 - \Delta\omega$ and $\omega_2 + \Delta\omega$, respectively. There is a conjugate relationship between the load impedances at positive and negative frequencies, such that $Z_L(-\Delta\omega) = {Z_L}^*(\Delta\omega)$, so there is an asymmetry between the bracketed terms in (15.12) for the upper and lower intermodulation products, which is often observed in measurements [25,26,29,30]. This asymmetry can be removed if $Z_L(\Delta\omega)$ is real. Also, a suitable choice of bias and load impedance at the difference-frequency, such that $g_{md}\, g'_m\, Z_L(\Delta\omega) = -g''_m$, can eliminate the intrinsic intermodulation altogether. There are useful operating regions where an optimal load can be realized in practice because $g''_m < 0$, as shown in Figure 15.13.

Note that the case above is simplified to illustrate a dominant intermodulation mechanism. It overlooks other intermediate second-order products that contribute to intermodulation [31–33].

15.7.2 Intermodulation from self-heating

Self-heating is a response to instantaneous power that has been described in Section 15.4.1. The power is the product of the drain current and drain potential, $i_D\, v_D$, which can be expressed in terms of the bias, I_D and V_D, and the signal components, i_d and v_d, to enable separation of bias and instantaneous power terms.

$$p_D = (I_D + i_d)(V_D + v_d) = I_D\, V_D + (i_d\, V_D + v_d\, I_D) + i_d\, v_d. \tag{15.13}$$

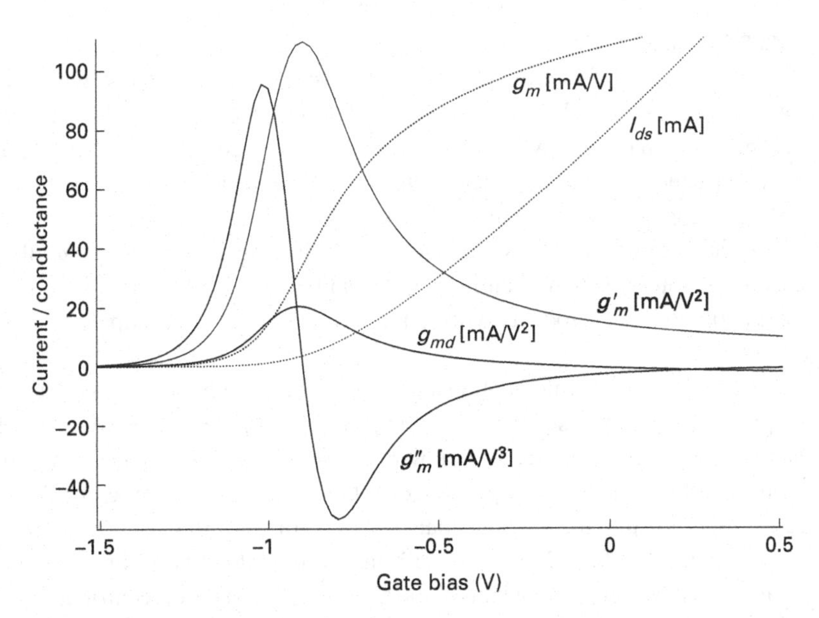

Fig. 15.13 Typical conductance and derivatives for a microwave transistor. This example is a model for a small-signal pHEMT.

In a weakly nonlinear scenario with two closely spaced tones, the signal currents and potentials will be dominated by the fundamental frequency components with frequencies near ω separated by $\Delta\omega$. The first term in (15.13) is the bias power component, which influences the bias temperature. The second, bracketed, term in (15.13) will be dominated by fundamental components with frequencies near ω. In most radio applications this frequency will be high, so there will be little, though not necessarily negligible, thermal response to this power component.

The last term in (15.13) is a product of fundamental tones, so will have components at frequencies $\pm\Delta\omega$ and $\pm 2\omega$. The latter is likely to be too high to excite a significant thermal response. However, for closely spaced tones, the former is capable of producing a significant thermal response.

The component of drain current that is directly affected by the thermal response through the self-heating mechanism is $-i_d \lambda p_D H_T$, which is derived from (15.1) and (15.2). Third-order intermodulation in the self-heating current comes from the product of the fundamental components in i_d with the difference-frequency components in p_D (from the last term in (15.13)). Expressing drain current and potential in terms of gate potential and load impedance gives the amplitude of the self-heating contribution to third-order intermodulation:

$$v_D^{<\mathrm{IMD}>} \approx -\frac{1}{4} {V_g}^3 {g_m}^3 Z_L^2(\omega) H_T(\pm\Delta\omega), \tag{15.14}$$

where $H_T(-\Delta\omega)$ and $H_T(+\Delta\omega)$ affect the upper and lower intermodulation products, respectively. The intermodulation product given by (15.14) is not negligible and can be easily observed at bias and load conditions that reduce the intrinsic nonlinearity.

15.7.3 Measuring heating response

The level of the thermal contribution to intermodulation relative to difference frequency mirrors the thermal response of the transistor. That is, $v_D^{\text{<IMD>}}$ versus $\Delta\omega$ is proportional to $H_T(\Delta\omega)$ given by (15.4). However, to observe the thermal response in a linearity measurement, other dynamic processes need to be reduced with a suitable bias condition and frequency-independent load. If the load is not constant then its frequency dependence can mask that of other slow-rate dynamics.

Intrinsic intermodulation, (15.12), varies with $Z_L(\Delta\omega)$ and the self-heating intermodulation (15.14), varies with $H_T(\Delta\omega)$. Presenting a constant-impedance load to the drain holds the intrinsic contribution constant, so any observed variation with frequency is from heating or other transistor dynamics.

The load shown in Figure 15.14 presents a 50 Ω load for all frequencies when $Z_L = 50\ \Omega$ and $L = 100C$. The capacitance should be large enough to allow the spectrum analyzer, Z_L, to observe the lowest frequency of interest; approximately 5 nF for 100 MHz. This is the frequency of the intermodulation product, which is high, and not the low difference-frequency.

A measurement of intermodulation generated by self-heating is shown in Figure 15.15 for a typical pHEMT. The transistor is biased near the zero crossing of g_m'', which occurs in the region of pinch-off. This zero crossing is illustrated in Figure 15.13. The intermodulation measurement that reveals the thermal response is found by choosing a bias that gives a minimal level at high difference-frequencies [3]. The high difference-frequency point has the lowest thermal contribution because $H_T(\omega)$ is small, so a bias that minimizes this point is one where the other dynamic processes are cancelled out. As the difference frequency is reduced, the thermal contribution increases.

In the example shown, the characteristic frequency of thermal response is 2.7 kHz and the order of the response is near 0.5, or only 10 dB per decade. The implication of this low order is that the magnitude of the heating contribution at 1 GHz is reduced by only 60 dB, which is not necessarily a negligible level. This dynamic process complicates the linearity of the broadband circuit that deals with a wide range of difference frequencies.

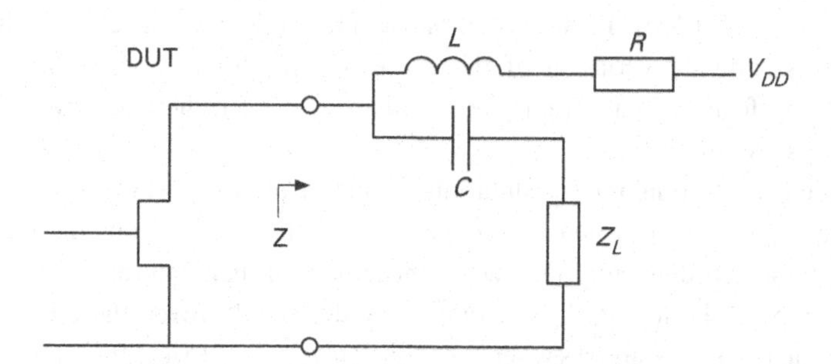

Fig. 15.14 Constant impedance bias network for measurement of slow dynamics within the DUT. The load presented to the drain is $Z = Z_L = 50\ \Omega$ for all frequencies when $L = 100C$.

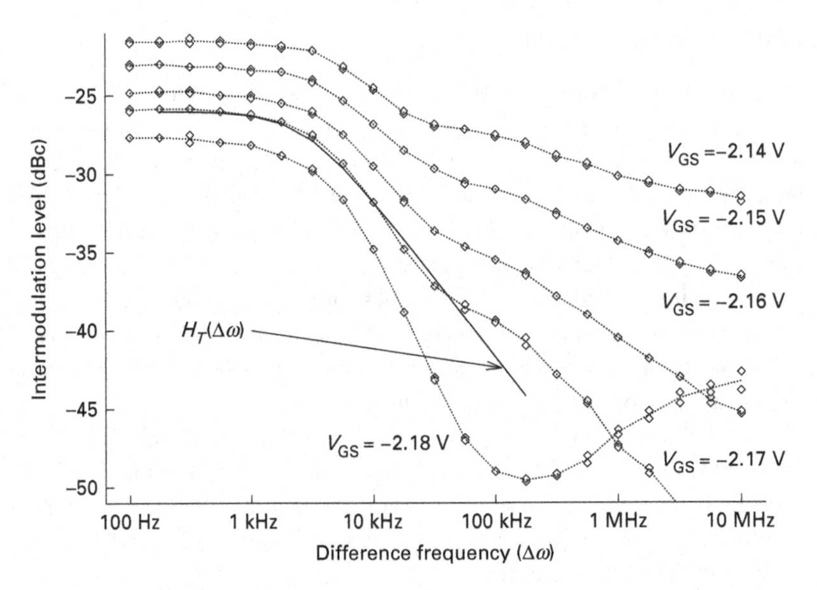

Fig. 15.15 Two-tone intermodulation measurement centered at 200 MHz of a typical pHEMT for $V_{DS} = 2.5$ V. A heating response, H_T, given by (15.4) is shown for $n_T = 0.5$ and $\omega_T = 2\pi \times 2700$ Hz. The load in Fig. 15.14 was used. Charge trapping accounts for the inflection near 80 kHz.

There is another dynamic process observed in Figure 15.15 as an inflection near 80 kHz. This process has a first-order response and a frequency that varies with drain bias, which would be expected if it were linked to a charge trapping process.

15.7.4 Measuring charge trapping response

Charge trapping provides a feedback path that can generate intermodulation in a mechanism similar to that for self-heating. The ionization potential of trap sites imparts an additional control over the channel current, which is influenced by drain and gate potentials. Since the trapping is a first-order low-pass phenomenon with characteristic frequency given by (15.7), it will respond readily to difference frequencies in the drain signal.

In principle, if other mechanisms are reduced, it is possible to observe a trap's frequency response, $H_E(\Delta\omega)$, in the variation of intermodulation level with difference frequency. A significant variation of the frequency of this response with bias is a clear identifying feature of a trapping mechanism. However, there are two issues that complicate this measurement.

Measurement of trapping intermodulation is an issue at biases where the trap is either fully ionized, $v_T \approx V_o$, or at neutral charge, $v_T \approx 0$. In these regions the trap cannot become more ionized than fully ionized or more neutral than neutral, so there is little change in trap potential to contribute to intermodulation. However, there is substantial trap related gate lag and hysteresis when the trap state is pulsed to and from these regions.

There is also an issue with the temperature of the trap site. Although the dependence of ω_E on the trap control potential is exponential for $v_I > 0$, this does not account for the frequency dependence of the intermodulation process because the trap potential is near zero in this region. Rather, the bias dependence of ω_E is mainly due to increasing temperature as power dissipation increases. As given by (15.7), the variation of ω_E with temperature is significant.

Pulse and transient step responses can be more suited to the characterization of trapping. Transient responses reveal trapping in regions where there is little trap-related intermodulation. The transients also include the effect of temperature change varying the trapping rate. This produces a variation with bias of the apparent order of the response, which can range from an order of one-half order if heating is coincident to a first-order response of trapping alone.

15.7.5 Measurement of impact ionization

Impact ionization works in conjunction with charge trapping to generate intermodulation that is highly dependent on bias. The response follows the combined transfer function of the first-order trap site and the second-order nonlinearity of impact ionization, derived from (15.8). This transfer function is proportional to the impact ionization rate, so the associated intermodulation contribution is only generated if the drain-source bias potential is high enough. It is a first-order response that depends on temperature and bias, so ω_I increases at a rate of about one decade per volt as drain bias increases. At moderate drain potentials, such as $v_{DS} = 7$ to 10 V, the response can be in the order of 1 to 10 GHz.

Clear evidence of impact ionization is seen in the striking variation of intermodulation with difference frequency and bias shown in Figure 15.16. To observe this, a bias in a

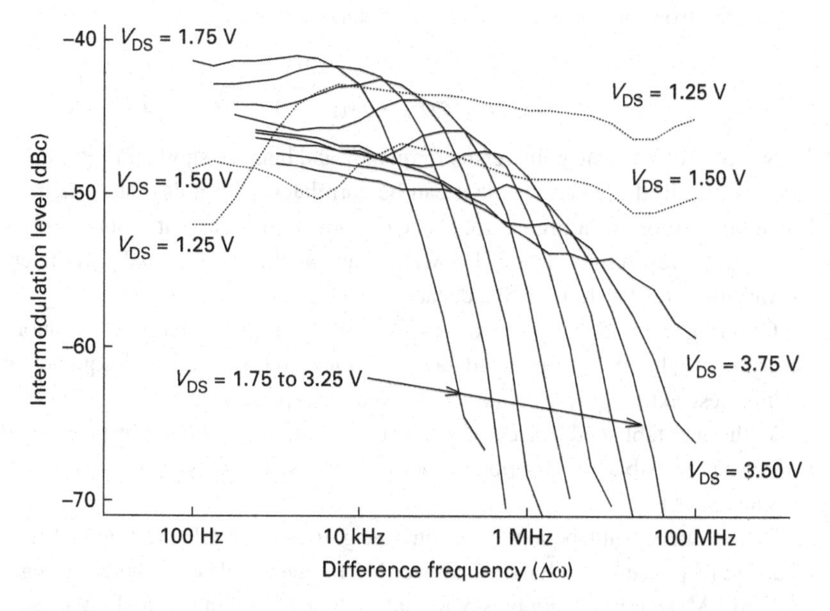

Fig. 15.16 Two-tone intermodulation centered at 500 MHz for a FET in common-source configuration with $V_{GS} = 0$ V. Note that the *waterfall* effect does not start till the drain-source potential reaches 1.75 V.

region of low g_m'' is required to reduce the intrinsic nonlinearity, such as a gate bias near zero as shown in Figure 15.13. This bias also produces substantial drain current, which is a prerequisite for impact ionization, while permitting hole tunneling to the surface states.

As the difference-frequency increases in Figure 15.16, the intermodulation reduces dramatically. This occurs at this gate bias because the intrinsic nonlinearity is canceled. This is a first-order response, which is consistent with charge trapping. The rate increases with drain potential at about one decade per volt due to the increased temperature, impact ionization rate, and reduced trap ionization. The rate of about 10 kHz at the onset of impact ionization ($V_{DS} = 1.75$ V) is slow enough to be observed in pulse measurements, such as in Figure 15.1. There is significantly less variation in intermodulation at low drain potentials in Figure 15.16 because the impact ionization rate is negligible.

15.8 Further investigation

Dynamic processes become faster with increasing drain bias, quickly falling to nanosecond scales. This renders pulsed techniques at fractions of a microsecond too slow to capture isodynamic characteristics. However, radio frequency measurements can bridge this gap.

Small-signal radio-frequency measurements cover a wide spectrum to more than 100 GHz. Network analyzers can provide small-signal isodynamic current and charge storage characteristics routinely. Extracting intrinsic gain from these measurements clearly reveals the frequency response of the dynamics processes in the FET [34].

Small-signal intrinsic gain, A_i, is voltage gain into an open-circuit load. It is easily calculated from measured network parameters:

$$A_i = \frac{y_{21}}{y_{22}} = \frac{2s_{21}}{s_{22} + s_{22}s_{11} - s_{11} - s_{12}s_{21} - 1}. \quad (15.15)$$

Surfaces of intrinsic gain versus frequency and bias are shown in Figure 15.17. There are features in these surfaces that can be correlated with pulse data, such as the time-evolution responses in Figure 15.9. The time-evolution starts at around 100 ns, which is a frequency resolution of only 1.6 MHz, whereas the small-signal parameters continue to vary for a further four or five decades.

Covering the wide spectral range of the intrinsic gain often requires more than one instrument. The low-frequency data can be measured with a low-frequency analyzer or similar test fixture [35], or could be derived from pulse data.

With an established correlation between the time-evolution response and intrinsic gain, it is possible to extrapolate the characteristics of dynamic processes to higher frequencies [15].

The frequency-independence of intrinsic gain above 1 GHz in Figure 15.17, suggests that the response is isodynamic above this frequency. The isodynamic region falls to around 1 MHz at low drain bias potentials, which is well in reach of pulse measurement. Above three volts, significantly higher frequencies are required for isodynamic characterization. This is evident in the time-evolution data of Figure 15.9, where the transients

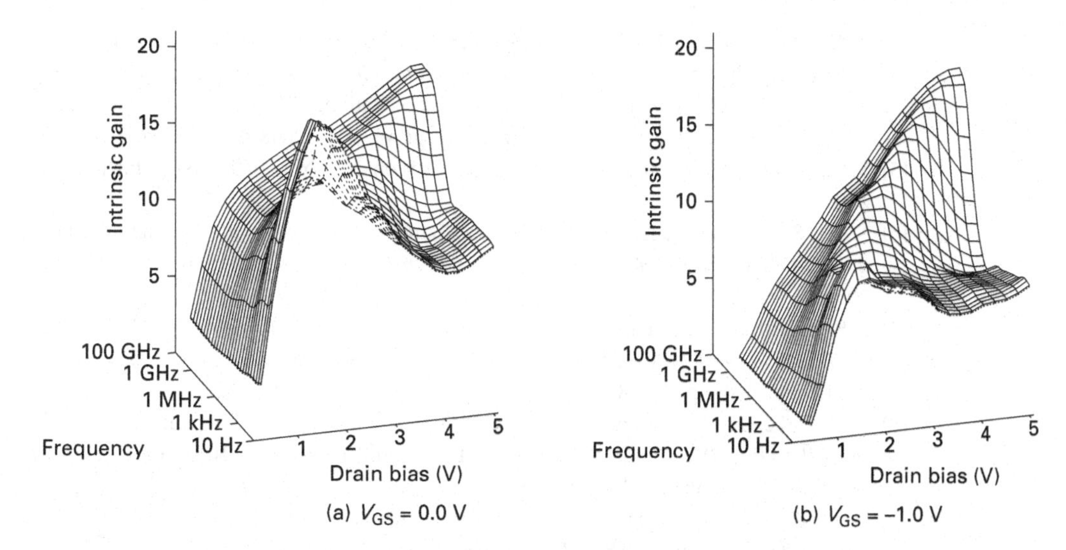

Fig. 15.17 Measured small-signal intrinsic gain surfaces versus frequency and drain-source bias for two gate-source biases. This is a pHEMT similar to that used for Fig. 15.9.

at high drain bias clearly start from higher points at times faster than those that were measured. At even higher drain potentials, the isodynamic region is pushed to beyond 10 GHz. This can have an impact on radio-frequency operation at these bias conditions.

A drop in gain at low frequencies is the most obvious feature of the intrinsic gain surface. This is caused by charge trapping and heating. Substrate trapping at low drain biases overlaps with impact ionization at higher drain biases. The kink in the time-evolution data at around three volts correlates to the fall in gain at around 1 MHz. This becomes exponentially faster with increasing drain bias as the emission rate increases with temperature and capture rates increase.

The peak at one volt comes from an interaction between substrate trapping and self-heating. These shift the knee of the drain current characteristic, which increases the drain conductance. A drop in gain occurs at extremely low frequencies, evident at 10 Hz in Figure 15.17(b), because the knee *walks out* to high drain potentials. In wideband-gap devices, the time constants can be so long that a significant reduction in gain is sustained for a very long period.

References

[1] A. E. Parker, J. G. Rathmell, and J. B. Scott, "Pulsed measurements," ch. 8, in *The RF and Microwave Handbook*, 2nd edn., ser. Electrical Engineering Handbook, M. Golio, Ed. New York, USA: CRC Press LLC, 2008, 40, pp. 8.1–8.30.

[2] A. E. Parker, J. G. Rathmell, and J. B. Scott, "Pulsed measurements," ch. 6, in *The RF and Microwave Handbook*, 1st edn., ser. Electrical Engineering Handbook, M. Golio, Ed. New York, USA: CRC Press LLC, 2000, 22, pp. 4–68–4–95.

[3] A. E. Parker and J. G. Rathmell, "Broad-band characterization of FET self-heating," *IEEE Trans. Microw. Theory Tech.*, 53, no. 7, pp. 2424–2429, July 2005.

[4] A. E. Parker and J. G. Rathmell, "Comprehensive model of microwave FET electro-thermal and trapping dynamics," in *Workshop on Applications in Radio Science*, ser. Commission D, G. James, Ed., URSI. Gold Coast, QLD, Australia: National Committee for Radio Science, 10–12 Feb. 2008, pp. 1–9, http://www.ncrs.org.au/wars/wars2008/sessions.htm.

[5] S. M. Sze, *Physics of Semiconductor Devices*, 3rd edn. New York, USA: Wiley InterScience, 2006.

[6] M. H. Somerville, A. Ernst, and J. A. del Alamo, "A physical model for the kink effect in InAlAs/InGaAs HEMT's," *IEEE Trans. Electron Devices*, 47, no. 5, pp. 993–930, May 2000.

[7] J. Brinkhoff and A. E. Parker, "Charge trapping and intermodulation in HEMTs," in *IEEE MTT-S IMS Digest*, K. Varian, Ed. Fort Worth, Texas, USA: IEEE, Inc., New York, NY, USA, 6–11 June 2004, pp. 799–802.

[8] A. E. Parker and J. G. Rathmell, "Contribution of self heating to intermodulation in FETs," in *IEEE MTT-S IMS Digest*, K. Varian, Ed. Fort Worth, Texas, USA: IEEE, Inc., New York, NY, USA, 6–11 June 2004, pp. 803–807.

[9] W. Shockley and W. T. Read, "Statistics of the recombinations of holes and electrons," *Physics Review*, 87, no. 5, pp. 835–842, 1 Sept. 1952.

[10] C.-T. Sha, "The equivalent circuit model in solid-state electronics—Part 1: The single energy level defect centers," *Proc. IEEE*, 55, no. 5, pp. 654–671, May 1967.

[11] J. G. Rathmell and A. E. Parker, "Circuit implementation of a theoretical model of trap centres in GaAs and GaN devices," in *Microelectronics: Design, Technology, and Packaging III*, ser. Proceedings of SPIE, V. K. V. Alex J. Hariz, Ed., 6798, Canberra, Australia, 4–7 Dec. 2007, pp. 67 980R (1–11).

[12] M. Yee, W. K. Ng, J. P. R. David, P. A. Houston, C. H. Tan, and A. Krysa, "Temperature dependence of breakdown and avalanche multiplication in $In_{0:53}Ga_{0:47}As$ diodes and heterojunction bipolar transistors," *IEEE Trans. Electron Devices*, 50, no. 10, pp. 2021–2026, Oct. 2003.

[13] C. Groves, R. Ghin, J. P. R. David, and G. J. Rees, "Temperature dependence of impact ionization in GaAs," *IEEE Trans. Electron Devices*, 50, no. 10, pp. 2027–2031, Oct. 2003.

[14] R. T. Webster, S. Wu, and A. F. M. Anwar, "Impact ionization in InAlAs/InGaAs/InAlAs HEMTs," *IEEE Electron Device Lett.*, 21, no. 5, pp. 193–195, May 2000.

[15] A. E. Parker and J. G. Rathmell, "Bias and frequency dependence of FET characteristics," *IEEE Trans. Microw. Theory Tech.*, 51, no. 2, pp. 588–592, Feb. 2003.

[16] J. G. Rathmell and A. E. Parker, "Characterization and modeling of substrate trapping in HEMTs," in *Proceedings of European Microwave Integrated Circuits Conference*, K. Beilenhoff, Ed. Munich, Germany: The European Microwave Association, 8–10 Oct. 2007, pp. 64–67.

[17] A. E. Parker and J. G. Rathmell, "Measurement and characterization of HEMT dynamics," *IEEE Trans. Microw. Theory Tech.*, 49, no. 11, pp. 2105–2111, Nov. 2001.

[18] J. Teyssier, J. Viaud, and R. Quere, "A new nonlinear I(V) model for FET devices including breakdown effects," *IEEE Trans. Microw. and Guid. Wave Lett.*, 4, no. 4, pp. 104–106, Apr 1994.

[19] J. B. Scott, J. G. Rathmell, A. E. Parker, and M. M. Sayed, "Pulsed device measurements and applications," *IEEE Trans. Microw. Theory Tech.*, 44, no. 12, pp. 2718–2723, Dec. 1996.

[20] J.-P. Teyssier, P. Bouysse, Z. Ouarch, D. Barataud, T. Peyretaillade, and R. Quere, "40-GHz/150-ns versatile pulsed measurement system for microwave transistor isothermal characterization," *IEEE Trans. Microw. Theory Tech.*, 46, no. 12, pp. 2043–2052, Dec. 1998.

[21] C. P. II Baylis, L. Dunleavy, and J. Daniel, "Direct measurement of thermal circuit parameters using pulsed IV and the normalized difference unit," *Microwave Symposium Digest, 2004 IEEE MTT-S International*, 2, pp. 1233–1236, Vol.2, June 2004.

[22] J. Scott, M. Sayed, P. Schmitz, and A. E. Parker, "Pulsed-bias/pulsed-RF device measurement system requirements," in *24th European Microwave Conference*, 1. Cannes, France: European Microwave Association, London, 5–8 Sept. 1994, pp. 951–961.

[23] A. N. Ernst, M. H. Somerville, and J. A. del Alamo, "Dynamics of the kink effect in InAlAs/InGAs HEMTs," *IEEE Electron Device Lett.*, 18, no. 12, pp. 613–615, Dec. 1997.

[24] S. A. Albahrani and A. E. Parker, "Impact of the pulse-amplifier slew-rate on the pulsed-IV measurement of GaN HEMTs," in *75th Microwave Measurements Conference (ARFTG)*, J. Wood, Ed. Anaheim, CA: IEEE Inc, 28 May 1994, pp. 1–7.

[25] N. B. de Carvalho and J. C. Pedro, "A comprehensive explanation of distortion sideband asymmetries," *IEEE Trans. Microw. Theory Tech.*, 50, no. 9, pp. 2090–2101, September 2002.

[26] J. Brinkhoff and A. E. Parker, "Effect of baseband impedance on FET intermodulation," *IEEE Trans. Microw. Theory Tech.*, 51, no. 3, pp. 1045–1051, Mar. 2003.

[27] S. A. Maas, "How to model intermodulation distortion," in *IEEE Trans. Microw. Theory Tech.*, 1, Boston, MA, 1991, pp. 149–151.

[28] S. A. Maas, *Nonlinear Microwave and RF Circuits,* 2nd edn. Norwood, MA: Artech House, 2003.

[29] G. Passiopoulos, D. R. Webster, A. E. Parker, D. G. Haigh, and I. D. Robertson, "Effect of bias and load on MESFET nonlinear characteristics," *Electronic Letters*, 32, no. 8, pp. 741–742, 11 Apr. 1996.

[30] D. R. Webster, A. E. Parker, D. G. Haigh, and J. B. Scott, "Effect of circuit parameters and topology on intermodulation in MESFET circuits," in *IEEE GaAs IC Symposium Digest*. San Jose, CA: IEEE, UK, 10–13 Oct. 1993, pp. 255–258.

[31] G. Qu and A. E. Parker, "Validation of a new HEMT model by intermodulation characterization," in *IEEE MTT-S IMS Digest*, ser. 1998 IEEE MTT-S International Microwave Symposium Digest, R. Meixner, Ed., 2. Baltimore, Maryland, USA: IEEE, Inc., New York, NY, USA, 7–12 June 1998, pp. 745–748.

[32] G. Qu and A. E. Parker, "New model extraction for predicting distortion in HEMT and MESFET circuits," *IEEE Trans. Microw. Guid. Wave Lett.*, 9, no. 9, pp. 363–365, Sept. 1999.

[33] A. E. Parker and G. Qu, "Intermodulation nulling in HEMT common source amplifiers," *IEEE Microw. Wireless Compon. Lett.*, 11, no. 3, pp. 109–111, Mar. 2001.

[34] A. E. Parker and J. G. Rathmell, "Novel technique for determining bias, temperature and frequency dependence of FET characteristics," in *IEEE MTT-S IMS Digest*, ser. 2002 IEEE MTT-S International Microwave Symposium Digest, R. Hamilton, Ed., 2. Seattle, Washington, USA: IEEE, Inc., New York, NY, USA, 2–7 June 2002, pp. 993–996.

[35] J. Brinkhoff and A. E. Parker, "Device characterization for distortion prediction including memory effects," *IEEE Microw. Wireless Compon. Lett.*, 12, no. 3, pp. 171–173, Mar. 2005.

Index

Printed in the United States
By Bookmasters